THE AMAZON FROM AN INTERNATIONAL LAW PERSPECTIVE

With a vast river network and rain forests extending over eight South American countries, the Amazon plays a vital role in maintaining biodiversity, regional hydrology, and terrestrial carbon storage. Due to its ecological characteristics, this region benefits not only the Amazon States, but also the international community at large. The rapid clearing of the Amazon forests, in particular, affects the balance of the global environment, exacerbating global warming trends and contributing to biodiversity loss. This book examines whether international law has an impact on the preservation of the Amazon and to what extent international cooperation can help protect this region. It examines both how the Amazon countries cooperate with each other, as well as how they cooperate with the broader international community. The book also explores whether the Amazon should be granted a special legal status and the possible implications in terms of international cooperation.

Beatriz Garcia is a Research Fellow at the Australian Centre for Climate and Environmental Law at Sydney Law School. She completed her PhD at the Graduate Institute of International and Development Studies in Geneva, Switzerland. She has worked at the Biodiversity and Climate Change Section of the United Nations Conference on Trade and Development, particularly on projects related to the Amazon. She has held positions at government agencies in Brazil and provided advice to institutions such as the Amazon Cooperation Treaty Organization.

The Amazon from an International Law Perspective

BEATRIZ GARCIA

Doctor in International Relations (International Law),
Graduate Institute of International and
Development Studies, Geneva

CAMBRIDGE
UNIVERSITY PRESS

CAMBRIDGE UNIVERSITY PRESS
Cambridge, New York, Melbourne, Madrid, Cape Town,
Singapore, São Paulo, Delhi, Tokyo, Mexico City

Cambridge University Press
32 Avenue of the Americas, New York, NY 10013-2473, USA

www.cambridge.org
Information on this title: www.cambridge.org/9780521769624

First published 2011

Printed in the United States of America

A catalog record for this publication is available from the British Library.

Library of Congress Cataloging in Publication data
Garcia, Beatriz, 1974–
The Amazon from an international law perspective / Beatriz Garcia.
p. cm.
Includes bibliographical references and index.
ISBN 978-0-521-76962-4 (hardback)
1. Natural resources – Law and legislation – Amazon River
Region. 2. Environmental law – Amazon River Region.
3. Environmental law, International. I. Title.
KH642.A46G37 2011
344.81′1046–dc22 2010039404

ISBN 978-0-521-76962-4 Hardback

For Laurent

Contents

List of Maps and Figures — *page* xi
List of Tables — xii
Preface — xv
Acknowledgments — xvii
Abbreviations and Acronyms — xix

1 **Introduction** — 1
- 1.1. International Cooperation in the Field of Environmental Protection — 3
- 1.2. The Context and Meaning of International Cooperation — 5
- 1.3. Limits to International Cooperation in the Environmental Field — 10
- 1.4. International Cooperation in the Context of the Present Study — 16

2 **The Characteristics of the Amazon Region** — 23
- 2.1. Physical Characteristics — 23
  - 2.1.1. Land Area — 23
  - 2.1.2. Hydrological Characteristics — 26
- 2.2. Socioeconomic Characteristics — 29
  - 2.2.1. Population — 29
  - 2.2.2. Economic Development and Trade — 32
- 2.3. Current Threats to the Amazonian Environment — 34
  - 2.3.1. Deforestation in the Amazon — 36
  - 2.3.2. Impacts of Deforestation on the Amazon — 46

3 **The Origins of Regional Cooperation in the Amazon** 50
3.1. International Boundaries Delimiting the Amazon Region 50
3.1.1. The *Uti Possidetis* as a Principle for Establishing International Boundaries 51
3.1.2. International Boundaries between Brazil and the Other Amazon States 54
3.2. Commerce and Navigation 60
3.3. Early Forms of Cooperation in the Amazon Region 66
3.3.1. Treaties of Friendship and Cooperation 66
3.3.2. Environmental Conservation 67
3.3.3. Toward Regional Cooperation 70
Conclusions 72

4 **The 1978 Amazon Cooperation Treaty** 74
A. Origins of the 1978 Amazon Cooperation Treaty 75
4.1. The Rationale of a Regional Cooperation Treaty 75
4.2. The Negotiation of a Regional Cooperation Treaty 82
B. An Overview of the 1978 Amazon Cooperation Treaty 85
4.3. Normative Framework 85
4.3.1. Spatial Domain of Validity 86
4.3.2. Principles and Obligations 88
4.3.3. Implementation Mechanisms 95
4.3.4. Normative Evolution 97
4.4. Institutional Framework 100
4.4.1. Meetings of the Presidents 100
4.4.2. Meetings of Ministers of Foreign Affairs 101
4.4.3. Amazon Cooperation Council 103
4.4.4. Permanent National Commissions 104
4.4.5. Special Commissions 105
4.4.6. From Pro Tempore Secretariats to a Permanent Structure 107
4.4.7. The Functions and Structure of the Permanent Secretariat 110
4.4.8. The Permanent Secretariat Today 113
4.4.9. Financial Mechanism 117
4.5. Assessing the Effectiveness of the 1978 Amazon Cooperation Treaty 118
Conclusions 123

5 **Regional and Subregional Organizations** 127
5.1. Organization of American States 129
5.2. Andean Community 132
5.3. Caribbean Community 137
5.4. Common Market of the South 140
5.5. Inter-American Development Bank 145
5.6. Andean Development Corporation 148
Conclusions 149

6 **Other Legal Instruments Adopted by the Amazon States *Inter Se*** 151
A. Cooperation among the Amazon States *Inter Se* 152
6.1. Freshwater Resources 153
6.2. Biodiversity 159
6.2.1. Genetic Resources 162
6.2.2. Wildlife Species of Fauna and Flora 164
6.3. Health, Food Safety, and Sanitation 166
6.4. Forests 170
6.5. Hazardous Substances and Activities 174
6.6. Disaster Prevention and Management 174
6.7. Indigenous Populations 175
B. Cooperation among the Amazon States and Other Countries in Latin America and the Caribbean 184
Conclusions 186

7 **Multilateral Treaties and Global Actors in the Amazon** 188
A. Multilateral Environmental Agreements 188
B. Global Actors 196
7.1. International Organizations 196
7.1.1. The United Nations Food and Agriculture Organization 196
7.1.2. United Nations Environment Programme 198
7.1.3. United Nations Development Programme 200
7.1.4. The World Bank and the Global Environment Facility 201
7.1.5. The Pilot Program to Conserve the Brazilian Rain Forest 207
7.2. International Nongovernmental Organizations 210
C. Synergies and Conflicts between Norms and Institutions 214
Conclusions 224

8 **Positive Incentives for Protecting the Amazon** 227
8.1. Carbon Trading 233
8.1.1. The Compliance Market 233
8.1.2. The Voluntary Carbon Market 236
8.2. Reducing Emissions from Deforestation and Forest Degradation 238
8.2.1. The International Negotiations on REDD 243
8.2.2. REDD in Practice 246
8.2.3. REDD: Pros and Cons 248
8.3. Payments for Environmental Services 253
8.3.1. Payments for Environmental Services in Practice 254
8.3.2. PES: Pros and Cons 256
Conclusions 260

9 **The Legal Status of the Amazon: Implications for International Cooperation** 266
9.1. The Amazon as a Common Heritage of Humankind? 270
9.2. The Amazon as an Area of Special Protection? 275
9.3. The Amazon as a Common Concern of Humankind 279
9.3.1. Consequences of a Common Concern Status 284
9.3.2. A Right of Surveillance in the Protection of the Amazon 294
Conclusions 304

10 **General Conclusions** 306
10.1. Cooperation among the Amazon States *Inter Se* 306
10.2. The Role of the 1978 Amazon Cooperation Treaty 311
10.3. Cooperation between the Amazon States and the International Community 319
10.4. International Legal Status for the Amazon and Possible Implications 323

Bibliography 327
Index 371

Maps and Figures

MAPS

2.1 The Amazon basin *page* 24
2.2 "Arc of Deforestation" in the Legal Amazon 41
3.1 Bolivia–Brazil 56
3.2 Brazil–Colombia–Venezuela 57
3.3 Brazil–Guyana–Suriname 58
3.4 Brazil–Peru 59

FIGURES

4.1 The structure of ACTO 116
4.2 The ACT structure prior to the establishment of the Permanent Secretariat 117

Tables

2.1 Brazil's borders with the Amazon States *page* 25
2.2 Amazon land area per country 26
2.3 Area of river basins 28
2.4 Population in the Amazon basin 31
2.5 Indigenous populations of the Amazon 32
2.6 General Information on the Amazon States 33
2.7 Annual GDP growth, 2006–2008 33
2.8 Trade between Brazil and other Amazon States, 1967–2007 35
2.9 Forest distribution by region 36
2.10 Deforestation in the Amazon region 37
2.11 Drivers of deforestation by country 39
2.12 Legal Amazon: annual rate of deforestation, 2000–2008 40
2.13 Cattle herds in the Legal Amazon 44
2.14 Protected areas in the Amazon States 48
2.15 Examples of transboundary terrestrial ecosystems among the Andean States 49
4.1 National policies for the occupation of the Brazilian Amazon 80
4.2 ACT normative framework 100
4.3 Meetings of Ministers of Foreign Affairs 103
4.4 Meetings of the Amazon Cooperation Council 105
4.5 Special Commissions 106
4.6 Pro Tempore Secretariats 108

4.7 Meetings of the ad hoc working group on the establishment of a Permanent Secretariat 110
4.8 Strategic plan 2004–2012 115
6.1 Some shared river basins in the Amazon 154
8.1 The state of the carbon market in the world 235

Preface

This book covers international treaties adopted by the Amazon States *inter se* since the end of the nineteenth century, when most of them attained independence from European colonial powers, until the present. It also includes nonbinding instruments, such as decisions, declarations, and plans of action, adopted under the aegis of subregional and regional organizations in which one or more Amazon States participate.

Most of the primary information related to the 1978 Amazon Cooperation Treaty (ACT) (for example, amendments, regulations, resolutions, presidential and ministerial declarations, minutes of meetings, etc.) can be found in successive publications of a document published by the Pro Tempore Secretariats of the ACT entitled "The Legal Basis of the Amazon Cooperation Treaty." Judicial decisions on matters related to the environment or indigenous communities, issued by tribunals established under subregional and regional organizations (i.e. the Court of Justice of the Andean Community, Mercosur's Permanent Court of Appeal and Ad Hoc Tribunals, and the Inter-American Court of Human Rights) are also examined here.

Apart from the information gathered at the library resources in Geneva, particularly at the United Nations library, other sources such as early bilateral treaties on boundaries and fluvial navigation among the Amazon States, or speeches delivered at the time of the adoption of the ACT, were collected at the Brazilian Ministry of Foreign Affairs and Federal Council of Justice (Conselho da Justiça Federal) in Brasília. Information was also collected during the author's short stay at the Amazon Cooperation Treaty Organization (January 2007) and at the Núcleo de Altos Estudos Amazônicos at the Federal University of Pará in Belém (December 2006). On both these occasions, interviews were conducted with various organizations and government agencies.

Acknowledgments

This book is based on my PhD thesis concluded at the Graduate Institute of International and Development Studies, Geneva. Therefore, I am first and foremost grateful to my thesis supervisors, Laurence Boisson de Chazournes and Marcelo Kohen, for their attentive and skilful guidance.

My thanks also go to Lucas Assunção, chief of the Biodiversity and Climate Change Section of the United Nations Conference on Trade and Development, and Claudia Mayer, at the German Technical Cooperation, whose encouragement and inspiration I will always treasure. I also express my gratitude to Márcia Mazo and Andresa Duarte at the Brazilian Federal Justice Tribunal (Tribunal Regional Federal da 1ª Região), who helped me collect essential information, particularly early treaties adopted among the Amazon States. I am greatly indebted to the Swiss Foundation Ernst et Lucie Schmidheiny for the financial support given both to the conclusion of my doctoral thesis and to the publication of this book, and personally to Annemarie S. Geckeler and Johanne Patenaude for their caring and kindness.

I also wish to thank a few institutions with which I have been associated over the past years that allowed me to conduct my research: the Amazon Cooperation Treaty Organization, Núcleo de Altos Estudos Amazônicos (University of Pará), the Centre for Comparative and Public Law (University of Hong Kong), and the Australian Centre for Climate and Environmental Law (University of Sydney).

I wish to thank Cambridge University Press, especially John Berger, for his support in the publication of this book, and the anonymous reviewers for their very useful comments and ideas. I also thank my English editor, Alex Potter, for his commitment and careful reading.

I am sincerely grateful to dear friends and fellow researchers who reviewed some of my drafts and were always ready to exchange ideas: Martha Balby Gandra, Daniela Guimarães Goulart, Yasmin Q. Naqvi, Gláucia Boyer, Makane Mbengue, Mara Tignino, and Haroldo Machado Filho. For their friendship and constant support, I thank Andrea Nascimento-Müller, Olimpia Fernandez Pordomingo, Sérgio

Saba, Melania Di Vara, and Teljya Oka-Pregel. A heartfelt thank you to my parents and sister, whose love and example encouraged me to conclude this project, especially my father, José Garcia Gasques, for his help and advice as an experienced researcher. Finally, the one to whom my deepest gratitude goes is my husband, Laurent Pauwels, for his love, understanding, and encouragement. This book would not have been possible without him.

Abbreviations and Acronyms

AAPAD	Andes Amazon Protected Areas Database
ACC	Amazon Cooperation Council
ACT	Amazon Cooperation Treaty
ACTO	Amazon Cooperation Treaty Organization
AEPS	Arctic Environmental Protection Strategy
ALFA	Application of Forest Legislation in the Amazon
AMAP	Arctic Monitoring and Assessment Program
ANEC	Associação Nacional dos Exportadores de Cereais
ASEAN	Association of Southeast Asian Nations
AWG-KP	Ad Hoc Working Group on Further Commitments for Annex 1 Parties under the Kyoto Protocol
AWG-LCA	Ad Hoc Working Group on Long-Term Cooperative Action under the Convention
BJ	Legal Basis of the Amazon Cooperation Treaty
BNDES	Brazilian National Economic and Social Development Bank
CAF	Andean Development Corporation
CAN	Andean Community
Caricom	Caribbean Community
CBD	Convention on Biological Diversity
CCBA	Climate Community and Biodiversity Alliance
CCJ	Caribbean Court of Justice
CCM	Mercosur's Council of the Common Market
CCOOR	Coordination Commission of the Amazon Cooperation Council
CDM	Clean Development Mechanism
CER	certified emission reduction
CERFLOR	Programa Brasileiro de Certifi cação Florestal
CIC	Intergovernmental Coordinating Committee of La Plata Basin Countries

CIF	Climate Investment Fund
CITES	Convention on International Trade in Endangered Species of Wild Fauna and Flora
CJAC	Court of Justice of the Andean Community
CMG	Mercosur's Common Market Group
CO_2	carbon dioxide
CO_2e	CO_2 equivalent
COFO	Committee on Forestry
COP	Conference of the Parties
DRC	Democratic Republic of Congo
EC	European Community
ECJ	European Court of Justice
ECLAC/CEPAL	Economic Commission for Latin America and the Caribbean
ECOSOC	United Nations Economic and Social Council
FAO	Food and Agriculture Organization of the United Nations
FCPF	Forest Carbon Partnership Facility
FIP	Forest Investment Programme
FSC	Forest Stewardship Council
FTAA	Free Trade Area of the Americas
FUNAI	Brazilian National Foundation of Indigenous People
G7	Group of Seven
GATT	General Agreement on Tariffs and Trade
GDP	gross domestic product
GEF	Global Environment Facility
GEO-Amazonia	Global Environmental Outlook
GHG	greenhouse gas
GPG	Good Practice Guidance
GTS	Grupo de Trabalho da Soja
GTZ	German Technical Cooperation
IABIN	Inter-American Biodiversity Information Network
IACHR	Inter-American Commission on Human Rights
IBAMA	Brazilian Institute of Environment and Renewable Natural Resources
IBGE	Brazilian Institute of Geography and Statistics
IBRD	International Bank for Reconstruction and Development
ICJ	International Court of Justice
IDB	Inter-American Development Bank
IDESAM	The Institute for Conservation and Sustainable Development of Amazonas

IFC	International Finance Corporation
IFOAM	International Federation of Organic Agriculture Movements
IIED	International Institute for Environment and Development
ILA	International Law Association
ILC	International Law Commission
ILM	International Legal Materials
ILO	International Labour Organization
INPE	Brazilian National Institute for Space Research (Instituto Nacional de Pesquisas Espaciais)
IPAM	Instituto de Pesquisa da Amazonia
IPCC	Intergovernmental Panel on Climate Change
IPEA	Brazilian Institute for Economic Research (Instituto de Pesquisas Econômicas Aplicadas)
ISA	Instituto Socioambiental
ITTA	International Tropical Timber Agreement
ITTO	International Tropical Timber Organization
IUCN	International Union for Conservation of Nature
J. Air L. & Com	Journal of Air Law and Commerce
JI	Joint Implementation
JRC	Joint Research Centre
LBA	Large Scale Biosphere-Atmosphere Experiment in Amazonia
MDG	Millennium Declaration Goals
MEA	multilateral environmental agreement
Mercosurl	Common Market of the South
MIGA	Multilateral Investment Guarantee Agency
MMA	Brazilian Ministry of Environment
MMFA	Meeting of Ministers of Foreign Affairs
MOP	Meeting of the Parties
MOU	memorandum of understanding
MRV	monitoring, reporting, and verification
MTC	Mercosur's Trade Commission
NGO	nongovernmental organization
NKCAP	Noel Kempff Climate Action Project
OAS	Organization of American States
ODA	official development assistance
OECD	Organization for Economic Cooperation and Development
OTC	"over-the-counter" offset market
PAC	Growth Acceleration Program
PAS	Sustainable Amazon Plan (Plano Amazônia Sustentável)
PES	payment for environmental services

PNC	Permanent National Commissions
POP	Protocol of Ouro Preto
PPCR	Pilot Program for Climate Resilience
PPG7	Pilot Program to Conserve the Brazilian Rain Forest
ppm	parts per million
PPTAL	Protection of Indigenous Peoples and Land in the Amazon
Proambiente	Program for the Socio-Environmental Development of Rural Family Production
PSA	Pagos por Servicios Ambientales
REDD	Reducing Emissions from Deforestation and Forest Degradation in Developing Countries
REMA	Reunión Especializada de Medio Ambiente
RESEX	Extractive Reserves Project
RFA	Rainforest Alliance
RTC	Revised Treaty of Chaguaramas Establishing the Caribbean Community
SAN	Sustainable Agriculture Network
SBSTA	Subsidiary Body for Scientific and Technological Advice
SC	Special Commissions
SECCI	Sustainable Energy and Climate Change Initiative
Sinmetro	National System of Metrology, Standardization, and Industrial Quality
Sipam	Amazonian Protection System (Sistema de Proteção da Amazônia)
SIS	Species Information Service
SIVAM	System for the Vigilance of the Amazon (Sistema de Vigilância da Amazônia)
Stan. Envtl. L.J.	Stanford Environmental Law Journal
TCCJ	Treaty Creating the Court of Justice of the Andean Community
TCO	Indigenous Ancestral Territory
TNC	Nature Conservancy
TRIPS	Trade-Related Aspects of Intellectual Property Rights
UN	United Nations
UNASUR	Union of South American Nations
UNCED	United Nations Conference on Environment and Development
UNCLOS	United Nations Convention on the Law of the Sea
UNCTAD	United Nations Conference on Trade and Development

UNDP	United Nations Development Programme
UNEP	United Nations Environment Programme
UNESCO	United Nations Educational, Scientific and Cultural Organization
UNFCCC	United Nations Framework Convention on Climate Change
UNFF	United Nations Forum on Forests
UNGA	United Nations General Assembly
UNTS	United Nations Treaty Series
v.	versus
VERs	verified emission reductions
WCED	World Commission on Environment and Development
WCMC	World Conservation Monitoring Centre
WHC	World Heritage Committee
WHO	World Health Organization
WRI	World Resources Institute
WTO	World Trade Organization
WWF	World Wildlife Fund for Nature

1

Introduction

The Amazon (also referred to as continental Amazon or pan-Amazônia) extends its limits beyond the Amazon River basin, including part of the Tocantins and Orinoco river basins and some other small basins.[1] Its territory includes eight South American countries (Bolivia, Brazil, Colombia, Ecuador, Guyana, Peru, Venezuela, and Suriname) over an area of approximately 7.5 million square kilometers (most of which is located in Brazil), comprising approximately 44 percent of the territory of South America.[2] The Amazon contains the largest freshwater reserve and forest cover on the planet, sheltering a mosaic of ecosystems with a great variety of species of fauna and flora, some of which are still unknown to science, and it is home to diverse indigenous communities, holders of ancestral knowledge and cultures. This region provides sustenance to its inhabitants, such as food, building materials, medicines, and other products, and has great potential as a field for scientific research and the development of the pharmaceutical, cosmetic, and food industries. The Amazon has recognized global importance because it plays a vital role in maintaining biodiversity, regional hydrology, and terrestrial carbon storage capacity.[3] Due to its conspicuous ecological characteristics, the Amazon benefits not only the basin States, but also the international community at large.

Despite the unique natural wealth and economic potential, the Amazon forests are being rapidly cleared, with a consequent loss of biodiversity and impact on climate change. Latin America and the Caribbean have already lost approximately 64 million hectares of their original forests,[4] and some estimates reveal that between

[1] For a general description of the Amazon River, see A. Biswas et al., *Management of Latin America River Basins: Amazon, Plata and São Francisco*, Tokyo: United Nations University, 1999.

[2] M. Goulding, R.B. Barthem, and E. Ferreira, *The Smithsonian Atlas of the Amazon*, Washington, DC: Smithsonian Books, 2003.

[3] W.F. Laurance et al., "The future of the Brazilian Amazon," *Science*, 2001, pp. 438–39, at p. 1.

[4] According to the Food and Agriculture Organisation, Latin America and the Caribbean lost about sixty-four million hectares of forest, FAO, *State of the World's Forests from 1990 to 2005, Oxford: Worlds and Publications*, 2007 report, p. 37.

1990 and 2020, tropical deforestation might extinguish 15 percent of the world's species.[5] The Brazilian Amazon alone has the world's highest absolute rate of forest destruction, averaging millions of hectares per year.[6] It is known that tropical forests store and process, via photosynthesis and respiration, large quantities of carbon dioxide released into the atmosphere through the burning of fossil fuels, and that small changes within tropical forest biomes can thus lead to major global impacts.[7] Although it is difficult to determine the exact number of species living in tropical forests, it is known that the majority of the world's species are located there, of which approximately one million species are "committed to extinction."[8] Whereas the clearing of tropical forests has global impacts, factors adversely affecting forests are also produced globally. Tropical forests now face dual pressure: direct deforestation and degradation, on the one hand, and climate change affecting, for example, patterns of tree growth and mortality, on the other. It is recognized that tropical forests affect climate, and vice versa.[9] This indicates that not only actions taken by States owning tropical forests might affect them, but also those taken by the wider community of States, which may lead to further climate change. Therefore, although the wider community of States might have a legitimate concern in the preservation of tropical forests, specifically the Amazon, they also share a responsibility to ensure their protection. In view of the situation of rapid deforestation and environmental degradation in the Amazon, this study looks at how the basin States have reacted to the problem, by assessing how they have been cooperating *inter se* and by examining what role the international community plays in reversing the current situation of environmental degradation. It is here assumed that at least some environmental problems have an "international dimension"[10] and require cooperative actions by a number of States.

The protection of the Amazon is an example *par excellence*, illustrating that cooperation is required at different levels in order to effectively handle some environmental problems. The extensive river network and rainforests extending over the territories of eight countries, where countless interdependent ecological processes

[5] K. R. Miller, W. V. Reid, and C. V. Barber, "Deforestation and species loss," in Jessica Tuchman Mathews (ed.), *Preserving the Global Environment: The Challenge of Shared Leadership*, New York & London: Norton, 1991, p. 84.

[6] Laurance et al., op. cit., p. 1.

[7] S.L. Lewis, "Tropical forests and the changing earth system," *Philosophical Transactions of the Royal Society*, 2006, pp. 195–210, at p. 196.

[8] As Lewis (ibid., p. 20) explains, tropical forests cover 10 percent of the Earth's land surface and the Brazilian Amazon alone contains about 40 percent of the world's remaining tropical rainforests. According to Lewis, although it is difficult to determine the exact number of species living in the tropics, it is estimated than about 40 percent of a total of 3.3 million species reside in tropical forests, from which approximately 1 million species are "committed to extinction."

[9] Ibid., p. 196.

[10] UNGA Res. 40/200, December 17, 1985.

occur, reveal the ecological unity of the Amazon and the fact that all basin States are naturally connected. Faced with similar ecological conditions, they have similar difficulties in dealing with certain environmental problems, and also common interests, particularly in exploiting the region's economic potential. Since those countries share that vast biome, regional cooperation is a predictable (and suitable) way they have at hand to tackle common environmental problems. In other areas of the world, the existence of shared ecosystems has also triggered international cooperation. For example, the adoption of the Arctic Environmental Protection Strategy in 1991, involving the eight Arctic States, was motivated by the fact that certain environmental issues could only be properly dealt with through international cooperation and recognition that the Arctic is a particular region in need of distinctive responses.[11] As a matter of fact, the interconnectedness of the natural environment, particularly observed in the case of shared ecosystems, has influenced recent developments in international environmental law based on a belief that some common values can only be safeguarded through international cooperation.[12] In this light, the scope of this book is to examine what forms of cooperation exist among the Amazon countries *inter se*, as well as between them and the international community, and finally to what extent international cooperation can help reverse the region's current situation of environmental degradation.

1.1. INTERNATIONAL COOPERATION IN THE FIELD OF ENVIRONMENTAL PROTECTION

A recognition that certain problems faced by the international community as a whole (e.g. human rights violations, environmental degradation, or threats to national security) require international cooperation among various actors (e.g. governments, international organizations, nongovernmental organizations (NGOs), and other sectors of civil society) has influenced the evolution of international law. In principle, matters such as these cannot be dealt with, or at least not adequately, by one State *uti singuli*, but demand international regulation at different levels.[13]

11 D.R. Rothwell, "The Arctic environmental protection strategy and international environmental cooperation in the far north," *Yearbook of International Environmental Law*, vol. 6, 1995, pp. 65–105, at p. 81.

12 J. Brunnée, "Common interests, echoes of an empty shell? Some thoughts on common interests and international environmental law," *Heidelberg Journal of International Law*, vol. 49, 1989, pp. 791–808, at p. 797.

13 G. Abi-Saab, "The changing world order and the international legal order: The structural evolution of international law beyond the state-centric model," in Y. Sakamoto (ed.), *Global Transformation: Challenges to the State System*, Tokyo: United Nations University Press, 1994, pp. 439–61, at p. 454.

A notable change in the international legal order in the last years, prompted by a recognition of common interests and values, is the evolution from a state-centered model derived from the Peace of Westphalia, which sanctioned an "international law of coexistence," to a more cooperative approach or, as Wolfgang Friedmann puts it, toward an "international law of cooperation."[14]

In comparing the international law of "coexistence" and that of "cooperation," Georges Abi-Saab concludes that they are in fact two different techniques of legal regulation differing in several respects.[15] According to him, these two approaches are distinct firstly from a normative perspective. Legal obligations in the realm of coexistence are mostly of abstention, whereas the international law of cooperation involves performance obligations and assumes that some tasks cannot be carried out by States individually, but require joint efforts. Whereas the international law of coexistence considers sovereign States as equal, at least formally, equality in the law of cooperation means equality of participation. Therefore, in this latter case, States are expected to perform and achieve certain results. However, the *de facto* disparities among them, in terms of their capacities and needs, are taken into account to ensure that performance obligations assumed by them can be met. Secondly, as far as the instruments of lawmaking are concerned, the international law of cooperation has witnessed a tendency toward more malleable or less constraining legal instruments designated as "soft law"[16] (e.g. resolutions, codes of conduct, declarations, etc.) and innovative types of multilateral treaties ("lawmaking treaties") that are particularly observed in the field of international environmental law. Thirdly, with regard to mechanisms of implementation, Abi-Saab notes that the law of coexistence envisages one single technique, that of autoregulation, whereas the law of cooperation is essentially "institutional," requiring a basic structure in charge of ensuring a certain "division of labor" among its participants and the fulfillment of performance obligations.

This change in the focus and structure of the international legal system observed since the postwar period has been particularly influenced by human rights movements, and later, from a slightly different perspective, by the recognition of new environmental protection concerns.[17] It has become common to affirm

14 Ibid., p. 440.

15 G. Abi-Saab, *Cours général de droit international public*, The Hague: Martinus Nijhoff, 1987, pp. 322–23.

16 According to Victor, soft law is used to set voluntary standards and to declare principles and aspirations (D.G. Victor, "The use and effectiveness of non-binding instruments in the management of complex international environmental problems," *American Society of International Law, Proceedings of the Annual Meeting*, 1997, pp. 241–50, at p. 241). Obligations contained in a treaty may also have a "soft-law" content.

17 M. Bowman, "The nature, development and philosophical foundations of the biodiversity concept in international law," in M. Bowman and C. Redgwell (eds.), *International Law and the Conservation of Biological Diversity*, The Hague: Kluwer Law International, 1996, pp. 5–31, at p. 12.

that the environment knows no political boundaries and that traditional regimes of resource exploitation grounded primarily on the notion of territorial sovereignty require more collectivist approaches.[18] Particularly after the 1972 United Nations Conference on the Human Environment (hereinafter: 1972 Stockholm Conference), environmental protection treaties adopted at the bilateral, regional, and global levels have reflected in their normative content and institutional framework the characteristics of the international law of cooperation, as will be observed in the subsequent chapters.[19]

As noted, the international law of cooperation is founded on the recognition of "common interests" or a "community."[20] In this light, this book assumes that the Amazon States and the international community share a "common interest" in the protection of the Amazon, which justifies further cooperation among the basin States, as well as between them and the international community. As Abi-Saab suggests, whereas a central question within the international law of coexistence has been that of how to keep States peacefully apart, the issue under the international law of cooperation is that of how to bring States together.[21] Taking into account that approach introduced by the international law of cooperation, this study looks at ways of bringing the Amazon States closer together, in order to address more effectively common environmental problems, and at forms of enhancing cooperation between them and the international community.

1.2. THE CONTEXT AND MEANING OF INTERNATIONAL COOPERATION

International cooperation is one of the main principles of the United Nations. Under Article 1(3) of the UN Charter, States are required to achieve international cooperation in order to solve problems of an economic, cultural, or humanitarian character. The whole of Chapter 9 is dedicated to the issue of "international economic and social co-operation." Article 55 (and subsequent provisions under Chapter 9 of the UN Charter) encourages States to cooperate in raising living standards; creating conditions of economic and social development; and providing solutions for socioeconomic, health, and related problems.[22] An array of issues in the socioeconomic field can be objects of international cooperation, including, for example, those related to education, public health, employment conditions, and also environmental protection.

18 Ibid.
19 Abi-Saab, "The changing world," op. cit., p. 445.
20 Abi-Saab, *Cours général*, op. cit., p. 321.
21 Ibid.
22 UN Charter, Article 55(b).

Under the 1970 Declaration on Principles of International Law Concerning Friendly Relations and Cooperation among States, the principle of international cooperation was enunciated as a "duty to cooperate."[23] According to this declaration, States "have the duty to cooperate with one another, irrespective of the differences in their political, economic and social systems in the various spheres of international relations."[24] This duty has been translated into obligations to negotiate, conduct consultations, exchange information, and notify other States in certain cases. As Franz Xaver Perrez notes, the duty to cooperate is now a central element of a sovereign State's authority and responsibility; it is inherently included in the principle of sovereignty and in that of a permanent sovereignty over natural resources.[25] Perrez proposes the idea of a "cooperative sovereignty," by which the authority of a State to decide on the utilization of natural resources also involves the responsibility to coordinate their use with other concerned States. This general duty to cooperate is made concrete, for example, by requiring a State to assess possible impacts and risks that its activities may have in other States, to notify and to inform other States, to enter into consultations, and to apply the precautionary principle in cases of scientific uncertainties.

Since the 1970s, the principle of international cooperation, previously invoked under Chapter 9 of the UN Charter as "international economic and social co-operation," has been advanced by developing countries, as they demanded higher levels of cooperation, notably in the socioeconomic field. Their claims were crystallized in the 1974 Declaration on the Establishment of a New International Economic Order[26] and its Program of Action,[27] whereby States committed to strengthen the role of the UN in a worldwide collaboration for economic and social development. In 1977, the notion of a "right to development" appeared for the first time, under the UN Human Rights Commission, reflecting developing countries' aspirations for a new international order.[28] A few years later, Article 9 of the 1984 Charter of Economic Rights and Duties of States[29] framed the general goal of international economic and social cooperation as a legal duty, by affirming that States

[23] UNGA Res. 2625 (XXV), October 24, 1970.

[24] Ibid.

[25] F. Xaver Perrez, *Cooperative Sovereignty: From Independence to Interdependence in the Structure of International Environmental Law*, The Hague: Kluwer Law International, 2000, p. 109. For a review of different opinions in the scholarship with respect to the duty to cooperate, see O. McIntyre, "The role of customary rules and principles of international environmental law in the protection of shared international freshwater resources," *Nature Resources Journal*, vol. 46, 2006, pp. 157–210.

[26] UNGA Res. 3201 (S-VI), May 1, 1974.

[27] Ibid., Chapter IX.

[28] Abi-Saab, "The changing world," op. cit., p. 455.

[29] GA Res.3281 (XXIX), December 13, 1974.

"have the responsibility to cooperate in the economic, social, cultural, scientific and technological fields for the promotion of economic and social progress." References to international cooperation in the social and economic fields have ever since been the object of many others UN General Assembly (UNGA) resolutions.[30]

However, the question of whether there exists a duty of international cooperation in the socioeconomic field and its exact content is controversial. According to Abi-Saab, the only real obligation in the domain of social and economic cooperation is that contained in Article 56 of the UN Charter, in which "all Members pledge themselves to take joint and separate action in co-operation with the Organization" for achieving the objectives of socioeconomic progress and development set forth in Article 55. The 1970 Declaration also requires cooperation in the economic, social, and cultural fields, as well as in the domain of science and technology. However, Abi-Saab argues that this Declaration does not clarify whether States recognize a legal duty to cooperate in the socioeconomic field, because under the heading: "The Duty of States to Cooperate with one another in accordance with the Charter," the 1970 Declaration uses the hortatory "should" in the economic, social, and cultural fields, as opposed to the "shall" contained in the remainder of the listing under that heading. Conversely, other commentators not only support the existence of a duty of international cooperation in the economic and social domains, particularly with respect to the protection of the atmosphere, but also claim that this duty has achieved customary law status.[31]

In 1992, the United Nations Conference on Environment and Development (UNCED) shifted the debate on economic development by addressing the environment-development relationship and firmly establishing the notion of sustainable development.[32] UNCED has also crystallized other ideas in the realm of the international law of cooperation, for example, that of an equal participation of States in the international legal system, which presupposes the recognition of their different responsibilities and tasks, as enunciated in Principle 7 of the 1992 Rio Declaration on Environment and Development (hereinafter: 1992 Rio Declaration).[33] Further international cooperation in the economic and social fields, for example, in the form of financial, scientific, and technological assistance required to help developing

[30] For example UNGA Res. 3362 (S-VII), September 16, 1975; Declaration on International Economic Cooperation, UNGA Res. S-18, May 1, 1990; International Co-operation for Economic Growth and Development, UNGA Res. 47/152, December 18, 1992.

[31] F. Biermann, "Common heritage of humankind: The emergence of a new concept of international environmental law," in *Archiv des Völkerrechts*, vol. 34, Hamburg, 1996, pp. 426–81, at pp. 462 and 465.

[32] B. Simma (ed.), *The Charter of the United Nations: A Commentary*, Second Edition, New York: Oxford University Press, 2002, p. 904.

[33] 1992 Rio Declaration on Environment and Development, United Nations Conference on Environment and Development, UN doc. A/CONF.151/6/Rev.1, 1992, 31 ILM 874, 1992.

countries meet certain objectives and ultimately achieve sustainable development, is now a common feature of environmental treaties.[34]

International cooperation is one of those notions that everyone understands but finds difficult to define precisely. In the environmental field, the principle of international cooperation is contained in virtually all multilateral environmental agreements (MEAs),[35] referred to in awards of international tribunals and generally supported by state practice, especially in areas involving the management of hazardous substances and environmental emergencies.[36] Since the early 1980s, UNGA resolutions[37] have addressed the issue of environmental cooperation and recalled the "historical responsibility" of States for the preservation of nature, attaching importance to "planned and constructive international co-operation in solving the problems of preserving nature."[38] Subsequent instruments have emphasized the crucial role of international cooperation in the prevention, reduction, and elimination of adverse environmental effects, as embodied in Principle 24 of the 1972 Stockholm Declaration[39] and Principle 27 of the 1992 Rio Declaration.[40] The Preamble of the 1982 World Charter for Nature[41] also stresses the importance of achieving cooperation in order to protect

[34] For example, Principle 9 of the 1992 Rio Declaration requests further international cooperation in the form of financial, scientific, and technological support for developing countries. References to cooperation in the form of financial and technical assistance are, for example, made in Articles 8, 9, and 12 of the 1992 Convention on Biological Diversity.

[35] Including for instance, the 1992 Convention on Biological Diversity, the 1968 African Convention on the Conservation of Nature and Natural Resources, the 1985 ASEAN Agreement on the Conservation of Nature and Natural Resources, the 1979 Bern Convention on the Conservation of European Wildlife and Natural Habitats, and the 1986 Convention for the Protection of the Natural Resources and Environment in the South Pacific Region.

[36] P. Sands, *Principles of International Environmental Law*, Cambridge: Cambridge University Press, 2003, p. 249.

[37] UNGA Res. 3129 (XXVIII), December 13, 1973, on Cooperation in the Field of the Environment Concerning Natural Resources Shared by Two or More States; UNGA Res. 34/186, 18 December 1979, on Co-operation in the Field of the Environment Concerning Natural Resources Shared by Two or More States; UNGA Res. 34/188, 18 December 1979, on International Cooperation in the Field of the Environment; and UNGA Res. 35/74, 5 December 1980, on International Co-operation in the Field of the Environment.

[38] UNGA Res. 36/7, October 27, 1981.

[39] 1972 Declaration of the United Nations Conference on the Human Environment, 11 ILM 1416, 1972.

[40] UN doc. A/CONF.151/26, vol. I. Various provisions in this declaration refer to cooperation, for example, Principle 5: "All States and all people shall *cooperate* in the essential task of eradicating poverty"; Principle 7: "States shall *cooperate* in a spirit of global partnership to conserve, protect and restore the health and integrity of the Earth's ecosystem"; Principle 12: "States should *cooperate* to promote a supportive and open international economic system that would lead to economic growth and sustainable development;" and Principle 13: "States shall also *cooperate* in an expeditious and more determined manner to develop further international law regarding liability and compensation for adverse effects of environmental damage."

[41] UNGA Res. 37/7, 28 October 1982, reprinted in 23 ILM 455, 1983. According to this declaration "States should effectively *cooperate* to discourage or prevent the relocation and transfer to other States of any

and safeguard the balance and quality of nature. The principle of international cooperation was framed under Principle 24 of the 1972 Stockholm Declaration as follows:

> International matters concerning the protection and improvement of the environment should be handled in a cooperative spirit by all countries, big and small, on an equal footing.
>
> Cooperation through multilateral or bilateral arrangements or other appropriate means is essential to effectively control, prevent, reduce and eliminate adverse environmental effects resulting from activities conducted in all spheres, in such a way that due account is taken of the sovereignty and interests of all States.

The vague content of the principle of international cooperation is given more definition in other declarations and environmental treaties. In particular, Principle 7 of the 1978 United Nations Environment Programme (UNEP) Draft Principles of Conduct in the Field of the Environment for the Guidance of States in the Conservation and Harmonious Utilization of Natural Resources Shared by Two or More States (hereinafter: 1978 UNEP Draft Principles)[42] provides further guidance as to the meaning of that principle by enumerating ways in which States can cooperate, including "exchange of information, notification, consultation and other forms of cooperation carried out on the basis of the principle of good faith and in the spirit of good neighborliness."[43] For example, under the 1994 Convention on the Cooperation for the Sustainable Development of the Danube River,[44] this principle is translated into a duty to negotiate, conduct consultations, and exchange information. The duty to cooperate takes the form of specific measures, enumerated in Articles 5 and 6, involving, for example, the identification of groundwater resources and protection zones, or the recording of conditions of natural water resources within the Danube River, among others.[45]

Other environmental treaties also invoke the principle of international cooperation as a legal duty. For example, Article 6(1) of the 1972 Convention Concerning the Protection of the World Cultural and Natural Heritage recognizes the cultural and natural heritage as a world heritage for whose protection "it is the duty of international community as a whole to co-operate." According to Article 197 of the 1982

activities and substances that cause severe environmental degradation or are found to be harmful to human health."

42 The Draft Principles of Conduct in the Field of the Environment for the Guidance of States in the Conservation and Harmonious Utilization of Natural Resources Shared by two or More States, were approved by the UNEP Governing Council on May 19, 1978 by Decision 6/14 (UNEP/IG. 12/2), reprinted in 17 ILM 1091, 1093, 1978.

43 Ibid.

44 Article 4, 35 ILM 651, 1996.

45 Ibid., Articles 5 and 6.

United Nations Convention on the Law of the Sea (UNCLOS), States "shall cooperate on a global basis, and as appropriate, on a regional basis" for the protection of the marine environment.[46] Despite the imprecise content of the principle of international cooperation (or the "duty to cooperate"), Alexandre Kiss suggests that its repetition in innumerable treaties, judicial awards, and nonbinding instruments reflects a general acceptance of a duty to cooperate, as framed in Principle 24 of the Stockholm Declaration, which seems to have acquired the status of customary international law.[47] Whereas the duty to cooperate in the environmental field is generally recognized in relation to activities involving, *inter alia*, exchange of information, consultation, and notification, its acceptance in areas involving financial assistance or technology transfer finds more resistance, as discussed in the following sections.

1.3. LIMITS TO INTERNATIONAL COOPERATION IN THE ENVIRONMENTAL FIELD

Some skeptics may claim that international cooperation might be inadequate, at least for addressing some environmental issues. As Daniel Bodansky suggests, certain environmental problems may require a multilateral approach, for example, those that have sources in many countries or produce transboundary effects and involve multiple parties that then have a legitimate claim to take part in certain decision-making processes. However, he argues that unilateral actions taken by a State, for example, to prevent the pollution of its coastline, as envisaged under the 1969 International Convention on Civil Liability for Oil Pollution Damage,[48] are the norm in environmental policy, whereas international action is the exception, requiring special justification. Therefore, despite the growth of multilateral decision making, "international cooperation often remains unachievable or illusory."[49] In the particular case of the Amazon, David Goodman and Michael Redclif claim that the root causes of deforestation, especially in the Brazilian Amazon, are related to complex land ownership issues that ultimately only Brazil can handle: "the fate of the forest is inextricably linked with national issues largely immune to international public opinion and inter-governmental agreements." They note that the Amazon is not a nature reserve, but a region inhabited by millions of people whose needs

[46] Sands, op. cit., p. 244.

[47] A. Kiss and D. Shelton, *International Environmental Law*, New York & London: Transnational/Graham & Trotman, 2000, p. 43.

[48] 9 ILM 45 (1970).

[49] D. Bodansky, "What's so bad about unilateral action to protect the environment?", *European Journal of International Law*, 2000, vol. 11, no. 2, pp. 339–47, at p. 347. Other examples provided by Bodansky of "unilateral actions" include the port State power to enforce international pollution standards, as established under Article 218 of the 1982 United Nations Convention on the Law of the Sea.

may involve clearing the forest; therefore, "the point is to provide a socio-economic framework within which forest clearance can be managed and ultimately only the Brazilian government can do that."[50]

International cooperation is indeed not a panacea for the protection of the Amazon, and is probably not the solution to certain environmental problems in the Brazilian Amazon, such as unclear land tenure, low institutional capacity, and poor governance,[51] which much depend on improving the capacity of the Brazilian government to handle its own issues. However, certain areas do require international cooperation, for example, those related to the management of migratory species, international rivers, transboundary pollution, or the combating of illegal wildlife trade, among others. In practice, Anthony Hall notes that international cooperation has proven useful, if not essential, for carrying out conservation programs in the Brazilian Amazon.[52] A number of international initiatives, such as the Pilot Program to Conserve the Brazilian RainForest (PPG7), have been instrumental in promoting conservation planning and sustainable development.[53]

At the regional level, international cooperation among the Amazon States *inter se* has been limited due to factors such as the lack of financial and technical capacity to carry out cooperative measures and insufficient knowledge or political will, whereas international cooperation between these countries and the international community has been shaped by other factors, for example, due to their different views as to what cooperation means in practice. In their efforts to promote international cooperation, different groups of States have distinct interests. For example, in a North–South perspective, the extent of international cooperation is limited to what developed countries may want to give and what developing countries may be willing to renounce. This latter group tends to perceive international cooperation as involving basically technical and financial assistance provided by developed countries as a means for attaining environmental goals. For example, under the 1989 Declaration of Caracas, adopted by the ministers of the Group of 77,[54] developing countries stated that "the possibility for our countries to assume concrete commitments and participate effectively in international instruments, programs and projects in the field of environmental protection depends, to a large extent, on our having adequate technology on accessible

[50] D. Goodman and M. Redclif, *Environment and Development in Latin America: the Politics of Sustainability*, Manchester: Manchester University Press, 1991, p. 129.

[51] World Bank, *A More Equitable, Sustainable, and Competitive Brazil: Country Assistance Strategy 2004–2007*, Annex 8: "A strategic contribution to a more sustainable Amazon," p. 197.

[52] A. Hall, *Sustaining Amazonia: Grassroots Actions for Productive Conservation*, New York: Manchester University Press, 1998, p. 35.

[53] Laurance et al., op. cit.

[54] Adopted by the ministers of foreign affairs of the member countries of the Group of 77 on the occasion of the twenty-fifth anniversary of the creation of the Group, held in Caracas, Venezuela, 21–23 June 1989, http://www.g77.org/doc

terms and additional financial resources."[55] Likewise, pursuant to the 1992 Manaus Declaration,[56] adopted by the heads of States of the 1978 Amazon Cooperation Treaty (ACT), the solution to environmental problems requires international cooperation, which should take place through increased financial resources, technology transfer, and greater commercial flows.

Developing countries reject any form of cooperation perceived as interfering with their domestic affairs and infringing on their national sovereignty over natural resources. This is particularly the case in relation to natural resources or sites located within their national jurisdictions, as observed in the debate over the adoption of a global legally binding instrument on forests at the 1992 UNCED. In regional forums, developing countries continue to oppose the adoption of a global binding instrument, reflected, for example, in the 2005 Paracas Declaration adopted at the First Meeting of the Andean Community Council of Ministers of Environment and Sustainable Development, which stressed the "inadvisability of moving toward a legally binding international regime for the conservation and sustainable use of forests of all kinds."[57]

For obvious reasons, States owning rich natural resources will refuse any form of international cooperation viewed as threatening to their national sovereignty. If cooperation is perceived as entailing any form of interference in domestic affairs, for example, in the form of international schemes such as that announced by the British environment secretary (David Miliband) of a wholesale privatization of the Amazon, it will not prosper.[58] In response to this particular proposal, the Brazilian minister of foreign affairs (Celso Amorim), on the occasion of the Eighth Meeting of Ministers of Foreign Affairs, held in 2004 under the aegis of the Amazon Cooperation Treaty Organization, reacted by stating "Amazonia is ours."[59] The 2004 Declaration of Manaus, an outcome of that meeting, reaffirmed the parties' sovereignty over the Amazon. However, this same declaration also recognized that the Amazon States have a "sovereign responsibility" over that region (paragraph 2) and, as stated by the Brazilian minister of foreign affairs, they should be seen as its "sovereign guardians."[60]

55 H. Munoz, "The environment in inter-American relations," in H. Munoz, *Environment and Diplomacy in the Americas*, London: Lynne Rienner, 1992, p. 7.

56 1992 Manaus Declaration, at the Second Meeting of Presidents of the Parties of the 1978 Amazon Cooperation Treaty, held in Manaus, February 10–11, 1992, Pro Tempore Secretariat, *Base Jurídica del Tratado de Cooperación Amazónica: Antecedentes Constitutivos de la Organización del Tratado de Cooperación Amazónica*, La Paz, Bolivia, 2002, pp. 42–43.

57 Paragraph 15. Adopted at the First Meeting of the Andean Community Council of Ministers of the Environment and Sustainable Development, held in Paracas, Peru, March 31 to April 1, 2005, http://www.comunidadandina.org

58 *Sunday Telegraph*, October 1, 2006.

59 *Jornal Hora do Povo*, September 17, 2004, p. 3.

60 Ibid.

The fact that international cooperation may be limited in some ways, especially in relation to resources falling within national boundaries, does not mean that national sovereignty is an obstacle to international cooperation. It is sometimes suggested that national priorities always come first, even with respect to ecologically important life-support systems, such as the atmosphere or cultural and natural heritage sites.[61] In this sense, principles such as that of permanent sovereignty over natural resources could be viewed as impairing international cooperation. However, national sovereignty has not restrained the development of MEAs (and other international legal instruments) aimed to protect natural resources situated exclusively within national boundaries, such as the 1992 Convention on Biological Diversity (CBD)[62] and the 1973 Convention on International Trade in Endangered Species of Wild Fauna and Flora (CITES).[63] As the World Commission on Environment and Development noted,[64] within an emerging international legal framework under which international cooperation seeks to ensure equitable sharing and environmental conservation, national sovereignty over natural resources may well continue to serve as a basic principle.[65] In this respect, Abi-Saab explains that the changed composition of the international community has pushed the substantive development of international law in two apparently opposite, but in fact complementary, directions: While it has reinforced national sovereignty by developing the principles of the UN Charter and making explicit the "non-trespass" rule through UNGA resolutions,[66] it has also reinforced the idea of a "duty to cooperate," the roots of which are found in the principles of the UN Charter, enabling a more collaborative approach in international relations:[67]

> le concept de souveraineté vient renforcer celui d'égalité et lui confère une plus grande substance. Ainsi, loin de faire obstacle à la coopération internationale,

[61] C. de Klemm and C. Shine, "Biological diversity conservation and the law: Legal mechanisms for conserving species and ecosystems," World Conservation Union (IUCN), Environmental Policy and Law Paper, no. 29, Gland, Switzerland, 1993, p. 1.

[62] 31 ILM 818, 1992.

[63] 993 UNTS 243, 1974.

[64] The World Commission on Environment and Development, also known as the Brundtland Commission, was created by UNGA Res. 38/161, December 19, 1983, adopted at the Thirty-eighth Session of the United Nations. The Commission, chaired by Mrs. Gro Harlem Brundtland of Norway, produced a report entitled *Our Common Future*, which was submitted to the UNGA, during its forty-second session in 1987 (in UN World Commission on Environment and Development, *Our Common Future*, Oxford: Oxford University Press, 1988, p. 352).

[65] See N. Schrijver, *Sovereignty over Natural Resources: Balancing Rights and Duties*, Cambridge: Cambridge University Press, 1997, pp. 249–50.

[66] Notably the 1970 Declaration of Principles of International Law Relating to Friendly Relations and Cooperation Among States in accordance with the Charter of the UN, UNGA Res. 2625 (XXV), October 24 1970, and on the Definition of Aggression, UNGA Res. 3314 (XXIX), December 14 1974.

[67] Abi-Saab, *Cours général*, op. cit., p. 351.

> comme d'aucuns pourraient le craindre, le principe de la souveraineté permanente jette-t-il les bases d'une coopération mieux équilibrée, et par conséquent, plus fructueuse et plus stable.[68]

Whereas national sovereignty, and especially the principle of national sovereignty over natural resources, is not an obstacle to international cooperation, it does shape the form in which international cooperation may occur. As noted, States owning rich natural resources, such as tropical forests, reject cooperative arrangements related to resources located in their own national territories, if they are perceived as threatening to their sovereignty or involve decreased levels of independence. Therefore, the way in which they cooperate with other countries is shaped to a great extent by national interests. For example, while developed countries may favor cooperative measures aimed at the conservation of natural resources and sites, developing countries, which depend economically on such resources, tend to privilege approaches based on their economic exploitation. Although the Amazon States recognize their "sovereign responsibility" over that region, they reject the idea of protecting the Amazon as the world's park. Instead, according to the document prepared in 1990 under the aegis of the Latin America and Caribbean Commission on the Development and Environment, entitled *Our Own Agenda,* these countries stressed the particular role of the Amazon:[69]

> The Amazon cannot be considered simply an ecological treasure and an important regulator of global climate; it is a major development resource. The preservation of the tropical rain forest, and especially the Amazon forest, will depend on the mobilization of research, and development, technology, and financial resources for sustainable management of this area. A commitment of support from the international community and voluntary co-operation of the eight sovereign Amazonian states is needed to preserve the Amazon as an asset for the region and the world community.
>
> Preservation of the Amazon is of interest to the world community. However, it is of great interest to the Amazonian countries. The issues must therefore be debated by them on their own terms with the support of foreign scientists and enlightened international public opinion.

Developed countries, on the other hand, seem to recognize the crucial role of international cooperation in the environmental field, as they have been providing financial assistance to developing countries, for example, through multilateral organizations or MEAs. However, they have shown resistance to some developing countries' claims, for example, those related to access to and transfer of technology in

[68] G. Abi-Saab, "La souveraineté permanente sur les ressources naturelles," in M. Bedjaoui, *Droit International, Bilan et Perspectives*, Paris: Editions A. Pedone, 1991, p. 661.

[69] Latin America and Caribbean Commission on the Development and Environment, Inter-American Development Bank, United Nations Environmental Programme, *Our Own Agenda,* in Munoz, op. cit., pp. 88–89.

fields such as biotechnology and pharmaceuticals. In the context of negotiations on an international seabed regime, under Part XI of UNCLOS,[70] for example, it became apparent that the United States and other industrialized countries would not become parties to the Convention unless significant changes were made in Part XI, modifying provisions that could potentially interfere with the principles of the free trade, such as those imposing production limitations and obligatory transfer of technology to developing countries.[71] Therefore, to ensure the participation of developed countries a New Agreement Relating to the Implementation of Part XI of UNCLOS was adopted in 1994, prior to the entry into force of that Convention, to take account of the objections raised by those countries.[72] Among other things, this agreement, which is now interpreted together with UNCLOS as a single instrument,[73] replaced the mandatory transfer of technology related to activities in the "Area,"[74] originally foreseen under Article 144 of the Convention. It introduced a system where the Enterprise[75] and developing countries wishing to obtain deep seabed mining technology will seek it on fair and reasonable commercial terms through joint venture arrangements.

Despite those more contentious areas, financial assistance has been provided by developed countries in relation to things deemed essential to the survival of humankind,[76] such as biodiversity and the climate system, under the CBD and United Nations Framework Convention on Climate Change, respectively.[77] The need to help developing countries protect natural resources considered to be of "world significance," such as tropical forests, has been long recognized, as observed at the Seventh World Forestry Congress held in 1972,[78] under the auspices of the Food and Agriculture Organization of the United Nations. In view of the gap between developed and developing States, this meeting stressed the need to further assist the latter States:

> The congress has noted with concern that the gap between developed and developing nations continues to widen. Some of the largest forest resources of the

[70] 21 ILM 1261, (1983).

[71] M.A. Fitzmaurice, "International protection of the environment," *Collected Courses of The Hague Academy of International Law*, The Hague/Boston/London: Martinus Nijhoff, 2001, vol. 293, 2002, p. 156.

[72] Ibid.

[73] Article 2(1), 1994 Agreement relating to the Implementation of Part XI of the United Nations Convention of the Law of the Sea (UNCLOS), 33 ILM 1309, 1994.

[74] According to Article 1(1)(1) of UNCLOS, "Area means the seabed and the ocean floor and subsoil thereof beyond the limits of national jurisdiction."

[75] The "Enterprise" is the operating arm of the Authority, through which the operations in the Area are conducted (UNCLOS, Articles 170(1) and 153(2)(a) and Annex IV, Article 1(1)).

[76] G. Handl, "International law and the protection of the atmosphere," *American Society of International Law, Proceedings* 62, 1989, p. 61.

[77] 31 ILM 849/851, 1992.

[78] This conference was held in Buenos Aires, Argentina, in October 1972; official documents are available at http://www.fao.org/forestry/18507/en

> world lie in countries where economic and social development is lagging. Over and above considerations of social justice, the fact that these resources have a world significance requires greatly intensified efforts by and aid to less developed countries for the development and proper management of their forest resources, so that these resources may simultaneously serve to raise the living standards of their own peoples and contribute to the world's expanding need for forest products.

Technical and financial assistance to developing countries, however, is not a cure-all. Complex environmental problems such as deforestation probably require innovative international cooperation mechanisms at the national, regional, and global levels. In view of this need, the present study enquires what forms of international cooperation are required among the Amazon States *inter se*, and between them and the international community to help protect the Amazon more effectively. With respect to the countries sharing the Amazon, should cooperation be limited to the exchange of information and the undertaking of joint activities? Conversely, what role should the international community play in the protection of the Amazon? Should it be limited to providing financial and technological assistance to the Amazon States? More generally, what forms of international cooperation are required in order to protect the Amazon more effectively?

1.4. INTERNATIONAL COOPERATION IN THE CONTEXT OF THE PRESENT STUDY

Despite the heterogeneity of the Amazon and the disparities among the basin States, the network of international rivers and ecosystems spread over their territories connects all these countries and reveals the ecological unity of the Amazon. Therefore, activities undertaken in the territory of one country can affect another basin State, or the ecological integrity of the Amazon a whole. Likewise, environmental mismanagement in that region, leading, for example, to deforestation or biodiversity loss, may impact the entire international community. For example, the water draining out of the Amazonian forests into the Atlantic Ocean constitutes nearly 15 to 20 percent of the total global discharge of fresh river water; therefore, a modification in the amount of fresh water in the Amazonian hydrological cycles could be sufficient to influence some of the great ocean currents that are important regulators of the global climate.[79] On the other hand, global environmental

79 United Nations Environment Programme (UNEP), Amazon Cooperation Treaty Organization (ACTO) and Research Center of Universidad del Pacífico (CIUP), *Environment Outlook in Amazonia, Geo Amazonia*, 2009, p. 101, http://www.unep.org/pdf/GEOAMAZONIA.pdf

problems such as atmospheric pollution or climate change also affect the Amazon. A tendency in that region for increased vulnerability to floods, droughts, and climate change has already been noted. It is predicted that if the loss of forest exceeds 30 percent of vegetation cover, rainfall inhibition will get stronger, generating a vicious circle leading to forest fires, reducing the release of water vapor, and increasing the emission of smoke into the atmosphere with the consequent suppression of precipitation.[80]

This positive feedback loop illustrates the often-cited idea that the environment knows no boundaries. Despite eventual differences, the Amazon States share a common interest in the conservation and sustainable use of the Amazon, and also face similar challenges in attaining that goal. Also, given the role of the Amazon in helping maintain the global climate, the international community seems to have a legitimate interest in its protection. Certain environmental issues, such as transboundary water pollution, the depletion of migratory species or the loss of habitat across their migration range, and the illegal trade in wildlife species require joint solutions, as do issues of the magnitude and complexity of global warming. International cooperation is then required at all levels (bilateral, regional, and global) to address different types of problems, especially when shared ecosystems are concerned.

The Amazon States seem to value international cooperation, as manifested in bilateral declarations (Bolivia–Brazil,[81] Peru–Brazil,[82] Brazil–Colombia,[83] Suriname–Brazil,[84] Brazil–Guyana,[85] and Brazil–Venezuela[86]) that reaffirm the idea of a "sovereign responsibility" over the Amazon and the need to cooperate. At the Second Meeting of the South American Community of Nations in 2006, where the regional integration of the Amazon (also of the Andes, Altiplano,

80 Ibid., p. 24.

81 "Visita de Estado do Presidente Evo Morales Ayma ao Brasil, Comunicado Conjunto Brasil – Bolívia: bases para uma parceria estratégica" of February 15, 2007, http://www.otca.org.br/ep/noticia/noticia.php?idNoticia=1578&tipoN=6

82 "Visita de Estado del Presidente Alan García Pérez ao Brasil, Comunicado Conjunto Peru–Brasil: Ampliando los horizontes de una Alianza Estratégica," of November 9, 2006, http://www.otca.org.br/en/noticia/noticia.php?idNoticia=1593&tipoN=9

83 "Declaração Conjunta dos Presidentes da República Federativa do Brasil e da República da Colômbia," December 14, 2005, http://www.otca.org.br/ep/noticia/noticia.php?idNoticia=1587&tipoN=6

84 "Declaração Conjunta: Suriname – Brasil," February 22, 2005, http://www.otca.org.br/en/noticia/noticia.php?idNoticia=1599&tipoN=9

85 "Comunicado Conjunto por ocasião da visita de Estado à Guiana do Excelentíssimo Senhor Presidente da República Federativa do Brasil, Luiz Inácio Lula da Silva," February 15, 2005, http://www.otca.org.br/en/noticia/noticia.php?idNoticia=1600&tipoN=9

86 "Comunicado Conjunto, Alianza Estratégica Venezuela – Brasil," February 14, 2005, http://www.otca.org.br/en/noticia/noticia.php?idNoticia=1601&tipoN=9

Chaco, Pantanal, and the Plata River Basin) was discussed, the Brazilian president (Luiz Inácio Lula da Silva) stated that "in a more complex world, no country, no matter how big or economically powerful, can remain in isolation."[87] Other regional arrangements in different parts of the world, for example the Association of Southeast Asian Nations (ASEAN), have proven to be an indispensable link between and complement to national and global initiatives.[88] As noted above, international cooperation is one of the pillars of current environmental treaties and is deemed essential to protect and restore the health and integrity of the global environment (Principle 7, Rio Declaration). In view of this emphasis, this study assumes that international cooperation plays a crucial role in the protection of the Amazon and looks at the region from an international perspective in order to examine how the Amazon States have been cooperating *inter se* and with other States, and what role the international community plays in that regard.

To assess how international cooperation takes place in the Amazon, this book focuses on international legal instruments adopted by the Amazon States *inter se* and with other States at the bilateral, regional, and global levels, which are part of an "international legal system of the Amazon." A legal system can be understood as a (1) set of elements, (2) arranged in an order (characterized by the interaction between the different elements), (3) with a certain degree of unity or cohesion (which facilitates the description of the elements as part of a bigger role).[89] The elements of the proposed system include (legally binding and nonbinding) instruments related to the environment (including matters such as freshwater, biodiversity, health and sanitation, forests, and indigenous peoples, among others) adopted by the Amazon countries *inter se*, or between them and other States. The international legal instruments examined here are grouped under three categories. Firstly, they include (legally binding and nonbinding) instruments related to the environment adopted by the Amazon States *inter se*, bilaterally or under subregional and regional organizations in which they participate. Secondly, they include legal instruments signed between those States and other countries in the Latin American and the Caribbean regions[90] at the bilateral or regional levels, which, if not addressing the Amazon in

[87] "Discurso do Presidente do Brasil na II Reunião de Presidentes da Comunidade Sul-Americana de Nações (CASA)," December 8, 2006, http://www.otca.org.br/en/noticia/noticia.php?idNoticia=1592&tipoN=9. This speech was delivered by the Brazilian president at the Second Meeting of Presidents of the South America Community of Nations, held in Cochabamba-Bolívia, December 8, 2006.

[88] K.L. Koh and N.A. Robinson, "Regional environmental governance: Examining the Association of Southeast Asian Nations (ASEAN) Model," *Global Environmental Governance*, Yale Center for Environmental Law and Policy, 2002, pp. 1–21, at p. 18.

[89] Y. Shany, *The Competing Jurisdiction of International Courts and Tribunals*, New York: Oxford University Press, 2003, p. 89.

[90] The Caribbean region is considered here because some Amazon countries, specifically Guyana and Suriname, are also Caribbean. Both countries participate in the Caribbean Community.

particular, are applicable to this region. Thirdly, they include MEAs to which some or all Amazon States are parties, which then apply to their respective Amazonian territories.

In practice, those instruments coexist without a particular order or hierarchy, but they still have a degree of coherence because they all relate to the environment, are applicable to the Amazon (or to parts of it), and involve some or all Amazon States. The selection of international legal instruments stretches from the time of the political independence of the Amazon States from the European colonial powers (mostly by the end of the nineteenth century) up to the present. Apart from these international legal instruments, this study also presents some of the major organizations working toward the protection of the Amazon at different levels (subregional, regional, and global) in which the Amazon countries participate.

A central element in the international legal system of the Amazon that will be closely scrutinized is the 1978 ACT.[91] The relevance of this treaty lies in the fact that it is the only regional agreement involving all and exclusively Amazon States (Bolivia, Brazil, Ecuador, Colombia, Guyana, Peru, Suriname, and Venezuela) intended to develop the parties' respective Amazonian territories and achieve the "preservation of the environment, and the conservation and rational utilization of the natural resources of those territories" (Article 1). French Guyana, an overseas department of France, although part of the Amazon basin, was excluded from the ACT due to its dependent status *vis-à-vis* a European State.

The ACT takes the form of a framework agreement and, as such, contains guiding principles and institutional bodies in charge of overseeing its implementation.[92] As Kiss notes, a framework agreement seems at first to be "soft law," but in fact it is a sort of *pactum de contrahendo*, by which the parties manifest their willingness to adopt more precise legal obligations, usually negotiated under meetings or conferences of the parties, and introduced through amendments, protocols, or annexes. Likewise, the ACT contains *lege feranda* provisions rather than specific legal obligations.[93] In any case, the evolution of this treaty, by a gradual introduction of more precise commitments, is foreseen under Article 1, in the form of "operational agreements and understandings" or other "pertinent legal instruments." The evolution of this treaty's normative and institutional frameworks over time is examined in detail in this study, which claims that the ACT can play a key role in improving the effectiveness of the international legal system of the Amazon as a factor of cohesion among the different legal instruments and

91 17 ILM 1045.

92 A. Kiss, "Les Traités-Cadres: une technique juridique caractéristique du droit international de l'environnement," *Annuaire Français de Droit International*, vol. 39, 1993, pp. 792–97.

93 C.G. Caubet, "Le Traité de Coopération Amazonienne: régionalisation et développement de l' Amazonie," *Annuaire Français de Droit International*, vol. 30, 1984, pp. 803–18.

institutions existing at different levels. The study assumes that a "stronger" ACT, able to change the parties' behavior in such a way as to eliminate or ameliorate existing environmental problems,[94] should involve a set of precise commitments, in the form of additional (legally binding or nonbinding) legal instruments, and more effective institutional bodies able to oversee this treaty's implementation, so that its unique mandate can be fulfilled.

In addition to those international legal instruments, comprising the international legal system of the Amazon, this book also examines certain "mechanisms" or "instruments" currently used as conservation tools in the Amazon and elsewhere, under the modality of "positive incentives," which include both market-based and nonmarket-based mechanisms. As more "traditional" techniques, such as that of command-and-control or technical cooperation, have showed certain limitations in dealing with environmental problems, other types of responses are being tested, especially involving market-based mechanisms, such as forest certification, payments for environmental services, and carbon trading. This book claims that such mechanisms can be effective and agile tools, especially for tackling environmental problems requiring urgent action, such as deforestation.

In view of the role of the Amazon in maintaining the balance of the global environment, this book also enquires whether that region should be granted a special legal status and, if so, the possible implications of this in terms of international cooperation. To assess whether the Amazon can be attributed a particular international legal status, the study looks at other areas in the world that have a particular status, from which some elements of comparison could be drawn.

The legal status of natural resources or sites varies, depending on whether they are located within the territory of one State, are shared by a group of States, or lie outside the territorial jurisdiction of a State. These categorizations have different impacts on the freedom of States to exploit such resources or certain areas.[95] For example, natural resources falling exclusively within the territory of one State remain the property of that State and are subject to its permanent national sovereignty.[96] In this case, States have considerable discretion to define how those resources should be used. However, the unsustainable use of certain natural

[94] O.R. Young and M.A. Levy (eds.), *The Effectiveness of International Environmental Regimes: Causal Connections and Behavioral Mechanisms*, Cambridge, MA and London: MIT Press, 1999, p. 2; D.G. Victor, Kal Raustiala, and E.B. Skolnikoff (eds.), *The Implementation and Effectiveness of International Environmental Commitments: Theory and Practice*, Cambridge MA, London: MIT Press, 1998, p. 1. Effectiveness is defined as the "degree to which international environmental accords lead to changes in behavior that held solve environmental problems."

[95] P. Birnie and A. Boyle, *International Law and the Environment*, Oxford: Oxford University Press, 2002, p. 137.

[96] J.A. Barberis, *Los recursos naturales compartidos entre estados y el derecho internacional*, Madrid: Editorial Tecnos, 1979, p. 145.

resources or sites, even when confined to precise territorial limits of one State, may adversely affect other States or, more broadly, the environmental equilibrium of a region or the global environment. For that reason, the protection of certain natural resources or areas, especially those essential for maintaining the ecological balance of the biosphere, may be a matter of international concern, as in the case of the Earth's climate or biodiversity. Nonetheless, the physical location of natural resources and sites, even in cases where their preservation can be considered a common concern, determines to some extent how much discretion States may have in their exploitation and also the form in which international cooperation takes place. States might be more reluctant to take part in cooperative arrangements when it comes to natural resources or sites located within their national jurisdictions. As Patricia Birnie and Alan Boyle note, whereas a reasonably comprehensive pattern of international cooperation now exists for protecting common areas such as the high seas, the deep seabed, or the Antarctic, it cannot be assumed that comparable obligations apply to areas that fall wholly within the boundaries of national States.[97]

In any case, certain natural resources or sites, due to their ecological importance, are internationally protected even when located within national borders, such as endangered species or natural sites of outstanding universal value, protected under CITES and the 1972 Convention Concerning the Protection of the World Cultural and Natural Heritage, respectively. In some cases, areas of high ecological importance are given a special international legal status, given the interest of the international community in their preservation, for example the Antarctic, the seabed beyond the limits of national jurisdiction, and the Moon. The particular legal status of such areas allows a group of States or the international community as a whole to take certain cooperative measures that would not otherwise be permitted. Therefore, not only the physical location of certain natural resources or sites, but also their particular international legal status, has implications in terms of international cooperation. This study claims that the protection of the Amazon in particular, due to its role in maintaining global ecological conditions, can be considered a common concern of humankind.

The chapters of this book move from the particular regional context of the Amazon – which includes an analysis of the socioeconomic characterization of the region, of international legal instruments adopted at different levels (bilateral, subregional, and regional) and an overview of some subregional and regional organizations in which the Amazon States participate (Chapters 2–6) – to a broader perspective that takes into account the interactions among those States, and between them and the international community. Whereas those initial chapters focus on the Amazon

97 Birnie and Boyle, op. cit., p. 89.

and on how the basin States interact among themselves and with other countries in the region, the three remaining chapters (Chapters 7–9) look at how international treaties and mechanisms negotiated at the global level apply in the Amazon (and may contribute to its protection) and also examine the policies of major global institutions in relation to this region. Building on the arguments in these chapters, the last chapter enquires whether the Amazon benefits from a particular international legal status and the possible implications of such a categorization.

2

The Characteristics of the Amazon Region

2.1. PHYSICAL CHARACTERISTICS

2.1.1. *Land Area*

The Amazon[1] is the name given to a vast region located in the northern portion of South America, drained by a vast network of rivers and their tributaries that discharge into the Atlantic at the mouth of the Amazon River. This region (also known as pan-Amazônia) is the planet's largest forest ecosystem and water basin.[2]

The Amazon extends over eight South American countries: Brazil, Bolivia, Ecuador, Colombia, Guyana, Peru, Suriname, and Venezuela (as well as French Guyana, an overseas department of France) with a total area of approximately 7.5 million square km.[3] Of this total, Brazil owns 63 percent (or 4 million square km). The remaining 37 percent (2.4 million square km) are distributed among Peru

[1] In 1541, the Spanish commander Gonzalo Pizarro led a great expedition from Quito in the Andes westward into the Amazon rainforest in search of gold. Pizarro eventually returned to Quito in 1543. A small advance detachment led by Francisco de Orellana, which had been designated to explore the Napo River, never returned to meet Pizarro or to Quito. Disregarding orders, Orellana's troops continued downstream on the Napo until they encountered a vast river, later named "River of the Amazons." The name derives from Greek mythology and, allegedly, was given after an attack on the expedition by Indian warrior women. According to Greek mythology, the Amazons were a legendary nation of tall, strong, and fierce warrior women (probably of Iranian origin, but whose location has also been attributed variously to the Black Sea, Scythia, and Libya). In their lands there were no men. To improve their archery, they disfigured their bodies, amputating one breast. In accordance with popular Greek etymology, the word *amazonas* derives from "*mazos*" meaning breast, with the prefix "a" meaning without, in other words, without breast(s) (E.E. Miranda, *Quando o Amazonas Corria para o Pacífico: Uma história desconhecida da Amazônia*, Petrópolis: Editora Vozes, 2007, pp. 138–39).

[2] United Nations Environment Programme (UNEP), Amazon Cooperation Treaty Organization (ACTO) and Research Center of Universidad del Pacífico (CIUP), *Environment Outlook in Amazonia, Geo Amazonia*, 2009, p. 148, http://www.unep.org/pdf/GEOAMAZONIA.pdf

[3] M.Goulding, R.B. Barthem, and E. Ferreira, *The Smithsonian Atlas of the Amazon*, Washington, DC: Smithsonian Books, 2003. Some studies indicate other figures for the total surface of the Amazon, see UNEP, ACTO, CIUP, op. cit., p. 41.

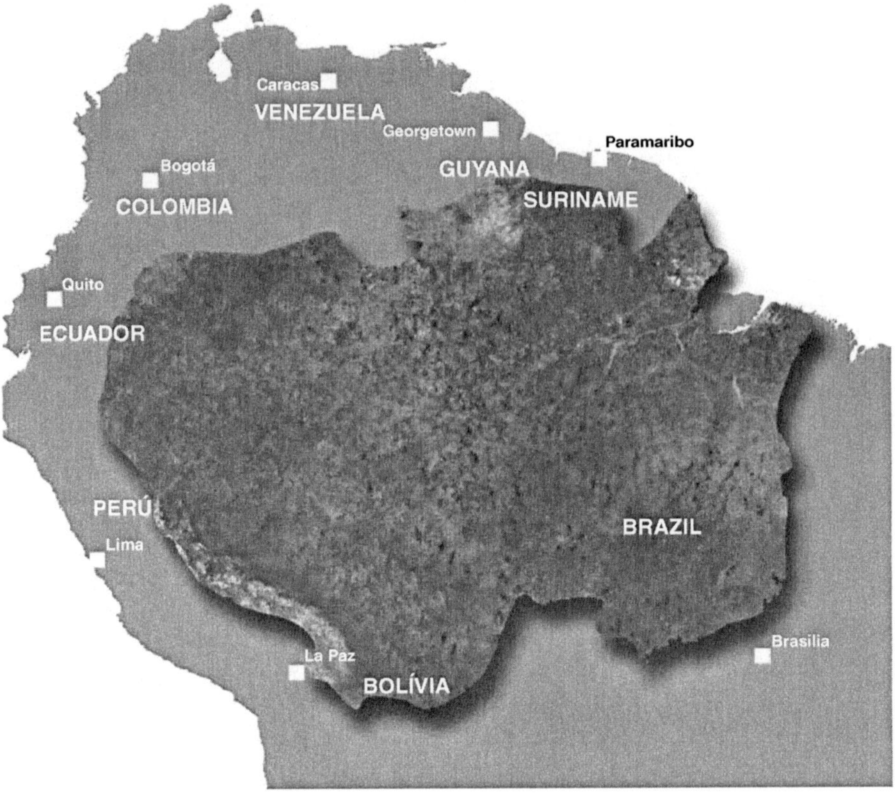

MAP 2.1. The Amazon basin.

Source: Informe de de Gestión 2004–2005, Amazon Cooperation Treaty Organization (ACTO), p. 6.

(10 percent), Colombia (7 percent), Bolivia (6 percent), Venezuela (6 percent), Guyana (3 percent), Suriname (2 percent), Ecuador (1.5 percent), and French Guyana (1.5 percent).[4] Including 7,408 km of Atlantic coastline, Brazil's borders extend for a total of 23,127 km and its land borders with the other seven Amazon countries extend 11,449 km (see Table 2.1). The longest of these borders are with Bolivia, Peru, and Colombia. Ecuador is the only Amazon State with which Brazil does not share a land border.

In Brazil, Article 2 of Law 1806 of January 6, 1953, created the concept of the "Legal Amazon" (Amazônia Legal), an administrative unit encompassing various states (Acre, Amazonas, Amapá, Pará, Rondônia, Roraima and Tocantins, Mato Grosso, part of

[4] M. Lentini et al., *Fatos Florestais da Amazônia,* Belém: Instituto do Homem e Meio Ambiente da Amazônia (Imazon), 2005, p. 23.

TABLE 2.1. *Brazil's borders with the Amazon States (km)*

Country	Land border (km)
Venezuela	1,495
Guyana	1,606
Suriname	583
Bolivia	3,126
Peru	2,995
Colombia	1,644
Ecuador	0
TOTAL	11,449

Source: Brazilian Institute of Geography and Statistics (IBGE), *Anuário Estatístico do Brasil*, 1980.

Maranhão, and some municipalities of Goiás). Amazônia Legal is not exactly equivalent to the Brazilian Amazon, since portions of some of these states lie outside the Amazon Basin watershed.[5] The Legal Amazon covers an area of approximately 5 million square kilometers, which represents 59 percent of Brazil's total land area.[6] Most of the existing official statistics available refer to the Legal Amazon (Table 2.2).[7]

The Amazon is home to a wide variety of species of flora and fauna and is an important area for endemism, making it a genetic reserve of global importance. It is also a culturally diverse region with 420 distinct indigenous peoples, 86 languages, and 650 dialects.[8] Although the Amazon is known for its abundance of natural resources, such as forests, minerals, petroleum, and natural gas, its inhabitants live at levels of poverty far worse than the national averages. The Amazon is also changing rapidly, with profound changes to its ecosystems, due to factors such as the growth of economic activities, the construction of infrastructure, and the establishment of human settlements. By 2005, accumulated deforestation in the Amazon had reached 857,666 km^2, which means that the vegetation cover of the region had been reduced by approximately 17 percent. The consequent loss of biodiversity is reflected in the increasing number of threatened species. The Amazon has been distanced for a long time from the basin States' "active space"; however, they are progressively integrating the region into their political-administrative systems and into their national economies, particularly Brazil. In spite of environmental changes, the Amazon still

5 P. Barreto et al., *Human Pressure on the Brazilian Amazon Forests*, Belem: IMAZON, Global Forest Watch and World Resources Institute, 2006, p. 23.

6 Lentini et al., op. cit., p. 24.

7 Barreto et al., op. cit., p. 23.

8 Information and reported conclusions extracted from UNEP, ACTO, and CIUP, op. cit.

TABLE 2.2. *Amazon land area per country*

	Size of the country (km²)	Size of the Amazon hydro-graphic area (km²)	Size of the Amazon ecological area (km²)	Size of the Amazon political area (km²)	Regional importance of national Amazon (country/total, percent)	National importance of the Amazon (percent)
Bolivia	1,098,581	724,000	567,303	724,000	9.8	65.9
Brazil	8,514,876	3,869,953	4,196,943	5,034,740	67.9	59.1
Colombia	1,141,748	345,293	452,572	477,274	6.4	41.8
Ecuador	283,561	146,688	76,761	115,613	1.6	40.8
Guyana	214,960	12,224	214,960	214,960	2.9	100.0
Peru	1,285,216	967,176	782,786	651,440	8.8	50.7
Suriname	142,800	–	142,800	142,800	1.9	100.0
Venezuela	916,445	53,000	391,296	53,000	0.7	5.8
TOTAL	13,598,187			7,413,827	100	

Source: UNEP, ACTO, and CIUP, Environment Outlook in Amazonia, GEO Amazonia, 2009, p.41.

contains areas that are untouched or show few signs of intervention. This fact should be an additional stimulus for joint conservation actions by all basin States.

2.1.2. *Hydrological Characteristics*

The Amazon River basin extends over an area of approximately 7 million square kilometers (25 percent of South America's total land area), with more than a thousand rivers and tributaries, which together constitute 15 percent of the world's non-frozen, surface freshwater.[9] The network of deep and fast-flowing rivers that drains the Amazon basin comprises the largest water basin on the planet and encompasses navigable stretches amounting to some 25,000 km.[10] Most of the Amazon basin (nearly 60 percent) lies within Brazil's boundaries.[11] Although 86 percent of the original vegetation of the Brazilian Amazon consists of dense forests, it also includes open forests in the Amazon-Cerrado transition and savanna-shrub vegetation (*campinaranas*) in the upper part of the Negro River.[12]

9 Lentini et al., op. cit., p. 23.

10 The Brazilian Institute of Geography and Statistics (IBGE) and the Brazilian Ministry of the Environment publish thematic maps of the Brazilian Amazon (IBGE, *Recursos Hídricos: Anuário Estatístico do Brasil*, Brasilia, 2000, http://www.ibge.gov.br, accessed January 26, 2007).

11 Barreto et al., op. cit., p. 23.

12 Ibid.

Annual mean flows discharged from the Amazon basin at the river's mouth into the Atlantic Ocean amount to some 209,000 cubic meters per second,[13] which represents approximately 18 percent of all freshwater discharged into the oceans in the world.[14] The Amazon and its many tributaries are of great importance as transport arteries, since inland navigation is the principal (and often the only) means of transport in the region. The Amazon River receives the flows of large tributaries, such as the Javari, Juruá, Purus, Madeira, Tapajós, and Xingu rivers on its south bank, and of the Içá, Japurá, Negro, Trombetas, Paru, and Jari rivers on its north bank.The most distant source of the Amazon is Lake Lauri, high in the Andes in Peru. From there, the river flows from south to north to Pongo de Manseriche, after which it runs east to west toward the river's mouth on the Atlantic Ocean. When it enters Brazil at Benjamim Constant (at an altitude of 65 meters above sea level), the river is already of formidable size. From there, it flows a further 3,000 kilometers before discharging into the Atlantic Ocean.[15]

The banks of the Amazon River itself are shared by Peru, Colombia, and Brazil, whereas Bolivia, Ecuador, Venezuela, and Guyana contain sources of its tributaries. As noted earlier, most of the Amazon basin is within Brazil's territory, whereas a smaller share is within the other basin States: Peru (17.0 percent), Bolivia (11.0 percent), Colombia (5.8 percent), Ecuador (2.2 percent), Venezuela (0.7 percent), and Guyana (0.2 percent).[16] The main river is known by various different names along its course. In Peru, the upper portion is known as "the Tunguragua"; then, above its confluence with the Ucayali, it is known as "the Marañón" (or "Maraftón"). Once the river enters Brazil, the upper stretch between its confluences with the Javari River (on the Peruvian border) and the Negro River (above the town of Manaus) is known as "the Rio Solimões," whereas from there to its mouth it is known as "the Amazonas."[17]

The Amazon River is navigable for sea-going ships traveling upriver to the town of Iquitos, 3,700 kilometers from the river mouth. Among the principal tributaries that flow into the Amazon on its south bank are the Huallaga and Ucayali (in Peru); and the Javari, Juruá, Tefé, Purus, Madeira, Tapajós, and Xingu (in Brazil); whereas its principal north bank tributaries are the Pastaza and Napo (in Peru); and the Içá, Japurá, Negro, Trombetas, Paru, and Jari (in Brazil).[18] Inland shipping is

[13] Brazilian Ministry of the Environment, *O Estado das Águas no Brasil 2001–2002*, Brasília, 2003, p. 494.

[14] J.A. Marengo, *Mudanças climáticas globais e seus efeito sobre a biodiversidade, caracterização do clima atual e definição das alterações climáticas para o territorio brasileiro ao longo do século XXI*, Brasília: Brazilian Ministry of the Environment, 2006, p. 41.

[15] Ibid.

[16] Casa Civil da Presidência da República et al., *Plano Amazônia Sustentável Diretrizes para o Desenvolvimento Sustentável da Amazônia Brasileira*, Brasília, 2008, p. 21.

[17] IBGE, op. cit.

[18] Ibid.

TABLE 2.3. *Area of river basins*

River basin	Area (km^2)
Amazon	3,836,529
Tocantins	932,070
Parnaíba	325,888
São Francisco	645,720
La Plata	1,320,275
North coastal	83,380
Western northeast coastal	682,449
Eastern northeast coastal	244,437
Southeast coastal	203,015
South coastal	273,612

Source: IBGE, *Anuário Estatístico do Brasil* 2000.

important to many living on Amazon tributaries like the Madeira, Negro, Tocantins, and Araguaia (in the state of Tocantins) in Brazil, and on the Guaporé River on the Brazilian border with Bolivia.[19] The importance of the Amazon basin for Brazil is illustrated in Table 2.3, which shows that the area covered by the Amazon River basin is three times larger than that of the La Plata River, South America's second-largest basin. Due to the vastness of this "water planet," the Amazon River basin provides large amounts of freshwater per capita, particularly when compared to other water basins. In Brazil's Amazon basin in particular, a very high annual water availability per capita of 984,000 cubic meters per person per year is observed.[20]

Two basic groups of activities are responsible for pollution of the Amazon River basin's aquatic environment: mining (on both an industrial and small scale, especially gold mining) and forest burning. To a lesser degree and as yet less investigated, the uncontrolled use of pesticides, especially in agro-industrial activities, will probably be one of the most serious environmental problem facing the Amazon in the coming years.[21]

Studies on the pollution of natural water resources in the Amazon have largely focused on mercury and arsenic. Forest burning, in particular, is a significant source of mercury in the aquatic ecosystems of the Amazon. According to Norbert Fenzl

[19] Ibid.

[20] N. Fenzl and A. Mathis, "Pollution of natural water resources in Amazonia: Sources, risks and consequences," in M. Clüsener-Godt and L.E. Aragón (eds.), *Water from the Amazon: Issues of Local and Global Use*, Montevideo: Núcleo de Altos Estudos Amazônicos, Universidade Federal do Pará and UNESCO, 2004, pp. 51–75.

[21] IBGE, op. cit.

and Armin Mathis, mercury (Hg) is a toxic metal occurring naturally at low concentration in the Earth's crust. The problem is that, even in low concentrations, mercury is very toxic, and thus even small increases in the environmental levels of the metal can have serious impacts on the environment and human health. While trace amounts of mercury have always been present in the environment, concentrations have been increasing to dangerous levels due to such human activities, such as coal burning and burning of forests, the latter being the main cause in the Amazon. When coal or forests are burned, mercury is released into the air and may be carried by winds over large areas before being deposited on the soil and in bodies of water. Aquatic microorganisms absorb such mercury, which then enters the food chain when these microorganisms are consumed by fish.[22]

Apart from mercury and arsenic found in water and sediments, other trace elements contribute to pollution of the Amazonian rivers. For example, various heavy metals in dissolved and suspended matter have been found in surface waters in the Mamore and Beni rivers in the Bolivian part of the Amazon basin. Harmful pesticides are also widely used in the Brazilian Amazon, but scientific information on environmental and human pesticide contamination is scarce.[23] In particular, the use of agrochemicals in large-scale soy production in the Brazilian Amazon leads to significant soil, air, and water pollution, affecting fish stocks.[24] In fact, a plethora of anthropogenic activities have decreased the quality of surface waters in the Amazon basin: mining washouts, hydrocarbon spills, the use of agrochemicals for agriculture, solid waste from cities, and waste from the transformation of illegal crops such as coca.[25]

2.2. SOCIOECONOMIC CHARACTERISTICS

2.2.1. *Population*

The Amazon's population is increasing.[26] While in the 1970s, Amazonian inhabitants numbered little more than 5 million, by 2007, the population was 33.5 million,

[22] Fenzl and Mathis, op. cit.

[23] Research on pesticide contamination in Santarém (in the state of Pará, Brazil) in 1997, 1998, and 2001 revealed the use of several insecticides and of various types of herbicides and fungicides for agriculture (N. Farella et al., "Deforestation modifying terrestrial organic transport in the Rio Tapajós," *Organic Geochemistry*, vol. 32, 2001, pp. 1443–58; M. Roulet et al., "Spatio-temporal geochemistry of Hg in waters of the Tapajós and Amazon rivers, Brazil," *Limnology and Oceanography*, vol. 46, no. 5, 2001, pp. 1158–70; J. Dolbec et al., "Sequential analysis of hair mercury levels in relation to fish diet of an Amazonian population, Brazil," *The Science of the Total Environment*, vol. 271, nos. 1–3, pp. 87–97).

[24] P. Verweij et al., *Keeping the Amazon Forests Standing: A Matter of Values*, report commissioned by WWF Netherlands, 2009, p. 42.

[25] UNEP, ACTO, and CIUP, op. cit., p. 22.

[26] Ibid., p. 21.

representing approximately 11 percent of the total population of the Amazon countries.[27] The population in the Amazon grows at an average annual rate that is greater than the population growth average of these countries, spurned by spontaneous migrations and state policies of colonization and settlement. As a result, the population density has grown from 3.4 inhabitants/km^2 (in the 1990s) to 4.2 inhabitants/km^2 (from 2000 to 2007). The Amazon has also undergone a process of accelerated unplanned urbanization that has led to approximately 62.8 percent of its population living in cities (or some 21 million inhabitants).[28]

The total population of the Legal Amazon is 24.9 million (13 percent of Brazil's total population), most of whom (68.9 percent) live in urban areas.[29] The population in the region has increased from 3.8 million to 23.55 million inhabitants (between 1950 to 2007), which represents a 516 percent population growth, much higher than the national average of 254 percent for the same period. Population growth in the Legal Amazon remains 40 percent higher than the national average. The Brazilian Institute of Geography and Statistics (IBGE) predicts that the population in the Legal Amazon will grow to 29.79 million inhabitants by 2020.[30]

For centuries, human settlement in the Brazilian Amazon occurred along the principal navigable rivers of the region. This pattern began to change in the 1960s owing to a number of factors, such as infrastructure build-up (e.g. roads, rural settlements, airports, and hydroelectric dams); subsidized credit channeled primarily to large-scale ranching; and the establishment of a free port in the city of Manaus. These factors provided an incentive for immigration, opening up extensive areas for settlement mostly along the southern margins of the region and in the major urban centers. Starting in the mid-1990s, infrastructure investments by the state and federal governments have targeted the paving of existing roads, developing ports, and building pipelines for hydrocarbons.[31]

Those socioeconomic processes, involving among other things, population growth, expansion of economic activities, and infrastructure development, have brought about significant changes in the Amazon as a whole, leading to ecosystem fragmentation, deforestation, and biodiversity loss. In particular, migratory agriculture and livestock production in the Amazon have generated accumulated deforestation of 857,666 square kilometers (as of 2005). In the Brazilian Amazon, over a period of thirty years (1975–2005), the highway network has multiplied tenfold,

[27] Ibid., p. 22.
[28] Ibid., p. 24.
[29] IBGE, op. cit.
[30] Casa Civil da Presidência da República et al., op. cit., p. 22.
[31] Barreto et al., op. cit., p. 25.

TABLE 2.4. *Population in the Amazon basin*

	Number of inhabitants			Average growth (%)	
Brazil	1980 11,015,363	1991 16,146,059	2007 24,970,600	1980–1991 3.5	1991–2007 2.8
Colombia	1985 1,607,093	1993 658,723	2005 960,239	1985–1993 −10.5	1993–2005 3.2
Ecuador	1982 263,797	1990 372,533	2005 629,373	1982–1990 4.4	1982–2005 3.6
Guyana	1980 759,568	1991 723,673	2002 751,223	1980–1991 −0.4	1991–2002 0.3
Peru	1981 1,253,355	1993 3,542,391	2005 4,361,858	1981–1993 9.0	1993–2005 1.38
Suriname	1980 354,860	1993 n.i.	2004 492,823	1980–1990 n.i.	1980–2004 1.38
Venezuela	1981 45,667	1990 55,717	2001 70,464	1981–1990 2.2	1990–2001 2.16

Source: UNEP, ACTO and CIUP, op. cit, p. 68.
Note: n.i. = no information.

stimulating the development of human settlements.[32] Those changes in the economic and social dynamics of the Amazon have affected the size of its native communities and weakened their culture and tradition practices (Table 2.4).[33]

Millennia of adaptation have endowed the indigenous peoples of the Amazon with knowledge of the various ecosystems, plants, and wildlife. This knowledge extends to terrestrial, arboreal, and aquatic wildlife, and to their habits, seasonal migrations, and availability at different seasons of the year. The natives of the Amazon also possess knowledge of medicinal plants and hallucinogens, plant-based poisons used for hunting, and various ways of preparing and storing foods.[34]

In the Brazilian Amazon, indigenous people constitute approximately 0.43 percent of the national population, somewhere between 700,000 and 750,000 people.[35] Despite a history of invasion and ongoing threats, indigenous peoples

[32] Information extracted from UNEP, ACTO, and CIUP, op. cit., p. 21.
[33] Ibid., p. 22.
[34] Commission on the Environment and Development for Amazonia, *Amazonia without Myths*, Inter-American Development Bank, United Nations Development Program, and ACTO, 1992, p. 26.
[35] James Anaya, *Report of the Special Rapporteur on the Situation of Human Rights and Fundamental Freedoms of Indigenous People: Report on the Situation of Human Rights of Indigenous Peoples in Brazil*, A/HRC/12/34/Add.2, Human Rights Council, 12th session, August 26, 2009, p. 5.

TABLE 2.5. *Indigenous populations of the Amazon*

Country	Number of inhabitants	Number of ethnic groups	Number of linguistic families
Bolivia	48,123 (2001)	25	18
Brazil	300,000 (2007)	175	34
Colombia	107,231 (2005)	62	n.i.
Ecuador	369,810 (2006)	10	n.i.
Peru	300,000 (2005)	59	15
Guyana	n.i.	n.i.	n.i.
Suriname	12,000	n.i.	n.i.
Venezuela	37,362 (2001)	17	n.i.
TOTAL			

Source: UNEP, ACTO and CIUP, op. cit., p. 72.
Note: n.i. = no information.

are still characterized by immense diversity and cultural wealth.[36] There are at least 225 indigenous peoples speaking some 180 different languages throughout Brazil, living both on traditional indigenous lands and in urban centers. The states comprising the legal Amazon have the highest concentrations of indigenous communities. It is estimated that there were around 5 million indigenous people living in the territory that is now Brazil (and 7 million in pan-Amazônia),[37] speaking as many as 1,300 languages, when Europeans first arrived. Due to factors common to the history experienced throughout the Americas, such as introduced diseases, forced displacement, and violent confrontation with colonizers, the indigenous population decreased dramatically, and numerous ethnically or linguistically distinct indigenous groups disappeared. In 1991, 76.1 percent of Brazil's indigenous population lived in rural areas, but a decade later, the majority of this population (52 percent) were living in urban areas. More than a thousand indigenous languages are spoken in Brazil, but only very few have been properly compiled in written form (Table 2.5).[38]

2.2.2. *Economic Development and Trade*

A comparison of the Amazon States shows considerable contrasts, especially in terms of land area and population. Guyana and Suriname are the smallest member

36 Ibid.
37 Commission on the Environment and Development for Amazonia, op. cit., p. 27.
38 Ibid.

TABLE 2.6. *General Information on the Amazon States*

Country	Capital	Language	Total population (2005)	Currency
Brazil	Brasília	Portuguese	184,184,264	Real
Venezuela	Caracas	Spanish	26,749,000	Bolivar
Bolivia	La Paz	Spanish, Quechua, Aymara	9,182,000	Boliviano
Peru	Lima	Spanish	27,968,000	Novo sol
Colombia	Bogotá	Spanish	45,600,000	Peso
Ecuador	Quito	Spanish	13,228,000	Sucre
Guyana	Georgetown	English	751,000	Dollar
Suriname	Paramaribo	Dutch	449,000	Dollar

Source: IBGE, http://www.ibge.gov.br

TABLE 2.7. *Annual GDP growth, 2006–2008 (%)*

Country	Annual GDP growth		
	2006	2007	2008
Brazil	4.0	5.7	5.9
Venezuela	10.3	8.4	4.8
Bolivia	4.8	4.6	5.8
Peru	7.6	8.9	9.4
Colombia	6.8	7.7	3.0
Ecuador	3.9	2.5	6.5
Guyana	5.1	5.4	4.8
Suriname	5.8	5.3	5.0

Source: Economic Commission for Latin America and the Caribbean (CEPAL), *Statistical Yearbook for Latin America and Caribbean 2008*, http://www.cepal.org

countries, both in terms of geographical size and population. In terms of land area, Brazil and Peru are the largest countries of the group, whereas Brazil and Colombia are the most populous (Table 2.6).

Economic growth rates vary considerably among the Amazon States. Between 2006 and 2007, Ecuador reported the lowest rates of gross domestic product (GDP) growth and Venezuela the highest rates. In 2008, GDP growth in Peru was the highest among the eight Amazon States (Table 2.7).

The overall level of economic development varies widely in the Amazon, with some areas such as Orellana (Ecuador) reporting a much higher GDP per capita than others areas, such as Putumayo (Colombia) of US$25,628.22 and US$705.33, respectively.[39] Despite population declines and the large amount of natural resources exploited, such as minerals, petroleum, and gas, the poverty indices of Amazonia are very high, due to the fact that in most cases the profits are not reinvested in the region.[40]

The volume of trade between Brazil and the other Amazon States (expressed in terms of the sum of imports and exports) has risen considerably in recent years. A comparison of sample trade figures for the past forty years, contrasting more recent values with those of the 1960s (1967–1969) and 1970s (1974–1976 and 1977–1979) shows that in this period the volume of trade between Brazil and the other Amazon States has increased more than ten-fold. As Table 2.8 shows, Brazilian exports have been the driving force behind this growth. The signing of the Amazon Cooperation Treaty in 1978 probably helped consolidate an existing trend of increased trade among member countries. However, trade relations with each of the partners have intensified at different rates, and the long-term trend has been for more countries to become engaged in these trade flows.

2.3. CURRENT THREATS TO THE AMAZONIAN ENVIRONMENT

Climate exerts a strong influence over the broad distribution of rainforests within the tropics and has a marked effect on regional patterns and structures of rainforest vegetation and habitat. Rainforests occur in hot, moist climates. They receive more solar radiation throughout the year than any other vegetation zone on Earth, which promotes the rich variety and luxurious character of their vegetation.[41]

As an integral component of the biosphere, forests contribute toward biological diversity and help maintain the stability of natural systems through the preservation of air, water, and soil quality. They influence biogeochemical processes by containing runoff and preserving groundwater, thereby reducing soil erosion. They enhance local climates, and reduce downstream sedimentation and flooding. As carbon sinks, forests sequester carbon dioxide from the atmosphere, thus reducing the greenhouse effect. Forests also have aesthetic value, provide recreational opportunities as well as commercial products such as timber, fibre, latex, fruits, and

[39] UNEP, ACTO, and CIUP, op. cit., p. 21.

[40] Ibid.

[41] C.C. Park, *Tropical Rainforests*, London & New York: Routledge, 1992.

TABLE 2.8. *Trade between Brazil and other Amazon States, 1967–2007 (US$ thousands)*

	Average 1967–1969	Average 1974–1976	Average 1977–1979	Average 1997–1999	Average 2006–2007
Exports					
Bolivia	3,479	101,547	133,824	613,137	156,047
Colombia	2,178	29,918	113,423	459,501	250,841
Equador	345	19,601	33,365	159,895	89,582
Peru	5,061	71,926	44,772	331,849	110,732
Venezuela	3,875	107,059	205,571	15,809	907,318
Guiana				6,299	3,391
Suriname		3,005	7,435	11,201	15,928
TOTAL	14,939	333,057	538,390	1,597,692	1,533,838
Imports					
Bolivia	473	21,642	49,949	20,728	21,505
Colombia	1,967	14,387	6,396	97,331	19,000
Equador	463	5,129	9,760	25,438	16,594
Peru	7,294	44,306	79,527	209,661	34,643
Venezuela	60,702	136,219	182,004	636,263	2,185
Guiana				4	826
Suriname		2,748	12,936	755	850
TOTAL	70,899	224,431	340,572	990,180	95,601
Trade (exp. + imp.)	85,838	557,488	878,962	2,587,871	1,629,440
Trade balance (exp. – imp.)	–55,961	108,625	197,818	607,512	1,438,237

Source: IBGE, *Statistical Yearbook*, 1970, 1977, 1980, 2000; Ministério do Desenvolvimento, Secretaria do Comercio Exterior, *Balança Comercial Dezembro de 2007*, Brasília, 2007.

medicinal plants, among others. For indigenous and traditional populations, forests are essential to livelihoods.[42]

The world has just less than 4 billion hectares of forests that cover about 30 percent of its land area.[43] Latin America and the Caribbean account for 22 percent of

[42] N.P. Sharma et al., "World forests in perspective," in N.P. Sharma (ed.), *Managing the Worlds Forests: Looking for Balance between Conservation and Development*, New York: World Bank, Kendall Hunt, 1992, pp. 17–31.

[43] Food and Agriculture Organization (FAO), *State of the World's Forests 2007*, Rome: FAO, 2007, p. ix.

TABLE 2.9. *Forest distribution by region (millions of hectares)*

Region	Forest area (2005)		
	Total forest (1,000 ha)	Land area (%)	Area per capita (ha)
Africa	635,412	21.4	0.7
Asia	571,577	18.5	0.1
Europe	1,001,394	44.3	1.4
Latin America and the Caribbean	859,925	47.3	1.9
North America	677,464	32.7	1.6
Oceania	206,254	24.3	6.3
TOTAL World	3,952,025	30.3	0.6

Source: FAO, State of the World's Forests 2007.

the global forest area, 14 percent of the global land area, and 7 percent of the world's population. The region contains the world's largest contiguous block of tropical moist forest – the Amazon basin (Table 2.9).[44]

2.3.1. *Deforestation in the Amazon*

Human activities have always modified forest environments, but not with the intensity and on the scale of the last decades. For example, from 1990 to 2005, the world lost 3 percent of its total forest area, an average of 0.2 percent per year.[45] During that period, Latin America and the Caribbean lost about 64 million hectares of forests, which represented a decline in forest area by 7 percent in that region.[46]

The depletion of forests has been most significant in the tropics and had been caused by socioeconomic and political factors that have created incentives for rapid exploitation of forests and intensified human pressure upon forest areas.[47] Deforestation – and its associated impacts, notably climate change and biodiversity loss – is the major current threat to the Amazon. Table 2.10 shows the average deforestation rates among the Amazon countries.

In economic terms, loss of environmental benefits from depletion of forests can be considerable, but such costs are difficult to quantify. When sectoral contributions

44 Ibid., p. 32.
45 Ibid., p. ix.
46 Ibid., p. 37.
47 Tropical forests are defined as forests lying between the tropics of Cancer and Capricorn, although it has proved difficult to apply in all cases. International Tropical Timber Organization (ITTO), *Status of Tropical Forest Management* 2005, 2006, p. 13, http://www.itto.int

TABLE 2.10. *Deforestation in the Amazon region*

Country	Total forest area (1) (1,000 ha)	Percent of Land area (2)	Amazon area (3) (ha)	Net deforestation rate, 2001 (4) (%)
Bolivia	58,740	54.2	688,450	0.57
Brazil	477,698	57.2	5,189,032	0.70
Colombia	60,728	58.5	445,085	0.24
Ecuador	10,853	39.2	116,947	0.41
Guyana	15,104	76.7	215,409	0.12
Peru	68,742	53.7	973,523	0.18
Suriname	14,776	94.7	147,479	0.18
Venezuela	47,713	54.1	184,265	0.35
TOTAL Amazon	754,354		8,045,491	0.54

Source: FAO, State of the World's Forests 2009, p. 114, (1), (2); Verweij et al., Keeping the Amazon Forests Standing: A Matter of Values, 2009, p. 28 (3), and (4).

to GDP are assessed, those of the forestry sector tend to be underestimated, because many forest products are noncommercial, and also because the environmental goods and services forests provide are not taken into account. Thus, in many countries, the contribution of the forestry sector to the economy is smaller than those of agriculture or industry. As a result, governments tend to assign low priority to the forestry sector and to make relatively little investment in forest management research and reforestation programs. Latin America and the Caribbean own 22 percent of the global forest area but only about 7 percent of the global forest sector's value. This region accounts for only 3 percent of the value added in wood-processing industries and 6 percent in the pulp and paper industry, which shows that the region is a major source of raw materials, but that much of the processing of these materials into finished products is done in other regions.[48] Both governments and the private sector do not always perceive the environmental benefits that can be derived from forests and the environmental costs associated with deforestation, and they usually fail to include local and native communities in discussions on the long-term management of forests.[49]

The causes of deforestation in the tropics are linked to the expansion of economic activities such as agriculture, ranching, commercial logging, mining, and oil extraction, which lead to infrastructure development and unplanned human settlement, among other things. Although the causes of deforestation vary among the Amazon countries, common drivers of deforestation include soybean cultivation, stock

[48] FAO, op. cit., p. 41.
[49] Sharma et al., op. cit., pp. 17–31.

farming, oil extraction, and the cultivation of illicit crops such as coca.[50] Natural hazards, for example the El Nino phenomenon and fires, have also affected the Amazon.[51] In general, both commercial and small-scale farming exert pressure on the Amazon biome, due to the use of slash-and-burn practices and other unsustainable techniques.[52] Nonetheless, some Amazon countries, such as Guyana and Suriname, have been less subject to human pressure leading to deforestation. These countries still maintain large intact forests with high conservation and ecological value (Table 2.11).[53]

Deforestation of the Brazilian Amazon in particular has been severe and has attracted worldwide attention. The Brazilian Amazon harbors about one-third of the world's remaining tropical forests, but land-use conversion occurs on an unprecedented scale and in a complex manner.[54] Brazil accounted for approximately 42 percent of global net forest loss (from 2000 to 2005), most of which occurred in the Amazon.[55]

While forests covered originally 75 percent of the Legal Amazon, they now cover 64 percent.[56] Nonforest formations (composed of *cerrados*, *campos naturais*, and *campinaranas*) represent 22 percent of the Legal Amazon land area. The remaining 14 percent of its vegetal cover has been deforested (up to 2004).[57] Apart from deforestation, large parts of primary forests are highly degraded due to factors such as habitat fragmentation, selective logging, surface fires, excessive hunting, and illegal gold mining, among others.[58]

Over the last decades, deforestation in the Brazilian Amazon has been a linear process, as thousands of square kilometers are lost every year. Since 1988, the Institute for Space Research (INPE) has been surveying deforestation of the Brazilian Amazon through satellite imaging. Results are reported in terms of annual deforestation indices, expressed in km^2 per year. In general, deforestation patterns are better monitored and documented in Brazil than in other Amazon countries.[59] The annual rates of deforestation of the Legal Amazon from 2000 to 2008 are indicated in Table 2.12.

Despite a decrease over the period 2006–2008 compared to previous years, deforestation in the Brazilian Amazon is still high. In 2008, the annual deforestation rate

50 ITTO, op. cit., pp. 200, 209, and 266.
51 Ibid., p. 217.
52 Description of causes of deforestation for each Amazon State can be found in ITTO, op. cit., p. 200.
53 Ibid., pp. 238 and 274.
54 Barreto et al., op. cit., p. 19.
55 Ibid.
56 Lentini et al., op. cit., p. 29.
57 Ibid.
58 Verweij et al., op. cit., p. 27.
59 Ibid.

TABLE 2.11. *Drivers of deforestation by country*

Country	Key drivers	Underlying causes
Bolivia	• Agricultural expansion (cattle ranching, soybeans, sorghum, sunflowers, and illicit crops)	• Public policies (the World Bank's structural adjustment program project in the 1990s, and fiscal and credit policies) • International demand for soybeans
Brazil	• Agricultural expansion (cattle ranching and soybeans) • Infrastructure extension • Selective logging	• Public policies (credits, roads, and resettlement) • International demand for soybeans • National, and recently global, beef demand • Institutional factors (weak law enforcement in frontier zones and weak property rights)
Colombia	• Agricultural expansion (cattle ranching, soybeans, and illicit crops) • Infrastructure extension (roads to forest frontier)	• Public policies (currency devaluation, low export tariffs, and easy access to land) • International demand for soybeans • Domestic demand for beef • Demographic factors (population pressure)
Ecuador	• Cattle ranching • Oil exploitation • Logging • Infrastructure extension	• Increased incomes raising demand for meat and dairy products • International demand for oil • Domestic timber demand for construction
Guyana	• Logging	• International timber demand
Peru	• Agricultural expansion (cattle ranching, soybeans, and shifting cultivation) • Logging	• Public policies (subsidized tax policies in frontier zones) • Institutional factors (land speculation)
Suriname	• Logging	• Public policies (logging concession policies)
Venezuela	• Cattle ranching • Logging • Mining • Oil exploitation	• Domestic demand for beef and timber • Weak governance with respect to private land appropriation

Source: S. Wertz-Kanounnikoff, M. Kongphan-Apirak, and S. Wunder, Reducing Forest Emissions in the Amazon Basin: A Review of Drivers of Land-use Change and How Payments for Environmental Services (PES) Schemes Can Affect Them, Bogor: Center for International Forestry Research (CIFOR), November 2008, p. 7.

TABLE 2.12. *Legal Amazon: annual rate of deforestation, 2000–2008 (km²/year)*

State	2000	2001	2002	2003	2004	2005	2006	2007	2008
Acre	547	419	883	1,078	728	592	398	184	254
Amazonas	612	634	885	1,558	1232	775	788	610	604
Amapá		7	0	25	46	33	3	39	100
Maranhão	1,065	958	1014	993	755	922	651	613	1,272
Mato Grosso	6,369	7,703	7,892	10,405	11,814	7,145	4,333	2,678	3,258
Pará	6,671	5,237	7,452	7,145	8,870	5,899	5,592	5,526	5,606
Rondônia	2,465	2,673	3,099	3,597	3,858	3,244	2,049	1,611	1,136
Roraima	253	345	84	439	311	133	231	309	574
Tocantins	244	189	212	156	158	271	124	63	107
TOTAL	18,226	18,165	21,523	25,396	27,772	19,014	14,196	11,633	12,911

Source: INPE, http://www.inpe.gov.br, accessed February 20, 2010.

has slightly increased when compared to 2007. In the ranking of deforestation by state, Mato Grosso, Pará, and Rondônia have usually been the leaders.[60]

Deforestation in the Brazilian Amazon is mostly concentrated in a narrow strip in the southern extremity of the region, stretching from the south of Maranhão to Rondônia, which is often referred to as the "Arc of Deforestation" (Arco do Desmatamento). Changes currently taking place in different parts of the Legal Amazon reflect the expansion of agricultural and stock-farming activity, as well as human population growth.

The most recent wave of agricultural expansion in the Legal Amazon owes much to new technologies.[61] Increases in production of mainly soybeans, but also corn, rice, and cotton, are the result of new varieties adapted to production in the savannahs (*cerrado*), using modern farming technologies. The areas in which such agricultural activities have most expanded are located in the savannahs of the states of Mato Grosso and Tocantins, and in the south of Maranhão, all of which are encompassed by the Legal Amazon.[62]

60 Brazilian Ministry of the Environment, *Ministério do Meio Ambiente apresenta estudos sobre mudanças climáticas e biodiversidade*, Brasília, 2007, http://www.meioambiente.gov.br

61 IBGE, *Report on the Agricultural Frontier in Amazônia Legal*, Brasília: IBGE, 2007.

62 Ibid. The main new grain production area is the region cut by the BR-163 highway, centered on the municipalities of Sorriso (which currently accounts for over 10 percent of Brazil's soybean production) and Sinop, where the large-scale grain production region meets the rainforest. Further north, rice is the main crop produced on recently cleared land. Much recently cleared land is also transformed into pasture for livestock production.

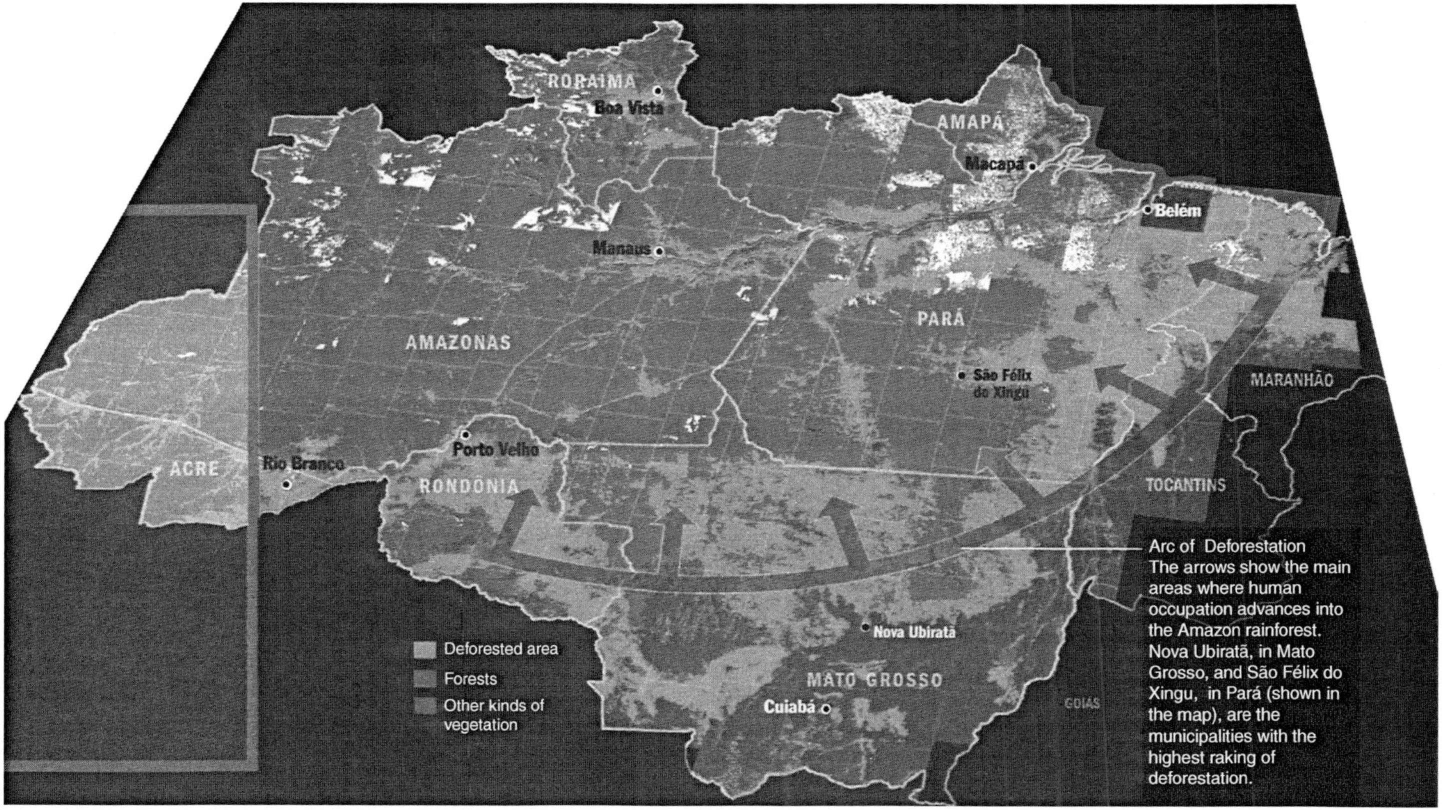

MAP 2.2. "Arc of Deforestation" in the Legal Amazon (text translated by the author).

Source: Revista Veja, *Veja: Especial Amazônia*, ed. 2053, vol. 41, no. 12, 26 March 2008.

Currently, Brazil is a leading exporter of soybeans.[63] Global commodities traders and Brazilian large producers have driven the expansion of soybean farming northwards initially into the savannahs on the edge of the rainforest. Within the space of a few years, Mato Grosso (one of the states in the Legal Amazon) has become the largest soybean-producing state in Brazil.[64] Soybean expansion in the states within Legal Amazon totals 14.1 percent a year (from 1990 to 2005).[65] In general, soya cultivation farms in the Legal Amazon are large, mechanized, and use significant chemical inputs.

New transport routes play an essential role in the spread of agriculture and livestock operations, and the expansion of areas of grazing land closely accompanies the construction of highways, notably the BR-364, the BR-163, the Trans-Amazonian (Transamazônica – BR-230), the BR-010 (Belém–Brasília), and such state highways as the PA-150 and MT-128. The building of new roads is also a factor in the increase of logging activity in the Brazilian Amazon.[66]

The most notable impact of the expansion of the agricultural frontier is deforestation, accompanied by degradation and fragmentation of ecosystems. Intensified use of agrochemicals and an increase in soil erosion resulting from the clearing of large tracts of land has affected river systems that provide drinking water to forest communities and habitats for innumerable wildlife species, many of them central to native livelihoods. For example, the Xingu River basin is being used as a waste drain by the soybean industry.[67] At the heart of the catchment is a near-pristine area of rainforest, the Xingu Indigenous Park, which is home to various indigenous tribes. The expanding agricultural frontier in the Brazilian Amazon has resulted in the establishment of very large ranches and a very unequal distribution of landholdings. Disputes over land are all too common in the Amazonian, though especially prevalent in the south of Pará and along the Trans-Amazonian highway; however, such conflicts also occur in other states, where indigenous communities, miners, and squatters tend to be involved.[68]

In addition to soybean cultivation, the other farming activity that has had the most impact on the Brazilian Amazon is cattle ranching.[69] Of these two activities,

63 Greenpeace International, *Eating up the Amazon*, Amsterdam: Greenpeace International, 2006, p. 13.

64 Ibid., p. 13.

65 Verweij et al., op. cit., p. 36.

66 IBGE, *Recursos Hídricos*, op. cit.

67 Greenpeace International, op. cit., p. 22.

68 IBGE, *Recursos Hídricos*, op. cit.

69 The Brazilian nongovernmental organization (NGO) Instituto Socioambiental (ISA) presents evidence of links between deforestation in the Legal Amazon and the expansion of farming and livestock activities (ISA, *Instrumentos econômicos e financeiros para a conservação ambiental no Brasil: Uma análise do estado da arte no Brasil e no Mato Grosso, Desafios e perspectivas*, 2007, http://www.socioambiental.org).

stock farming has had the greatest impact on the Brazilian Amazon in recent years.[70] Despite the decrease of fiscal incentives and technological improvements that have led to increased productivity, cattle raising by large and medium producers is responsible for nearly 80 percent of the total deforested area in the Legal Amazon.[71]

Brazil has the world's largest commercial cattle herd and is the world's largest beef exporter.[72] Between 2003 and 2006 there was hardly any growth outside the Legal Amazon: 96 percent of all growth in cattle raising registered in this period in the country (equivalent to nearly 10 million head) was from the Amazon.[73] Contrary to the nationwide tendency, in the Legal Amazon pastures have been expanding: over a ten-year period (1997 to 2007), the cattle herd in the states making up the Legal Amazon grew 77.4 percent, compared to a 23.7 percent growth in the Brazilian herd.[74] Tocantins (western), Pará (southern and southeastern), Rondônia (southeastern), and Mato Grosso are the states with the largest concentration of the cattle herd in the Legal Amazon (Table 2.13).[75]

In Brazil, stock farming is mainly financed by public institutions. In 2008, according to the Central Bank, more than 50 percent of financing for the ranching activity was provided by government banks (almost exclusively federal), and the other half by private banks and credit cooperatives.[76] In the Legal Amazon, federal government banks have financed approximately 81 percent of ranching activities.[77] For example, in 2008, direct operations by the National Economic and Social Development Bank with the meat industry reached the amount of R$6

[70] Agricultural and livestock expansion are currently two of the most representative examples of economic growth at the expense of the Amazon forest. Soy is the most important indirect driver of deforestation, and cattle farming the major direct cause (Verweij et al., op. cit., p. 28); "Whatever factors may be involved in the transformation and displacement of agricultural activities, land use change in the Amazon is first and foremost a product of ranching" (R. Smeraldi and Peter H. May, *The Cattle Realm: A New Phase in the Livestock Colonization of the Brazilian Amazon*, São Paulo: Amigos da Terra, 2008, p. 1).

[71] Ministério da Integração Nacional et al., *Plano de Ação para a Prevenção e Controle do Desmatamento na Amazônia Legal*, Brasília, 2004, p. 10.

[72] M.M.C. Bustamante, C.A. Nobre, and R. Smeraldi, *Estimating Recent Greenhouse Gas Emissions from Cattle Raising in Brazil*, 2008, http://www.amazonia.org.br/english/; E. Arima, P. Barreto, and M. Brito, *Cattle Ranching in the Amazon: Trends and Implications for Environmental Conservation*, Belém: Instituto do Homem e Meio Ambiente da Amazônia (Imazon), 2006, p. 38. According to this study, with more than 200 million head of cattle, Brazil possesses the world second largest herd, surpassed only by India, where bovine livestock farming is not done for commercial ends. More than one-third of the herd is located in the Central-West Region. More specifically, the state of Mato Grosso makes up 13 percent of the total Brazilian herd.

[73] Smeraldi and May, op. cit., pp. 1–2.

[74] Friends of the Amazon, Brazilian Amazon, *Time to Pay the Bill: The Current Situation of Cattle Ranching in the Amazon*, São Paulo: Friends of the Amazon, Brazilian Amazon, 2009, p. 11.

[75] Ibid., p. 15.

[76] Ibid., p. 18.

[77] Ibid.

TABLE 2.13. *Cattle herds in the Legal Amazon*

States	Total head (2006)	% in relation to Brazil
Rondônia	8,490,822	4.95
Acre	1,721,660	1.00
Amazonas	1,154,269	0.67
Roraima	480,704	0.28
Pará	13,354,858	7.78
Amapá	57,728	0.03
Tocantins	6,076,249	3.54
TOTAL	31,336,290	18.26

Source: IBGE, *Censo agropecuário 2006*, Rio de Janeiro: IBGE, 2006.

billion, equal to all of the bank's other investments in the industrial area, from the automobile sector to ethanol mills.[78]

The agricultural expansion in the Amazon has benefited from several factors, including the availability of cheap lands, weak enforcement of environmental and labor legislation,[79] easy occupation of public lands, land tenure policies, and good climate conditions, among others.[80] In particular, the expansion of stock farming since the 1970s has been initially linked to subsidized rural credit, tax incentives, and other facilities granted by the federal and state governments, which were gradually withdrawn. In any case, stock farming in the Brazilian Amazon currently produces satisfactory rates of return on investment, and is now the driving force behind the expansion of cattle ranches in the region.[81] In general, the production costs of raising beef cattle in the Amazon are lower than in the remainder of the country mainly due to lower land prices, which is the main component of production costs in extensive cattle raising. For example, between 1977 and 2000, planted pasture prices in Pará were on average only 11 percent of pasture prices in São Paulo.[82]

As a major driver of deforestation in the Brazilian Amazon, stock farming, apart from other environmental impacts, contributes to climate change. It is estimated that the emissions of greenhouse gases (GHGs) accumulating deforestation and burning for the formation of pastures (from 2003 to 2008) were about 3,416 million tonnes of CO_2 equivalent. The contribution of pasture to the land use change emissions in the Amazon is substantial; it accounts for more than 75 percent of GHG

78 Ibid., p. 6.

79 Ministério da Integração Nacional et al., op. cit., p. 10.

80 Ibid.

81 ISA, op. cit.

82 Friends of the Amazon, Brazilian Amazon, op. cit., p. 26.

emissions from enteric fermentation related to cattle herd in the *cerrado* and the Amazon, which correspond to approximately 39 percent and 24 percent of the emissions from the total Brazilian herd, respectively.[83]

Other economic activities, such as logging, also exert pressure on the Amazon. Logging is one of the most important and traditional activities particularly in the Brazilian Amazon. Brazil is the world's largest producer (and consumer) of tropical timber, with productive chains directly based in forest timber products representing 3.5 percent of its GDP.[84] Logging is a key driver of deforestation in other Amazon States, such as Ecuador, Guyana, Peru, and Venezuela. Other activities that have caused the disappearance of large tracts of rainforest and the emission of large volumes of GHGs include the building of infrastructure, for example hydroelectric projects. In particular, hydroelectric reservoirs cause high emissions of methane due to the decomposition of plant material. In the Amazon basin, the Balbina dam (in the River Uatumã), for example, was responsible for the loss of 2,400 km^2 of forest, with a total area flooded of more than 3,000 km^2 and annual emissions initially equivalent to about 30.2 million tonnes of CO_2. The Tucurui reservoir (on the Tocantins River) caused the flooding of 2,430 square kilometers of forest and emitted 51.5 million tons of CO_2 equivalent per year.[85]

As suggested, Brazil has two major opportunities to end the clearing of the Amazon and reduce GHG emissions.[86] The first is the formal announcement in 2008 at the Conference of the Parties to the UN Convention on Climate Change in Posnan (Poland) to end the net loss of forest cover in Brazil under its National Climate Change Plan,[87] which prompted Norway to commit US$ 1 billion if progress is made toward this target. Under this Plan, Brazil established a voluntary emission reduction target between 36.1 to 38.9 percent until 2020 (Article 12).[89] In order to attract international funds to achieve that target, Brazil created the Amazon Fund (Fundo Amazônia) in August 2008. A first donation agreement (Decree 6527/08, 25.03.2009) has been signed between the government of Norway and the Brazilian National Development Bank, the administrator of the fund. The second opportunity is provided by a widespread marketplace transition within the beef and soybean industries to exclude Amazon deforesters from commercial supply chains.[88] In this regard, a two-year soy moratorium has been established in 2006 (Termo de Compromisso da Moratória da Soja,

[83] Bustamante et al., op. cit., p. 4.

[84] Comitê Interministerial sobre Mudança do Clima, *Plano Amazônia Sustentável Diretrizes para o Desenvolvimento Sustentável da Amazônia Brasileira*, Brasília, 2008, p. 39.

[85] Verweij et al., op. cit., p. 29.

[86] Greenpeace International, *Slaughtering the Amazon*, Amsterdam: Greenpeace International, 2009, p. 43.

[87] Comitê Interministerial sobre Mudança do Clima, *Plano Nacional de Mudanca do Clima*, 2008 (Decree 6.263, 21.11.2007), Brasília, 2007, p. 28.

[88] D. Nepstad et al., "The end of deforestation in the Brazilian Amazon," *Science*, vol. 326, no. 5958, 2009, pp. 1350–51.

29 July 2009),[89] later extended until 2010, to prevent the purchase of soy from newly deforested areas in the Amazon. With regard to the meat industry, the Ministry of the Environment banned authorities from granting permits for deforestation to unregistered ranches within "embargoed municipalities" within the Legal Amazon (Decree 6.321/2007 and Portaria 28/2008 of the Ministry of the Environment). Moreover, a national certification program (Programa de Certificação de Produção Responsável na Cadeia Bovina) was launched by the Brazilian Association of Supermarkets (Associação Brasileira dos Supermercados – Abras) in December 2009 to attest the origins of the meat produced and commercialized in Brazil.[90]

2.3.2. *Impacts of Deforestation on the Amazon*

Some of the most notable impacts of deforestation include climate change and biodiversity loss. According to the Intergovernmental Panel on Climate Change (IPCC), deforestation accounts for 17 percent of global GHG emissions, which is more than the global transport sector.[91] Brazil, in particular, has one of the world's fastest rate of deforestation and is the fifth largest producer of GHG emissions among developing countries.[92] It is precisely the deforestation of the Brazilian Amazon that makes the largest contribution to the total amount of Brazil's carbon emissions, making up more than 60 percent of the country's total GHG emissions.[93] As a result of deforestation, forests' carbon absorption capacity is affected and they end up being themselves a source of carbon emissions through the process of burning, which also affects air quality.[94]

Due to the abundant forestlands and variety of ecosystems, South America has a rich biological diversity, encompassing six megadiverse countries: Colombia, Brazil, Ecuador, Venezuela, Peru, and Bolivia.[95] In terms of numbers of species, South

[89] http://www.greenpeace.org/brasil/amazonia/moratoria-da-soja

[90] Abras, "Abras adere a desmatamento zero," December 7, 2009, http://www.greenpeace.org/brasil/amazonia/noticias/abras-adere-desmatamento-zero; "Carne Bovina: Garantias de ponta a ponta," January 18, 2010, http://www.abrasnet.com.br/fundacao/sustentabilidade/noticias/?materia=593

[91] Intergovernmental Panel on Climate Change (IPCC), "The physical science basis," in *Climate Change 2007*, 2007, chap. 9, http://www.ipcc.ch/publications_and_data/publications_ipcc_ fourth_ assessment_report_wg1_report_the_physical_science_basis.htm

[92] After China, India, Mexico, and South Africa (United Nations Conference on Trade and Development (UNCTAD), *Developing Country Interests in Climate Change Actions and Implications for a Post-2012 Climate Change Regime*, Rome: UNCTAD, 2009, p. 19).

[93] The deforestation of the Brazilian Amazon makes the largest contribution to the total amount of emissions in that country (Brazilian Ministry of External Relations et al., *Brazil's Contribution to Prevent Climate Change*, White Paper, Brasília, 2007, p. 36; Comitê Interministerial sobre Mudança do Clima, 2007, op. cit., p. 66).

[94] UNEP, ACTO and CIUP, op. cit., p. 23.

[95] Brazilian Ministry of the Environment, *Estratégias Nacionais de Biodiversidade na América do Sul: Perspectivas para a Cooperação Regional*, Brasília: Secretaria de Biodiversidade e Florestas, 2004.

America has more than 100,000 species of higher plants, and more species of birds, amphibians, mammals, and butterflies than anywhere else on the planet.[96] This remarkable biological diversity is matched by exceptional cultural diversity, as represented by numerous native ethnic groups.[97]

The Amazon has been considered as one of the planet's most biologically diverse areas, and it is estimated that around 10 percent of all species of plants are found in this region.[98] All Amazon States have a variety of ecosystems and great richness of species with high levels of endemism. For example, Bolivia contains at least 18,000 species of plants and many undisturbed ecosystems.[99] Colombia has one of the highest levels of biodiversity in the world, boasting some 55,000 plant species (of which one-third are endemic), 1,721 bird species, and 205 reptile species.[100] Both Guyana and Suriname have large areas of intact forest ecosystems with a high conservation and ecological value.[101] Despite this biological wealth, much of this heritage is under threat due to deforestation and consequent habitat loss, pollution, and the overall degradation and fragmentation of ecosystems. Deforestation is a major threat to the survival of those species, many of which inhabit the forests of the Amazon basin. Latin America and the Caribbean have the world's highest number of tree species considered endangered or vulnerable to extinction,[102] and many of the species occurring in the Amazon are listed as critically endangered or vulnerable in the International Union for Conservation of Nature Red List,[103] the threat being greatest in Brazil, Ecuador, Peru, and Colombia.[104] Biodiversity loss is not the only casualty of deforestation, which also causes soil erosion, nutrient depletion, loss of watershed regulation functions, and climate change, among others.[105]

In response to the alarming loss of biodiversity and ecosystem degradation, the Amazon countries have formulated National Biodiversity Strategies, one of the commitments assumed under the 1992 Convention on Biological Diversity, and have made efforts to strengthen their national systems of protected areas. Different categories of national protected areas (e.g. national parks, nature reserves, sanctuaries, etc.) have been created by national law to provide *in situ* conservation of wildlife species of fauna and flora and ecosystems. For example, in Brazil 33 percent of the Amazon area is legally protected in the form of conservation units and indigenous lands

96 Ibid.
97 Ibid.
98 UNEP, ACTO, and CIUP, op. cit., p. 110.
99 ITTO, op. cit., p. 204.
100 Ibid., p. 221.
101 Ibid., pp. 243, 277, respectively.
102 FAO, op. cit., p. 37.
103 See ITTO, op. cit. for information on each Amazon State.
104 Brazilian Ministry of the Environment, op. cit.
105 Barreto et al., op. cit., p. 19.

TABLE 2.14. *Protected areas in the Amazon States*

	Brazil	Venezuela	Bolivia	Peru	Colombia	Ecuador	Guyana	Suriname
Protected areas (% in 2004)	3.9	34.2	11.1	3.1	8.6	9.3	2.3	12.7
Areas under cultivation (% in 2002)	6.97	2.95	2.67	2.89	2.21	5.85	2.44	0.37
Permanent grazing land (% in 2002)	23.29	20.68	31.2	21.17	40.49	18.39	6.25	0.13
CO_2 emissions (tonnes per person, in 2000)	1.6	5.9	1.2	1.1	1.3	1.9	2.1	4.7

Source: IBGE, http://www.ibge.gov.br, accessed February 16, 2007.

(whereas 24 percent of the Brazilian Amazon is private land).[106] The Ecuadorian Amazon, for example, is classified as a "region of special attention" under the National Biodiversity Strategy, which are areas of major strategic value, in which serious environmental problems are most evident (Table 2.14).

At the regional level, the Amazon States, together with other South American countries, have discussed ways of enhancing regional cooperation especially in transboundary protected areas.[107] For example, among the main transboundary ecosystems of the Andean countries are the forests of the Amazon floodplain, the humid mountain forests of the eastern slopes of the Andes and the forests of the Chocó region on the Pacific coast; however, only few joint initiatives exist to protect those transboundary ecosystems. Under the initiative of the Brazilian Ministry of the Environment all South American countries gathered in 2003 to identify those areas and try to come up with strategies for joint activities. As a result of this consultation process, these countries proposed, among other things, the formulation of common policies for development and cooperation between countries sharing transboundary

[106] Approximately 10 percent can be considered special areas, including military lands (*terras militares*), areas of environmental protection (*áreas de proteção ambiental*), and rural settlements. The remaining 33 percent of the Brazilian Amazon can be considered "terras devolutas" or disputed private lands (Lentini et al., op. cit., p. 32).

[107] Brazilian Ministry of the Environment, op. cit. Funding was provided by the Global Environment Facility, the United Nations Development Program, the United Nations Environment Program, and the Brazilian Cooperation Agency.

TABLE 2.15. *Examples of transboundary terrestrial ecosystems among the Andean States*

Ecosystems	Bolivia	Peru	Ecuador	Colombia	Venezuela
Amazon rainforest (Hylaea)					
Moist forests of the southeast (Acre, Madre de Dios)		X	X		
Moist forests of the Napo (western Hylaea)		X	X	X	
Moist forests of the apurá-Rio Negro area (northwestern Hylaea)		X		X	X
Meadow forests and floodable forests	(X)	X	(X)	X	X

Note: (X): occurring over smaller areas.
Source: National Biodiversity Strategies in South America, Perspectives for Regional Cooperation, Brazilian Ministry of the Environment, 2004.

protected areas, including, for example, the monitoring and zoning of these areas, security and basic services in border areas, and capacity building of local agencies for biodiversity conservation (Table 2.15).

While deforestation of the Amazon is a cause of biodiversity loss and contributes to global warming, global climate change also affects the Amazon biome. It is noted that the warming of the climate system is already affecting Latin America's climate. Temperatures in Latin America increased by about 1°C during the twentieth century, while sea level rise has reached 2–3 mm/yr since the 1980s.[108] Apart from some possible positive effects on crop yields in the southern hemisphere, the impacts of climate change have been profoundly negative, already affecting some of the unique features and ecosystems of the region, including the risk of forest dieback in the Amazon basin.[109] The IPCC predicts that already by midcentury, increases in temperature and associated decreases in soil water may lead to a gradual replacement of tropical forest by savannah in eastern Amazon.[110]

[108] A. de la Torre, P. Fajnzylber, and J. Nash, *Low Carbon High Growth Latin America Responses to Climate Change: An Overview*, Washington, DC: World Bank Latin America and Caribbean Studies, 2009, p. 1.

[109] Ibid., p. 2.

[110] IPCC, *Climate Change 2007: Synthesis Report*, 2007, http://www.aussmc.org/IPCC_Synthesis_Report.php

3

The Origins of Regional Cooperation in the Amazon

This chapter provides an overview of the first treaties adopted by the eight Amazon States, from the time they attained independence from the European colonial powers, mostly by the end of the nineteenth century (except for Guyana and Suriname) until 1978, the year of the adoption of the Amazon Cooperation Treaty (ACT).[1] Signed by two or more Amazon States, those instruments were aimed mainly at establishing their common boundaries and setting up rules on commerce and navigation. Some few agreements and resolutions concluded, for instance at inter-American conferences, had the objective of promoting technical, scientific, and cultural cooperation, particularly in border areas. Such instruments paved the way for other forms of cooperation, of which the 1969 La Plata River Basin Treaty and the 1978 ACT are examples.

3.1. INTERNATIONAL BOUNDARIES DELIMITING THE AMAZON REGION

The papal bull *Inter Caetera* of 1493[2] issued by Alexander VI provided a first delimitation line between the Spanish and Portuguese possessions in the New World. Portugal, however, challenged this bull as being favorable to the Kingdom of

1 Brazil became an independent kingdom in 1822; Bolivia gained independence in 1825; Ecuador in 1822; the Republic of Gran Colombia (which included today's Colombia, Ecuador, Panama, and Venezuela) in 1819; and Venezuela in 1830. Guyana, a Dutch territory from the mid-seventeenth century onward, was acquired by Britain in 1799 and became independent in 1966. Suriname, founded by British settlers, who established the town of Paramaribo in 1610, by the mid-seventeenth century became Dutch territory. It was again ruled by Britain between 1799 and 1815, and finally returned to the Dutch after the 1815 Congress of Vienna. Suriname achieved independence in 1975 (G. Biger (ed.), *The Encyclopedia of International Boundaries*, Jerusalem: Jerusalem Publishing House, 1995, pp. 78–80, 91–115, and 169–72).

2 Bulls *Inter Caetera* of May 4, 1493 and May 3, 1493 (J.C. Macedo Soares, *Fronteiras do Brasil no regime colonial*, Rio de Janeiro: Livraria José Olimpio, 1939, pp. 25–32).

Castile. As a result, negotiations began for achieving a new boundary line and a year later the two kingdoms agreed on a boundary treaty signed on June 7, 1494 in the Spanish town of Tordesillas.[3] This treaty drew a line running from north to south, 370 leagues west of the Cape Verde Islands, and provided that everything west of this line should belong to Spain and everything east of it to Portugal. Subsequently, the treaties of Madrid of January 13, 1750[4] and of San Ildefonso of October 1, 1777[5] redefined the dividing line between Spanish and Portuguese possessions. This latter treaty placed the boundary from the Javary (or Yavarí) and Amazon rivers to the confluence of the latter with the Yapurá (or Caquetá) River,[6] placing the majority of the Amazon basin under Portuguese control. From then onward, Portugal took decisive steps to ensure control over the Amazon River by establishing fortifications at strategic points and creating *capitanias* in the region of the Cabo Norte (today the state of Amapá) and Negro rivers, pushing the English, French, and Dutch colonizers toward the north.[7]

3.1.1. *The* Uti Possidetis *as a Principle for Establishing International Boundaries*

Following independence from the European colonial powers, mostly by the 1820s, the newly created States in the Amazon basin began to negotiate and adopt international treaties delimitating their common boundaries, which were in general established by peaceful means.[8] The *uti possidetis* principle was at the basis of the boundary treaties adopted among the Amazon States and was also invoked in the settlement of boundary disputes among those countries. Article 7 of the 1851 Treaty on Commerce, Navigation, and Boundaries[9] between Brazil and Peru, for

3 G.F. Martens, *Suppléments au recueil des traités*, vol. 1, Gottingue: Dieterich, 1802–1808, pp. 372–89.

4 A. Cantillo, *Tratados, convenios y declaraciones de paz y de comercio*, Madrid: Imprenta de Alegria y Charlain, 1843, p. 400.

5 C. Parry, *The Consolidated Treaty Series*, vol. 46 (1775–1778), London: Oceana, 1979.

6 A map indicating this is available in W. Boggs, *International Boundaries: A Study of Boundary Functions and Problems*, New York: Columbia University Press, 1940, p. 78.

7 J.T. Calasans, *Le concept de ressource naturelle partagée: Application aux ressources en eau: l'exemple de l'Amérique du Sud*, thèse de doctorat en droit, Université Paris I, Panthéon Sorbonne, Septentrion Presses Universitaires, 1996, p. 299.

8 P. Motta Pinto Coelho, *Fronteiras na Amazônia, um espaço integrado*, Brasília: Fundação Alexandre Gusmão, Instituto de Pesquisa de Relações Internacionais, 1992, p. 58.

9 J.M. de Oliveira, *Actos diplomáticos do Brasil*, vol. 1, Rio de Janeiro: Jornal do Commercio, 1912, pp. 166–67; R. Aranda, *Colección de tratados del Perú*, vol. 2, Lima: Imprenta del Estado, 1843, pp. 514–20; UK, *British and Foreign State Papers*, vol. 42, London: William Ridgway, pp. 1308–312. Ratified by Brazil on March 18, 1852; ratified by Peru on December 1, 1851; ratifications exchanged at Rio de Janeiro on October 18, 1852 (G. Ireland, *Boundaries, Possessions and Conflicts in South America*, Cambridge, MA: Harvard University Press, 1938, p. 125).

instance, expressly recognized that the principle of *uti possidetis* should guide the drawing of their common boundary. Also, the 1867 Treaty of Friendship, Commerce, Navigation, Boundaries, and Extradition between Brazil and Bolivia affirmed that the parties agreed "in recognizing as a basis on which to determine the boundaries between their respective territories, the *uti possidetis*" and to define their boundary "in conformity with this principle" (Article 2). The *uti possidetis* was also reflected in the 1848 Treaty of Confederation signed by Colombia, Peru, Chile, Bolivia, and Ecuador at the First Congress of Lima, between 1847–1848 (Article 7):[10]

> The confederated republics declare that they have a perfect right to the preservation of their territorial boundaries as they existed at the time of the independence from Spain in their respective vice-royalties, *capitanias generales*, or presidencies in which Spanish America was divided.... Republics which, having been part of a same state when their independence was proclaimed, separated from each other after 1810, shall preserve the boundaries that they had recognized, without prejudice to the treaties that they have concluded or may conclude to change them or validate them pursuant to the present article.

A definition of the *uti possidetis ita possideatis* principle, which literarily means "as you possess you shall continue to possess,"[11] can be found in the award by the Swiss Federal Council in the boundary dispute between Colombia and Venezuela. According to this arbitral award, the boundary between the disputant parties should be identical with the frontiers laid down by the Spanish authorities between the respective territorial units prior to the independence of the Latin American States.[12]

This principle was founded on two main ideas: the absence of *terra nullius* in Latin America, and the recognition that boundaries of newly independent States should follow the boundaries of the old colonial territories from which they emerged.[13] Although sometimes neglected, that first element, which denies the existence of *terra nullius*, is a main feature of the *uti possidetis* principle as conceived in Latin America.[14] The

[10] Archivo Diplomático del Peru, *Congressos Americanos de Lima: recopilación de documentos*, vol. 1, Lima, 1938, p. 301. This principle was tacitly accepted by the Hispanic American States, but was formally ennunciated in Colombia in 1819 and later at the Congress of Panama of 1826 and at the Congress of Lima of 1847–1848 (J.L. Suárez, "El uti possidetis y los límites americanos," *Diplomacia Universitaria Americana, Argentina en el Brasil, Ciclo de Conferencias*, Buenos Aires: Escoffier, Caracciolo, 1918, p. 46).

[11] C. Antonopoulos, "The principle of *uti possidetis iuris* in contemporary international law," *Revue Hellénique de Droit International*, vol. 49, 1996, pp. 29–88, at p. 29.

[12] 1922 Swiss Federal Council Arbitral Award (Colombia–Venezuela), United Nations, Reports of International Arbitral Awards, vol. I, pp. 223–98.

[13] J. Barberis, "Les règles spécifiques du droit international en Amérique Latine," *Recueil des cours*, vol. 235, 1992, p. 153.

[14] M. Kohen, "L'uti possidetis revisité: l'arrêt du 11 septembre 1992 dans l'affaire el Salvador/Honduras," *Revue Générale de Droit International Public*, vol. 97, no. 4, 1993, pp. 939–73, at p. 942. As Kohen notes, this aspect has been rightly highlighted by the International Court of Justice in the case El Salvador–Honduras of September 11, 1992.

rejection of *terra nullius* was, however, based on an assertion of "constructive possession or occupation," given that in the nineteenth century many unexplored or uninhabited regions still existed in Latin America.[15] This assumption served the purpose of preventing boundary disputes among the newly created States. As asserted by the Chamber of the International Court of Justice (ICJ) in the Burkina Faso–Mali case, the obvious purpose of the *uti possidetis* was precisely to "prevent the independence and stability of States being endangered by fratricidal struggles provoked by the challenging of frontiers following the withdrawal of the administering power."[16]

The *uti possidetis* appears in two manifestations in Latin America. The Hispanic American States, successor States of the Spanish Empire, supported the *uti possidetis iuris,* whereas Brazil, previously a Portuguese colony, claimed that the *uti possidetis de facto* should govern its relations with its Hispanic American neighbors. These different approaches were due to their disparate views regarding the applicability of the treaties of Madrid of 1750 and San Ildefonso of 1777, which defined the boundaries in Latin America. Brazil considered that these treaties were no longer applicable by virtue of the war between Spain and Portugal, which resulted in the signature of the 1801 Treaty of Badajoz, which put an end to the hostilities. Brazil supported the view that this latter treaty did not bring the treaties of 1750 and 1777 back into force, and that therefore they could not be used to delimitate the boundaries of the newly independent States in South America. In the absence of a treaty basis for defining their international boundaries, the principle of *uti possidetis de facto* should instead apply, by which rightful sovereignty should be based on occupation, utilization, and effective control of territory, rather than written or verbal claims. The Brazilian concept of *uti possidetis* is thus based on effective possession, and for this reason this principle is referred to as *uti possidetis de facto*.[17] The Hispanic American States, on the other hand, considered that the 1777 Treaty of San Ildefonso remained in force even after 1801, and, based on the notion of *uti possidetis iuris,* they claimed that this treaty continued to govern their international boundaries.[18]

Those two different approaches are not, as some suggest, contradictory or irreconcilable.[19] As Marcelo Kohen notes, Brazil has never denied the principle of

[15] M.N. Shaw, "The heritage of states: The principle of *uti possidetis juris* today," *British Yearbook of International Law,* 1996, pp. 75–154.

[16] ICJ, "Frontier Dispute (Burkina Faso v. Republic of Mali)," *ICJ Reports 1986,* p. 565, para. 20. According to Antonopoulos (op. cit.), "the *uti possidetis* has the result of (at least *a priori*) of dispensing boundaries or territorial disputes among newly independent entities."

[17] M.G. Kohen, "La contribución de América Latina al Desarrollo Progresivo del Derecho Internacional en Materia Territorial," *Anuario de Derecho Internacional,* Universidad de Navarra, vol. 16, 2001, pp. 57–77, at p. 63.

[18] M.G. Kohen, *Possession contestée et souveraineté territoriale,* Paris: Presses Universitaires de France, 1997, pp. 446–50.

[19] Kohen, "La contribución de América Latina," op. cit.

uti possidetis juris, which in practice continued to be used among the Hispanic American States, but only rejected the application of this principle in its relations with those States. Brazil made use of the notion of the *uti possidetis de facto* instead, as it took the view that there existed no other valid title at the time. According to Kohen, there existed a consensus among South American States, including Brazil, in relation to the principle of *uti possidetis*: They all acknowledged the absence of *terra nullius* in South America, agreeing that the newly independent States enjoyed a territorial sovereignty that previously pertained to Spain and Portugal, and also recognized that existing boundary treaties should be used to define their common boundaries.

The *uti possidetis* principle was also present in the modern decolonization process in Africa[20] to ensure that boundary disputes would be settled by respecting pre-existing borders rather than by resorting to force. In 1983, a Chamber of the ICJ, in examining the frontier dispute between Burkina Faso and the Republic of Mali, recalled the essence of this principle as follows:[21]

> the essence of the principle lies in its primary aim of securing respect for territorial boundaries at the moment when independence is achieved. Such territorial boundaries may be no more than delimitation between different administrative divisions or colonies all subject to the same sovereign. In that case, the application of the principle of *uti possidetis* resulted in administrative boundaries being transformed into international frontiers in the full sense of the term.

3.1.2. *International Boundaries between Brazil and the Other Amazon States*

After its independence from Portugal on September 7, 1822, Brazil put into practice a policy to promptly delimitate its extensive international boundaries,[22] which represented 77 percent of its territorial perimeter.[23] With the exception of Ecuador, Brazil has common boundaries with all other Amazon States (Bolivia, Colombia, Guyana, Suriname, Peru, and Venezuela), and also with French Guyana.

Negotiations to define a frontier with Bolivia began in 1837, which resulted in the conclusion of a boundary treaty signed at La Paz on March 27, 1867. By virtue of an agreement with the United States signed in London on July 11, 1901, Bolivia granted to an Anglo-American syndicate, incorporated under the name Bolivian Syndicate

[20] The *uti possidetis* principle has evolved significantly with circumstances, as seen in recent cases, such as in that of the breakup of the former Republic of Yugoslavia, the Soviet Union, and Czechoslovakia (J. Castellino and S. Allen, "The doctrine of *uti possidetis*: Crystallization of modern post-colonial identity," *German Book of International Law*, vol. 43, 2000, pp. 205–26).

[21] ICJ, op. cit., pp. 554, 586.

[22] Calasans, op. cit., p. 283.

[23] Motta Pinto Coelho, op. cit., p. 14.

of New York City, a concession of rights including powers of administration and government in the Bolivian territory of Acre. The Bolivian Syndicate of New York City was allowed to purchase lands in Acre, was exempted from paying taxes, and was given extensive rights, for example, to navigate freely on all rivers of the territory, and to maintain railroads, electric power plants, and a police force. In response to this concession, on August 8, 1902, Brazil closed the Amazon to commerce in transit to or from Bolivia.[24] This dispute was settled through a treaty signed at Petrópolis, Brazil on November 17, 1903,[25] by which Bolivia relinquished the Acre region to Brazil in exchange for a smaller territorial parcel and the payment of an indemnity by Brazil. A treaty signed in Rio de Janeiro on December 25, 1928 finally settled the present boundary between Brazil and Bolivia.[26] Brazil and Bolivia also signed an agreement through exchange of notes on March 11, 1997, in order to create boundary commissions.[27]

Brazil and the Republic of Gran Colombia established the northern section of their common boundary through a treaty on boundaries and navigation signed on April 24, 1907,[28] while its remaining part was subject to a later arrangement, once boundary claims between Colombia, Ecuador, and Peru, on the one hand, and between Brazil and Peru, on the other, had been settled. The treaty signed in 1907 envisaged the adoption of a treaty on commerce and navigation based on the fullest liberty of land transit and river navigation within the Brazil–Colombia border area, which was ultimately agreed to in Rio de Janeiro on August 21, 1908.[29] A treaty signed in Rio de Janeiro on November 15, 1928 finally completed the delimitation of the Brazil–Colombia boundary by defining its southern section.[30] Subsequently, Brazil and Colombia signed an agreement on March 12, 1930 with further instructions for the delimitation or their common boundary, followed by an agreement adopted through exchange of notes on June 10, 1937, establishing the final delimitation of their boundary.[31]

24 J.B. Moore, *Digest of International Law*, vol. 1, Washington, DC: Government Printing Office, 1906, p. 646.

25 *British and Foreign State Papers*, op. cit. vol. 96, pp. 383–87. Ratified by Bolivia on 6 January 1904; sanctioned by Brazil, Decree 1179 of February 18, 1904; ratifications exchanged at Rio de Janeiro on 10 March 1904; promulgated by Brazil, Decree 5161 of March 10, 1904 (Ireland, op. cit., p. 47).

26 *British and Foreign State Papers*, op. cit., vol. 128, pp. 677–78. Sanctioned by Brazil, Decree 5649 of January 8, 1929; ratified by Brazil on February 19, 1929; ratified by Bolivia June 7, 1929; ratifications exchanged at Rio de Janeiro, June 27, 1929; promulgated by Brazil, Decree 18838 of July 9, 1929 (Ireland, op. cit., p. 52).

27 Information on boundary treaties between Brazil and other South American countries is available on the Web site of the Brazilian Ministry of Foreign Affairs: www2.mre.gov.br/dai/bifront.htm.

28 G. Cavelier, *Tratados de Colombia 1811–1910*, vol. 1, Bogota: Editorial Kelly, 1982, pp. 682–85.

29 Ibid., pp. 710–14.

30 Cavelier, op. cit., vol. 2, 1982, 1984, pp. 509–12.

31 www2.mre.gov.br/dai/bifront.htm.

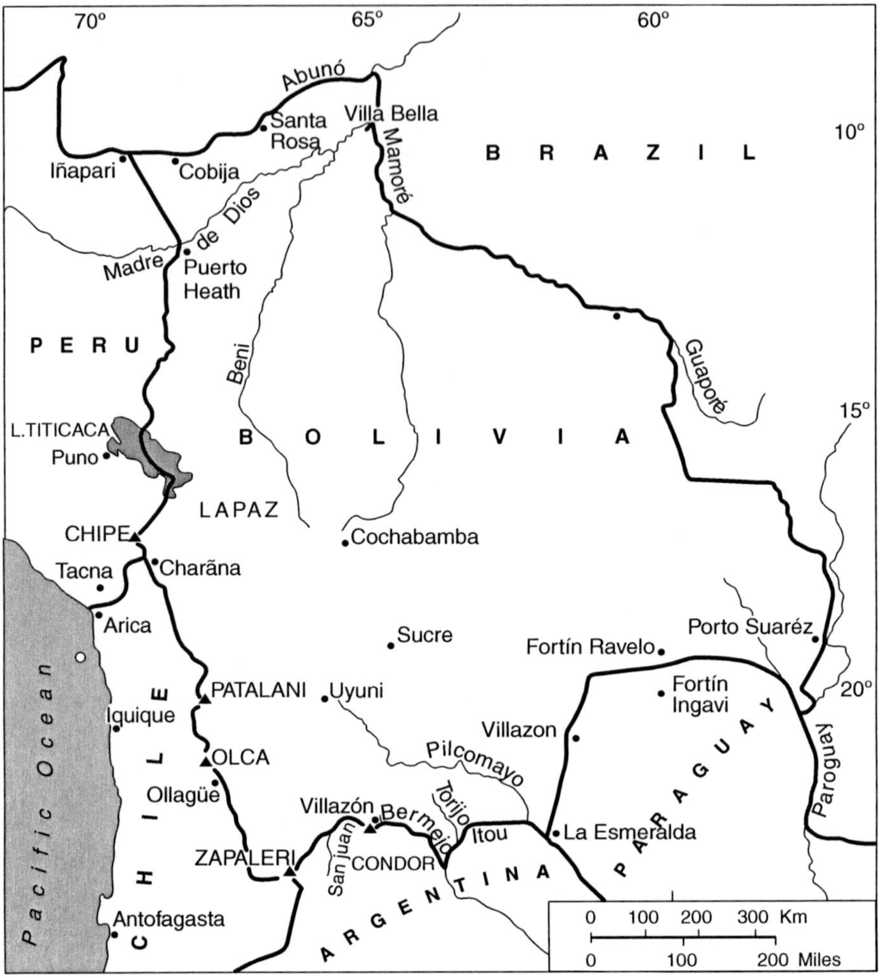

MAP 3.1. Bolivia–Brazil[32]

It is noteworthy that international boundaries in the Amazon basin were defined in the context of the prosperous period of the "rubber boom," which occurred mostly in the Brazilian Amazon. In the nineteenth century, the Industrial Revolution triggered an increased demand for rubber, which prompted an intensive and predatory exploitation of Amazonian rubber until approximately 1914, entailing disputes over ill-defined boundaries, such as that between Brazil and British Guyana. On November 6, 1901, Brazil and Britain convened in London a meeting envisaging

[32] Biger, op. cit., p. 79.

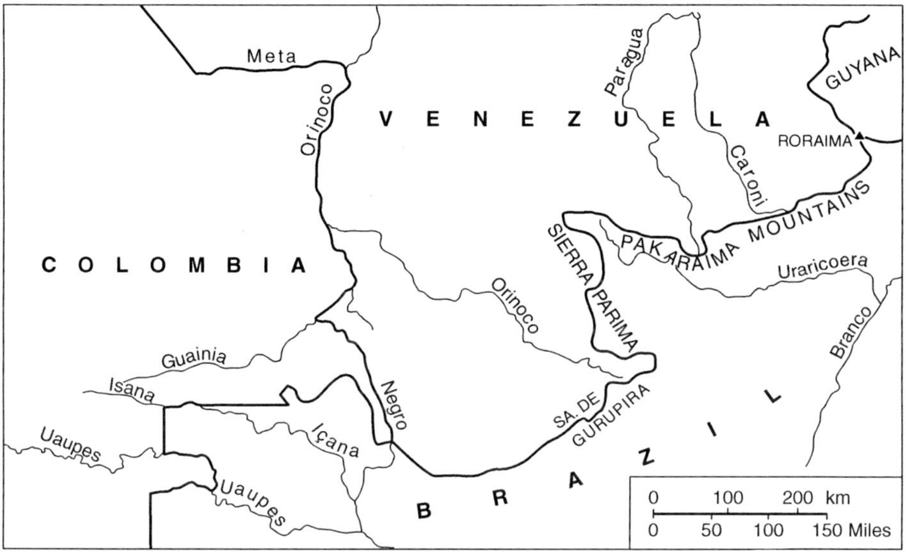

MAP 3.2. Brazil–Colombia–Venezuela[33]

the establishment of the boundary between Brazil and the colony of British Guyana through arbitration.[34] A few years later, an arbitral award by the king of Italy was delivered in Rome on June 6, 1904.[35] On April 22, 1926, Brazil and Great Britain adopted a boundary treaty to rectify certain inaccuracies in the award and to demarcate the boundary line along the watershed between the Amazon basin and the Essequibo and Courantyne rivers as far as the point of junction with Suriname.[36] This treaty was supplemented by a protocol of March 18, 1930,[37] which appointed a demarcation commission to give effect to the boundary treaty of April 22, 1926. An exchange of notes of March 15, 1940 ultimately approved the boundary demarcation.[38] The treaty signed at Rio de Janeiro on May 5, 1906 defined the Brazil–Dutch Guyana (today's Suriname) boundary.[39]

33 Biger, op. cit., p. 511.
34 *British and Foreign State Papers*, op. cit., vol. 94, pp. 23–29. Sanctioned by Brazil, Decree 824 of December 27, 1901; ratified by Great Britain on December 23, 1901; ratifications exchanged at Rio de Janeiro on January 28, 1902; promulgated by Brazil, Decree 4329 of January 28, 1902 (Ireland, op. cit., p. 155).
35 *British and Foreign State Papers*, op. cit., vol. 99, pp. 930–32.
36 Ibid., vol. 123, p. 468.
37 League of Nations, *Treaty Series*, vol. CI, Geneva, p. 401.
38 5 UNTS 71.
39 *British and Foreign State Papers*, op. cit., vol. 99, pp. 932–33. Sanctioned by Brazil, Decree 1659 of June 25, 1907; ratified by the Netherlands on September 12, 1908; ratifications exchanged at The Hague on September 15 1908; promulgated by Brazil, Decree 7133 of September 24, 1908 (Ireland, op. cit., p. 159).

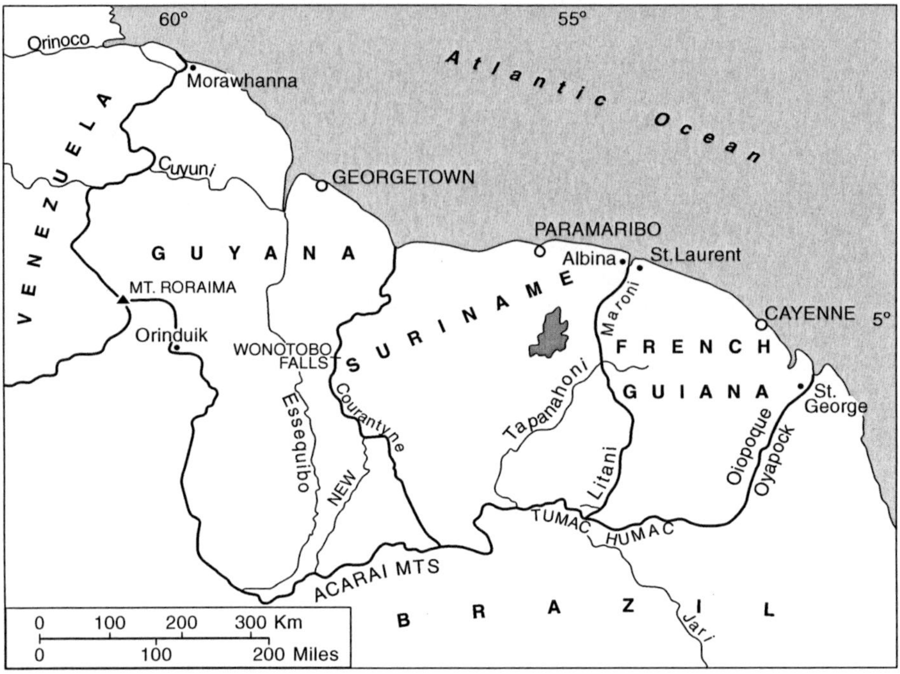

MAP 3.3. Brazil–Guyana–Suriname[40]

Since its independence in 1824, Peru has claimed lands east to the Javary (or Yavarí) River and resisted the application of the San Ildefonso line. A treaty of friendship between Brazil and Peru signed in Lima on July 8, 1841 called for the establishment of a precise boundary based on the *uti possidetis* principle. This treaty has never been ratified by Brazil, because war between Bolivia and Peru intervened.[41] Once peace was restored, Brazil and Peru signed a treaty on commerce and river navigation in Lima on October 23, 1851,[42] delimitating their boundary. Subsequently, Brazil and Peru signed a river convention in Lima on October 23, 1858[43] providing that, within twelve months from the exchange of ratifications, the parties would name a mixed commission to demarcate the frontier, as provided by Article 7 of the treaty of October 23, 1851. Subsequently, Brazil and Peru signed a treaty on September 8, 1909 to complement the delimitation of their common boundary and establish

[40] Biger, op. cit., p. 259.

[41] Aranda, op. cit., pp. 507–14.

[42] De Oliveira, op. cit., pp. 166–67. Ratified by Brazil on March 18, 1852; ratified by Peru on December 1, 1851; ratifications exchanged at Rio de Janeiro on October 18, 1852.

[43] Aranda, op. cit., pp. 520–26. Ratified by Brazil on April 5, 1859; ratified by Peru on April 1, 1859; ratifications exchanged at Paris on March 27, 1859; promulgated by Brazil, Decree 2442 of July 16, 1859.

MAP 3.4. Brazil–Peru[44]

general principles on commerce and navigation in the Amazon basin, followed by an agreement adopted on July 21, 1999, through exchange of notes, to create boundary commissions.[45]

[44] www2.mre.gov.br/dai/bifront.htm.

[45] Biger, op. cit., p. 96.

Finally, with regard to the Brazil-Venezuela boundary a first treaty was signed on November 25, 1852. Some difficulties arose with regard to the western section of the boundary junction with Colombia located west of the Negro River, which was claimed by Colombia. As a result, a new boundary and navigation treaty between Brazil and Venezuela was signed in Caracas on May 5, 1859.[46] A protocol was signed in Rio de Janeiro on July 24, 1928[47] that provided for the creation of a mixed commission to complete the boundary demarcations. A demarcation agreement to establish the Brazil-Venezuela common boundary was finally signed on November 7, 1929.[48]

3.2. COMMERCE AND NAVIGATION

Under the Treaties of Madrid of January 13, 1750[49] and the 1777 Treaty of San Ildefonso[50] between Portugal and Spain, the colonial powers were granted exclusive navigation rights over international rivers under their jurisdiction and adjacent territories.[51] Following independence, the newly created States sharing the Amazon basin continued to exercise such exclusive navigation rights over their rivers, which were closed to international navigation.[52] Nonriparian States, notably Great Britain, France, and the United Sates, however, began to demand the opening of the Amazon to international navigation. Fearing any forms of foreign intervention at a time when international boundaries were yet to be established, the then Empire of Brazil undertook a policy of reinforcing control over the Amazon River particularly through the Companhias de Navegação, which was given the monopoly to undertake navigation and commerce on the Amazon and to colonize areas under its influence.[53]

Brazil's protectionist position with respect to navigation of the Amazon River and tributaries was, however, challenged by increased pressure from third States and also from the other riparian States. On July 26, 1851, the United States

46 Venezuela, *Colección de tratados públicos de Venezuela*, Caracas: Imprenta Nacional, 1910, pp. 144–48, 343–45. Ratified by Brazil on September 6, 1858; ratified by Venezuela on July 9, 1860; ratifications exchanged at Caracas on July 31, 1860; promulgated by Brazil, Decree 2726 of January 12, 1861 (Ireland, op. cit., p. 140).

47 *British and Foreign State Papers*, op. cit., vol. 130, pp. 447–49. Approved by Brazil, Decree 5664 of January 12, 1929; ratified by Venezuelan legislature on May 31, 1929; ratified by Venezuelan Executive on June 24, 1929; ratifications exchanged at Rio de Janeiro on August 31, 1929; promulgated by Brazil, Decree 18905 of September 17, 1929 (Ireland, op. cit., p. 142).

48 www2.mre.gov.br/dai/bifront.htm.

49 Cantillo, op. cit. p. 400.

50 Ibid., pp. 534–44.

51 C. Sosa-Rodriguez, *Le droit fluvial international et les fleuves de l'Amérique Latine*, Paris: Pedone, 1935, pp. 98–99.

52 Calasans, op. cit., p. 301.

53 Ibid., p. 302.

concluded with Peru the Treaty of Friendship, Commerce, and Navigation,[54] which conferred on American citizens full liberty to trade in all parts of the Peruvian territory and secured most-favored nation treatment in matters of commerce and navigation.[55] Brazil responded by concluding with Peru a few months later the Treaty on Commerce and River Navigation, signed in Lima on October 23, 1851, which provided for exclusive navigation rights over the Amazon to the "respective States owning its banks."[56]

By the Decree of January 27, 1853, Bolivia declared that the rivers flowing through its territory discharging into the Amazon were free to the commerce and navigation of all nations.[57] At that same year, Peru issued the Decree of April 15, 1853 by which the towns of Loreto and Nauto were made ports of entry for international vessels and the privileges conferred on Brazil by the Treaty on Commerce and River Navigation signed in Lima on October 23, 1851 were extended to all most-favored nations.[58] Based on the Decree of April 15, 1853, the United States sought recognition of the right to navigate in Peruvian waters.

In reaction to those developments, Brazil invited New Granada (the modern Colombia), Ecuador, and Venezuela to negotiate an international agreement for closing the Amazon River to international navigation and conferring exclusive navigation rights on its riparian States. Subsequently, Peru reconsidered its position and, through the Decree of January 4, 1854, modified the prior Decree of April 15, 1853 by allowing the use of the Amazon only to Brazilian vessels.[59] The 1851 Treaty on Commerce and River Navigation between Brazil and Peru expired on October 23, 1858, and, in the same year, they adopted a river convention on October 22, 1858 by which Peru and the Empire of Brazil allowed, as a "special concession," reciprocal navigation of the Amazon paying only charges of lightning, pilotage, and police.[60] In 1862, the Peruvian government gave notice of the termination of the 1851 Treaty of Friendship, Commerce, and Navigation with the United States, which came[61] to an end on December 9, 1863.

[54] W.M. Malloy, *Treaties, Conventions, International Acts, Protocols and Agreements between the United States and Other Powers, 1776–1909*, vol. 2, Washington, DC: Government Printing Office, 1910, pp. 1388–410.

[55] Ibid., pp. 1388–401.

[56] Article 2 (De Oliveira, op. cit., pp. 166–67).

[57] *British and Foreign State Papers*, op. cit., vol. 55, p. 505.

[58] Moore, op. cit., p. 641.

[59] Ibid., p. 645.

[60] Fluvial Convention between the Empire of Brazil and the Republic of Peru, signed on October 22, 1858 (Aranda, op. cit., pp. 520–25), Article 2: "Su Majestad el Emperador de Brasil, conviene en permitir como concesión especial, que las embarcaciones peruanas, registradas en forma, puedan pasar libremente del Perú al Brasil, y vice-versa, por el río Amazonas o Marañón, y salir por el mismo rió al Océano y vice-versa, siempre que se sujeten a los reglamentos fiscales y de policía establecidos por la autoridad superior brasilera."

[61] Moore, op. cit., p. 645.

The American government, however, renewed efforts to attain the opening of the Amazon River and its tributaries by encouraging the Amazon upper riparian States to open their rivers to international navigation and also obtain the removal of all restrictions from Brazil. The American secretary of state instructed US diplomatic representatives in Colombia, Bolivia, and Ecuador to seek support from these governments,[62] as evidenced in a letter sent by the American secretary of state to the US *chargé d'affaires* in Ecuador:[63]

> The Republics of Venezuela, New Granada, Ecuador, Peru and Bolivia are all interested in procuring the removal of the restrictions which now practically deprive them in a considerable degree of the advantages of the portion of their respective territories which would be (by the opening of the Amazon) opened to an extensive and profitable foreign commerce ... Bolivia and Peru have already taken steps for testing the extent to which the pretensions of Brazil may be maintained.... It would be well for you ... to present to the Government the advantages that would accrue to Ecuador in following the example of Bolivia and Peru. For, when all the five States whose fertile regions are watered by the Amazon and its tributaries shall have thrown open their rivers and ports to foreign commerce it is though that Brazil will not be able long to withstand the moral power which will thus be arrayed against her selfish and restrictive policy.

Basically, the American government claimed that Brazil's position obstructed trade among the Amazon upper riparian States, which injured their interests to use the navigable watercourses of the Amazon for commercial purposes, as proposed in a note of August 8, 1853 from the American secretary of state to the United States minister to Brazil:[64]

> The most important object of our mission – an object to which you will devote your early and earnest effort – is to secure the citizens of the United States the free use of the Amazon. There are several republics with which our countrymen have commercial intercourse situated on the upper waters and tributaries of that great river. With these States they would carry on an extensive trade were not our vessels excluded from approaching their internal ports by the selfish and unjustifiable policy of the Brazilian Government that claims and has hitherto exercised the right to obstruct the trade to the countries bordering upon and contiguous to the Amazon

62 Secretary of State Marcy instructed the United States ministers in Colombia (letter of October 15, 1853 to Mr. Green, minister of the United States in Colombia), Ecuador (letter of August 20, 1853 to Mr. White, United States *chargé d'affaires étrangers*) and Bolivia (letter of November 1, 1853 to Mr. Dana, United States *chargé d'affaires étrangers*) to gain support from these governments for the opening of the Amazon; see Sosa-Rodriguez, op. cit., pp. 160–61.

63 Letter of August 20, 1853 from Mr. Marcy, United States secretary of state, to Mr. White, *chargé d'affaires* to Ecuador (Moore, op. cit., p. 644).

64 W. R. Manning, Diplomatic Correspondence of USA – Inter-American Affairs 1831–1860, vol. 2, Bolivia and Brazil, Rumford Press, Concord N.H., p. 170.

> with foreign nations through this great natural highway. The assumption and exercise of this right is not only injurious to the interests of the states on the navigable waters of the Amazon, but to all other nations wishing to use these waters for the purpose of commercial intercourse.... This restricted policy, which it is understood Brazil still persists in maintaining, in regard to the navigable rivers passing though her territories is the relic of an age less enlightened than the present.

From 1853, the Amazon upper riparian States finally began to open their rivers to international navigation: Bolivia, by the Decree of January 27, 1853; Ecuador, by the Decree of November 26, 1853; and Colombia, by the Law of May 24, 1856.[65] Bolivia concluded a treaty of commerce and navigation with the United States on May 13, 1858[66] by which the Amazon and the La Plata Rivers were regarded as "highways or channels open by nature for the commerce of all nations."[67] Therefore, whereas international navigation was secured by Ecuador, Bolivia, and Colombia in the upper Amazon and its tributaries, Brazil and Peru adopted and maintained a restrictive navigation regime. In practice, the navigation of the Amazon was impeded by Brazil, which possessed the Amazon's entire downstream course. Finally, a few years later, these two countries opened the Amazon to international navigation: Brazil by Decree 3.749 of December 7, 1866 followed by the regulatory Decree of July 31, 1867, and subsequently, Peru through Decree of December 17, 1868. International navigation was then secured on the entire Amazon.

It should be noted that freedom of navigation in Latin America has a different connotation than in other regions of the world. A distinctive feature of international watercourses law in Latin America is that navigation rights are based on a "gracious concession" exercised by the territorial State as an expression of its sovereign rights.[68] In general, foreign vessels can only navigate international rivers if they have been given prior authorization by a unilateral act (constitution, law, or decree) or an international treaty conferring such rights.[69] Such a distinctive feature was explained by the Chilean jurist, Alejandro Alvarez, on the occasion of the Barcelona Conference on the Convention on the Regime of Navigable Waterways of International Concern of 1921 at the League of Nations:[70]

> Sur le continent Américain, le principe de la libre navigation de fleuves n'a pas suivi la même évolution [qu'en Europe ou Afrique]: s'il a été admis, c'est non pas

65 Sosa-Rodrigues, op. cit., p. 11.

66 Malloy, op. cit., vol. 1, pp. 113–25.

67 Article 24 (Malloy, op. cit., vol. 2, pp. 113–25).

68 Sosa-Rodriguez, op. cit., pp. 109–14; E.J. Rey Caro, *El derecho de los cursos de agua internacionales: la navegación*, Córdoba: Universidad Nacional, 1986, pp. 61–62; Barberis, op. cit., pp. 180–84.

69 Barberis, op. cit., p. 184.

70 Transcribed in L. Caflisch, "Règles générales du droit des cours d'eau internationaux," in *Collected Courses of the Hague Academy of International Law*, vol. II, Dordrecht: Nijhoff, 1989, pp. 117–18.

> par extension du principe européen, mais comme une concession que les Etats riverains ont volontairement octroyée dans des accords inter partes ou dans des actes législatifs.

That particular contention with regard to freedom of navigation was no exception in the Amazon basin. Bilateral treaties on commerce and navigation adopted at the time reveal that the Amazon basin States permitted the free transit of foreign vessels as a concession or special favor. For example, Article 2 of the Treaty of Friendship, Commerce and Navigation between Brazil and Peru signed on October 22, 1858[71] reads as follows: "The Emperor of Brazil permits, *as a special concession*, free transit to registered Peruvian vessels, from Peru to Brazil and vice versa, through the Amazon River or Marañón and exit from that River to the Ocean and vice versa" (emphasis added). Similarly, according to Article 7 of the Treaty of Friendship, Commerce, Navigation, Boundaries, and Extradition between Brazil and Bolivia signed on March 27, 1867:[72] "His Majesty the Emperor of Brazil permits, *as a special favor*, the water of the navigable rivers running through Brazilian territory to the ocean, to be free to the commercial navigation of the Republic" (emphasis added).

The Amazon basin States not only viewed freedom of navigation as a special concession, but would grant it on a reciprocal basis, as reflected in Article 4 of the Treaty of Friendship, Commerce, and Navigation between Colombia and Ecuador signed on August 10, 1905:[73] "there will exist between the two contracting Republics *reciprocal freedom* of commerce and navigation" (emphasis added); or in Article 2 of the Treaty on the Navigation of River Yapurá or Caquetá between Brazil and Peru signed on April 15, 1908:[74]

> navigation of the Peruvian waters if the Yapurá or Caquetá is free to Brazilian vessels which may be proceeding to any conterminous State and vice-versa, *in reciprocity for the equivalent concession* which Brazil makes to Peruvian merchant vessels to navigate the principal stream of the Brazilian Yapurá. (emphasis added)

Similarly, according to Article 3 of the ACT, commercial navigation among the riparian States is conferred on a "reciprocal basis" and shall be mutually guaranteed:

> In accordance with and without prejudice to the rights granted by unilateral acts, to the provisions of bilateral treaties among the Parties and to the principles and rules of International Law, the Contracting Parties mutually guarantee, on a

[71] "Su Majestad el Emperador del Brasil, conviene en permitir como concesión especial que las embarcaciones peruanas registradas en forma, puedan pasar libremente del Perú al Brasil, y viceversa, por el río Amazonas o Marañón, y salir por el mismo río al Océano y viceversa" (Aranda, op. cit., pp. 520–29).

[72] *British and Foreign State Papers*, op. cit., vol. 59, 1868–1869, pp. 1161–169.

[73] "Habrá entre las dos Repúblicas contratantes reciproca libertad de comercio y navegación" (Cavelier, op. cit., vol. 1, pp. 568–75).

[74] *British and Foreign State Papers*, op. cit., vol. 102, 1908–1909, p. 912.

> reciprocal basis, that there shall be complete freedom of commercial navigation on the Amazon and other international Amazonian rivers, observing the fiscal and police regulations in force now or in the future within the territory of each. Such regulations should, insofar as possible, be uniform and favor said navigation and trade.

The freedom of navigation of the Amazon River under this provision is restricted to the riparian States, as only they can confer on a "reciprocal basis" similar rights to the other riparian States.[75]

The navigation of the Amazon and its tributaries, as noted above, is dependent on existing treaties or national laws granting this concession both to riparian States and foreign vessels. For example, according to Article 3 of the Brazilian Decree 3.749 of December 7, 1866, which opened the Amazon River to merchant vessels of all nations (and also the Tocantins, Tapajós, Madera, Negro, and San Francisco rivers), navigation of the Amazon River would depend on "previous conventions with other Riparian States, with regard to respective limits, fiscal and police regulations."[76]

Following the opening of the Amazon River and its tributaries, the riparian States began to adopt international treaties on commerce and fluvial navigation.[77] Brazil and Peru adopted a provisional protocol on navigation of the River Iça (or Putumayo) in Lima on September 29, 1876, which provided for free navigation of commercial vessels on a reciprocal basis, subject to fiscal and police regulations. Navigation of war vessels was subject to a number limit. These countries have also concluded an agreement on April 15, 1908[78] regulating the navigation of the Yapurá River.[79] The Treaty of Petropolis, adopted by Brazil and Bolivia on November 17, 1903, pledged for the "most ample freedom of transit and river navigation to both countries."[80] Similarly, Colombia and Ecuador concluded on August 10, 1905 a treaty ensuring perpetual freedom of

[75] Caflisch, op. cit., p. 122.

[76] "La navegación de los afluentes del Amazonas en la parte en que una sola de sus márgenes pertenece al Brasil, queda dependiente de Convenciones previas con los otros Estados ribereños acerca de los respectivos limites y reglamentos de policía y fiscales" (Aranda, op. cit., pp. 529–30).

[77] Examples of bilateral agreements on friendship, commerce, and navigation include: the Treaty of Friendship, Commerce and Navigation between Ecuador and Nueva Granada adopted on July 9, 1856 (Cavalier, op. cit., vol. 1, pp. 251–58); the Treaty of Friendship, Commerce and Navigation between Brazil and Peru adopted on October 22, 1858 (Aranda, op. cit., pp. 520–29); the Treaty of Friendship, Commerce and Navigation between Colombia and Ecuador adopted in Quito on August 10, 1905 (Cavelier, op. cit., vol. 1, pp. 568–75); the Treaty on Commerce and Navigation between Brazil and Colombia adopted in Rio de Janeiro on August 21, 1908 (Cavelier, op. cit., vol. 1, pp. 710–13); and the Treaty on Commerce and River Navigation signed on August 12, 1910 between Brazil and Bolivia (699 UNTS, 270).

[78] *British and Foreign State Papers*, op. cit., vol. 102, p. 912.

[79] Aranda, op. cit., pp. 619–25, 631–34.

[80] *British and Foreign State Papers*, op. cit., vol. 96, pp. 383–87.

navigation of the Napo and Putumayo rivers.[81] Also, the Treaty on Commerce and Navigation of April 24, 1907 between Brazil and Colombia recognized ample freedom of terrestrial transit and fluvial navigation. These countries agreed to regulate matters concerning trade and navigation in a special treaty, which was done by a treaty signed in Rio de Janeiro on August 21, 1908.[82] A few years later, an agreement on limits and fluvial navigation between Brazil and Colombia signed on November 15, 1928 restated the right to free navigation on the Amazon, Japurá (or Caquetá), and Içа (or Putumayo) rivers and all their tributaries.

In general, the freedom of navigation on the Amazon and its tributaries granted on a reciprocal basis was subject to some restrictions: river-coasting trade was not permitted; merchant vessels were subject to the fiscal and police regulations established by the respective competent authorities, which should in general be the most favorable to the navigation and commerce between the contracting parties; and warships reciprocally enjoyed the privilege of free passage and entry, but this liberty was in general dependent on prior notification or agreement between the parties. The early bilateral treaties did not establish river commissions or any similar institutions.

3.3. EARLY FORMS OF COOPERATION IN THE AMAZON REGION

3.3.1. *Treaties of Friendship and Cooperation*

After independence and up to the adoption of the Amazon Cooperation Treaty in the 1970s, the Amazon States committed to cooperate by signing bilateral friendship agreements, such as the 1976 Treaty of Friendship, Cooperation, and Trade between Brazil and Suriname[83] and the 1977 Friendship and Cooperation Treaty between Brazil and Venezuela.[84] These treaties normally created joint commissions to draw up guidelines on matters of common interest such as trade, transport, communications, science and technology, and envisaged regular meetings between the parties to oversee the treaty implementation. Specific areas of cooperation foreseen under these bilateral treaties included cultural exchange, and the dissemination of the other party's language and culture,[85] or technical and scientific cooperation to be promoted by joint scientific

[81] Cavelier, op. cit., vol 1, pp. 568–75.

[82] Ibid., pp. 710–14.

[83] Brazilian Ministry of Foreign Affairs, Divisão de Atos Internacionais, *Coleção de Atos Internacionais*, vol. 875, Brasília, 1978.

[84] Ibid., vol. 941, 1978.

[85] Treaties on cultural exchange signed between the Amazon countries include the cultural agreement between Brazil and Ecuador signed on May 24, 1944 (73 UNTS 223) and on July 12, 1973 (975 UNTS 283); and the agreement on cultural exchange between Brazil and Colombia on March 20, 1963 (975 UNTS 269).

research, training, and regular exchange of information. Public health and sanitation was also a matter of common concern and subject of various bilateral treaties adopted by the Amazon States, *inter alia* the 1972 cooperation agreement on sanitation in the Amazon region between Brazil and Colombia,[86] the 1972 cooperation agreement on sanitation between Brazil and Bolivia,[87] and the 1976 sanitary agreement for the tropical environment[88] between Brazil and Peru. These treaties envisaged coordinated activities of epidemiological control, including the building of vaccination facilities in border areas and unified methods for disease control. Notification of transmittable diseases involving risks to the parties' respective populations was often mandatory.

3.3.2. *Environmental Conservation*

Except for a handful of examples, the environment was not a policy priority for the Amazon basin States for some time. This situation began to change in the 1970s, probably due to an increased awareness of the environmental degradation observed worldwide, which culminated in the United Nations Conference on the Human Environment held in Stockholm in 1972.[89] In particular, Brazil seemed convinced that States should define their own environmental policies *vis-à-vis* the international community and began to place greater emphasis on elaborating its national policies in that domain.[90] This country would later become the main instigator of a regional treaty in the Amazon and take the leadership in its preparatory process.

Until the 1970s, the Amazon States adopted only some few bilateral agreements with the objective of preserving the environment, which were mostly aimed at the protection of species of fauna and flora, such as the 1973 and 1975 agreements on the conservation of fauna and flora in the Amazonian territories between Brazil and Colombia,[91] and Brazil and Peru,[92] respectively. With very similar objectives, these agreements intended to draw joint guidelines through technical bilateral meetings on areas such as hunting, fishing, or on the management of chemicals and the introduction of alien species. They also envisaged the creation of nurseries in boundary areas to protect species of fauna and flora of scientific and economic interest, and the establishment of biological reserves representative of particular ecosystems and geographical unities.

[86] Brazilian Ministry of Foreign Affairs, Divisão de Atos Internacionais, op. cit., vol. 847, 1976.
[87] Ibid., vol. 936, 1978.
[88] Ibid., vol. 887, 1978.
[89] H. Muñoz, *Environment and Diplomacy in the Americas*, London: Lynne Rienner, 1992, p. 1.
[90] D. Ware, "The Amazon Treaty: a turning point in Latin American cooperation?", *Texas International Law Journal*, vol. 15, 1980, pp. 118–37, at p. 127.
[91] Brazil, *Diário Oficial da União* of July 12, 1976, Decree 78016, p. 9367.
[92] Brazil, *Diário Oficial da União* of November 24, 1976, Decree 78802, p. 15382.

Under the framework of early inter-American conferences started in Washington, DC since 1889,[93] which are at the origins of the inter-American system under the aegis of the modern Organization of American States (OAS), the Amazon States, among others, have also adopted resolutions on environmental protection. For example, Resolution LXXII on industrial and agricultural uses of international rivers, adopted in 1933 at the Seventh Inter-American Meeting held in Montevideo,[94] urged States to avoid damages to other riparian States and seek their prior consent before altering the course of international rivers. Those inter-American conferences also addressed other issues connected to the environment, such as international sanitary policies (Mexico, 1901–1902; Rio de Janeiro, 1906; Buenos Aires, 1910; Santiago, 1923),[95] technical and cultural cooperation (Washington, DC, 1951; Caracas, 1954)[96] and indigenous populations (Montevideo, 1933; Lima, 1938).[97]

Of particular interest is Resolution XXXVIII, adopted at the Eighth Inter-American Conference held in Lima, which, as early as 1938, recommended the drafting of a convention on environmental protection under the aegis of the Pan American Union[98] involving the American States. This resolution led to the adoption of the 1940 Convention on Nature Protection and Wildlife Preservation in the Western Hemisphere[99] (to which most of the Amazon States were parties except Guyana).

93 OAS, *The Inter-American System: Treaties Conventions and Other Documents* (compilation annotated by F.V. García-Amador), vol. 1, Washington, DC, Sub-Secretary of Legal Affairs, General Secretary of the OAS, 1982, p. 59. The objective was to discuss methods for preventing the use of force among the American nations. This conference initiated a series of inter-American conferences, which shaped through a number of resolutions, declarations, and treaties the inter-American system under the auspices of the OAS.

94 Ibid., pp. 380–82. Similarly, the OAS economic conference held in Buenos Aires in 1957 adopted Resolution XXIII on the *Aprovechamiento de sistemas fluviales y facilidades a los Estados Mediterráneos*.

95 Pan American Union, *Manual of Inter-American Relations, a Systematic Classification of the Treaties, Conventions, Resolutions, Declarations and Recommendations Adopted at Inter-American Conferences and Meetings of Consultation*, Washington, DC: Division of Conferences and Organizations, Department of International Law, Pan American Union, 1956, pp. 258–66.

96 Ibid., pp. 277–78.

97 Resolution XCIII on the Conference on Indian Life, Seventh International Conference of American States, Montevideo, December 3–26, 1933; Declaration XI on the Protection of the American Indigenous Populations; Resolution XIII on the Problem of the Indigenous Woman, Conference of Experts on Indian Life in the Americas, Eighth International Conference of American States, Lima, December 9–27, 1938 (Pan American Union, op. cit., p. 239).

98 In 1889–1890, the First International Conference of American States held in Washington, DC founded the International Union of American Republics, served by a permanent secretariat called the Commercial Bureau of the American Republics (renamed the International Commercial Bureau at the Second International Conference, which took place in Mexico City in 1901–1902). At the Fourth International Conference of American States, held in Buenos Aires in 1910, the name of the organization was changed to the Union of American Republics and it became the Pan American Union.

99 161 UNTS 193.

This treaty was opened for signature by the American states at the Pan American Union on October 12, 1940 and entered into force on May 1, 1942. Despite some interesting mechanisms aimed at promoting *in situ* conservation of native flora and fauna (including migratory birds) and protecting regions of aesthetic, historic, and scientific value, this convention has never been well implemented.[100]

A few years after the adoption of the 1940 Convention on Nature Protection and Wild Life Preservation, an initiative under the aegis of the UN Educational Scientific and Cultural Organization's General Conference (UNESCO), involving several States within and outside the region, addressed the Amazon in particular. In 1946, the first session of UNESCO's General Conference started the debate on the creation of an international research institute in the Amazon region.[101] UNESCO's director-general was then instructed to convene a meeting among the concerned States and international organizations to discuss the proposal.[102] A commission involving a group of States and some organizations met in Belém, Brazil on August 12–18, 1947 to draft a proposal for defining the objectives and structure of an International Institute of the Hylean Amazon. This proposal was later presented to UNESCO's director-general and submitted to the Second Session of the General Conference of UNESCO held in Mexico City in 1947.

In 1948, the UNESCO's General Assembly approved a resolution authorizing the setting up of the institute[103] and proposed an international conference to reach an agreement on the International Hylean Amazon Institute. After four sessions (August 1947, in Belem, Brazil; December 1947, in Mexico City; April 1948, in Tingo Maria,

[100] P. Birnie and A. Boyle, *International Law and the Environment*, Oxford: Oxford University Press, 2002, p. 202.

[101] UNESCO, *Provisional Programme of Action of the Hylean Amazon Project during 1948*, IIHA/7, Nat. Sci./52, Paris: UNESCO, April 7, 1948, p. 1.

[102] As affirmed at UNESCO's Conference for the Establishment of the International Institute of the Hylean Amazon: "The project for the creation of an international research institute in the Hylean Amazon has been inspired by the past history of the region. Since the discovery of Amazonia it has been continuously explored by scientific missions of many nationalities with the aim of drawing up its botanical and zoological inventory, of becoming acquainted with the state of social development and organization of its native tribes, of determining the essential characteristics of its climate and soil, of carrying out archaeological excavations, and finally of opening up the economic wealth and exploring the demographic possibilities of this vast area. These long and difficult tasks have too often remained fruitless for want of a permanent centre of body, to coordinate them, follow them up and pass them on to succeeding generations.... Only an international body jointly maintained by the countries of the Amazon region and those which, although not a part of it, are specially interested in the problems of a natural and social sciences peculiar to it, is capable of ensuring lasting results of such an undertaking" (UNESCO, *General Information on the Conference for the Establishment of the International Institute of the Hylean Amazon*, IIHA/1, Nat.Sci./42, Paris: UNESCO, February 12, 1948, p. 7). The German naturalist von Humboldt (1769–1859) coined the expression *hylaea* (from the Greek for wood) to describe the Amazonian rainforest.

[103] Ibid., pp. 2–6.

Peru; and May 1948, in Iquitos, Peru), a group of experts appointed by UNESCO concluded a draft agreement on the International Institute of the Hylean Amazon,[104] which was approved between April 30 and May 10, 1948 in Iquitos and opened to ratification. Meanwhile, a provisional commission composed of the signatories States was established in Manaus, Brazil. However, in the following year the Brazilian government refused to ratify the agreement, which hindered the formal opening of the institute.[105]

Brazil objected in particular to the autonomy conferred on the institute *vis-à-vis* its member States and opposed the fact that non-Amazon countries and international organizations were granted the same rights and privileges as those conferred to Amazon countries.[106] The UNESCO initiative has then failed, but it brought to light the need to produce further scientific knowledge on the Amazon. A few years later the president of the Brazilian Council for National Research proposed the creation of a national research institute on the Amazon. This institution was established in 1952 (in Portuguese, Instituto Nacional de Pesquisas da Amazônia) and is to date one of the leading research institutions in the Brazilian Amazon.[107]

3.3.3. *Toward Regional Cooperation*

At the late 1960s some South American States engaged in a new form of cooperation, which was introduced by the 1969 La Plata River Basin Treaty and later by the Amazon Cooperation Treaty. In both cases, a group of States connected by a shared ecosystems recognized the need to cooperate and take joint actions required for their preservation and sustainable development. Article 1 of the 1969 La Plata River Basin Treaty calls the parties to "unite efforts" and promote its "harmonious development." Similarly, the Preamble of the ACT sees this treaty as the start of a "process of cooperation" that should benefit "their respective countries and the Amazon region as a whole."

The idea of jointly developing Latin America's largest river basins was put forward by the OAS secretary-general (José A. Mora) and the president

[104] UNESCO, *Final Act of the Conference Creating the International Institute of the Hylean Amazon*, Nat.Sci./IIHA/10, Paris: UNESCO, June 14, 1948, Annex 1, p. 1. According to this document, the region of the Hylean Amazon included the "great forest areas of (a) the basin drained by the Amazon, extending from the Andes to the Atlantic; (b) the basin drained by the right basin of Orinoco; (c) British, Dutch and French Guyana; (d) the lower Tocantins and the Maritime cost of Para and (e) the north-western part of the State of Maranhao" (Article 1).

[105] UNESCO, *Report on the Measures Undertaken by UNESCO with a View to Creating the International Institute of the Hylean Amazon*, doc. NS/IIHA/12, Paris: UNESCO, October 18, 1948, p. 2.

[106] A.C. Reis, *A Amazonia e a cobiça internacional*, Rio de Janeiro: Editora Nova, 1965, p. 9.

[107] Brazil, *Diário Oficial da União* of October 29, 1952, Decree 31672, p. 16906 (translated by the author).

of Argentina (Arturo Illia) in 1966 on the occasion of the Fourth Ministerial Meeting of the OAS.[108] On that occasion, Argentina suggested a meeting of ministers of foreign affairs to discuss the possibility of enhancing subregional cooperation in the La Plata River Basin. This river system covers more than 3.1 million square kilometers, second in size only to the Amazon basin, and is composed of major rivers: the Paraguay, Paraná, Uruguay, and La Plata.[109] Following Argentina's proposal, the riparian States' ministers of foreign affairs (Argentina, Bolivia, Brazil, Paraguay, and Uruguay) met in Buenos Aires in 1967. They agreed to create the Intergovernmental Coordinating Committee of the La Plata Basin Countries (CIC) to be based in Buenos Aires and composed of their ambassadors serving in Argentina. In 1968, they met again in Santa Cruz de la Sierra, Bolivia, to approve the statute of the CIC and entrusted this institution to draft a treaty involving all riparian States of the La Plata River basin. A year later, in an extraordinary meeting held on April 23, 1969 in Brasília, the La Plata River Basin Treaty was signed and entered into force on August 14, 1970.[110]

In line with the experience in the La Plata, the Amazon States negotiated a regional agreement some years later. Both instruments are multiparty treaties covering the two greatest drainage basins in South America[111] shared by a group of States. Both treaties have the objective of achieving the "harmonious development" of each region through joint actions (as stated in Article 1 of both treaties). In both cases, the parties agree to cooperate in similar areas, such as navigation,[112] water resource preservation,[113] health and sanitation,[114] among others. These treaties also have analogous institutional bodies charged with similar functions, such as the Meetings of Ministers of Foreign Affairs[115] or National Commissions.[116]

[108] Calasans, op. cit., pp. 281–82.

[109] OAS, *La Plata River Basin: A Framework for the Sustainable Management of Its Water Resources with Respect to the Hydrolocial Effects of Climate Variability and Change*, OAS Water Project Series, no. 6, October 2005, p. 1. Most of the basin is situated in Brazil (1.4 million km^2, or 45 percent), another 30 percent lies in Argentina, 13 percent in Paraguay, 7 percent in Bolivia, and 5 percent in Uruguay (Brazil and Bolivia are riparians both of the La Plata River basin and the Amazon) (Newton V. Cordeiro, "Environmental management issues in the Plata basin," in Asit K. Biswas, Newton V. Cordeiro, Benedito P.F. Braga, and C. Tortajada (eds.), *Management of Latin America River Basins: Amazon, Plata and Sao Francisco*, Tokyo: United Nations University Press, 1998, pp. 148–73).

[110] 875 UNTS 3.

[111] On the differences and analogies between the two treaties, see J.H. Greño Velasco, "Pacto Amazónico y Tratado de la Cuenca del Plata: analogías y diferencia," *Revista de Política Internacional*, vol. 165, 1979, pp. 75–92, at p. 75.

[112] 1969 La Plata River Basin Treaty (La Plata Treaty), Article 1(a); 1978 Amazon Cooperation Treaty (ACT), Articles 3, 6.

[113] La Plata Treaty, Article 1(b); ACT, Article 5.

[114] La Plata Treaty, Article 1(g); ACT, Article 8.

[115] La Plata Treaty, Article 2; ACT, Article 20.

[116] La Plata Treaty, Article 4; ACT, Article 23.

Most importantly, both treaties are based on the idea that each of those regions should be viewed as an ecological unity and therefore managed jointly. As Jorge Thierry Calasans notes, those treaties were first manifestations of a willingness to cooperate on a broader scale.[117] The 1969 La Plata River Basin Treaty calls the parties to promote the physical integration of this River basin (Article 1) and generate an "integral knowledge," and the ACT aims at the "harmonious development" and conservation of the Amazon region. However, whereas the La Plata River Basin Treaty focuses on the rational use and conservation of this water basin, the ACT looks at the development of the parties' "respective Amazonian territories" (Articles 1, 4, 10, 11, 13, 16, 17) or the "Amazon region" (Preamble; Articles 7, 14), not limited to the management of water resources. In the Amazon, as discussed in subsequent chapters, a major concern was to ensure sovereignty over the parties' respective territories, one of the major reasons behind the adoption of the ACT. In part due to those different approaches, the La Plata River basin has witnessed greater development compared to the Amazon basin, in the adoption of treaties, notably on hydropower use, and the creation of river commissions, which did facilitate the physical integration of that region.[118]

CONCLUSIONS

Following independence from the European colonial powers, mostly by the end of the nineteenth century, the Amazon countries began to adopt bilateral treaties to establish their common international boundaries and lay down rules on commerce and navigation. Particularly from the 1970s, they concluded bilateral friendship and cooperation treaties in areas such as culture, science, health, and sanitation. Some few bilateral agreements and resolutions concluded at inter-American conferences addressed matters related to environmental protection. Of particular note is the 1940 Convention on Nature Protection and Wild Life Preservation in the Western Hemisphere, which is the only example of a regional treaty prior to 1978 dealing with environmental matters. In this domain, until the 1970s, the Amazon States adopted mostly bilateral treaties for the protection of species of fauna and flora. The adoption of the 1969 La Plata River Basin Treaty

[117] J.H. Greño Velasco, op. cit., p. 278.

[118] Such as the 1973 agreement between Brazil and Paraguay creating the binational Itaipú dam, the 1973 Treaty of the Plata River between Argentina and Uruguay, which created two commissions charged respectively with river administration and maritime affairs; and the 1979 meeting between the governments of Argentina, Brazil, and Paraguay that established the Tripartite Agreement on Corpus and Itaipú, both dams located on the Paraná River, among others (Calasans, op. cit., p. 298).

and, a few years later, of the 1978 ACT introduced a new form of cooperation among the respective basin States. In both cases, a particular region is viewed as encompassing a coherent ecosystem and the States sharing its natural resources committed to taking joint actions for its conservation and sustainable use. The following chapter presents the normative and institutional framework of the 1978 ACT in particular and enquires into how this treaty regime has evolved from its adoption until the present.

4

The 1978 Amazon Cooperation Treaty

The 1978 Amazon Cooperation Treaty[1] (ACT) is an international agreement with regional application adopted by the eight Amazon countries (Bolivia, Brazil, Colombia, Guyana, Ecuador, Peru, Suriname, and Venezuela) with a view to promoting the harmonious development of their respective Amazonian territories in such a way as to "achieve also the preservation of the environment, and the conservation and rational utilization of the natural resources of those territories."[2] As a framework agreement, the ACT sets forth general guidelines for enhancing regional cooperation on a wide range of issues such as the utilization of natural resources, transport and communications, science and technology research, health, and tourism.[3] The parties to this treaty intentionally devised a flexible and simple institutional framework, comprised of few organs and only temporary secretariats, in order to avoid excessive financial costs and burdensome bureaucratic procedures.[4] However, they have soon realized the limitations of such a structure and as early as 1989 began to conceive ways of institutionally strengthening this treaty.[5]

1 1978 Amazon Cooperation Treaty, 17 ILM 1045.

2 ACT, Article 1.

3 According to P. Sands (*Principles of International Environmental Law*, Cambridge: Cambridge University Press, 2003, p. 502), international law for the conservation of biodiversity may be arranged in three categories. The first category of treaties is applicable to all species and habitats on the planet, such as the 1973 Convention on International Trade and Endangered Species and the 1992 Convention on Biological Diversity. The second category contains obligations that are applicable to all species and habitats located within a particular region. Finally, the third category includes treaties applicable at the regional or global level for the protection of a particular species types or habitats, such as birds, polar bears, and migratory species, as well as wetlands and forests. The ACT falls under the second category.

4 Pro Tempore Secretariat, *Base Jurídica del Tratado de Cooperación Amazónica: Antecedentes Constitutivos de la Organización del Tratado de Cooperación Amazónica* (BJ), SPT-TCA-PER-37, La Paz, Bolivia, 2002, p. 63.

5 1989 San Francisco de Quito Declaration, adopted at the Third Meeting of Ministers of Foreign Affairs, held in Quito, Ecuador on March 6–7, 1989 (BJ, pp. 73–79, at p. 74).

The normative and institutional evolution of the ACT is envisaged in Article 1, which requires the parties to undertake "operational agreements and understandings, as well as the pertinent legal instruments" in order to attain the treaty's objectives.[6] Under this provision, ACT signatories are encouraged to negotiate more specific obligations to give effect to the treaty's general objectives. The question of whether the ACT's normative and institutional frameworks have indeed evolved to a more robust legal regime since the ACT's adoption in 1978 is the subject of the present chapter.

A. ORIGINS OF THE 1978 AMAZON COOPERATION TREATY

4.1. THE RATIONALE OF A REGIONAL COOPERATION TREATY

On July 3, 1978 the eight Amazon basin States adopted the Amazon Cooperation Treaty in Brasília, following a speedy negotiation process. In 1977 the Brazilian Ministry of Foreign Affairs circulated a draft regional treaty among the other Amazon States, which was discussed and adopted one year later. On February 2, 1980, the ACT entered into force.[7]

The reasons why Brazil took the initiative of proposing a regional treaty and why its neighbors finally welcomed it have been a subject of debate.[8] It is often suggested that the Amazon States agreed to adopt a regional cooperation treaty in order to ensure national sovereignty over their respective Amazonian territories against risks (real or not) of foreign interference in the region.[9] Some commentators even suggest that it was not a spirit of cooperation that motivated the adoption of the ACT, but

6 ACT, Article 1.

7 The ministers of foreign affairs present at the ACT signing ceremony were the following: Oscar Adriázola Valda (Bolivia); Antonio Azeredo da Silveira (Brazi); Indalécio Liévano Aguirre (Colombia); José Ayala Lasso (Ecuador); Rasleigh Esmond Jackson (Guyana); José de la Puente Radbill (Peru); Henck Arron (Suriname); and Simón Alberto Consalvi (Venezuela).

8 D. Ware, "The Amazon treaty: A turning point in Latin America co-operation?", *Texas International Law Journal*, vol. 15, 1980, pp. 118–37, at p. 127; E. Ferris, "The Andean Pact and the Amazon Treaty: Reflections of changing Latin American relations," *Journal of Inter-American Studies on World Affairs*, vol. 23, 1981, pp. 147–75.

9 According to C. Ossa Henao ("Aspectos de la implementación del Tratado de Cooperación Amazónica en las relaciones bilaterales entre Brasil y Colombia: El plan modelo para el desarrollo integrado de las comunidades vecinas del Eje Tabatinga-Apaporis," *Boletim da Sociedade Brasileira de Direito Internacional*, vols. 79/80, 1992, pp. 81–89, at p. 82): "Los principios y propósitos que orientan el Tratado, son claro reflejo de la posición predominante en la época, cuando se hacia indispensable, a la luz de la política interna, reafirmar la soberanía de cada uno de los Estados sobre sus respectivos territorios, y de la necesidad de realizar actos de presencia en tan vastas extensiones, que, entre otras cosas, asegurasen dicha soberanía, frente a una amenaza, real o no, de internacionalización de la Amazonia, provocada por el creciente interés, a nivel mundial, en los temas de conservación del medio ambiente a raíz de la Conferencia de Estocolmo de 1972."

the willingness to reaffirm national sovereignty over the Amazon against attempts of its "internationalization."[10]

The Amazon countries feared that third-party States motivated by what is often called "international eagerness" (in Portuguese, "*cobiça international*") could interfere in some way and strive for the internationalization of the Amazon. These concerns were intensified by episodes such as the controversy between Brazil and Bolivia over the Acre region or the attempt at setting up an International Institute of the Hylean Amazon (both described in Chapter 3). Proposals such as that of the American magnate Daniel Ludwig to set up a large-scale pulp and cellulose production in the Brazilian Amazon (known as the Jari project)[11] or that of the American Hudson Institute to build a massive dam across the Amazon River[12] were interpreted as evidence of foreign economic interests over the Amazon.[13] Some quotations from the international media are often mentioned to suggest a broad interest in the Amazon, for example, citations such as: "the Amazon is the lungs of the world" (editorial, *Houston Post*, 1989), "Brazil must accept a relative sovereignty over the Amazon" (François Mitterrand, 1989), "Brazil should delegate some of its rights [over the Amazon] to competent international organizations" (Mikail Gorbachov, 1989), "Contrarily to what the Brazilians think, the Amazon is not theirs, but for everyone" (Al Gore, 1989).[14] Despite the common references to a possible "internationalization" of the Amazon, this term is not well defined. It seems to involve the

[10] C. Caubet, "Le Traité de Coopération Amazonienne: régionalisation et développement de L'Amazonie," *Annuaire Français de droit international*, vol. 30, 1984, pp. 803–18, at p. 812, quoting M. Medina, "Algunos aspectos políticos en el proceso de integración," *Universo Amazónico y la Integración Latino-americana*, Caracas: Universidad Simón Bolívar, 1983, p. 261.

[11] The shipping magnate Daniel Ludwig originally owned the Jari project located in Amapá and Pará, aimed at the establishment of monocultures of fast-growing trees, for large-scale pulp and cellulose production occupying in 1969 an area of approximately 3.6 million hectares (J. Sautchuk, *Projeto Jari: A invasão Americana*, São Paulo: Editora Brasil Debates, 1979, p. 14; see also L.F. Pinto, *Amazônia Brasileira no rastro do saque*, São Paulo: Editora Hucitec, 1980; and A. Meira Mattos, "Pacto Amazônico: cooperação e integração," *Revista Brasileira de estudos políticos*, vol. 53, 1981, pp. 91–152).

[12] In the 1960s, the Hudson Institute of New York proposed creating several "Great Lakes" in the Amazon basin: three in Colombia (the Chocó project), one in Peru (a projected dam on the River Ucayali), one on the border of Colombia and Venezuela (a projected dam on the Atabapo and Guainia rivers) and one in Brazil (a projected dam on the Amazon River near Monte Alegre). This project, which would have allowed the navigation of 20,000-ton vessels and facilitated the exports of resources in the region, was the object of strong criticism and was never implemented (Caubet, op. cit.)

[13] B. Kucinski, "A Amazônia e a geopolítica do mundo," *Encontros com a civilização Brasileira*, vol. 11, 1979, pp. 12–19, at p. 18. According to Kucinski, the idea of an Institute of the Hylean Amazon and the "Great Lakes" project were attempts to internationalize the Amazon. Currently, it is still being claimed that Brazil must ensure sovereignty over the Amazon against attempts at internationalization (C. Meira Mattos, in a debate hosted by the Brazilian Superior War College, "Pacto Amazônico precisa ser reativado," *Cadernos do terceiro mundo*, vol. 19, 2000, pp. 24–25).

[14] Quotations found in G. Fregapani, *Amazônia e a grande cobiça internacional*, Brasília: Editora Thesaurus, 2000, pp. 94–95.

transfer of land ownership or administrative powers to foreign States, international organizations, or individuals.[15]

Undoubtedly, the ACT was based on a spirit of defense and a main concern was to ensure sovereignty over a vast and underdeveloped region against any form of foreign interference. A regional treaty was regarded as an "antidote" against attempts toward the internationalization of the Amazon, as asserted by the then Brazilian president (Ernesto Geisel) on the occasion of the ACT signing ceremony:[16]

> Against the attempts of internationalization, in the abrupt form of the past or the subtle interference of the present, the best antidote is the regionalization and the protection of the area. In *Amazonia*, I repeat, it is the countries, which share it, and only them, who must have exclusive responsibility for its development.

The need to ensure sovereignty over the Amazon was invoked several times during the ACT negotiations and was eventually reflected in its final text, particularly in Article 4, in which the parties declare the exclusive use of natural resources as a right inherent to the sovereignty of each State. The ACT's "Monrovian" overtones, reflected in the often-cited formula of the time, "*Amazonia* for the Amazonians,"[17] indicated that the Amazon States alone should develop this region and benefit from its exploitation. Through a regional treaty they would create an "Amazon club" (an expression coined by the then Brazilian minister of foreign affairs, Azeredo da Silveira), based on a system of unanimous votes able to ensure equality among the member States and aimed at the defense and integration of this region.[18]

[15] A. Kolk, "From conflict to cooperation: International policies to protect the Brazilian Amazon," *World Development*, vol. 26, 1998, p. 1483. According to Kolk, the "'internationalization' was seen as a threat to the Brazilian sovereignty and the legitimate right to use and manage the Amazon. This could imply not only the creation of a large Amazon reserve to protect the environment, but also a further 'internationalization' and exploitation of the large mineral reserves in the Amazon by international forces under the pretext of the environment." Allegations of internationalization, especially during the rule of Brazil's military governments, were largely rhetorical, as the entrance of foreign capital was part and parcel of the economic model under the dictatorship, as M.H. Moreira Alves indicates (*State and Opposition in Military Brazil*, Austin: University of Texas Press, 1985).

[16] Speech of the Brazilian president (Ernesto Geisel) at the ACT signing ceremony in Brasília on July 3, 1978 "Tratado de Cooperação Amazônica," Brazilian Ministry of Foreign Affairs, *Resenha de Política Exterior do Brasil*, Brasília, 1978, pp. 59–65, at p. 64). In Portuguese: "Contra os ensaios de internacionalização, sob a forma abrupta do passado ou da interferência sutil do presente, o melhor antídoto é a regionalização da proteção da área. Na Amazônia, repito, compete aos países que a partilham, e a eles só, a responsabilidade exclusiva pelo seu desenvolvimento."

[17] D. Landau, "The Treaty for Amazonian Cooperation: A bold new instrument for development," *Georgia Journal of International Law and Comparative Law*, vol. 10, 1980, pp. 463–89, at p. 480.

[18] On the idea of an "Amazon club" ("*clube Amazônico*"), see the interview of the Brazilian minister of foreign affairs (Antonio Azeredo da Silveira) with Venezuelan television on November 14, 1977, "Silveira explica política brasiliera à tv da Venezuela," Brazilian Ministry of Foreign Affairs, *Resenha de Política Exterior do Brasil*, vol. 15, 1977, pp. 11–15, at p. 12). In Portuguese: "E tenho certeza de que vamos iniciar entre nós – uma espécie de Clube Amazônico – nada será imposto a nenhum país,

To ensure national sovereignty over the Amazon, the ACT advocated the "full incorporation" of the Amazon into its member States' respective territories. This was, as enunciated in Article 10, the ACT's "priority goal." It should be noted that this goal envisaged the integration of the ACT parties' respective Amazon territories into their territories, more than an overall integration of the Amazon basin. As Thierry Calasans observes when examining the scope of Article 10:[19] "nous comprenons par là qu'il est question de l'intégration du territoire amazonien à l'intérieur du territoire de chaque Partie et non de l'intégration du basin amazonien en tant que tel."

At the root of the full incorporation goal were national security doctrines prevailing during the rule of Brazil's military regime (1964–1985), which supported the integration and occupation of the Amazon as a way of limiting the likelihood of foreign interference in the region.[20] Theorists within the Brazilian Superior War College supported the view that a regional treaty would give a regional dimension to a challenge faced by all Amazon States, that of "conquering" its "vast open spaces."[21] The Brazilian government intended to create a so-called "protective barrier" around the Amazon ("*cordão protetivo*") against any form of external interference, a concern that was incorporated in official maxims such as "to integrate in order not to forfeit" ("*integrar para não entegrar*").[22]

In fact, the Brazilian government's attempts to "incorporate" the Amazon in its mainstream economy began in the 1950s with policies aimed at the human occupation of that region. In 1953, the Superintendência do Desenvolvimento Econômico da Amazônia was created, later transformed into the Superintendência para o Desenvolvimento da Amazônia. During the 1960s, the Brazilian government introduced a policy aimed at providing fiscal and financial incentives for attracting development projects in the Amazon. In 1974, those incentives were regulated with the creation of the Fundo de Investimento da Amazônia.[23] The construction of Brasília in the 1960s was also an attempt to decentralize the political and economic power concentrated in the south of Brazil toward the north of the country. In particular,

porque nós, mesmo nas proposições que fizemos, declaramos que pretendíamos que o processo funcionasse à base de votos unânimes. Como conseqüência de um 'clube' de defesa de uma região e de integração física, nascem as estruturas que constituem a infra-estrutura das economias de escala."

19 J.T. Calasans, *Le concept de ressource naturelle partagée. Application aux ressources en eau: l'exemple de l'Amérique du Sud*, thèse de doctorat en droit, Université Paris I, Panthéon Sorbonne, Septentrion Presses Universitaires, 1996, pp. 342–43.

20 Kolk, op. cit., p. 8.

21 A. Teixeira Soares, "Tratado de Cooperação Amazônica," *Segurança de desenvolvimento, Associação dos diplomados da Escola Superior de Guerra*, vol. 188, Rio de Janeiro, 1981, pp. 6–14.

22 A. Hall, *Sustaining Amazonia: Grassroots Actions for Productive Conservation*, New York: Manchester University Press, 1998, p. 48.

23 By 1986, fiscal incentives were practically over; see J.E. Gasques and C. Yokomizo, *Avaliação dos incentivos fiscais na Amazônia*, Brasília: Instituto de Pesquisas Econômicas Aplicadas, 1987.

the construction of the highway Transamâzonica launched in 1972 (BR 230) was followed by a series of major infrastructure works, including roads, dams, and mining sites. It is also suggested that the "colonization" of the Amazon served the purpose of alleviating growing pressures for land reform by resettling "excess" rural populations in other regions and reducing land conflict-induced tensions (Table 4.1).[24]

From the 1960s, the other Amazon States also began to formulate national strategies for integrating their respective Amazonian territories. A regional treaty would then complement national development policies and help reduce the isolation of these countries' respective Amazonian territories *vis-à-vis* their own national economies.[25] As suggested by the then Brazilian president (Ernesto Geisel), "once dominated by man, the Amazon would unite its basin countries instead of separating them."[26]

To say that national sovereignty and defense were central concerns behind the adoption of a regional treaty does not exclude other important motivations, notably that of enhancing regional cooperation. All Amazon States faced the challenge of exploiting a highly complex and rich ecosystem and at the same time experienced similar difficulties; therefore, regional cooperation was a way of attaining joint solutions.[27] As noted by the then Brazilian minister of foreign affairs (Antonio Azeredo da Silveira) at the opening session of the First Preparatory Meeting of the ACT, "the very physical reality of the Amazon motivated and justified further regional cooperation."[28] The treaty negotiations evidenced a belief that ways of adequately

[24] Hall, op. cit., p. 47.

[25] M. Montenegro, "Política externa e cooperação amazônica: a negociação do Tratado de Cooperação Amazônica," *Sessenta anos de política externa*, São Paulo: Fundação Getulio Vargas, 2000, pp. 355–80, at p. 367.

[26] Speech of the Brazilian president (Ernesto Geisel) on the occasion of the official dinner offered to the Venezuelan president (Carlos Andrés Perez) in Brasília on November 16, 1977, Brazilian Ministry of Foreign Affairs, "Entre Brasil e Venezuela, idêntico espírito de cooperação," *Resenha de Política Exterior do Brasil*, vol. 15, 1977, pp. 15–17, at p. 17. In Portuguese: "Agora, as iniciativas que pretendemos tomar em conjunto no plano sub-regional ainda mais nos aproximarão entre outras razões pelo simples fato de que a floresta amazônica, que nos separa, será, dominada pelo homem, um traço de união."

[27] This was stated in the interview of the Brazilian minister of foreign affairs (Antonio Azeredo da Silveira) with Venezuelan television, op. cit., p. 12. According to d Silveira, the development of the Amazon should be conducted by cooperative means, given the common problems faced by the countries sharing the region. In Portuguese: "Cremos que o Pacto Amazônico tem uma virtude muito especial. Em primeiro lugar é uma grande aventura regional. Nada pode ser mais atraente que considerar o desenvolvimento da região Amazônica. Além disso, esse desenvolvimento não pode ser feito de maneira egoísta. Deve ser feito de forma cooperativa, porque todos os países, que tem suas regiões amazônicas, têm problemas comuns."

[28] Speech of the Brazilian minister of foreign affairs (Antonio Azeredo da Silveira) at the opening session of the First Preparatory Meeting of the ACT, "Silveira abre a primeira reunião preparatória do pacto amazônico," Brazilian Ministry of Foreign Affairs, *Resenha de Política Externa*, vol. 15, 1977, pp. 53–55, at p. 55. In Portuguese: "para o processo amazônico, o elemento catalizador é a própria realidade física de uma imensa região que é preciso vincular, povoar e desenvolver."

TABLE 4.1. *National policies for the occupation of the Brazilian Amazon*

Year	Programs and executing agencies	Objectives
1953	Plano de Valorização Econômica da Amazônia, (SPVEA), Brazilian Presidency	Elaborate a five-year plan for enhancing the economic value of the Amazon
1958	Belém–Brasília highway, BR-010, Ministry of Transport	Create a pioneer axis to integrate the eastern portion of the Amazon with the rest of the country
1960	Cuiabá–Porto Velho highway, BR-364, Ministry of Transport	Create a pioneer axis to integrate the southern portion of the Amazon with the rest of the country
1966	Superintendência de Desenvolvimento da Amazônia (SUDAM), Ministry of the Interior	Coordinate and monitor programs and regional plans; decide on the redistribution of tax incentives
1967	Superintendência da Zona Franca de Manaus, Ministry of the Interior	Integrate the western portion of the Amazon through the creation of an industrial, agricultural, and stock-raising center for tax exemptions
1968	Committee on Energy Studies of the Amazon, Ministry of Mines and Energy	Monitor studies related to the use of the energy potential of the Amazon
1968	Fiscal incentives (SUDAM)	Promote investments in the region through significant tax deductions
1970	Programa de Integração Nacional (PIN)	Extend the highway network and implement programs for official colonization
1970	Programa de Redistribuição de terras e de Estimulo a agroindústria do Norte e Nordeste (Proterra)	Promote rural capitalization
1970	Instituto Nacional de Colonização e Reforma Agrária (INCRA), Ministry of the Interior	Implement strategies for the controlled distribution of land
1974	Programa de Pólos Agropecuários e Agrominerais da Amazônia (Polamazônia), Ministries of the Interior, Agriculture, and Transport	Concentrate resources in selected areas with a view to encouraging migratory flows, increasing cattle farming and improving urban infrastructure
1978	Adoption of the Amazon Cooperation Treaty	

Source: B. Becker, *Amazônia*, São Paulo: Atica, 1998 (translated by the author).

exploiting the Amazon should come from the basin States. As stated during the treaty negotiations, the Amazon countries should seek out their own methods instead of relying on foreign States unaware of the ecological functioning and characteristics of life in the tropics.[29]

Traditionally, States have sought cooperation for managing shared ecosystems, notably water basins, by adopting regional agreements for instance, with respect to the Danube-Rhine river systems in Europe, the lower Mekong in Asia, the Senegal in Africa, and the La Plata River basin in Latin America.[30] Likewise, in the Amazon basin, States also intended to achieve common objectives through regional cooperation, for example improving the navigability of the Amazon and its tributaries,[31] preserving wildlife species and maintaining the region's ecological balance,[32] and creating infrastructure,[33] among others. Some countries also expected to have their access to the sea facilitated, such as the land-locked Bolivia,[34] or to the Atlantic Ocean, such as Peru and Ecuador (which are Pacific States).[35]

Furthermore, a contributing factor leading to a regional treaty was an increase in commercial relations among the Amazon countries from the 1960s that intensified diplomatic relations among them and reinforced a process of regional cooperation, which until then had been sporadic and discontinuous.[36] This fact can be observed by the significant increase of exports and imports between Brazil and the other Amazon countries from 1971 to 1980.[37] At the signing ceremony of the ACT, the

[29] Speech of the Brazilian president (Ernesto Geisel) at the ACT signing ceremony, op. cit. In Portuguese: "Não serão os centros do Norte industrializado, de diversa ecologia, que nos hão de facilitar, pronta e acabada, a chave da civilização nos Trópicos."

[30] E.M. Jarrin, "Análisis de los aspectos políticos relacionados con el Pacto Amazónico y la integración Latinoamericana," *El universo Amazónico y la integración Latinoamericana*, 1981, pp. 237–47, at p. 7. In this regard, see the speech of the Brazilian president (Ernesto Geisel) at the ACT signing ceremony (op. cit., p. 61). In Portuguese: "A necessidade de mudança decorre igualmente de outro fator: a tendência universalmente consagrada de dar tratamento mais amplo e de conjunto, através de acordos de cooperação, aos problemas das grandes regiões geográficas comuns a vários países".

[31] ACT, Article 6.

[32] ACT, Article 7.

[33] ACT, Article 10.

[34] Speech of the Bolivian minister of foreign affairs (Oscar Adriázola Valda) at a celebration ceremony provided by the Brazilian government to the ACT ministers of foreign affairs on July 4, 1978, "Governo brasileiro condecora chanceleres dos países amazônicos," Brazilian Ministry of Foreign Affairs, *Resenha de Politica Exterior*, vol. 18, 1978, pp. 23–24, at p. 24. In Portuguese: "A facilidades para habilitar de forma permanente e fácil as vias fluviais consagradas no Tratado Amazônico, só encontrarão seu verdadeiro e transcendente sentido quando a Bolívia, uma pátria enclausurada no coração do continente, puder chegar até o Atlântico, livre de qualquer obstáculo natural, jurídico ou político."

[35] G. Sousa Franco Filho, *O Pacto Amazônico: idéias e conceitos*, Belém: Falangola, 1979.

[36] Jarrin, op. cit.; R. Ricúpero, "O Tratado de Cooperação Amazônica," *Revista de informação legislativa*, no. 81, Senado Federal, 1984, pp. 177–96, at p. 177.

[37] Montenegro, op. cit., p. 363.

then Brazilian president (Ernesto Geisel) made reference to that fact, by stating that diplomatic relations among the Amazon States had been intensified by virtue of more dynamic commercial exchanges among them.[38] As a result in the 1970s, the Amazon countries concluded various bilateral and multilateral treaties creating international bodies in different domains, such as the Technical Intergovernmental Committee for the Protection of the Amazonian Fauna and Flora set up in 1975 by Bolivia, Brazil, Colombia, Ecuador, Peru, and Venezuela.[39] Moreover, the 1972 UN Conference on the Human Environment brought about an international debate over the protection of natural resources, which also influenced the adoption of the ACT.[40]

4.2. THE NEGOTIATION OF A REGIONAL COOPERATION TREATY

In 1976, the then minister of foreign affairs (Azeredo da Silveira), in a speech delivered to the Brazilian National Congress, signaled the Brazilian government's intention to seek further cooperation with the other States sharing the Amazon basin through a regional arrangement.[41] After some informal consultations with its neighboring States, the Brazilian Ministry of Foreign Affairs proposed in 1977 a draft text of a regional cooperation treaty, which was circulated among the other seven Amazon States and served as the basic text for the treaty negotiations.[42] Brazil played a leading role during this treaty's negotiating process, which, according to David

[38] Speech of the Brazilian President (Ernesto Geisel) at the ACT signing ceremony, op. cit. In Portuguese: "Hoje, dez milhões de pessoas e uma economia em expansão já não permitem tratar a região apenas como uma expressão geográfica. A dinamização da vida social e econômica não podia deixar de contagiar a diplomacia para a Amazônia.... Multiplicam-se, em todos os campos, os contatos entre os países da área e a intensificação das relações já não cabe mais dentro do estreito molde tradicional, cordial e amistoso, no plano político, mas de pouca expressão econômica e social."

[39] Other examples cited by Ricúpero (op. cit.), include the Ata do Rio Branco signed between Brazil, Bolivia and Peru with a view to plan and interconnect the highways in the area where the three borders converge, in addition to the Intergovernmental Committee for the Protection and Management of the Amazonian Fauna and Flora, between Brazil, Bolivia, Colombia, Ecuador, Peru, and Venezuela.

[40] J.F. Magalhães Pimenta, *The Treaty of Amazonian Cooperation: An Analysis of the Brazilian Proposal in the Light of Brazil's Regional and International Constraints*, master's thesis, George Washington University, 1982, p. 65.

[41] Speech by the Brazilian minister of foreign affairs (Antonio Azeredo da Silveira) to the Brazilian National Congress, "Silveira no Congresso expõe bases da diplomacia brasileira," Brazilian Ministry of Foreign Affairs, *Resenha de Política Exterior*, vol. 10, 1976, pp. 57–65, at p. 61. In Portuguese: "se manifesta nosso interesse cada vez maior pela Amazônia. Inspira-nos ainda, nesses contatos, a consideração de que uma cooperação ativa com os nossos vizinhos amazônicos pode ter um efeito salutar para o desenvolvimento da região, contribuindo para assegurar na área o exercício efetivo das soberanias de cada Estado e para dissipar eventuais idéias de internacionalização."

[42] C. Apesteguy, G. Martiniere and H. Thery, "Frontières en Amazonie: la politique du Brésil et l'intégration de l'Amérique du Sud," *Problèmes de l'Amérique Latine*, vols. 4533–34, 1979, pp. 76–98, at p. 89.

Ware, was not surprising, considering that it had "the most to gain and the least to lose by any development, cooperative or otherwise, of the Amazon Basin since it is geographically and economically best situated to exploit the Basin."[43]

Initial reactions to the Brazilian proposal were in general positive, but cautious. It is suggested that the history of the Portuguese conquest in South America and the later affirmation of Brazil's sovereignty over the Amazon in particular by resort to the *uti possidetis de facto* principle justified some apprehension by its neighbors, who feared Brazil's potential hidden intentions of expanding its influence over the region.[44] The reason for initial suspicions by the five Andean countries (Bolivia, Colombia, Ecuador, Venezuela, and Peru) was above all connected with concerns over a possible "disintegrating effect" that an Amazon treaty could have on the 1969 Andean Subregional Integration Agreement, concluded years earlier by these countries for the creation of a subregional common market. Venezuela, in particular, feared that its increasingly significant position in Latin American as an oil exporter could be challenged by Brazil with the adoption of a regional treaty in the Amazon.[45] However, without Venezuela, a regional treaty was unlikely to prosper. For the other Andean countries, Venezuela's participation was fundamental, as it was the only country able to counterbalance the influence of Brazil in the region.[46]

Intensive, behind-the-scenes diplomacy by the Brazilian Ministry of Foreign Affairs succeeded in dissipating the initial suspicions of the Andean countries. In 1977, the Venezuelan president was invited to pay a formal visit to Brasília that opened the way for formal negotiations on a regional cooperation treaty. The first round of negotiations took place in Brasília on November 28–30, 1977. Brazil's counterparts took the position that the draft treaty presented by the Brazilian Ministry of Foreign Affairs was too incisive in some respects and proposed a number of changes.[47] The second preparatory meeting, also held in Brasília, opened on March 27, 1978, reexamined the draft treaty, but it did not reach consensus on a final text. At a third meeting held in Caracas on June 15–16, 1978, the final text was approved.[48]

[43] Ware, op. cit., p. 127. According to Bertha Becker, from 1966 to 1985 Brazil started a process of regional planning of the Amazon. Brazil had two main concerns with respect to the continental Amazon. The first was the migration of the populations in neighboring States from urban centers to their respective Amazonian territories, and the second was the construction of the highway Carretera Bolivariana Marginal de la Selva, extending to the Pacific, which could "capture" the Amazon into the orbit of the Caribbean and Pacific, reducing the influence of Brazil in that area of the continent (B. Becker, *Amazônia: Geopolítica na virada do III milênio*, Rio de Janeiro: Garamond Universitária, 2007, p. 26).

[44] Calasans, op. cit., p. 341.

[45] E.J. Amaral Souza Neto, "O Tratado de Cooperação Amazônica," in S. França Danese (ed.), *Ensaios de historia diplomática do Brasil (1930–1986): cadernos do Instituto de Pesquisa de Relações Internacionais*, vol. 2, Brasília: Fundação Alexandre Gusmão, 1989, pp. 165–75, at p. 169.

[46] Apesteguy, Martiniere and Thery, op. cit., p. 90.

[47] Landau, op. cit., p. 475.

[48] G.D. Landau, "Tratado de Cooperación Amazónica: nuevo ensayo de cooperación," *Integración Latinoamericana*, no. 27, 1978, pp. 3–10, at p. 5.

The draft treaty presented by the Brazilian Ministry of Foreign Affairs – which, as noted earlier in this study, took inspiration from the prior 1969 La Plata River Basin Treaty[49] adopted by Argentina, Bolivia, Brazil, Paraguay, and Uruguay,[50] – contained nineteen articles that partially changed during the treaty negotiations to accommodate the interests of the Andean countries. Basically, these States advocated the removal of terms suggesting any form of economic integration among the parties, likely to conflict with the purposes of the 1969 Andean Subregional Integration Agreement.[51]

Article 8 of the original draft treaty text referred to the "physical integration" of the Amazon through the setting up of transport and communications infrastructure, which would facilitate the flow of people and goods, and also urged States to consider ways of enhancing a process of "regional economic complementation."[52] Draft Article 10 required the parties to eliminate customs duties, fiscal taxes, and burdensome administrative requirements related to the transit of their border populations.[53] Instead of making reference to a "physical integration" of the Amazon, the ACT final text points to the advisability of creating a suitable "physical infrastructure" among the parties, especially in relation to transport and communications.[54] In addition, rather than aiming at eliminating customs duties, fiscal taxes, and burdensome administrative requirements, as envisaged in Draft Article 10, the ACT final text only acknowledges in Article 12 the benefits deriving from retail trade in products for local consumption among their border populations.[55]

With respect to international rivers, Draft Article 5(b) asserted that States were allowed to use their successive rivers according to their own needs, insofar as no significant harm was caused to other States.[56] Final Article 5 of the ACT removed those references and simply states that the signatories should "make efforts aimed at achieving rational utilization of water resources." Draft Article 7(b) requested the parties to promote scientific research on wildlife species of fauna and flora, jointly monitor their management, and assess the effectiveness of conservation measures adopted by each State. Final Article 7(b) makes no reference to joint monitoring of wildlife

49 875 UNTS 3.

50 G. Sousa Franco Filho, "Aspectos controversos do Pacto Amazônico," *Revista do Tribunal de Justiça do Estado do Pará*, vol. 21, 1980, pp. 35–53, at p. 36.

51 Ware, op. cit., p. 123.

52 Draft Article 9.

53 Draft Article 10. In Portuguese: "As Partes contratantes se comprometem a livrar de direitos, impostos aduaneiros, disposições cambiais e consulares, e de todo gravame fiscal, existentes ou por serem criados no futuro, o tráfego trasfonteiriço que se realiza entre as populações fronteiriças de seus respectivos territórios amazônicos, e a reduzir ao mínimo os tramites administrativos imprescindíveis."

54 ACT, Article 10.

55 ACT, Article 12.

56 Draft Article 5(b).

species and only provides for regular exchanges of information on conservation measures.[57]

After the introduction of those few changes, the Amazon Cooperation Treaty (or Amazon Pact) was signed in Brasília on July 3, 1978 and entered into force in 1980. It was deposited with the Brazilian Ministry of Foreign Affairs in the four official languages (English, Dutch, Portuguese, and Spanish). This treaty was received with great enthusiasm by all of its member States and praised as an "exceptional example of good political will,"[58] an "expression of a generous will put in the service of a promising cause"[59] and an event that "will mark the history of Latin America for its significance."[60]

The product of that negotiation process was a treaty containing a total of twenty-eight articles. In terms of substance, it contains principles and norms addressing various areas, such as fluvial navigation, health and sanitation, and the preservation of wildlife species, among others (Articles 1–19). Among these provisions are some procedural obligations or mechanisms intended to fulfill the treaty's objectives. The final articles set up the treaty's institutional structure by creating its organs and specifying their functions (Articles 20–25). The subsequent sections examine the ACT normative framework, including principles, norms, and mechanisms, as well as its institutional structure, as foreseen in its original text and as they have evolved over time.

B. AN OVERVIEW OF THE 1978 AMAZON COOPERATION TREATY

4.3. NORMATIVE FRAMEWORK

The 1978 Amazon Cooperation Treaty was not conceived as an instrument aimed at integrating national economies and markets;[61] neither was it devised as a conservation agreement, although the conservation and rational use of natural resources is one of its main purposes. The treaty was perceived rather as a mechanism to allow

57 ACT, Article 7(b).

58 Brazilian Ministry of Foreign Affairs, "Pacto vai evitar interferência na regiao amazônica," *Tratado de Cooperação Amazônica*, Brasília, 1978, pp. 115–17, at p. 115 (originally published in *Correio Braziliense*, Brasília, July 4, 1978).

59 Brazilian Ministry of Foreign Affairs, "El Tratado organizará vínculos entre los 8 países Amazónicos," *Tratado de Cooperação Amazônica*, Brasília, 1978, pp. 131–35, at p. 132 (originally published in *La Prensa*, Lima, July 5, 1978).

60 Brazilian Ministry of Foreign Affairs, "Silveira revives ideal of Shuman in Latin America," *Tratado de Cooperação Amazônica*, Brasília, 1978, pp. 144–45, at p. 144 (originally published in *Sunday Chronicle*, Georgetown, July 23, 1978).

61 According to Meira Mattos, the Amazon Cooperation Treaty was the beginning of a process of cooperation (in the socioeconomic field) that would move toward regional integration (in the political and economic domains) (A. Meira Mattos, op. cit., p. 151).

further cooperation of a noneconomic nature among its parties.[62] Cooperation is foreseen in a wide array of areas by resorting to mechanisms that would enable a regular exchange of information and experiences. This section enquires into what principles and substantive obligations are laid down by the treaty and what procedural obligations (here referred to as "mechanisms" or "techniques") are foreseen in order to meet its objectives.

Given the nature of the ACT, the evolution of its principles and obligations was not only likely to occur over time, as it is often the case with respect to framework treaties, but was expressly envisaged in Article 1, according to which its aims should be attained through the adoption of "operational agreements and understandings, as well as pertinent legal instruments." The ACT normative framework was then expected to develop substantive and procedural norms over time, for example through separate treaties, amendments, protocols, and resolutions adopted by the ACT institutional organs, which would be gradually introduced under this treaty. The present section examines whether more specific substantive and procedural obligations among the parties have been introduced through additional "operational agreements and understandings."

4.3.1 *Spatial Domain of Validity*

The ACT spatial domain of validity, as defined in Article 2, is not limited to the States forming part of the Amazon River's drainage basin, but also extends to countries that are not Amazonians from a hydrological standpoint, (such as Suriname), yet are within the area of influence of the Amazon basin. French Guyana, an overseas department of France, although reached by the Amazon River's drainage basin, was excluded from this treaty due to its dependent status *vis-à-vis* a European power. Article 2 reads as follows:

> This treaty shall be in force in the territories of the contracting parties in the Amazon River basin as well as in any territory of a contracting party which, by virtue of its geographical, ecological or economic characteristics, is considered closely connected with that basin.

As defined in the above provision, the ACT applies in the territories of the contracting parties in the Amazon River basin or of a contracting party closely connected with that basin. Several references are made within the ACT to the "Amazon region" or to the parties' respective "Amazonian regions" or "territories."[63] However, their exact

[62] L. Amado Cervo, *Relações Internacionais na América Latina: velhos e novos paradigmas*, Brasília: Instituto Brasileiro de Relações Internacionais, Universidade de Brasília, 2001.

[63] For example, "*respective Amazonian regions*" (Preamble); "to promote the harmonious development of the *Amazon region*" (Preamble); "to promote the harmonious development of their respective

limits are not defined in Article 2 or elsewhere in the treaty. Theoretically, the limits of the Amazon region could be determined by assembling each State's respective Amazon territories, according to each party's definition of its own Amazonian area. This solution would, however, present a methodological problem, as each Amazon State applies different criteria and thresholds for defining its Amazon territory, for example, physical, socioeconomic, and political-administrative criteria.[64]

To obtain clarification on the precise geographical limits of the Amazon region, the current Amazon Cooperation Treaty Organization (ACTO) requested support from the European Commission's Joint Research Centre (JRC) in 2004. As noted by the ACTO secretary-general at the time, that would enhance the understanding of current political, socioeconomic, and ecological processes and their impacts in the Amazon region, and also help in devising better tools for sustainable use and land management.[65]

The European Commission agreed to carry out a study on the geographical limits of the Amazon, in collaboration with Latin American experts, based on the parameters contained in Article 2 of the ACT. According to the JRC, the difficulty in defining the limits of the "Amazon region" lay in the fact that the Amazon eco-region is not necessarily identical to the Amazon River basin area, and both also face permanent natural and man-made changes.[66] The JRC findings were compiled in a document entitled *A Proposal for Defining the Geographical Boundaries of Amazonia*. According to this study, the total area of the "Amazon region" is actually larger than that covered by the ACT, whereas the area considered *Amazonia stricto sensu* is an

Amazonian territories" (Article 1); "flora and fauna of the Amazon region"; "maintain the ecological balance within the region" (Article 7).

64 European Commission Joint Research Centre, *A Proposal for Defining the Geographical Boundaries Amazônia*, Luxembourg: European Commission, Directorate General, JRC, ACTO, Office for Official Publications of the European Communities, 2005, p. 1.

65 According to this report, this initiative had its basis in the ACTO Strategic Plan 2004 2012, which states: "In order to obtain the best planning tools to strengthen the ability of ACTO to analyze change-processes undergoing in the Amazon, which highly contributes to the decision-making process of Amazon countries' governments, ACTO will coordinate the use of the Geo-referenced Information, Monitoring and Appraisal Systems that have already been developed. This will be coordinated with the support of national and regional agencies such as SIVAM, IIRSA among others" (ACTO, Strategic Plan 2004–12, Brasília, 2004, p. 68).

66 The director of the Institute for Environment and Sustainability of the European Commission's JRC (Manfred Grasserbauer) comments on the difficulties in achieving a definition of the Amazon (JRC, op. cit., p. vii): "at first glance, defining Amazonia seems to be an easy task. However, when the Institute for Environment and Sustainability received through European Commission President José Manuel Barroso the request from the Amazon Cooperation Treaty Organization (ACTO) to support the development of a definition for the geographical delimitation of Amazonia, this proved to be a scientifically highly challenging exercise. The reason for this is the fact that the Amazon eco-region is not necessarily identical with the Amazon River basin, both of them being areas facing permanent natural and man-made change."

"area of typical Amazon lowland rainforest (or *Hylaea*)."[67] The criteria for spatially delimitating this subunit include the overall presence of the typical Amazon lowland rainforest biome extending within the hydrographical limits of the Amazon basin. This area also comprises the extensions of recently cleared forest lands, or in other words, areas covered by forest in the past and identifiable as such on earlier maps.[68] In conclusion, this study proposed that areas such as the state of Bolívar (Venezuela); the headwaters of the Amazon in the Andes of Bolivia, Peru, Ecuador, and Colombia; the region to the southeast of Santa Cruz (Bolivia); and the northern part of Goiás (Brazil) be included in those States' national definitions of the Amazon.Currently, the Amazon countries have still not reached consensus on such a definition. Meanwhile, the ACT's domain of validity remains imprecise.

4.3.2 *Principles and Obligations*

The principle of national sovereignty over natural resources embodied in Article 4 of the ACT is one of the treaty's main pillars. According to this provision, the ACT parties have the right of exclusive use of natural resources, as follows:

> The Contracting Parties declare that the exclusive use and utilization of natural resources within their respective territories is a right inherent to the sovereignty of each state and that the exercise of this right shall not be subject to any restrictions other than those arising from International Law.

During the ACT negotiations, the Amazon States made clear that a regional treaty should not hamper their national sovereignty in any way. As affirmed by the Venezuelan minister of foreign affairs (Alberto Consalvi) on the occasion of the ACT signing ceremony, this treaty could not be interpreted in such a way as to restrict the member States' exclusive and sovereign rights over their natural resources:

> On the one hand, we have made clear that the treaty does not interfere in any way with the sovereign jurisdiction lying over each country's Amazonian territory. We did not intend in any way to replace the exclusive right and responsibility of each country to ensure the development of its Amazonian territory as part of its national setting. On the contrary, we have created a cooperation mechanism so that the

67 The *Hylaea* is defined as follows: "all natural forest types growing in the macrothermic lowlands (Mean Annual Temperature MAT > 24°C) of the Amazon River basin and on sub-montane slopes and hill-lands of the piedmont belonging to the surrounding mountain systems up to an elevation of 700 m asl (MAT ≈ 20°C), or forests of evergreen to sub-evergreen phenology growing under medium to high rainfall regimes (Medium Annual Precipitation of > 1.400mm), therefore often called "rainforest" or "forests growing on a wide variety of tropical soil types with different chemical and physical properties, on both flooded or non-flooded terrain, including the riparian habitats of várzea and igapó" (ibid., pp. 16–17).

68 Ibid., p. 17.

> knowledge and experiences of each of us can help and support all the others, we have also created the basis to jointly deepen studies, research and projects of common interest, under the principle that we will equitably benefit from the results we may achieve.[69]

This treaty should not create any kind of institutional structure superseding existing national institutions, and should only address matters concerning a larger group of States, without interfering in their domestic affairs, as noted very early on in the negotiations by the then Brazilian minister of foreign affairs (Antonio Azeredo da Silveira) at the opening session of the First Preparatory Meeting of the ACT on November 28,1977. According to him, the treaty was not intended to:

> interfere in national spheres or create a multinational entity that would superpose other existing institutions, but only deal with matters that involved the interests of a larger number of countries if not all ... it is not intended to superpose multilateral criteria over national ones. On the contrary, national efforts will continue to be the fundamental and main factor in the development of the Amazon territories.[70]

Although the ACT parties declared sovereign rights to exclusively use their natural resources, they also committed to the "preservation of the environment and the conservation and rational utilization of natural resources" (Article 1). This treaty embraced both the principles of the preservation of the environment and sustainable development. The parties recognized the need to rationally exploit their fauna and flora – as well as their water resources (Article 5) – in such a way as to "maintain the ecological balance within the region" and "preserve the species" (Article 7). Importantly, they acknowledged the need to "maintain a balance between economic growth and conservation of the environment" (Preamble) and assumed that "both the socio-economic development and conservation of the environment

[69] Speech of the Venezuelan minister of foreign affairs (Alberto Consalvi) at the ACT signing ceremony in Brasília on July 3, 1978, "A assinatura do tratado de cooperação multilateral na Amazônia," Brazilian Ministry of Foreign Affairs, *Resenha de Política Exterior do Brasil*, vol. 18, 1978, pp. 11–13, at p. 12. In Portuguese: "Por uma parte, consagramos com claridade meridiana que o Tratado não interfere de nenhuma maneira na jurisdição soberana que a cada pais corresponde em seus próprios territórios amazônicos. Não pretendemos de maneira nenhuma substituir o direito e a responsabilidade exclusivos de cada Estado em assegurar o desenvolvimento de sua própria Amazônia como uma parte integrante de seu conjunto nacional."

[70] Speech of the Brazilian minister of foreign affairs (Antonio Azeredo da Silveira) at the opening session of the First Preparatory Meeting of the ACT, op. cit.. In Portuguese: "O primeiro é o que visa a reservar com exclusividade as nações da área a responsabilidade pelo desenvolvimento da Amazônia. Ao mesmo tempo, teve-se o cuidado de claramente reconhecer no projeto o contorno das esferas de soberania. Não se tenciona superpor critérios multilaterais aos nacionais. Ao contrario, proclama-se que o esforço interno continuará a ser o fator fundamental e prioritário no desenvolvimento dos territorios amazônicos."

are responsibilities *inherent* to the sovereignty of each State" (Preamble). In other instances, for example, in the 2004 Declaration of Manaus adopted at the Eighth Meeting of Ministers of Foreign Affairs (MMFA), the ACT parties reaffirmed their sovereignty over the Amazon (para. 3) but also recognized a "sovereign responsibility" with respect to its protection (para. 2). The principle of national sovereignty over the region's natural resources under this treaty is balanced with an "inherent responsibility" to preserve the Amazon.[71]

In terms of its legal status the concept of a "sovereign responsibility" is not an established principle of international environmental law, and has not been reflected in a regional State practice. It was declared only a few times in ministerial declarations (e.g. 2004 Declaration of Manaus), adopted under the framework of the ACT and in some official statements (e.g. speech of the Brazilian Foreign Affairs Minister, at the Eighth MMFA, in 2004).[72] Its normative content can be inferred from the context in which it has been used. According to the Preamble of the ACT, the "conservation of the environment is a responsibility inherent to each State." At the Eighth MMFA, the then Brazilian minister of foreign affairs declared that the ACT States are "sovereign guardians" of the Amazon, which was reiterated in the subsequent 2004 Declaration of Manaus. The 2004 Declaration of Manaus reaffirmed the "sovereign responsibility" of the States Members of the ACTO in relation to the sustainable development of the Amazon. The notion of a "sovereign responsibility," as invoked under the ACT, seems to indicate that sovereignty over the Amazon involves a duty to preserve the environment.

In that sense, this concept is not entirely new. The duty to preserve a State's own environment is contained in the principle of environmental preservation, which is foundational to international environmental law. This duty is reflected, for example, in Article 192 of the 1982 United Nations Convention on the Law of the Sea (UNCLOS),[73] which affirms that States have an obligation to protect and preserve the marine environment. The novelty of the notion of "sovereign responsibility" is probably in assembling the principle of environmental preservation and that of a permanent sovereignty over natural resources, in the particular context of the Amazon.

Other principles that can be derived from the provisions of the ACT relate to freedom of navigation and absolute equality among the member States. Under Article 3, the parties guarantee, on a reciprocal basis, the complete freedom of navigation of the Amazon and its tributaries. The equality of member States in ensured by a

[71] Declaration of Manaus, 2004, available at http://www.otca.org.br/ep/Institucional/index.php?id=1230

[72] The minutes of this meeting are available at: http://www.otca.org.br

[73] 21 ILM 1261 (1983).

system of unanimous votes, as provided for in Article 25, the aim of which is to avoid the hegemony of one State over the others.[74]

As regards substantive obligations, the ACT sets forth a general duty to cooperate. This obligation is embodied in Article 1 of the treaty, under which the parties commit to promote a harmonious development of the Amazon and rationally use its natural resources through "joint actions and efforts," as follows:

> The Contracting Parties agree to undertake joint actions and efforts to promote the harmonious development of their respective Amazonian territories in such a way that these joint actions produce equitable and mutually beneficial results and achieve also the preservation of the environment, and the conservation and rational utilization of the natural resources of those territories.

This general obligation to cooperate is manifested in particular areas covered by the ACT. For example, the parties commit to make efforts to achieve a rational utilization of water resources (Article 5), undertake measures aimed at improving navigation of the Amazonian rivers (Article 6), maintain an "ecological balance in the region and preserve species" (Article 7), coordinate health services to improve sanitary conditions and combat epidemic diseases (Article 8), jointly conduct scientific research (Article 9), and cooperate in order to increasing the flow of tourists in their respective Amazonian territories (Article 13).

With respect to environmental protection, the ACT contains two general obligations, embodied in Articles 1 and 7, which provide guidance as to how the parties should conduct economic and other activities in their respective Amazonian territories. Firstly, according to Article 1, the Amazon States are to seek the sustainable development of the Amazon (although these terms are not referred to as such); in other words, in their efforts to promote the harmonious development of their respective Amazonian territories, they shall achieve the preservation of the environment and the rational utilization of natural resources. Secondly, as envisaged in Article 7, the ACT parties are expected to maintain the ecological balance of the region, and especially preserve species of fauna and flora.

[74] J.E. Greño Velasco identifies three principles: (1) the principle of equality of the parties; (2) the principle of free navigation; and (3) the right to exclusively use natural resources (J.E. Greño Velasco, "Pacto Amazónico y Tratado de la Cuenca del Plata: Analogías y Diferencias," *Revista de política internacional*, vol. 165, 1979, pp. 75–92, at pp. 83–84). According to Rubens Ricúpero (op. cit., p. 6), five main principles can be extracted from the ACT: (1) the member States' exclusive competence to develop the Amazon; (2) national sovereignty over natural resources; (3) regional cooperation as a means of achieving the treaty's objectives; (4) a balance between development and ecological protection; and (5) absolute equality among the member states, which is ensured by a system of unanimous vote. These five principles are also identified by Jarrin (op. cit., p. 8): "sus principios generales son: desarrollo económico de la Amazonia, incorporación de los territorios amazónicos a sus respectivas economías, el equilibrio entre crecimiento económico y preservación del medio ambiente, a la reafirmación de la soberanía nacional y a consagrar el interés regional frente a pretensiones de internacionalización de la Amazonía."

Although not conceived as a conservation agreement,[75] the ACT does embrace environmental protection concerns and places the conservation and rational use of natural resources as one of its main objectives. As noted above, this treaty calls on the parties to achieve a balance between economic growth and environmental conservation, as well as the ecological balance in the region, and recognizes the parties' "responsibility to conserve the environment" (Preamble). As Calasans observes, environmental protection is one of the treaty's main features: "malgré son caractère éminemment politique, un des traits le plus caractéristiques du Traité est celui d'un réel engagement envers le respect de l'environnement."[76] At present, the environment is manifestly a central concern within the framework of ACTO. Three of the six "programmatic areas" identified under this organization's Strategic Plan 2004–2012 relate to the environment: waters; forests, soils, and protected natural areas; and biological diversity, biotechnology, and biotrade.

More specifically, the ACT provides for the protection of water resources, fauna and flora, and for the conservation of the region's ethnological and archeological wealth. The parties shall make efforts to achieve a "rational utilization of the hydro-resources" (Article 5), strive to rationally plan the exploitation of species of fauna and flora (Article 7) and cooperate in the conservation of the ethnological and archeological wealth of the Amazon (Article 14). This latter provision was included specifically to protect the archeological and ethnological relics in the Peruvian Amazon inherited from the Incan civilization.[77] The ACT is silent in relation to other environmental matters of relevance in the conservation and management of the Amazonian environment, such as the protection of forests (the word "forests" is nowhere mentioned), the prevention of transboundary pollution, and the protection of indigenous populations. Article 13 makes some reference to this issue, by stating that tourism might occur "without prejudice to national regulations for the protection of indigenous cultures and natural resources."[78]

As one could expect, some areas not originally covered by the ACT have been gradually introduced by subsequent resolutions and nonbinding instruments adopted under the framework of the ACT. The lack of reference to some environmental issues in the ACT original text is in part due to the fact that, at the time of its adoption, international environmental law was in its infancy and some of its concepts and principles were yet being developed. For example, the term "biodiversity" was used under the aegis of the ACT only several years after its adoption.

[75] According to A. Meira Mattos (op. cit., p. 151), the ACT was the beginning of a process of cooperation (in the socioeconomic field), which would move toward regional integration (in the political and economic domains).

[76] Calasans, op. cit., p. 346.

[77] Sousa Franco Filho, op. cit., p. 42.

[78] ACT, Article 13.

In fact, the foundation of this concept worldwide was laid down a few years after the adoption of the ACT in the 1980 World Conservation Strategy,[79] and did not become familiar to the general public until the 1992 United Nations Conference on Environment and Development (UNCED),[80] which proposed a definition of biological diversity (or biodiversity) in Article 2 of the Convention on Biological Diversity (CBD).[81]

The ACT does not contain the term "biodiversity" and uses words such as "environment" and "natural resources" (without defining them).[82] Article 1 proclaims that the parties agree to achieve the "preservation of the environment" and the "conservation of the natural resources of those territories." Under the aegis of the ACT meetings, the term "biological diversity" was first used at the Second Meeting of the Presidents of the Amazon countries, held in Manaus, Brazil in 1992. This meeting adopted the 1992 Manaus Declaration on the United Nations Conference on Environment and Development,[83] which included a whole section on "Biological Diversity and Biotechnology." Section 2(1) of the 1992 Manaus Declaration affirms that biological resources are "unmistakably natural resources of each State" and "greater emphasis should be placed on the use and sustainable development of these resources." Subsequently, these terms have been included in instruments adopted under the ACT.[84]

79 Prepared by the World Conservation Union, UN Environment Program and World Wildlife Fund, in 1980 in collaboration with the UN Food and Agriculture Organization and UN Educational, Scientific and Cultural Organization, and the revised program *Caring for the Earth: A Strategy for Sustainable Living* (Gland, 1991). See M. Bowman, "The nature, development and philosophical foundations of the biodiversity concept in international law," in M. Bowman and C. Redgwell (eds.), *International Law and the Conservation of Biological Diversity*, The Hague: Kluwer Law International, 1996, pp. 5–31, at p. 8.

80 R. Paehlke (ed.), *Conservation and Environmentalism: An Encyclopedia*, New York & London: Garland, 1995, p. 81. The notion of biological diversity first entered scientific literature in the 1980s: The term "biodiversity" was used in the title of a symposium held in Washington, DC in 1986. The papers presented at that meeting were compiled in a landmark book entitled *Biodiversity* edited by Edward O. Wilson.

81 "Biological diversity" is defined in Article 2 of the 1992 CBD as the "variability among living organisms from all sources including *inter alia*, terrestrial, marine, and other aquatic ecosystems and the ecological complexes which they are part; this includes diversity within species, between species and ecosystems" (31 ILM 818).

82 Natural resources are used to satisfy human needs. The term "natural" means that there is no human intervention. Resources produced or cultivated by man, such as cattle, wheat, and corn, are not considered "natural resources." Lastly, as stated in Principle 3 of the Stockholm Declaration, natural resources can be differentiated as being renewable (such as fauna, flora, and water) or nonrenewable (such as mineral and oil deposits) (J.S. Barberis, "Introduction," in *Los recursos naturales compartidos entre estados y el derecho internacional*, Madrid: Editorial Tecnos, 1979, p. 145, note 72.

83 BJ, pp. 42–43.

84 The Third MMFA first used the term "biodiversity," which was included in the Declaration of San Francisco de Quito in 1989 (BJ, pp. 73–79). In 1995, the Lima Declaration adopted at the Fifth MMFA,

Similarly, matters related to forests have been gradually introduced under the aegis of the treaty. In the 1995 Lima Declaration adopted at the Fifth MMFA, the first ministerial meeting following the 1992 UNCED, the ACT parties expressly recognized the importance of the Amazonian forests.[85] In 1995 they held a first regional meeting in Tarapoto, Peru, on the subject of common criteria and indicators for managing the Amazonian forests (an issue also addressed in subsequent meetings). Some of those areas not originally foreseen under the ACT, notably forests or indigenous populations, have now been included as "priority areas" within the ACTO Strategic Plan 2004–2012.

As observed earlier, environmental protection, albeit important, is but one area covered by the ACT. This treaty also addresses issues such as fluvial navigation, by calling the riparian States to adopt (national, bilateral, or multilateral) measures aimed at improving the navigability of the Amazonian rivers (Article 6). Article 3 provides for the complete freedom of commercial navigation of the Amazon and its tributaries, on a reciprocal basis, excluding cabotage. The parties are urged to coordinate health services in their respective Amazonian territories and to take measures to improve the sanitary conditions in the region for preventing and combating epidemics (Article 8); to study ways of creating a suitable physical infrastructure among their respective countries, especially in relation to transportation and communications (Article 10); to recognize the benefits of retail trade in products for local consumption among Amazonian border populations, by means of suitable bilateral or multilateral agreements (Article 12); and also to increase the flow of tourists in their respective Amazonian territories (Article 13). Cooperation is particularly sought in the fields of scientific and technological research, where the parties are requested to cooperate in order to accelerate the economic and social development of the region (Article 9) (references to research and joint studies are made in other provisions, *inter alia*, Articles 7, 10, and 24).

which was the first ministerial meeting following the 1992 UNCED, acknowledged the importance of the Amazonian forests for the preservation of global and environmental processes and the biological megadiversity in the region (ACT, *Amazon Legal Framework of the Amazon Cooperation Treaty*, SPT-TCA-PER-37, 1996). The ACT parties reaffirmed in the Declaration of Caracas signed at the Sixth MMFA in 2000 (BJ, pp. 117–21) that the region's biodiversity and biotechnology offered great opportunities for the sustainable development of the Amazon countries. The ACT parties reaffirmed at the Eighth MMFA held in Santa Cruz de la Sierra, Bolivia in 2004 (http://www.otca.org.br/ep/Institucional/index.php?id=1285) their responsibility to manage their natural resources and rationally use their biological diversity, and to equitably share the benefits deriving from its use. The 2004 Manaus Declaration ratified the commitment to harmonize policies in order to guarantee the protection of biodiversity in the Amazon and restated the need for implementing the provisions of the CBD concerning the protection of the genetic heritage of Amazonian species and the sharing of the benefits arising from its utilization.

85 ACT, op. cit., see declaration in http://www.otca.org.br

The general duty to cooperate in such areas covered by the ACT is framed in vague language, where the parties "declare that,"[86] "shall make efforts,"[87] "agree on the advisability of,"[88] "agree to encourage,"[89] "seek to maintain,"[90] and "give special attention to."[91] Rather than imposing specific legal obligations, these provisions contain best-endeavor commitments to achieving further cooperation in different areas. The obligations contained in the ACT can be said to have a "soft" content, albeit inserted in a legally binding instrument. Some of the consequences deriving from the imprecise nature of legal obligations, as Abi-Saab notes, is that States have large discretion as to what needs to be done and by which means; moreover, their execution can hardly be assessed by an external objective control mechanism, such as a process of judicial control.[92] Despite the vagueness of the ACT provisions, they do have legal effects. The "soft" or "hard" nature of a norm does not affect its normative character: A treaty or customary law may be vague, but will not cease to be a legal norm and produce legal effects.[93] The provisions of the ACT have provided guidance for the adoption of further resolutions and ministerial declarations in the context of the ACT, and, in practice, to the undertaking of field projects and to the process of institutional strengthening of this treaty. What is examined subsequently is whether this general duty to cooperate has evolved in the sense of creating more specific obligations.

4.3.3. *Implementation Mechanisms*

The mechanisms foreseen to fulfill the ACT objectives involve no more than the generation of knowledge and exchange information.[94] Exchange of information is expected to occur, notably with respect to scientific research (Article 7(a)), conservation measures adopted by each party in its respective Amazonian territories (Article 7(b)), and also between the ACT parties and Latin American cooperation

86 ACT, Article 4.
87 ACT, Article 5.
88 ACT, Article 10.
89 ACT, Article 11.
90 ACT, Article 15.
91 ACT, Article 17.
92 G. Abi-Saab, "Cours général de droit international public," *Recueil des Cours de l'Académie de Droit International*, vol. 207, The Hague: Martinus Nijhoff, 1987, p. 212: "La gradation de la marge de discrétion laissée aux sujets dans l'exécution des obligations de moyen rend cette exécution, au delà d'un certain point, invérifiable par un contrôle objective externe, tel le contrôle judiciaire. De sorte que mêmes quand de telles obligations sont portées par des instruments contraignants comme des traités, elles ne se prêtent pas à ce type de vérification. Nous parlons alors d'un contenu soft."
93 Prosper Weil, "Towards relative normativity in international law?", *American Journal of International Law*, vol. 77, 1983, pp. 413–42, at p. 414.
94 ACT, Article 15.

agencies (Article 15). Much emphasis is put on the generation of knowledge through the undertaking of joint studies, for example, on means for improving fluvial navigation (Articles 6 and 11), in coordinated implementation of scientific research (Article 9) and programs of common interest (Article 17), or by the organization of seminars and conferences (Article 9(c)). In practice, information exchange also occurs through the regular meetings of the ACT parties, notably the MMFA and the Amazon Cooperation Council (ACC), and by the release of materials produced by the now Permanent Secretariat, such as bulletins and progress reports. An obligation to report on conservation measures is foreseen in Article 7(b), by which the parties are under an obligation to present annual reports on existing or future measures aimed at the protection of the region's wildlife. The fragile mechanisms under the ACT involving merely information exchange have been criticized. As Ware asserted, despite its ambitious objectives, this treaty's "twenty-eight articles fail to provide a mechanism capable of any substantial action."[95] Others criticize the fact that the treaty relies mostly on national undertakings instead of creating joint mechanisms involving some or all of the parties.[96] Apparently, "light" mechanisms and only few institutional bodies were a conscious choice of the ACT signatories, who intented to create a flexible structure that could adapt to changing circumstances. At the ACT signing ceremony, the then Venezuelan minister of foreign affairs (Alberto Consalvi) emphasized that the ACT was intended to leave enough flexibility to allow the ACT parties to adapt this treaty to their own national circumstances.[97] Also on the occasion of its adoption, the then Brazilian minister of foreign affairs stressed the fact that the ACT was an instrument "devised for the future" and an example of a "diplomacia de projeto."[98] In that light, the subsequent sections enquire whether the ACT has managed to develop both its normative and institutional frameworks over the years.

[95] Ware, op. cit., p. 117.

[96] According to Landau, the ACT encourages the exchange of experiences and promotion of scientific cooperation, but achieving the treaty goals remains a matter of each country's sovereign prerogatives (G.D. Landau, "The Treaty for Amazonian Cooperation: A bold new instrument for development," *Georgia Journal of International and Comparative Law*, vol. 10, 1980, pp. 463–89, at p. 479).

[97] Speech of the Venezuelan minister of foreign affairs (Alberto Consalvi) at the ACT signing ceremony, op. cit.

[98] Speech of the Brazilian minister of foreign affairs (Azeredo da Silveira) on the occasion of the official dinner offered to the ACT ministers of foreign affairs, "Tratado amazônico reafirma unidade fundamental dos países latino-americanos," Brazilian Ministery of Foreign Affairs, *Resenha de Política Exterior do Brasil*, vol. 18, 1978, pp. 13–15, at p. 15. In Portuguese: "Essa obra de criação do futuro, essa diplomacia de projeto é a que exige e reclama atuação pioneira, onde não valem métodos de outras terras e tudo tem de ser redescoberto e reinventado."

4.3.4. *Normative Evolution*

The idea of a treaty adaptable to national circumstances and factual changes, put forward during the ACT negotiations, was eventually incorporated into its final text. As noted earlier, according to Article 1, the aims of this treaty should be attained through the adoption of "operational agreements and understandings, as well as pertinent legal instruments," from which more precise rights and obligations should derive. This provision calls on the ACT parties to make the hortatory provisions contained in its original text more operational. However, up to the present, little evolution has been noted toward more precise and compelling norms, although-some progress occurred with respect to the ACT's institutional structure.

Until present, the ACT has been modified only once by the Protocol of Amendment of the Amazon Cooperation Treaty,[99] adopted on December 14, 1998, which modified Article 22 of the ACT, dealing with the functions of the treaty's Secretariat, in order to create ACTO. Most developments introduced under the ACT occurred through instruments adopted in the context of regular meetings of the treaty's institutional bodies, which include resolutions and declarations approved at the MMFA, the ACT's highest institutional body, and resolutions taken by the ACC. On two occasions, the presidents of the ACT member States also adopted declarations: the 1989 Declaration of Manaus[100] and the 1992 Declaration of Manaus on the 1992 UNCED.[101] The ACT institutional bodies have also adopted regulations governing their functions, which have conferred more precision on the attributions given to these organs laid down in the ACT original text. In addition to those instruments, a variety of nonbinding instruments have been produced within the framework of this treaty, such as work plans, guidelines, and memoranda of understanding.

Most resolutions adopted by the MMFA and the ACC have addressed institutional matters only. The declarations adopted at the MMFA have provided guidance and political backing for measures and projects conducted under the aegis of the ACT, and the resolutions have addressed mostly administrative matters, for example they have created institutional bodies,[102] defined meetings dates,[103] approved the

[99] 2199 UNTS, A-19194.

[100] BJ, pp. 39–41.

[101] Ibid., pp. 42–49.

[102] RES/V MRE-TCA/1 creates a permanent secretariat (BJ, p. 90); RES/V MRE-TCA/4 creates the Special Commission for Education (BJ, pp. 91–92); RES/V MRE-TCA/2 creates a financial mechanism (BJ, pp. 90–91); and RES/VI CCA 2 instituted the ad hoc working group to evaluate the administrative, legal, technical, and financial implications of establishing a permanent secretariat (BJ, pp. 200–1).

[103] RES/VII MRE-TCA/6 requests a hemispheric meeting on sustainable development (BJ, p. 161); RES/IX CCA-TCA/6 requests the Pro Tempore Secretariat to organize a regional workshop on an Amazonian Institute for the Research and Protection of Genetic Resources (BJ, p. 239).

treaty institutional bodies' regulations[104] (or proposed changes in existing ones),[105] and authorized activities and projects.[106] Likewise, resolutions taken by the ACC have addressed institutional matters, for example they have approved projects[107] and have extended the lifetimes of institutional bodies.[108]

In practice, the resolutions and regulations adopted under the ACT have contributed to its institutional strengthening, as they have created new bodies and specified their functions, but they have not tackled more substantive matters, for example by providing a list of protected species, identifying sites to be legally protected, or restricting the use of certain pollutants. The ministerial declarations have introduced common principles, such as those contained in the 1992 Declaration of Manaus on the 1992 UNCED, adopted at the Second Presidential Meeting of ACT parties,[109] and provided guidance to the parties to undertake activities under the treaty.

It should be noted that the legal status of resolutions adopted by the ACT organs is not defined by their constituent instruments. In general, resolutions are not all of a kind, and the approach to their interpretation may vary depending on their nature: Some deal with internal administrative matters of a certain organization or are purely recommendatory, whereas others address substantive matters and may have a mandatory content, for example some UN Security Council resolutions.[110] The legal effect of resolutions depends to some extent to how they are drafted. In examining the status of Security Council resolutions, in the Namibia Advisory Opinion, the International Court of Justice (ICJ) held that "the language of a resolution of the Security Council should be carefully analyzed before a conclusion can be made as to its binding effect."[111] As the binding character of resolutions taken

104 RES/VI MRE-TCA/1 approves the Regulation of ACTO in 2000 (BJ, p. 113); RES/VII MRE-TCA/4 adopts the headquarters agreement between Brazil and ACTO in 2002 (BJ, p. 160); RES/VII MRE-TCA/5 approves the staff regulation of ACTO in 2002 (BJ, p. 160).

105 RES/IX CCA-TCA/8 initiates a study to evaluate whether the legal basis of the ACT should be changed with regard to the presentation, approval, financing, and implementation of projects (BJ, p. 240).

106 RES/V MRE-TCA/6 encourages the adoption of a regional document on criteria for and indicators of the sustainability of the Amazonian forests (BJ, p. 93); RES/V MRE-TCA/10 requests the Special Commission on Transport Communications and Infrastructure of the Amazon to undertake a study on the navigability of Amazonian rivers (BJ, p. 95).

107 RES/XI CCA-TCA/2 approves the project on the combating of malaria and requests the Permanent Secretariat to seek funding for developing a "sanitary surveillance network" (BJ, p. 301).

108 RES/VI CCA/3 extends the exercise of the Pro Tempore Secretariat in Lima for a three-year period, given that the annual period for the exercise of Pro Tempore Secretariats was considered insufficient to ensure its effective functioning (BJ, p. 201).

109 BJ, pp. 42–49.

110 M.C. Wood, "The interpretation of Security Council Resolutions," *Max Planck Yearbook of United Nations Law*, vol. 2, The Hague: Kluwer Law International, 1997, pp. 73–95, at p. 79.

111 *ICJ Reports* 1971, p. 53.

by the ACT organs is not defined under the ACT or subsequent instruments, the analysis of their actual content and language might be useful for assessing their normative character. In general, even if resolutions are not binding in law, they often command strong moral obligations. In the context of the ACT, resolutions, independently of their binding character, have produced significant legal effects, notably that of strengthening its institutional structure. The question of the normative character of ACT resolutions might deserve more attention if, rather than dealing with administrative matters, they give rise to more precise legal commitments.

Some attempts have been made toward the adoption of additional agreements under the aegis of the ACT, in particular with respect to commercial navigation[112] and the prevention of water contamination,[113] but no final agreement has been reached so far. Since the Ninth MMFA, the ACT parties have been negotiating an "Amazon charter," proposed by the government of Peru to address matters related to the political and physical integration of the member States. The 2005 Declaration of Iquitos adopted on that occasion made reference to the commitment of the ACT parties to draft the charter in order to provide guidance to the countries in the region.[114] According to the foreign minister of Peru, in his inaugural speech at the Ninth MMFA, the adoption of a charter would reaffirm the principles and common values among the ACT parties in the light of the new challenges faced in the twenty-first century.[115] In 2006, the ACC held three extraordinary meetings with the aim of discussing the content of an Amazon charter. At the third extraordinary meeting, held in Lima on May 23, 2006, this organ approved the start of its drafting, which would then be submitted to the consideration of the ministers of foreign affairs.[116] The Amazon charter has not yet been adopted, but will probably be instrumental in introducing new principles and guidelines in face of the new circumstances in the region (Table 4.2).

In order to fulfill the objectives laid down by the ACT and make its mechanisms operative, the treaty set up a few institutional bodies. The next section examines

[112] Proposed at the third meeting of ACTO, held in Brasília in 1988 (BJ, p. 185). The proposal of a regulation for the free commercial navigation of the international Amazon rivers was contained in the document TCA/III CCA/SPT BR/W.P. 1/88.

[113] In 1995, at the Fifth MMFA, the ACT parties requested the Pro Tempore Secretariat through resolution RES/V MRE-TCA/7 to draft, in consultation with the parties, and to begin to formulate an agreement aimed at preventing and controlling contamination of water resources in the Amazonian rivers and its harmful effects on human health (BJ, p. 93). A proposal for a regional plan of action for preventing contamination of the Amazonian rivers by mercury was elaborated in 2006 (ACTO, *Integrando la Amazonía Continental*, report, November 2005–October 2006, Brasília, 2007).

[114] 2005 Declaration of Iquitos, para. 5, http://www.otca.org.br/ep/Institucional/index.php?id=1247

[115] Speech of the minister of foreign affairs of Peru, available at: http://www.otca.org.br/ep/Institucional/index.php?id=1368&tipoN=

[116] ACTO, *Integrando la Amazonia Continental*, op. cit., pp. 24–25, 42–43.

TABLE 4.2. *ACT normative framework*

Agreements
1978 Amazon Cooperation Treaty
1998 Amendment Protocol to the Amazon Cooperation Treaty
2002 Headquarters' Agreement between ACTO and Brazil
Declarations and resolutions
Meetings of Presidents of ACT Member States (1989, 1992, and 2009)
Meetings of Ministers of Foreign Affairs (1980 –)
Amazon Cooperation Council (1983 –)
Regulations of the ACT institutional bodies
Meetings of Ministers of Foreign Affairs
Amazon Cooperation Council
Permanent Secretariat of ACTO
Regulation of the Personnel of the Secretariat of ACTO

what these institutional bodies and their main functions are, and assesses how they have operated and evolved during the past few years.

4.4. INSTITUTIONAL FRAMEWORK

The institutional structure foreseen under the ACT has evolved more significantly than its normative framework. This section presents the organs created under the ACT and examines what has been modified in its original institutional design. A major change occurred in 2003, as a result of the establishment of a Permanent Secretariat in Brasília with the creation of ACTO.

4.4.1. *Meetings of the Presidents*

The presidents of the ACT member States have met three times under the aegis of the ACT, in Manaus, Brazil, on May 6, 1989, February 10–11, 1992, and on November 26, 2009. These presidential meetings, however, are not a regular instance within the treaty's institutional structure. The first of those meetings discussed the future of regional cooperation in the Amazon and expressed support for the 1989 Quito Declaration,[117] by which the ACT ministers of foreign affairs affirmed their intention

[117] 1989 San Francisco de Quito Declaration, BJ, pp. 73–79.

to institutionally strengthen the ACT. At this meeting, the ACT signatories agreed to meet annually,[118] but this commitment has not been observed.

The second presidential meeting was held in preparation for the 1992 UNCED. This meeting produced a document entitled "Joint Positions of the Amazon Countries with a View to the United Nations Conference on the Environment and Development"[119] containing their joint positions on issues such as climate change, biodiversity, forests, freshwater, indigenous communities, financial resources, and technology transfer. The 1989 Amazon Declaration[120] and the 1992 Manaus Declaration on the UNCED are the outcomes of those two presidential meetings.[121] More recently, the Heads of States of the ACTO member countries gathered prior to the Fifteenth Conference of the Parties (COP-15) to the United Nations Framework Convention on Climate Change (UNFCCC), held in Copenhagen in December 2009. The outcome of this meeting was a declaration, signed on November 26, 2009, by which they committed to consolidate regional cooperation and endow ACTO with an active cooperative role[122]

4.4.2. *Meetings of Ministers of Foreign Affairs*

The MMFA is the highest body within the ACT institutional structure. As defined in Article 20 of the ACT, this body shall establish common policy guidelines, assess the development of the process of Amazonian cooperation, and take decisions to carry out the aims of the treaty. The MMFA adopts declarations and resolutions. The regulation of the MMFA[123] provides guidance as to how those meetings occur, but it does not detail the general functions contained in Article 20 of the ACT. The first MMFA was held in Belém, Brazil on October 23–24, 1980, in the same year of the ACT's entry into force; therefore, it came earlier than was envisaged in Article 22(2) of the treaty, according to which the first MMFA should occur "within a period of two years following the date of entry into force of this Treaty," or, in other words, by 1982.

As originally foreseen in Article 20 of the ACT, MMFA would be convened upon the request of any party with the support of at least four other member states.[124] Otherwise, the ACC, the second highest body within the ACT institutional

[118] 1989 Declaration of Manaus, para. 10, BJ, pp. 39–41.

[119] "Joint Positions of the Amazon Countries with a view to the United Nations Conference on the Environment and Development," BJ, pp. 44–49.

[120] 1989 Amazon Declaration, BJ, pp. 39–41.

[121] 1992 Manaus Declaration, BJ, pp. 42–43.

[122] Declaration of Heads of States of the Amazon Cooperation Treaty Organization, November 26, 2009, available at: http://www.otca.org.br

[123] BJ, pp. 305–8.

[124] ACT, Article 20.

framework, could also make recommendations on the advisability of convening such meetings. Later on, Article 3 of the MMFA Regulation redefined these meetings' frequency: Ordinary meetings should be convened every two years, whereas extraordinary meetings should be held upon the request of any party with the support of at least four other parties or upon the recommendation of the ACC.[125] Debate has been raised on whether to convene annual ordinary meetings instead of biannual meetings, in order to speed up decision-making processes at the ministerial level.[126]

The MMFA plenary sessions are public, whereas the sessions of commissions and subcommissions,[127] as well as those of working groups, are private. In addition to the delegations, only the secretarial staff and invited observers can participate in private sessions. Article 6 of the MMFA Regulation contains a list of States and organizations allowed to participate as observers, namely interested States, the UN and its specialized agencies, the OAS, the Association of Latin American Integration, the Latin American Economic System, and any other international (governmental or nongovernmental) organizations.[128]

Up to the present, nine regular MMFA have been held in total, seven of which convened prior to the establishment of the ACT's Permanent Secretariat in 2003. The Eighth MMFA, held in Manaus, Brazil in September 2004, was the first ministerial meeting after the establishment of the Permanent Secretariat. Of those nine ministerial meetings, the ACT parties have met extraordinarily only once, in Brasília on December 6, 2002, in order to elect ACTO's interim secretary-general.[129]

[125] Approved at the First MMFA, held in Belém, Brazil in 1980 (BJ, pp. 305–8).

[126] Proposed at the Eighth MMFA held in Manaus, Brazil in 2004; see the minutes of the meeting available at: http://www.otca.org.br/ep/Institucional/index.php?id=1230

[127] According to Article 13 of the Regulation of the MMFA, the meeting was conducted by a General Commission composed of the heads of delegations, in addition to Subcommissions on (1) Credentials, (2) Technical and Scientific Cooperation, (3) Natural Resources and Infrastructure, (4) Social, Cultural and Economic Affairs, and (5) Style (BJ, pp. 307–6).

[128] In 2004, at the Eighth MMFA, Brazil and Peru proposed inviting third countries and organizations to participate as observers, and specifically granting observer status to French Guyana. The Bolivian delegation also proposed the participation of local governments in the treaty's regular meetings. The coordination commission of the ACC and the Permanent Secretariat were then requested to define criteria for the participation of observers in future meetings (2004 Manaus Declaration, para. 39, adopted on that occasion, http://www.otca.org.br/ep/Institucional/index.php?id=1230). At the Ninth MMFA in 2005, the French delegate participating in the meeting affirmed that cooperation with the Amazon countries could be strengthened, for example, via technical assistance particularly in developing surveillance systems for the Amazon region, taking as an example that created in Cayanne, French Guyana, for capturing high-resolution images. On that occasion, the ACT member States requested the Coordination Committee of the ACC to define, in collaboration with the Permanent Secretariat, criteria for the possible participation of observers, through RES/IX MRE-OTCA/07 (http://www.otca.org.br/ep/Institucional/index.php?id=1248). From the Eighth MMFA onward, minutes of the meeting and resolutions are only available on ACTO's Web site.

[129] Extraordinary MMFA, BJ, p. 166.

TABLE 4.3. *Meetings of Ministers of Foreign Affairs*

	Type of meeting	Venue	Date
1st	Ordinary	Belém, Brazil	October 23–24, 1980
2nd	Ordinary	Santiago de Cali, Colombia	December 7–8, 1983
3rd	Ordinary	Quito, Ecuador	March 6–7, 1989
4th	Ordinary	Santa Cruz de la Sierra, Bolivia	November 7–8, 1991
5th	Ordinary	Lima, Peru	December 4–5, 1995
6th	Ordinary	Caracas, Venezuela	April 6, 2000
7th	Ordinary	Santa Cruz de la Sierra, Bolivia	November 22, 2002
1st	Extraordinary	Brasília, Brazil	December 6, 2002
8th	Ordinary	Manaus, Brazil	September 14, 2004
9th	Ordinary	Iquitos, Peru	November 25, 2005

Over the years, the MMFA have taken place in a discontinuous way (Table 4.3). For example, six years elapsed between the second and third MMFA, four years between the fourth and fifth meetings, and five years between the fifth and sixth meetings. Since 2000, those meetings have taken place on a more regular basis. The last MMFA took place in 2005.

In 2004, the ACT signatories decided to convene, in addition to ordinary (and extraordinary) meetings, "thematic ministerial meetings" on specific areas (e.g. national security, social development, physical integration, trade, science and technology, biodiversity, and intellectual property).[130] These meetings have been useful in gathering actors such as governments, national institutions, and nongovernmental organizations (NGOs), and allowing the exchange of information among ACTO member States.

4.4.3. *Amazon Cooperation Council*

The Amazon Cooperation Council, the second highest body within the ACT hierarchy, is composed of high-level diplomatic representatives of each member State. According to Article 21 of the ACT, this body shall ensure compliance with the aims of the treaty, carry out decisions adopted at the MMFA, consider initiatives presented by the parties, and decide on the undertaking of bilateral or multilateral

[130] For example, meetings of ministers of health (2006), the environment (2006 and 2008), tourism (2007 and 2008), science and technology, (2005) and intellectual property (2005) (ACTO, *Management Report 2007–2008*, Brasília: ACTO Permament Secretariat, 2008, p. 13). The thematic ministerial meetings were approved at the Eighth MMFA in 2004 (2004 Manaus Declaration, para. 10, http://www.otca.org.br/ep/Institucional/index.php?id=1285).

studies and technical cooperation projects. It must also evaluate the implementation of bilateral or multilateral plans. The Regulation of the ACC, similarly to that of the MMFA, specifies how the ACC meetings shall occur, but does not detail the ACC functions contained in Article 21.[131] The ACC meetings adopt resolutions. Decisions adopted both at the MMFA and the ACC meetings require a unanimous vote, as established in Article 25 of the ACT.

According to Article 3 of the ACC Regulation (and Article 21 of the ACT), ordinary meetings should be held annually, whereas extraordinary meetings should be convened upon the request of any of the parties with the support of the majority of the other parties.[132] The ACC plenary sessions are public, but they may be made private upon the request of any party.[133] Article 8 of the ACC Regulation contains a listing of States and institutions that may participate as observers in ACC meetings (the same as that contained in Article 6 of the MMFA Regulation).

Up to the present, the ACC has convened thirteen ordinary meetings in total. This latter meeting was held in Quito, Ecuador in July 1993 to discuss ways of institutionally strengthening the ACT. Those ACC meetings took place prior to the establishment of the ACT Permanent Secretariat. ACC meetings have been convened mostly every two years, thus disregarding the one-year frequency foreseen in Article 21 of the ACT and Article 3 of the ACC Regulation. The last ACC meeting was held in 2007.

4.4.4. *Permanent National Commissions*

Article 23 of the ACT requests the parties to create permanent national commissions (PNCs) that should operate in their respective territories. The PNCs should enforce the ACT provisions domestically and carry out decisions taken at the MMFA and ACC meetings. In principle, each State can adopt national regulations to govern their respective national commissions. The PNCs are composed of ministries, governmental and nongovernmental national institutions, and headed by each State's ministry of foreign affairs.

Although formally created in most of the ACT member States, the PNCs have not been operative or have functioned on a very limited scale. In view of this, the ACT Permanent Secretariat is now exploring ways of strengthening the PNCs. After the establishment of the ACT Permanent Secretariat, two international meetings of the parties' PNCs were convened in Brasília on July 1–2, 2004 and May 10–12, 2005 to evaluate their current state, future role, and ways of reinforcing such commissions. The Ninth MMFA held in 2005 suggested that meetings assembling each party's

[131] BJ, pp. 309–11.

[132] ACT, Article 21.

[133] Regulation of the ACC, Article 20.

TABLE 4.4. *Meetings of the Amazon Cooperation Council*

	Type of meeting	Venue	Date
1st	Ordinary	Lima, Peru	July 5–8, 1983
2nd	Ordinary	La Paz, Bolivia	September 25–26, 1986
3rd	Ordinary	Brasília, Brazil	March 16–18, 1988
4th	Ordinary	Bogota, Colombia	May 2–5, 1990
1st	Extraordinary	Quito, Ecuador	July 19–22, 1993
5th	Ordinary		
6th	Ordinary	Lima, Peru	October 10–11, 1994
7th	Ordinary	Lima, Peru	November 30–December 1, 1995
8th	Ordinary	Caracas, Venezuela	March 10–11, 1997
9th	Ordinary	Caracas, Venezuela	October 5–6, 1998
10th	Ordinary	Caracas, Venezuela	April 4–5, 2000
11th	Ordinary	Santa Cruz de la Sierra, Bolivia	November 20–21, 2002
12nd	Ordinary	Manaus, Brazil	September 15–16, 2004
13th	Ordinary	Iquitos, Peru	November 24, 2005
2nd	Extraordinary	Brasilia, Brazil	July 2, 2007
3rd	Extraordinary	Brasilia, Brazil	August 29, 2007
4th	Extraordinary	Brasilia, Brazil	September 26, 2007

PNC should be held annually and also instructed the ACT Permanent Secretariat to prepare a work program for strengthening the PNCs. The latest declaration of the heads of States of ACTO member countries signed on the occasion of the preparatory meeting to the UNFCCC COP-15 in 2009, reaffirming the importance of establishing and reactivating the permanent national commissions.[134]

4.4.5. *Special Commissions*

Article 24 of the ACT envisaged the creation of special commissions (SCs) by the contracting parties aimed at studying specific problems.[135] To date, seven SCs have been created in the following areas: 1) environment, 2) science and technology, 3) transport, communication and infrastructure, 4) health, 5) indigenous affairs, 6) tourism, and 7) education.

[134] Declaration of Heads of States of the ACTO, op. cit.
[135] ACT, Article 24.

TABLE 4.5. *Special Commissions*

Special commission	Date of constitution
Science and Technology	March 18, 1988
Health	March 18, 1988
Environment	March 7, 1989
Indigenous Affairs	March 7, 1989
Tourism	May 5, 1990
Transport, Communications, and Infrastructure*	May 5, 1990
Education	December 5, 1995

* The original SC on Transport was designated SC on Transport, Communications, and Infrastructure on November 8, 1991.

In 1994, the Sixth ACC meeting held in Lima on October 10–11, 1994 requested the Pro Tempore Secretariat to draft a new regulation for the SCs and replace the executive (or technical) secretary at the head of each SC by a coordinator designated by the secretary *pro tempore* in four domains: 1) the environment; 2) health and indigenous affairs; 3) transport, communications, and infrastructure; tourism; and 4) education and science and technology.[136] The special commissions are now grouped in five Coordination offices (i.e. Environment; Health; Transport, Infrastructure, Communication and Tourism; Indigenous Affairs; and Science, Technology and Education.)[137]

Both the SCs and the Pro Tempore Secretariats can propose programs that may be implemented with the approval of each concerned party (member States not participating in a specific program can also present observations).[138] The Pro Tempore Secretariats, which functioned on a temporary basis until 2002, were expected to follow up and examine the activities carried out under the ACT.[139] Some special commissions have been more active than others in terms of number of activities and meetings.

Certain SCs, notably those on Science and Technology and on the Environment, have set up criteria to prioritize projects and evaluate their own activities.[140] These criteria were intended to reduce the number of projects approved by each SC without the corresponding human and financial capacity to put them into practice,

[136] RES/VI CCA/4 proposed the preparation of a Standard Regulation for SCs.

[137] ACTO, *Management Report*, op. cit., p. 54.

[138] Regulation of the Pro Tempore Secretariat, Article 7. This provision makes reference to the activities and programs in general, not only those undertaken by the SCs (BJ, 1992).

[139] Regulation of the Pro Tempore Secretariat, Article 10, ibid.

[140] BJ, 2002, p. 109.

and also to facilitate the management of the funds and technical cooperation required. The ACT parties debated whether such criteria should be extended to all SCs, and requested the Pro Tempore Secretariat to develop criteria for setting up bilateral and multilateral projects under the framework of the ACT.[141] Through Resolution RES/IX CCA-TCA/8, the ACC requested a study on whether the legal basis of the ACT should be changed with regards to the presentation, approval, financing, and implementation of projects, and whether an *ad hoc* working group should be created to undertake this study.[142]

4.4.6. *From Pro Tempore Secretariats to a Permanent Structure*

The ACT was intentionally conceived, as noted earlier, with a flexible and simple structure, comprising few institutions and a temporary Secretariat, involving little costs and bureaucratic work.[143] Until recently, each State's ministry of foreign affairs functioned as the treaty's secretariat, which operated on a temporary basis and rotated among the parties in alphabetical order. The Pro Tempore Secretariats were established in the country where the next ACC regular meeting was scheduled to occur.[144] The Pro Tempore Secretariats had originally a very limited role of sending "the pertinent documentation to the Parties," as envisaged in Article 22 of the ACT.

A few years later, on the occasion of the Fourth MMFA in 1991, the parties approved the Regulation of the Pro Tempore Secretariat, which was intended to "confer to the Pro Tempore Secretariat a wide range of attributions to ensure the fulfillment of the tasks assigned to this body."[145] In addition to secretarial tasks defined in Article 5 of the 1991 Regulation of the Pro Tempore Secretariat – for example, preparing and circulating documents or keeping the member States informed via diplomatic channels (information was normally circulated among the embassies located in the country in which the Pro Tempore Secretariat was located), that body was also responsible for ensuring compliance with the treaty objectives, and MMFA and ACC resolutions.[146] The activities of the Pro Tempore Secretariats should be undertaken in conformity with specific mandates conferred by the MMFA, ACC, and SCs, pursuant to Article 5(6) of the Regulation of the Pro Tempore Secretariats.

[141] Third MMFA held in Quito, Ecuador in 1989 (BJ, 2002, pp. 70–79).
[142] BJ, 2002, p. 240.
[143] BJ, 2002, p. 63.
[144] ACT, Article 22.
[145] At the Fourth MMFA (BJ, pp. 82–83).
[146] The functions of the Pro Tempore Secretariat are defined in Article 5 of the Regulation of the Pro Tempore Secretariat, adopted at Santa Cruz in 1991 (BJ, 1992).

TABLE 4.6. *Pro Tempore Secretariats*

Host Country	Time period
Peru	October 1980–July 1983
Bolivia	July 1983–September 1986
Brazil	September 1986–March 1988
Colombia	March 1988–May 1990
Ecuador	May 1990–January 1994
Peru	February 1994–March 1997
Venezuela	March 1997–July 2000
Bolivia	August 2000–February 2003

Source: L. Carrera de la Torre, *Contribuciones para la definición de una propuesta de trabajo técnico para la Organización del Tratado de Cooperación Amazónica: 2004–5*, report prepared for the ACTO Permanent Secretariat, *Brasília*, 2003, p. 31.

Originally, each Pro Tempore Secretariat was to function for a period of one year, pursuant to Article 22 of the ACT, as they would be instituted at every regular annual meeting of the ACC. In 1994, the mandate of the Pro Tempore Secretariat was extended to a three-year period.[147] Until the establishment of the Permanent Secretariat in Brasília, in total, eight Pro Tempore Secretariats were instituted (from 1980 to 2003). Peru and Bolivia have performed secretariat functions twice, whereas Guyana and Suriname have never hosted Pro Tempore Secretariats (Table 4.6).

At the suggestion of the government of Colombia put forward at the Third MMFA in 1989,[148] an *ad hoc* mechanism was created to reinforce and assist the work of the Pro Tempore Secretariats, which was welcomed by Article 4 of the 1991 Pro Tempore Secretariat Regulation. According to this provision, an ad hoc consultation committee should be formed, composed of heads of diplomatic missions serving in the Pro Tempore Secretariat's host country. This body should support the activities of the Pro Tempore Secretariats, coordinate activities, and facilitate the adoption of common positions by the ACT parties.[149] Up to 2002, the Ad Hoc Consultation Committee convened nineteen meetings.[150]

[147] With the RES/VI CCA 3 adopted at the sixth meeting of the ACC, held in Lima in 1994 (BJ, p. 201). The changes to the Regulation of the Pro Tempore Secretariat agreed upon at the sixth and seventh meetings of the ACC were latter approved by RES/V MRE-TCA/3 at the Fifth MMFA (BJ, 2002, p. 91).

[148] BJ, 2002, p. 63.

[149] Regulation of the Pro Tempore Secretariat, Article 4.

[150] Carrera de la Torre, op. cit., p. 32.

A process of institutional strengthening of the ACT that led to the approval of the various regulations of the ACT bodies (and the creation of new ones), was in fact initiated back in 1989, when the ACT parties started discussions on ways of reinforcing the treaty's institutional structure.[151] In 1991, the Fourth MMFA approved a resolution by which the Pro Tempore Secretariat was instructed to set up a working group to formulate a proposal for the institutional strengthening of the ACT.[152] Following this resolution, a major change occurred in 1995 at the Fifth MMFA. On that occasion, a permanent secretariat was formally created through RES/V MRE-TCA 1[153] to be based in Brasília.

Following the formal creation of the Permanent Secretariat in 1995, an *ad hoc* working group composed of representatives of all member states was created to prepare a proposal defining the structure and objectives of a permanent secretariat.[154] After eight meetings that took place between 1996 and 2002, the ad hoc working group recommended the setting up of a permanent secretariat with greater management capacities relative to the prior Pro Tempore Secretariats, and endowed with a legal personality to represent the ACT member States.[155] As emphasized by the government of Venezuela at the Second Meeting of the Ad Hoc Working Group in 1996, this body should be, however, auxiliary to the MMFA and ACC, with functional and administrative autonomy, but should not function as a political organ.[156]

A Protocol of Amendment of the ACT[157] modifying Article 22 on the Pro Tempore Secretariats was adopted on December 14, 1998, in order to create the Amazon Cooperation Treaty Organization, to be permanently established in Brasília.

The Seventh MMFA in 2002, the last meeting convened by the treaty's Pro Tempore Secretariat prior to the establishment of ACTO, approved a headquarters agreement between Brazil and this newly created organization,[158] its staff

[151] 1989 San Francisco de Quito Declaration, adopted at the Third MMFA, held in Quito, Ecuador in 1989 (BJ, 2002, pp. 73–79, at p. 74).

[152] Fourth MMFA held in Santa Cruz de la Sierra, Bolivia in 1991 (BJ, 2002, p. 82).

[153] BJ, 2002, p. 90.

[154] RES/V MRE-TCA/1, approved at the Fifth MMFA held in Lima, Peru in 1995 (BJ, 2002, p. 90).

[155] BJ, 2002, p. 605. As the Peruvian delegation asserted, "la Secretaria Permanente debería desempeñar un rol substancial y cualitativamente mayor que el actualmente asignado a la Secretaria Pro Tempore, debería robustecer al Tratado, de acuerdo al desarrollo registrado gracias al esfuerzo desplegado por las ultimas Secretarias Pro Tempore, debería ser dotada de mayor flexibilidad ante los nuevos retos a enfrentar y ante la mayor complejidad del trabajo conjunto; debía favorecer una mayor coordinación y participación de las Partes, así como incrementar su capacidad de interlocución ante organismos internacionales, gobiernos cooperantes y ONGs."

[156] BJ, 2002, p. 605.

[157] UNTS, vol. 2199, A-19194.

[158] RES/VII MRE-TCA/4, BJ, 2002, p. 160.

TABLE 4.7. *Meetings of the ad hoc working group on the establishment of a Permanent Secretariat*

Meeting Venue	Date
Brasília, Brazil	May 13, 1996
Lima, Peru	June 3–4, 1996
Brasília, Brazil	July 31–August 1, 1996
Brasília, Brazil	January 21–23, 1998
Caracas, Venezuela	June 9–10, 1998
Brasília, Brazil	September 10–11, 1998
Caracas, Venezuela	July 7–8, 1999

regulation,[159] and institutional structure.[160] In 2002, the ACT started a new phase, as ACTO's interim secretary-general pointed out:

> the ministerial meeting that bring us together today is of particular symbolism as it closes a long phase of the functioning of pro-tempore secretariats and that new organization should confer vitality and potential as a catalyst of the initiatives carried out by its member countries in their search for solutions to problems common to the Amazon region.[161]

4.4.7. *The Functions and Structure of the Permanent Secretariat*

With the establishment of ACTO in Brasília in 2003,[162] the period of temporary Secretariats came to an end and a new phase began of greater institutional stability. The ACT member States perceived the creation of this organization as a way of opening new possibilities for regional cooperation. As noted by the then Brazilian president (Fernando Henrique Cardoso) at the signing ceremony of the headquarters agreement between Brazil and ACTO, with the creation of a Permanent Secretariat the Amazon States were taking a step forward and conferring more stability on the organization.[163]

[159] RES/VII MRE-TCA/5, BJ, 2002, pp. 160–61.

[160] RES/VII MRE-TCA/8, BJ, 2002, p. 162.

[161] BJ, 2002, p. 134: "El encuentro ministerial que hoy nos reúne se reviste de particular simbolismo toda vez que en el se cierra una larga etapa de ejercicio de secretaria pro-tempore y se establece en la capital brasileña la secretaria permanente de la ACTO. La efectiva transformación del tratado en una organización internacional dotada de personería jurídicas, le confiere la necesaria vitalidad y a potencia como catalizador de las iniciativas de sus estados miembros en el propósito de avanzar en el tratamiento y búsqueda de soluciones para los problemas comunes de la región amazónica."

[162] ACTO, *Management Report*, op. cit., p. 7.

[163] Speech of the then Brazilian president (Fernando Henrique Cardoso) at the signing ceremony of the headquarters agreement between Brazil and ACTO, "Discurso do presidente da República, Fernando

With the establishment of ACTO, the Regulation of the Permanent Secretariat was approved, defining its functions (Articles 4 and 5).[164] In general, they are similar to those of the Pro Tempore Secretariats. Differently from the latter, the Permanent Secretariat of ACTO, as stated in Article 1, has legal personality and is competent to enter into agreements with member countries, third parties, and international organizations in conformity with specific mandates conferred on it by the MMFA and the ACC.

The powers conferred on the Permanent Secretariat are dependent on specific mandates given by the ACT's highest organs. Therefore, it can only represent the member States and enter into agreements on their behalf if it has been given specific mandates conferred by the MMFA or the ACC in accordance with its work plan, as defined in Articles 3 and 4 of the Regulation of the Permanent Secretariat of ACTO. To perform other functions the Secretariat shall also observe the decisions and guidelines of the MMFA or ACC. As established in Article 1, the Permanent Secretariat is an auxiliary body of the political organs of the treaty and should undertake the activities foreseen under the ACT, according to the guidance of those two bodies.

ACTO's institutional structure was approved in 2002 at the Seventh MMFA.[165] It is headed by a secretary-general, who is a national of one of the member States, unanimously elected by a MMFA for a three-year period.[166] The ACTO Permanent Secretariat has a directive board formed by the secretary-general and an administrative director. Five sectoral coordinating offices are currently in operation: Environment; Health; Transport, Infrastructure, Communication and Tourism; Indigenous Affairs; and Science, Technology and Education. The secretary-general has the support of a technical staff composed of nationals of the eight member States. ACTO's programs are developed by teams of consultants under contract for the duration of the programs.[167] ACTO maintained the original bodies as foreseen in the ACT, described above, included one more: the coordination commission of the ACC.

With a similar role to that of the *ad hoc* consultation committee to the Pro Tempore Secretariat,[168] Brazil suggested at the Third Meeting of the Ad Hoc Working Group on the Establishment of a Permanent Secretariat to ACTO in 1996 the creation of a

Henrique Cardoso, na cerimônia de assinatura do Acordo de Sede entre o Brasil e a Organização do Tratado de Cooperação Amazônica," http://www.radiobras.gov.br. In Portuguese: "Isso tem, a meu ver, um efeito muito maior do que um simples tratado, porque diz respeito a algo que é vital para a humanidade e para nós, que é a região amazônica. E acredito que o fato de nós criarmos esta secretaria pro tempore, que já deu um passo. Mas agora, estamos dando um passo adiante, que é de dar uma estabilidade maior na organização do tratado amazônico."

164 Regulation of ACTO, Article 5.

165 Adopted by RES/VII MRE-TCA/8, BJ, 2002, p. 162.

166 Regulation of the Permanent Secretariat of the Amazon Cooperation Treaty Organization, BJ, Article 2, para.1, p. 314.

167 ACTO, *Management Report*, op. cit., p. 54.

168 As suggested the Fourth Meeting of the Ad Hoc Working Group on the Establishment of a Permanent Secretariat held in Brasília in 1998 (BJ, 2002, p. 615).

consultative organ auxiliary to the Permanent Secretariat responsible for the coordination among the governments and the ACT's institutional bodies.[169] According to Brazil, the institutional strengthening of the ACT required not only a permanent secretariat, but also further intergovernmental coordination, which could be facilitated by that body.[170] The other ACT parties welcomed this proposal, but emphasis was put on the fact that this organ should only have consultative and liaising functions, without decision-making power, as this lay within the competence of the ACC.[171] Brazil presented a draft regulation for the Coordination Commission of the ACC (Spanish acronym: CCOOR), which was discussed within the framework of the Ad Hoc Working Group on the Establishment of a Permanent Secretariat.

The CCOOR was formally created in 2000 with its respective regulation.[172] It is a consultative organ auxiliary to the ACC, composed of diplomatic officers of each country's diplomatic representations in Brasília. It monitors the planning and functioning of the Permanent Secretariat, evaluates activities carried out by ACTO, and prepares recommendations to ACC.[173] Article 3 of the CCOOR Regulation provides a detailed description of its functions. This body can meet as many times as necessary upon the request of any of the member States, but must at least twice a year.[174] In practice, these meetings have been held more regularly (usually on a monthly basis) at ACTO headquarters in Brasilia.[175] CCOOR has mostly discussed issues related to the Permanent Secretariat's administration and personnel, including staff travel expenses, salary policy and scales, and the evaluation of applications for vacant posts.[176] The CCOOR meetings are attended by the directive board of ACTO, ambassadors of the member States in Brazil, and representatives of the Brazilian chancellery.[177] These meetings have allowed a regular exchange of information and greater participation of ACTO member States in the Permanent Secretariat's activities.

[169] At the Third Meeting of the Ad Hoc Working Group on the Establishment of a Permanent Secretariat held in Brasília in August 1996 (BJ, 2002, p. 610).

[170] Brazil proposed the creation of this organ, which was discussed in that Third Meeting of the Ad Hoc Working Group on the Establishment of a Permanent Secretariat (BJ, 2002, pp. 611–12).

[171] As argued by Colombia, Ecuador, Bolivia, and Peru at the Fourth Meeting of the Ad Hoc Working Group on the Establishment of a Permanent Secretariat held in Brasília in 1998 (BJ, 2002, p. 615).

[172] Though RES/VI MRE-TCA/2, approved at the Sixth MMFA held in Caracas, Venezuela in 2000 (BJ, 2002, p. 113).

[173] Additional functions are established in Article 3 of the Regulation of the CCOOR (BJ, 2002, pp. 312–16.)

[174] Article 4 of the Regulation of the CCOOR (BJ, 2002, pp. 312–16).

[175] ACTO, *Management Report*, op. cit., p. 13. About the frequency of CCOOR meetings, see ACTO, *Integrando la Amazonía Continental*, op. cit., p. 24.

[176] For example, ACTO's first secretary-general (Rosalia Arteaga) was nominated in one of the CCOOR meetings for a period of three years starting May 1, 2004, RES/VIII CCOOR-OTCA/02/04 adopted at the eighth meeting of the CCOOR held in Brasília on April 28, 2004.

[177] ACTO, *Management Report*, op. cit., p. 13.

4.4.8. *The Permanent Secretariat Today*

Once established, ACTO requested a study on the performance of the ACT's institutional bodies over twenty-five years, from 1978 to 2003. In a document entitled *Contribuciones para la definición de una propuesta de trabajo técnico para la Organización del Tratado de Cooperación Amazónica,*[178] the results of the activities conducted during the time of Pro Tempore Secretariats were examined. According to this study, the great majority of projects approved by the ACT's SCs have not been executed. With regards to the PNCs, although formally created at the national level, they have not been operative. The little practical impact of the ACT, as concluded by the study, was due to a number of factors: the excessive number of projects approved by the SCs without required technical and financial capacity, the institutional weaknesses of the Pro Tempore Secretariats and the lack of financial resources.[179] Field research conducted in the Brazilian Amazon assessed whether members of parliament in each of the States comprising the Legal Amazon were aware of the ACT and ever considered it in their voting. A series of interviews indicated that the treaty was rarely or never considered in the context of these parliaments.[180] Of particular note, the ACT has not been mentioned in the jurisprudence of the Brazilian federal courts.[181]

To avoid some of the difficulties encountered in the past, in particular those related to the handling of too many projects in different areas, the Permanent Secretariat requested in 2003 a study intended to guide ACTO's future policies and highlight some areas on which ACTO should focus attention. A document entitled *Lineamientos estratégicos para la Organización del Tratado de Cooperación Amazónica*[182] was drawn up in the same year and provided the basis for the formulation of the Strategic Plan prepared by the Permanent Secretariat for the period 2004–2012, which would provide guidance to the ACT future activities. The objective was to select a few areas on which the Secretariat should focus, in order to avoid the overloading of issues and projects, which had proven counterproductive during the time of the Pro Tempore Secretariats.[183]

178 Carrera de la Torre, op. cit., p. 93.

179 A. Aranìbar Quiroga and J. Marcovitch, *Lineamientos estratégicos para la Organización del Tratado de Cooperación Amazónica: contribuciones para la reflexión a 25 años de suscripción del Tratado de Cooperación Amazónica,* report prepared for the ACTO Permanent Secretariat, 2003, p. 65.

180 E.D. da Silveira, *Direito ambiental: Tratado de Cooperação Amazônica*, Curitiba: Juruá Editora, 2005, pp. 141–47.

181 With respect to national courts, the author has searched for references to the ACT in the jurisprudence of Brazilian federal courts from 1980, the date of its entry into force, until the present, but no references were found.

182 Aranìbar Quiroga and Marcovitch, op. cit.

183 BJ, 2002, p. 109.

The work plan, formulated in consultation with the ACT member States, international organizations and civil society, was first presented at the First International Meeting of the PNCs held in Brasília on July 1–2, 2004. A revised version of the Strategic Plan 2004–2012 was submitted to the Twelfth Meeting of the ACC in 2004,[184] and finally, the Eighth MMFA adopted ACTO's Strategic Plan 2004–2012 on September 14, 2004.[185] The plan identifies four "strategic axes" and six "programmatic areas," as shown in Table 4.8.

The latest declaration signed by the heads of States of the ACTO member countries on November 26, 2009,[186] required the coordination commission of the ACC to draft a Strategic Agenda for ACTO to be approved by the Meetings of Foreign Affairs Ministers at the second half of 2010.[187]

Since 2003, ACTO's Permanent Secretariat has signed technical cooperation agreements and understandings with other international organizations in areas such as waters, forests, and biodiversity.[188] In general, projects carried out by ACTO are financed by international organizations and governments. ACTO carries out projects of regional scope, involving a group of States (e.g. Sala Andes-Amazonia, Destination Amazonia Year 2009, or the Global Environment Facility (GEF) Project on Integrated and Sustainable Management of Transboundary Water Resources in the Amazon River Basin) as well as "demonstration projects" implemented in particular countries.[189]

ACTO has being a forum where its member States have built consensus on issues of common interest. For example, under the aegis of ACTO, a joint position of the Amazon countries was agreed upon and presented at the Fifth Session of the UN Forum on Forests (UNFF) in 2005.[190] With the facilitation of ACTO, these countries have managed to negotiate as a block in subsequent UNFF sessions[191] and were identified as the "Amazon Group" in UNFF negotiations. At the Ninth MMFA in 2005, the ACT member States praised the Permanent Secretariat for this positive contribution. The 2005 Declaration of Iquitos signed on that occasion,

184 This meeting was held in Lima, Peru on October 10–11, 1994.

185 At the Eighth MMFA held in Manaus, Brazil on September 14, 2004, included in para. 6 of the 2003 Declaration of Manaus, http://www.otca.org.br/ep/Institucional/index.php?id=1285

186 Declaration of Heads of States of the ACTO, op. cit.

187 Ibid.

188 R. Arteaga, "Organização do Tratado de Cooperação Amazonica (OTCA): um desafio permanente," *Diplomacia, estratégia, politica*, vol. 4, 2006, pp. 87–119, at p. 95.

189 ACTO, *Management Report*, op. cit., pp. 45–46.

190 ACTO played a role of "facilitator" in the dialogue and interaction among member States during the 2005, 2006, and 2007 sessions of the UNFF (ibid., p. 19).

191 ACTO Permanent Secretariat, "OTCA busca posição comum amazônica sobre florestas," SP-OTCA Comunicado no. 26, Brasília, April 1, 2005.

TABLE 4.8. *Strategic plan 2004–2012*

Programmatic areas	Strategic axes			
	Conservation and sustainable use of biodiversity	Technological transfer and knowledge management	Regional integration	Institutional strengthening
	Intervention spaces			
Water	Integrated management of hydrobiological resources Standard agreement on measures toward preventing contamination			
Forests, soils, and protected natural areas	Sustainability of the Amazon as a standing forest Forestry management criteria (Tarapoto Process) Integral management of natural protected areas Ecotourism			
Biological diversity, biotechnology, and biotrade	Regional Amazon strategy on biodiversity Regional Amazon program on biotrade			
Territorial organization, human settlements, and indigenous affairs	Ecological and sustainable economic macrozoning Cultural diversity and respect for the rights of indigenous peoples Training for indigenous leaders			
Social infrastructure: Health and education	Network on epidemiological surveillance and the fight against malaria Support for local capacity-building Virtual education			
Transportation, energy, and communications infrastructure	Initiative on the Integration of the South American Regional Infrastructure (IIRSA) Free navigation in the Amazon rivers System of early warning aimed at preventing disasters			

Source: ACTO, Strategic Plan 2004–2012, Brasília, 2004, p. 24.

stressed the importance of adopting joint positions at international forums.[192] More recently, the Second Meeting of Ministers of Environment of ACTO member States, held on May 29, 2008 on the occasion of the Ninth Conference of the Parties to the 1992 Convention on Biological Diversity, produced the "Bonn

[192] Prior to the formation of ACTO, the Amazon countries had only once adopted a joint position, at the Second Meeting of Presidents of the Amazon States, the 1992 Declaration of Manaus on the UNCED and the "Document on a Common Position of the Amazon States" (BJ, 2002, pp. 42–49).

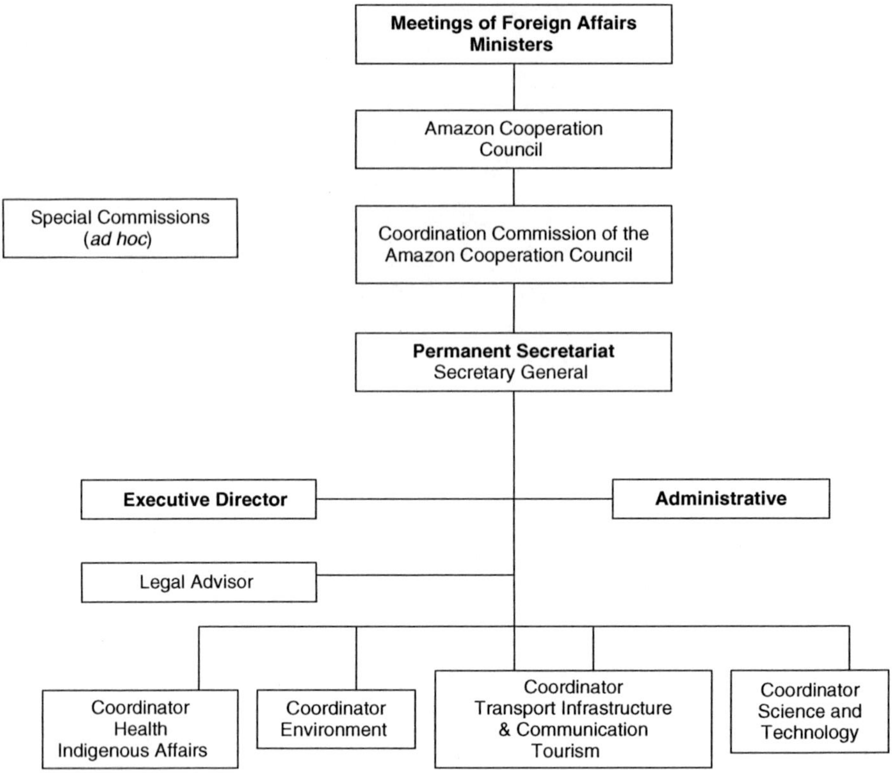

FIGURE 4.1. The structure of ACTO.
Note: Approved by RES/VII MRE-TCA/8, included in the ACTO Strategic Plan, p. 66.

Report," which compiles their common views on biodiversity.[193] At the declaration signed by the heads of States of ACTO member States on November 26, 2009[194] the parties reaffirmed the importance of the ACTO Permanent Secretariat and instructed their ministers of foreign affairs to ensure its institutional strengthening to permit compliance with the mandates conferred to the Secretariat by the member States.

[193] ACTO, *Management Report*, op. cit., p. 18. In recognition of the role it can play in the Amazon, ACTO was invited to participate in the Region-Led Initiative promoted by Switzerland and Australia in the city of Geneva in January of 2008. ACTO also participates in the Puembo Initiative, a dialogue and coordination platform led by FAO and other regional organizations and formed by national forest authorities (ibid., p. 19).

[194] Declaration of Heads of States of the ACTO, op. cit.

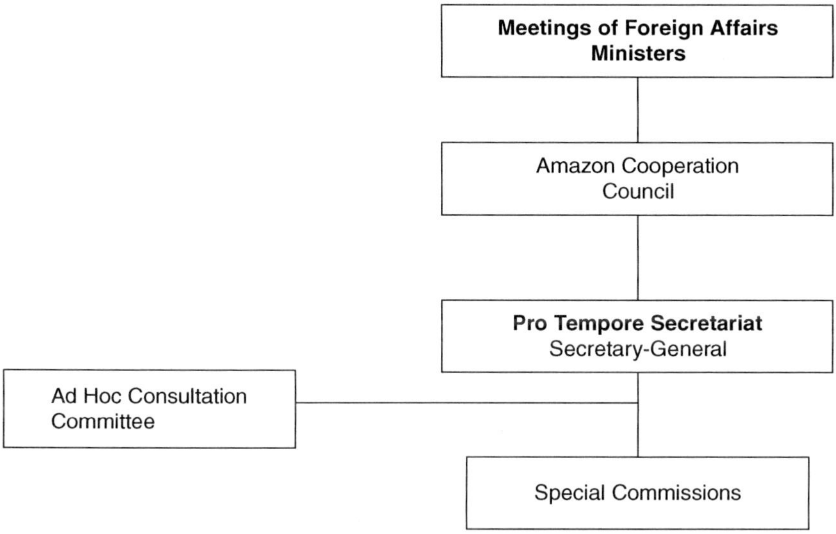

FIGURE 4.2. The ACT structure prior to the establishment of the Permanent Secretariat.

4.4.9. *Financial Mechanism*

Parallel to the debate on the mandate and structure of the Permanent Secretariat, the *ad hoc* working group emphasized that the institutional strengthening of the ACT should be accompanied by a financial mechanism able to support the work of the Permanent Secretariat. This mechanism should complement funds coming from international cooperation[195] and facilitate cooperation among the ACT member States, by allowing more expeditious and efficient interactions among these countries.[196]

As one could expect, the financing of the ACT's activities was a central issue in the discussions on the establishment of a Permanent Secretariat. During the existence of Pro Tempore Secretariats, the process of obtaining financial and technical support would take from one to three years. Consequently, most of the activities planned under the ACT have never evolved from the stage of pilot programs.[197] The ACT Pro Tempore Secretariats relied almost exclusively on international funding, coming from institutions such as the UN Food and Agriculture Organization (FAO),[198] the European Community, the UN Development Program (UNDP), and

[195] 1995 Lima Declaration, BJ, 2002, p. 96.

[196] Resolution RES/V MRE-TCA/1 and RES/V MRE-TCA/2 on the establishment of a Permanent Secretariat and financial mechanism, respectively (BJ, 2002, p. 90).

[197] ACTO, Strategic Plan 2004–2012, op. cit., p. 20.

[198] Ninth meeting of the ACC, BJ, 2002, p. 218.

governments of countries such as the Netherlands and Germany among others.[199] A Permanent Secretariat was expected to have more managerial capacity and be better placed to attract international cooperation and ultimately manage ACT projects more effectively.[200]

In addition to the decision to create a Permanent Secretariat, the ACT member States agreed to have a system of compulsory contributions to be regularly paid to ACTO. As proposed by Bolivia, the national contributions should be proportional to each country's level of development.[201] In view of this, a scale of contributions was devised, assembling the parties in three groups: 1) Brazil; 2) Colombia, Peru, and Venezuela; and 3) Bolivia, Ecuador, Guyana, and Suriname.[202] As Suriname stressed, despite these different levels of contributions, the parties should be equally represented in the Permanent Secretariat.[203]

In 2000, the ACT ministers of foreign affairs approved the Permanent Secretariat total budget (US$ 1,139,600) and agreed on a specific scale of contributions according to the following distribution: Bolivia, 6.5 percent; Brazil, 35 percent; Colombia, 16 percent; Ecuador, 6.5 percent; Guyana, 2 percent; Peru, 16 percent; Suriname, 2 percent; and Venezuela, 16 percent. It should be noted that the ACT parties' national contributions only cover costs related to the functioning of the Permanent Secretariat. Additional funding is still needed for undertaking projects.[204] This system of national contributions is now in place, and since the establishment of the Permanent Secretariat, the parties have met their financial obligations, paying (in the period of 2003–2008) 92 percent of their respective annual contributions.[205]According to the latest declaration signed by the heads of States of ACTO on November 26, 2009, the parties requested the Permanent Secretariat to conduct a study on possible sources of funding within the member States, in order to overcome ACTO's dependence on foreign financing.

4.5. ASSESSING THE EFFECTIVENESS OF THE 1978 AMAZON COOPERATION TREATY

Over the last few years, the ACT has been strengthened in some ways. The creation of a Permanent Secretariat and a compulsory financial mechanism indicates a political commitment to this treaty regime. ACTO has managed to build partnerships with other organizations, leveraging funds for field projects. One of its most notable

199 Ibid.

200 BJ, 2002, p. 605.

201 BJ, 2002, p. 620.

202 BJ, 2002, p. 612. Brazil proposed to contribute 30 percent, in addition to the costs it would cover for setting up the secretariat facilities in Brasília.

203 BJ, 2002, p. 620.

204 RES/VI MRE-TCA/6, adopted at the sixth MMFA, held in Caracas on April 6, 2000.

205 ACTO, *Management Report*, op. cit., p. 50. ACTO's annual functioning and maintenance budget is still of US$ 1,139,600 (ibid., p. 50). For a list of contributions per country, see ibid., p. 50.

achievements was that of providing a forum where the parties can agree on common positions, as occurred in the context of UNFF negotiations. This "coordinating" role is probably one of its potential strengths. As a matter of fact, under resolution RES/IX MRE-OTCA/05 adopted at the Ninth MMFA in 2005, the parties requested the Permanent Secretariat to continue playing the role of facilitator in preparations for international negotiations. The Permanent Secretariat has also facilitated the flow of information among the parties, through more regular meetings, especially of the coordination commission of the ACC, which has proven to be an effective communication channel between the Secretariat and the member States. ACTO has produced information (e.g. progress reports, bulletins, technical reports, etc.) that has brought more transparency and ensured better exchange among the parties. Despite its weaknesses, ACTO has managed to maintain some cohesion among its member States, without which the evolution toward a Permanent Secretariat would not have happened.

On different occasions, the Amazon States manifested their willingness to attain further regional cooperation and strengthen the ACT. For example, at the inter-American level, they committed, under the 1996 Plan of Action for the Sustainable Development of the Americas,[206] to promote regional cooperation, particularly for the use of genetic resources, and hemispheric cooperation at all levels, by adopting treaties for the conservation, management, and sustainable use of water resources and biodiversity. While reviewing this Plan, the First Inter-American Meeting of Ministers and High Level Authorities on Sustainable Development in 2006 reaffirmed the need to foster regional cooperation as a way of achieving sustainable development.[207]

In the context of the ACT, the parties have affirmed many times their willingness to make this treaty stronger. As early as 1983, only a few years after its entry into force, the parties initiated the debate on ways of strengthening the ACT.[208] At the Third Meeting of Ministers of Foreign Affairs, the then president of Ecuador (Rodrigo Borja Cevallos) stressed their "shared responsibility" over the Amazon and the fact that the ACT had introduced a new dimension for regional cooperation.[209] Subsequent ministerial declarations reaffirmed the need of enhanced regional cooperation under the ACT. In 1995 the Amazon States affirmed their

[206] Adopted at the Summit of the Americas on Sustainable Development, held in Santa Cruz de la Sierra, Bolivia, December 7–8, 1996, http://www.summit-americas.org

[207] Information available at: http://www.oas.org/dsd/ministerialmeeting/reunioninteram_eng_v1.htm

[208] BJ, pp. 63–64.

[209] Meeting Minutes, adopted at the third MMFA, held in San Francisco de Quito, Ecuador, March 6–7, 1989, BJ, pp. 70–72: "destacando la responsabilidad que asiste a los países Amazónicos, sin ingerencias extranjeras, en la defensa de los ecosistemas de la Amazonía. El Tratado de Cooperación Amazónica va a eso, a la defensa de la ecología, va a la preservación del medio ambiente, al respeto a las etnias preexistentes."

"historical responsibility" in affecting the future of the Amazon and formally created the Permanent Secretariat.[210] Ever since, other ministerial declarations have reiterated the need to adopt common policies for the Amazon[211] and forge a common vision in order to make regional cooperation more effective[212] in the context of the ACT – viewed as an adequate instrument for promoting regional cooperation.[213] In 2004, at the first MMFA held after the establishment of ACTO, the parties reaffirmed the importance of the ACT.[214] At the last MMFA, the then secretary-general (Rosalía Arteaga) stated that the newly created ACTO was the best and most suitable forum for integrating the Amazon countries. More recently, the heads of State of the ACTO member countries gathered in preparation for the UNFCCC COP-15 in 2009 and affirmed their commitment to consolidate cooperation among the parties and endow the Permanent Secretariat with an active role to this end.[215]

A first step has been taken by the Amazon States in creating the conditions for strengthening the ACT. However, much needs to be done to make this treaty regime effective. Among other things, this involves creating more precise legal obligations, and stronger mechanisms and institutions. What needs to be seen is whether the political statements in that direction will be translated into action and whether the ACT will ever be a cornerstone in the international legal protection of the Amazon. In practice, this treaty has not been able to address common problems, such as deforestation; it has had little practical impact and has operated with scarce human and financial resources. Some of the regular ACT bodies, in particular the MMFA and the ACC, have not met regularly or not at all. In view of this fact, one may question the chances of this treaty being strengthened further and receiving the necessary support to become an effective legal regime for the protection of the Amazon.

210 RES/V MRE-TCA/1 creates a Permanent Secretariat to the ACT with headquarters in Brasília, and RES/V MRE-TCA/2 establishes a financial mechanism for the ACT, both adopted at the Fifth MMFA, held in Lima, Peru, December 4–5, 1995, BJ, pp. 90–91.

211 The 2002 Declaration of Santa Cruz de la Sierra (ibid.) reiterated the importance of the ACT as the principal instrument able to put into effect concerted policies for the sustainable development of the Amazon. In Spanish: "Reafirman la relevancia del Tratado de Cooperación Amazónica como principal instrumento para llevar a efecto políticas concertadas para el desarrollo sostenible de lo territorios amazónicos de los países del TCA."

212 2000 Declaration of Caracas, adopted at the sixth MMFA, held in Caracas, Venezuela, April 6, 2000, BJ, pp. 117–21, at p. 117.

213 2002 Santa Cruz de la Sierra Declaration, adopted at the Seventh MMFA, held in Santa Cruz de la Sierra, Bolivia, November 22, 2002, BJ, pp. 163–65. On this occasion, see also the statement by the minister of Venezuela (Roy Chaderton Matos), BJ, pp. 133–35: "Surge así el Tratado de Cooperación Amazónica como un instrumento idóneo para la cooperación multilateral internacional en el esfuerzo de preservar y desarrollar sustentablemente la Cuenca Amazónica…."

214 Declaration of the Eighth MMFA, held in Santa Cruz de la Sierra, Bolivia, September 14, 2004, available at: http://www.otca.org.br

215 Declaration of the Heads of States of the Amazon Cooperation Treaty Organization, op. cit.

Could legal changes in the framework of the ACT bring about this support? What are the obstacles in designing a more effective treaty?

Before answering those questions, one may ask whether the ACT should be strengthened at all. Should this treaty be revitalized, or would it be better to start with a new treaty altogether? Is regional cooperation the best way of dealing with environmental problems in the Amazon, or would it be better to opt for a global legal regime?

Cooperation is needed at different levels for the protection of the Amazon. The ACT is one possible way of tackling common problems in the Amazon and should be strengthened for various reasons. The international legal system dealing with the Amazon involves norms and institutions in two dimensions: regional and global. At the regional level, the ACT is the only treaty to involve all Amazon States with the specific objective of promoting the conservation and rational use of natural resources. Therefore, it is an ideal forum where all basin States can negotiate, adopt, implement, and enforce norms and policies applicable basin-wide. The ACT has attempted to provide this forum, for example, in hosting negotiations on common criteria and indicators for the management of Amazonian forests (under the Tarapoto Process)[216] or on norms related to commercial navigation[217] and pollution of the Amazonian rivers.[218]

As the only international treaty involving all Amazon States with the purpose of protecting the Amazonian environment, the ACT could be an instrument of cohesion among the different instruments and institutions at all levels in matters related to the Amazon. Due to its unique mandate and position *vis-à-vis* other legal regimes and institutions, the ACT should be strengthened to help in attaining further cohesion and effectiveness in the efforts to protect the Amazon. However, this requires time, additional (financial and human) resources, and strong political support – as would be required in the negotiation of a new treaty, either of a regional or global scope. Nevertheless, unlike a new treaty, the ACT has the advantage of having a diplomatic history, institutional bodies created over the years, and now a permanent

[216] At the First Regional Meeting on the Proposal on Criteria and Sustainability Indicators for Amazon Forest Sustainability held in Tarapoto, Peru in 1995, the Amazon countries discussed possible criteria and indicators for managing Amazonian forests with a view to harmonizing forest and environmental policies. This meeting agreed on twelve criterions. The latest agreed draft adopted at the Second Regional Workshop held in Tarapoto, Peru in 2001 included indicators such as measures for the *in situ* conservation of endangered species and the conservation of genetic resources. This meeting launched the so-called Tarapoto Process under which these criteria will be discussed and validated by the ACT member States. Through Resolution RES/VII MRE–TCA/7, adopted at the Eighth MMFA held in Santa Cruz de la Sierra, Bolivia on November 22, 2002, the Tarapoto Process formally started.

[217] Third Meeting of the ACC, held in Brasília on March 16–18, 1988, BJ, 2002, p. 185. The proposal of a regulation for free commercial navigation in the international Amazon rivers was contained in the document TCA/III CCA/SPT BR/W.P. 1/88.

[218] RES/V MRE-TCA/7, BJ, 2002, pp. 93–94.

structure and financial mechanism. Therefore, it might be easier to build a more effective legal regime from this existing structure.

On the other hand, it is unlikely that the Amazon States would welcome a global treaty dealing with the Amazon, as it would probably represent a threat to their national sovereignty. The need to ensure sovereignty over the Amazon was often invoked during the ACT negotiations, and also the issue of whether a regional treaty would in any way affect the parties' sovereign rights. The Amazon States also made it clear that the ACT should not in any way supersede their national institutions and should only address matters of common interest, without interfering in any way in their domestic affairs.[219] This concern was ultimately reflected in Article 4 of the ACT, which states that the exclusive use of natural resources is a right inherent to the sovereignty of each State. In any case, a regional treaty is by no means the only answer to the problems of the Amazon; rather, it complements the efforts taken at the national and global levels. A regional treaty is not suitable for handling all types of problems either, but only those concerning two or more States. For example, problems related to land ownership and rights can hardly be dealt with at a regional level.

Despite the unique position of the ACT in the international legal protection of the Amazon, this treaty may eventually not have enough support to become an effective instrument. One may doubt the ability or willingness of the ACTO member States to make it an agile and effective regime, for instance, in view of the fact that its bodies have not met regularly or that it took the parties two long years to elect the current secretary-general (Manuel Ernesto Picasso Botto),[220] after the first one (Rosalia Arteaga) left ACTO in 2007. Furthermore, although the ACTO member States have diligently complied with their financial obligations *vis-à-vis* this organization, which is a sign of interest in the treaty, the resources allocated to ACTO are still fairly limited.[221]

Certain factors may make it more difficult for the ACT parties to achieve further regional cooperation. For example, Brazil has a dominant position *vis-à-vis* the other countries as a stronger economy, but also because it owns most of the Amazon and may be less willing to look for cooperation with its neighbors. In practice, Brazil and the other Amazon countries are very reliant on the notion of national sovereignty when assessing laws and policies related to the Amazon. Also, it seems that

[219] Speech of the Brazilian minister of foreign affairs (Antonio Azeredo da Silveira) at the opening session of the First Preparatory Meeting of the ACT, Brazilian Ministry of Foreign Affairs, *Resenha de Política Exterior do Brasil*, vol. 18, 1978, pp. 11–13.

[220] "Embajador Manuel Picasso es elegido al cargo de Secretario General de la OTCA," Electronic Bulletin no. 6/13, January 2010, http://www.otca.org.br

[221] Between 2003, when the Permanent Secretariat was established in Brasília, and 2008, the member countries paid 92 percent of their corresponding annual contributions (ACTO, *Management Report*, op. cit., p. 50).

the Amazon States have not developed a strong state practice in consulting one another before taking decisions with potential transboundary effects, as observed in the case of the construction of the Santo Antônio and Jirau hydropower plants on the Madeira River (one of the major tributaries of the Amazon River), where Brazil failed to hold proper consultations with Bolivia. Furthermore, Brazil has traditionally had fewer interactions with the Andean-Amazonian countries (Bolivia, Colombia, Ecuador, Peru, and Venezuela) due to historical, cultural, and linguistic reasons. This seems also to be the case for Guyana and Suriname *vis-à-vis* their Amazonian neighbors. For example, Guyana and Suriname have never hosted ACT Pro Tempore Secretariats.

Despite all this, the Amazon States continue to affirm their willingness to entrust ACTO with an prominent role as a cooperation forum, as noted at the 2009 Declaration of the Heads of the States of the ACTO member countries, which could indeed lead to a stronger regime. As noted earlier, the creation of the Permanent Secretariat and a compulsory financial mechanism also confers more stability and visibility on this treaty regime. In this context, can legal changes bring about further political support? It is probably the other way round, with effective legal changes depending on the political support of the member States, as they involve negotiations and the allocation of financial and human resources.

In short, the ACT parties have taken steps to strengthen this legal regime. Is this enough, or will it never be enough to handle problems in the Amazon, even if a stronger normative and institutional framework is built? Can one treaty alone solve complex and long-lasting problems, such as deforestation? Probably not. Despite its unique position, the ACT (even if strengthened) cannot single-handedly tackle the environmental problems of the Amazon. As noted earlier, it can be a factor for cohesion among the many efforts taken in that direction and can complement actions at the national and global levels.

CONCLUSIONS

As a framework agreement, the 1978 ACT contains general guidelines on ways in which its main objectives should be achieved. Despite some progress observed over the last few years, some normative and institutional weaknesses within this treaty regime still need to be addressed to allow the strengthening of the treaty.

With respect to the ACT normative framework, it can be observed that, throughout the years of existence of this treaty, the parties have not managed to give more precision to the general guidelines contained in its original text through the adoption of "operational agreements and understandings, as well as the pertinent legal instruments," as foreseen in Article 1. So far, there has been only one amendment to the ACT, which modified Article 22 by replacing the Pro Tempore Secretariat with the Permanent Secretariat. No other protocols or additional treaties have been concluded to specify the commitments under this treaty. Subsequent declarations adopted by the ACT's institutional bodies have introduced new concepts or principles, for example the conservation of biodiversity or forests, and the regulations on the functioning of the ACT organs have brought more clarity to their mandates and specific tasks. The resolutions taken by the MMFA and the ACC, however, have mostly addressed administrative or institutional matters, without giving further guidance as to how the general objectives of the ACT should be achieved. Some attempts to create substantive obligations in specific areas, through, for example, agreements on commercial navigation of the Amazon basin and the prevention of water contamination have not prospered. Other initiatives, such as the Amazon Charter, albeit relevant for putting some of the principles of the ACT into modern legal and factual perspective, are unlikely to entail further substantive obligations.

As far as legal obligations are concerned, the ACT embodies a general duty to cooperate that is manifested in various ways. For instance, States are encouraged to take joint studies and exchange information or coordinate health services, but the treaty has not developed specific obligations. This can be particularly observed in areas on which the treaty is silent and in which no significant progress has occurred, for example, in relation to forests, the prevention of transboundary harm and pollution, or the protection of indigenous populations. In addition, the procedural obligations or mechanisms devised for ensuring the fulfillment of the ACT's objectives involve no more than the exchange and dissemination of information. The obligation to submit national reports on conservation measures taken by each of the parties in the management of species of fauna and flora, envisaged in Article 7(b), has never been met, as can be observed in the practice of the ACT parties.

Despite the progress made in the institutional strengthening of the ACT, this treaty has not been very successful in achieving its objectives. During the last few years, it has been faced not only with the normative deficiencies outlined above, but also institutional inadequacies. Probably the major institutional weakness of the ACT lies in the fact that its executive organs (i.e. the PNCs and SCs, charged with ensuring compliance with the ACT objectives) have not been operative or have functioned on a very limited scale. Therefore, even if the ACT parties had come up with more specific substantive obligations through additional instruments, those organs would probably not have been able to put them into effect. The need to strengthen the ACT's institutional bodies was acknowledged at the Ninth MMFA in 2005, which requested the Permanent Secretariat to review its regulations (RES/IX MRE-OTCA/12) and prepare a program for the strengthening of the PNCs (RES/IX MRE-OTCA/04).

Moreover, the Pro Tempore Secretariats, which existed until the establishment of the Permanent Secretariat, could not ensure the continuity of the activities undertaken under the ACT or become a forum where long-term common policies could be negotiated and jointly implemented. It should also be noted that the Pro Tempore Secretariats operated under severe financial restrictions, depending almost exclusively on international funding, and that therefore most activities under the ACT failed to progress beyond the formulation stage. The regular meetings of the ACT institutional bodies, notably the MMFA and ACC, have occurred in a discontinuous way, which was in part due to the absence of a Permanent Secretariat. As a result, the adoption of norms and policies that could be negotiated in the framework of those meetings was postponed.

The legal and institutional weaknesses of the ACT, however, have not prevented the strengthening of the treaty in some respects. The establishment of a Permanent Secretariat with the creation of ACTO and of a financial mechanism involving regular national contributions by the member States is evidence of the willingness of the Amazon States to reinforce the ACT as an instrument for the conservation and management of the Amazon basin. Also, the Amazon States have agreed to an eight-year working plan to guide their future common actions. Since its establishment, ACTO has succeeded in creating partnerships with other international organizations and leveraging funds for implementing projects, and has been a forum where the parties agreed on common positions, such as in the context of the UNFF negotiations. The role of "coordinating" the interests and positions of those countries in different negotiation forums is one of the greatest potential strengths of ACTO. As a matter of fact, under resolution RES/IX MRE-OTCA/05 adopted at the Ninth MMFA in 2005, the parties requested the Permanent Secretariat to continue to play a "facilitator" role. A permanent institutional structure has also facilitated the flow of information among the parties, particularly in the framework of the coordination commission of the ACC, which has been an effective communication channel between the Permanent Secretariat and the ACT member States. The Permanent Secretariat has also attempted to bring transparency to its activities and ensure better communication among the parties and institutional bodies. According to the latest declaration signed by the heads of States of the ACTO on November 26, 2009, on the occasion of preparatory meeting for the United Nations Conference on Climate Change held in that year, the parties instructed the ACTO Permanent Secretariat to follow up international negotiations on issues of relevance to Amazonian cooperation, such as climate change, biological diversity, and forests.[222]

[222] Declaration of Heads of States of the ACTO, op. cit.

ACTO faces many challenges and there is anxiety as to whether this organization will ever become a major vehicle for the fulfillment of the objectives of the ACT, in particular one capable of achieving the conservation and sustainable use of the Amazonian natural resources. Some difficulties still faced by the Permanent Secretariat relate to the fact that the national financial contributions, which have been timely met by the parties, are only enough to cover the Secretariat's running costs, but not other projects and initiatives. This organization is still strongly dependent on international financial assistance. Other challenges associated with tackling the normative and institutional weaknesses of the ACT are still to be addressed: Having examined the role of the ACT in regional cooperation, the next chapter looks at the institutional framework in the Amazon involving some other regional and subregional organizations.

5

Regional and Subregional Organizations

The Amazon States participate in different regional and subregional organizations. Although they are all members of the 1978 Amazon Cooperation Treaty (ACT), they are also members of other regional or subregional organizations such as the Organization of American States (OAS), the Andean Community (CAN), and the Caribbean Community (Caricom). For example, although Brazil (and probably soon Venezuela) is part of the Common Market of the South (Mercosur), Guyana, and Suriname participate in Caricom, whereas the Andean countries (Bolivia, Colombia, Ecuador, and Peru) are members of CAN.

Over the last years, these organizations have developed their normative and institutional frameworks by adopting legal instruments, such as treaties, decisions, resolutions, and plans of action, and by creating permanent bodies, some of which are competent to adopt binding decisions that can be directly applicable in the member States' national legal systems. Virtually all of those organizations have developed bodies of law on different aspects of environmental protection, for example, climate change, forests, and freshwater. The legal instruments developed by such institutions are applicable in the Amazon States' territories, including their respective Amazonian territories, and are, therefore, part of the international legal system of the Amazon.

In addition to the regional and subregional organizations in which the Amazon States participate, other actors at the national level play a crucial role in the protection of the Amazon. For example, in Brazil the National Institute for Space Research within the Ministry of Science and Technology has monitored the deforestation of the Amazon using modern satellite technology since 1988. Apart from the ministries of the environment and related government agencies, other national stakeholders such as nongovernmental organizations (NGOs) and indigenous communities also have a prominent role in the Amazon. Indigenous communities are holders of ancestral knowledge about their natural environment, which can be used to inform conservation policies. Since 1984, national indigenous organizations in the Amazon have been gathered under a regional institution, the Coordinator of Indigenous Organizations of the Amazon Basin. In general, indigenous communities are granted certain rights to allow for the better protection of their traditional lands, usually defined under national law, such as the right to prior informed consent regarding measures that may affect their environment[1] and access to justice.[2] In countries such as Brazil and Colombia, the area designated as indigenous land is substantial. Due to their size and protected status they play a role in the protection of Amazonian ecosystems. For example, these (and other conservational areas) act as a barrier to deforestation along the arc of deforestation in the Brazilian Amazon.[3] International instruments to which the Amazon States are parties – such as the ILO Convention (No. 169) concerning Indigenous and Tribal Peoples in Independent Countries[4] or the 2007 United Nations Declaration on the Rights of Indigenous Peoples[5] – call for the participation of indigenous peoples in decisions affecting their rights. At the regional level, the jurisprudence of the Inter-American Court of Human Rights has also supported the view that indigenous peoples shall benefit from special protection.[6] NGOs, on the other hand, have produced knowledge and helped monitor the state of the Amazonian environment. For example, they have participated in the formulation of "positive agendas" to guide development and conservation policies in each state of the Brazilian Amazon.

Regional and subregional organizations to which the Amazon States are parties have also encouraged the participation of indigenous communities and NGOs in decision-making processes, such as the Andean Community Decision N. 524 (Article 1)[7] and the 2001 Mercosur Framework Agreement on the Environment (Article 6). Both actors may also have standing before subregional and regional tribunals, for example the Inter-American Court of Human Rights and the Court of

1 Article 84 (4), 1998 Constitution of Ecuador.

2 Article 232, 1988 Brazilian Constitution, granting indigenous communities standing to defend their rights and interests (the Public Prosecution intervening in all the procedural acts on their behalf).

3 P. Verweij et al., *Keeping the Amazon Forests Standing: A Matter of Values*, report commissioned by WWF Netherlands, 2009, p. 33.

4 Pursuant to Article 6 (1) of the ILO Convention (No. 169) concerning Indigenous and Tribal Peoples in Independent Countries, governments shall consult indigenous peoples through their representative institutions with regards to measures that may affect them and ensure their free participation in elective institutions.

5 According to Article 41 of the 2007 United Nations Declaration on the Rights of Indigenous Peoples, States shall provide ways and means for ensuring participation of indigenous peoples on matters that may affect their rights (Article 18), through representatives chosen by them in accordance with their own procedures. In particular, States are to obtain prior informed consent from those communities, before taking measures that may affect their lands (Articles 19 and Article 32 (2)).

6 Inter-American Court of Human Rights, case of the *Yakye Axa Indigenous Community v. Paraguay*, op cit., para. 54

7 The CAN Decision 524 creates a Working Group on the Rights of Indigenous Peoples to promote the active participation of indigenous peoples in the subregional integration process.

Justice of the Cartagena Agreement.[8] In the context of the ACT, the participation of NGOs and indigenous communities is still restricted. According to the Amazon Cooperation Treaty Organization (ACTO) regulations, international NGOs can participate in regular meetings with observer status[9] – a right that is not expressly recognized to indigenous communities.[10]

This chapter focuses on regional and subregional institutions relevant to the international legal protection of the Amazon. Due to their development, the institutional setting in the Amazon region is currently more complex than the one existing at the time of the adoption of the ACT in 1978. These institutions are now dealing with a broader array of issues, in particular the environment. For example, both CAN law and the ACT are applicable in the territories of Bolivia, Colombia, Ecuador, and Peru, which are parties to both the 1969 Andean Subregional Integration Agreement and the ACT. As a result, conflicts of law are more likely to arise among these different legal regimes. In practice, as these institutions operate in overlapping geographical areas, they may also compete for funding and duplicate efforts by working on similar projects in the same area. In order to show the greater complexity of this institutional arrangement, this chapter presents an overview of the major regional and subregional institutions in which two or more Amazon States participate, whose interactions will be further discussed in the next chapter.

5.1. ORGANIZATION OF AMERICAN STATES

The Organization of American States was formally created with the adoption of the 1948 Charter of the Organization of American States,[11] but it in fact resulted from a

[8] According to Article 44 of the 1969 American Convention on Human Rights, any person or group of persons, or nongovernmental entities legally recognized can submit petitions containing denunciations or complaints of violation of that treaty by any member State. Likewise, under the framework of the CAN noncompliance procedure, legal persons can present claims to the General Secretariat or, in some cases, to the Court of Justice of the Cartagena Agreement (Articles 24 and 37 of the 1979 Treaty Creating the Court of Justice of the Andean Community). Legal persons may include indigenous communities' representative institutions and NGOs. Under Mercosur, natural or legal persons can submit cases to its Arbitral Tribunal (Article 25, of the 1991 Protocol of Brasilia on dispute settlement), which may include entities representing indigenous communities or NGOs. Differently, under Caricom, in principle only disputes between States can be submitted to the Caribbean Court of Justice. Private entities may also have standing with special authorization (Article 24, 2001 Agreement Establishing the Caribbean Court of Justice).

[9] Article 6, MMFA Regulation, and Article 8 ACC Regulation, Pro Tempore Secretariat, Base Jurídica del Tratado de Cooperación Amazónica: Antecedentes Constitutivos de la Organización del Tratado de Cooperación Amazónica, La Paz, Bolivia, 2002.

[10] The list of institutions under Article 6 of the MMFA and Article 8 of the ACC do no include indigenous communities.

[11] 119 UNTS 3; entered into force on December 13, 1951.

series of conferences started in the nineteenth century, following the independence of most Latin American States.[12] The purpose of this organization, as enunciated in Article 1 of the Charter, is to promote solidarity and strengthen collaboration among the American States, as well as to defend their sovereignty, territorial integrity, and independence. The principles on which OAS is founded are embodied in Article 3 of the Charter, which privileges international law as the standard of conduct in the relations among member States and posits an international order based on respect for their personality, sovereignty, and independence, without external interference in each member State's political, economic, and social systems. This organization now comprises a large group of States, among which are the eight Amazon countries.[13]

OAS's supreme organ is the General Assembly, which holds annual regular sessions that issue resolutions, declarations, and recommendations, and adopt treaties.[14] Any OAS member State and organ can submit for the consideration of the General Assembly a draft treaty relating to any topic.[15] A Meeting of Consultation of Ministers of Foreign Affairs is held in order to consider problems of an urgent nature and of common interest to the American States. Any member State may request that a Meeting of Consultation be called. A Permanent Council, composed of one representative of each member State with the rank of ambassador, appointed by their governments, carries out decisions of the General Assembly and of the Meeting of Consultation of Ministers of Foreign Affairs, if their implementation has not been assigned to any other body.[16] Other organs include the General Secretariat; councils such as the Inter-American Council for Integral Development, as well as committees, specialized conferences and organizations, listed in Part II of the OAS

[12] The Congress of Panama was convened from June 22 to July 15, 1826 at the call of Simón Bolívar, "the liberator" of many Latin American States, who proposed creating a league of American republics with a common military, a mutual defense pact, and a supranational parliamentary assembly. In his circular of invitation, Bolívar made reference to a federation of the American republics (S.E. Baldwin, "The international congresses and conferences of the last century as forces working toward the solidarity of the world," *American Journal of International Law*, vol. 1, 1907, pp. 565–78, at p. 567). Following that meeting, the First Pan American Conference was held in 1889–1890 in Washington, DC, followed by a series of Pan American Conferences (Editorial comment, "The arbitral award in the Peru-Bolivia boundary controversy," *American Journal of International Law*, vol. 3, 1909, pp. 949–53, at p. 950).

[13] OAS member States are: Antigua and Barbuda, Argentina, the Bahamas, Barbados, Belize, Bolivia, Brazil, Canada, Chile, Colombia, Costa Rica, Dominica, the Dominican Republic, Ecuador, El Salvador, Grenada, Guatemala, Guyana, Haiti, Honduras, Jamaica, Mexico, Nicaragua, Panama, Paraguay, Peru, Saint Kitts and Nevis, Saint Lucia, Saint Vincent and the Grenadines, Suriname, Trinidad and Tobago, the United States of America, Uruguay, and Venezuela (http://www.oas.org/documents/eng/memberstates.asp).

[14] OAS, Amendments to the Rules of Procedure of the General Assembly, Resolution, adopted at the First Plenary Session, held on June 5, 2000, OEA/Ser.P AG/RES. 1737 (XXX-O/00).

[15] Ibid.

[16] OAS, Charter of the Organization of American States, 1948, Articles 80–92, p. 119 UNTS 3.

Charter. The OAS structure also encompasses autonomous institutions, such as the Inter-American Commission and Court of Human Rights.[17]

After the 1992 UN Conference on Environment and Development (UNCED), OAS put more emphasis on environmental protection, and created in 1996 the Inter-American Committee on Sustainable Development[18] as a subsidiary body of the Inter-American Council for Integral Development, which is responsible for formulating OAS' sustainable development policy. Two regional agreements adopted under the aegis of OAS relate to environmental protection, namely the 1940 Convention on Nature Protection and Wildlife Preservation in the Western Hemisphere,[19] and the 1976 Convention on the Protection of the Archeological, Historical and Artistic Heritage of American Nations.[20] This latter treaty aimed to set up an inter-American registry to protect cultural property of special value, but also fauna and flora related to such heritage.

Some of the projects OAS carries out, particularly in the area of biodiversity, include the Inter-American Biodiversity Information Network (IABIN)[21] and Andes Amazon Protected Areas Database (AAPAD), aimed at the standardization of national and subnational databases on species and specimens, ecosystems, and protected and pollination areas. The AAPAD (involving the eight Amazon States) created an information system to maintain an Internet database containing information on the situation of protected areas in the eight countries that make up the Andean-Amazon Basin region.[22] Also, under the Department of Sustainable Development, the program entitled Sustainable Development and Bio-Cultural Conservation in the Brazil-Suriname Border Region[23] aims at promoting conservation and

[17] The Inter-American System comprises the OAS, the Inter-American Development Bank (IDB), and other subregional organizations, including CAN and the Andean Development Corporation, the Inter-American Court of Human Rights, Mercosur, and the Summits of the Americas Process. A list of institutions is available at: http://www.oas.org/OASpage/Inter-American_System.asp

[18] OAS, AG/RES. 1440 (XXVI-O/96), Sustainable Development, adopted at the Eighth Plenary Session held on June 7, 1996.

[19] 161 UNTS 193. This treaty was signed on October 12, 1940 under the auspices of the Pan American Union and entered into force on May 1, 1942. The objective was to promote *in situ* conservation of native flora and fauna, including migratory birds, and regions of aesthetic, historic, and scientific value. The Pan American Union is the former name for the General Secretariat of the OAS. It was founded (1889–1890) at the first Inter-American Conference as the Commercial Bureau of the American Republics and changed its name to the International Bureau of the American Republics in 1902.

[20] 15 ILM 1350.

[21] Information is available at the OAS Web site: http://www.oas.org/dsd/Bio-Proj-Sum.htm

[22] Permanent Council of the Organization of American States, Committee on Hemispheric Security, OEA/Ser.G, CP/CSH-951/08 corr. 1, February 25, 2008.

[23] The Amazon Conservation Team (Brazil and Suriname Offices) is the coordinating institution, funded by the OAS Inter-American Council for Integral Development, http://www.oas.org/dsd/Working%20 Documents/Lan-Man-su-brproject.htm

sustainable development (primarily of nontimber forest products), and improving the healthcare and security of the indigenous communities in the Brazil-Suriname border region. In partnership with the UN Environment Program (UNEP) and ACTO, OAS undertakes a multiyear project on water management in the Amazon financed by the Global Environment Facility involving all Amazon countries.[24]

5.2. ANDEAN COMMUNITY

The Andean Community (CAN) is the subregional organization that has probably developed its normative and institutional frameworks the most during the past years. Decisions taken by its political organs are binding and directly applicable in the member States' national legal systems. In the environmental field, CAN has approved a number of decisions and created comprehensive subregional regimes in areas such as biodiversity, agriculture, industrial, and intellectual property rights. A number of institutions endowed with legislative, executive, and supervisory powers have also been set up since the adoption of the 1969 Subregional Integration Agreement (the Andean Pact or Cartagena Agreement),[25] originally signed by Bolivia, Colombia, Chile, Ecuador, and Peru with the objective of creating a common market among the Andean countries. Venezuela acceded to this agreement in February 1973, whereas Chile withdrew from it in October 1976. In 1996, the Protocol of Trujillo[26] modified the Cartagena Agreement to create CAN and the Andean Integration System. CAN is now composed of Bolivia, Colombia, Ecuador, and Peru (also Amazon States), as Venezuela withdrew from the organization in September 2006. Chile rejoined CAN in April 2006 as an associate member country (other associate countries include Argentina, Brazil, Chile, Paraguay, and Uruguay).[27]

A number of bodies and institutions make up the Andean Integration System, which is now composed of CAN's executive, legislative, and adjudicatory bodies; various councils; and research institutions created under the framework of the Andean

[24] This program, entitled "Integrated and Sustainable Management of the Trans-boundary Water Resources of the Amazon River Basin Considering Climate Variability and Change" aims to strengthen the institutional framework for the coordinated planning and execution of activities for the protection and sustainable management of land and water resources of the Amazon Basin. It was launched in 2005 with the support of the Global Environmental Facility (GEF), UNEP, and of the OAS. ACTO is the regional entity responsible for its management (ACTO, *Management Report 2007–2008*, Brasília: ACTO Permanent Secretariat, 2008, p. 45).

[25] 8 ILM 910.

[26] Protocol Modifying the Subregional Andean Integration Agreement, http://www.comunidadandina.org/normativa/tratprot/trujillo.htm

[27] Decision 645, September 20, 2006, *Otorgamiento de la condición de País Miembro Asociado de la Comunidad Andina a la República de Chile*, grants to Chile the status of associate member country (http://www.comunidadandina.org/normativa/dec/D645.htm). Mercosur member States are granted associate members status by Decision 613, July 7, 2005, http://www.comunidadandina.org/exterior/can_mercosur.htm

subregional integration process.[28] The major permanent institutional bodies comprise the Andean Presidential Council, which is the highest body in charge of establishing CAN's guidelines and policies, the Andean Council of Foreign Ministers,[29] and the Andean Community Commission.[30]

The Andean Community Parliament, which represents the nations of the member States, is the deliberating body under CAN. It was created in 1979 by a Treaty Constituting the Andean Parliament, signed by Bolivia, Colombia, Ecuador, Peru, and Venezuela, based in Bogota (Colombia).[31] A few years later, the 1997 Additional Protocol to the Treaty Establishing the Andean Parliament[32] replaced that treaty (Article 20) and envisaged a procedure for direct and universal vote for CAN parliamentary representatives.[33] The Andean Parliament provides advice to CAN member States and is now developing the required procedure for direct elections of its representatives.[34] A judicial body was set up by the 1979 Treaty Creating the Court of Justice of the Andean Community (TCCJ),[35] which is competent to declare the nullity of decisions, resolutions, and agreements adopted by the member States, to provide authoritative interpretations of CAN law and declare noncompliance with CAN legal obligations. The jurisdiction of the Court of Justice of the Andean Community (CJAC) is binding on CAN member States.[36]

28 According to Article 6 of the Cartagena Agreement, this system is comprised of the following bodies and institutions: the Andean Presidential Council, Andean Council of Foreign Affairs, Andean Community Commission, Andean Community General Secretariat, Andean Community Court of Justice, Andean Parliament, Business Consultative Council, Labor Consultative Council, Andean Development Corporation, Latin American Reserve Fund, Simón Rodríguez Agreement, Andean Health Organization, Simón Bolívar Andean University, and "all other bodies and institutions that are created within the framework of the Andean subregional integration."

29 This body comprises the ministers of foreign affairs of the Cartagena Agreement's member States and holds ordinary meetings twice a year (Articles 15–20 of the Cartagena Agreement).

30 Constituted by a plenipotentiary representative from each member country, this organ meets regularly three times a year (Articles 24–28 of the Cartagena Agreement).

31 The text of the 1979 Treaty Constituting the Andean Parliament is available at the Andean Parliament Web site: http://www.congreso.gob.pe/parlamento-andino

32 Sucre Protocol Establishing the Andean Parliament of June 25, 1997, http://www.comunidadandina.org/ingles/treaties.htm

33 1997 Additional Protocol to the Treaty Establishing the Andean Parliament http://www.comunidadandina.org/INGLES/normativa/ande_trie5.htm, Article 2. Direct and universal elections was foreseen under the Trujillo Act of 1996 (Article 42), http://www.comunidadandina.org/normativa/tratprot/trujillo.htm

34 http://www.congreso.gob.pe/parlamento-andino

35 Treaty Creating the Court of Justice of the Andean Community, 18 ILM 1203, amended by the Cochabamba Protocol of May 28, 1996, http://www.comunidadandina.org/ingles/normativa/ande_trie2.htm

36 In the case *Acción de incumplimiento interpuesta por la Secretaría General de la Comunidad Andina contra la República del Ecuador*: "Los Países Miembros no pueden dejar de reconocer que la jurisdicción del Tribunal Andino es obligatoria, ya que no se ha dejado al arbitrio de aquellos someter o no

Decisions, resolutions, and protocols adopted by CAN institutional bodies constitute the CAN legal system. According to Article 1 of the 1979 TCCJ (amended by the 1996 Cochabamba Protocol), such a system consists of this treaty, the Cartagena Agreement (including protocols and additional instruments), decisions of the Andean Council of Foreign Ministers and the Commission, resolutions of the general secretariat, and any other agreements adopted by the member States among themselves in the context of the Andean subregional integration process. Under the Andean Integration System, the Andean Council of Foreign Ministers and Andean Community Commission are the two organs endowed with lawmaking powers, whereas the general secretariat and member States have "initiating legislative powers," which means that they can submit legislative proposals to those two bodies.[37] The decisions adopted by the Andean Council of Foreign Affairs and the Commission are binding on member States.[38] Both the decisions of these organs and resolutions adopted by the General Secretariat are directly applicable in the territories of the member States, without requiring an express act for incorporation into domestic law.[39]

The binding nature and direct applicability of CAN law derives from treaty, provided respectively in Articles 2 and 3 of the 1996 TCCJ, and from the CJAC's jurisprudence.[40] On the direct applicability of the decisions by the Andean Council of Foreign Ministers or the Commission, the CJAC takes the view that no domestic legislative procedure is required to give effect to those decisions, because requiring such internal procedures would deny the very existence of CAN law.[41] The norms adopted under the aegis of CAN, according to the CJAC, have direct effect in the

sus controversias al Juez Comunitario sino que, según disposición expresa contemplada en el artículo 2 del Estatuto del Tribunal, éste es un 'Órgano Jurisdiccional instituido para asegurar el respeto al derecho en la aplicación e interpretación del ordenamiento jurídico del Acuerdo'" (PROCESO 2-AI-97, *Cartagena Agreement Official Gazette*, no. 391, December 11, 1998).

37 Cartagena Agreement, Article 27.

38 Decisions become binding as of the date they are approved by the Andean Council of Foreign Ministers or the Commission of the Andean Community, according to TCCJ, Article 2.

39 TCCJ, Article 3. According to this Article, decisions are directly applicable as of the date they are published in the *Cartagena Agreement Official Gazette*, unless they indicate another date. Only when their text so stipulates, "decisions must be incorporated into national law through an express act stipulating the date they will enter into effect in each Member Country."

40 On the direct applicability of CAN's decisions, see PROCESOS 1-IP-87, 6-IP-94, 10-IP-94, and 1-AI-96, published in the *Cartagena Agreement Official Gazette*, no. 28 of February 15, 1988, no. 170 of January 23, 1995, no. 177 of April 20, 1995, and no. 234 of November 21, 1996, respectively.

41 In the case *Acción de incumplimiento interpuesta por la Junta del Acuerdo de Cartagena contra la República de Venezuela*: "La sola suposición de que las Decisiones de la Comisión o las Resoluciones de la Junta, tuvieran que pasar por el tamiz legislativo de cada uno de los Países Miembros, antes de su aplicación interna, habría conducido a negar la existencia de un derecho comunitario andino" (PROCESO 3-AI-96, *Cartagena Agreement Official Gazette*, no. 261, April 29, 1997).

territories of its member States and generate rights and obligations, just as any norm issued within their domestic legal systems.[42]

When CAN member States fail to comply with obligations under CAN law, the General Secretariat can activate a noncompliance procedure, foreseen in the TCCJ (chapter 1 on "Actions of Nullification or Noncompliance"). This possibility of monitoring noncompliance puts CAN one step ahead of other subregional organizations in terms of reviewing member States' performance. In cases where the defaulting State continues with the behavior that motivated the noncompliance procedure, the General Secretariat can require, as a last recourse, a decision by the CJAC.[43] Interestingly, any member State, as well as natural or legal persons, can also present a claim of noncompliance to the General Secretariat, and even appeal directly to the CJAC, if the General Secretariat fails to do so in due time.[44] A verdict of noncompliance by the CJAC is to be executed within a ninety-day period, otherwise the CJAC may establish the limits within which the claimant country or any other state may restrict or suspend the benefits obtained by the party at fault under the Cartagena Agreement.[45] In cases involving natural and legal persons, the noncompliance decision issued by the Court constitutes sufficient grounds for the claimant to appeal to the competent national court to seek compensation for damages.[46]

In examining cases of noncompliance (*acción de incumplimiento*), the CJAC has given a wide interpretation of what compliance means. According to the Court, compliance with CAN law goes beyond the statement by a competent national authority that a country is observing the law, or by its introduction in a State's national registry or *Official Gazette*. Rather, compliance requires certain behavior or acts that give effect to what is prescribed by law or measures avoiding the production of a situation forbidden by law.[47] In addition, according to CJAC jurisprudence, a member

[42] In the case *Acción de incumplimiento interpuesta por la Junta del Acuerdo de Cartagena contra la República de Venezuela*: the legal effects of CAN law are discussed: effects "generan derechos y obligaciones para los particulares al igual que ocurre en las normas de los ordenamientos estatales, permitiendo la posibilidad de que aquellos puedan exigir directamente su observancia ante sus respectivos tribunales" (PROCESO 3-AI-96, op. cit.).

[43] TCCJ, Article 23.

[44] Within sixty days after the date of the General Secretariat's administrative ruling, according to TCCJ, Article 24.

[45] Article 27. In cases where the noncessation is likely to cause irreparable damage or damage difficult to repair, before handling down its final judgment the court may order the temporary suspension of the alleged violating measure, as stated in Article 28 of this treaty.

[46] TCCJ, Article 30.

[47] In the case *Acción de Incumplimiento interpuesta por la Secretaría General de la Comunidad Andina en contra de la República de Bolivia*: "Es igualmente necesario a los efectos de la resolución del caso planteado advertir que el cumplimiento de las obligaciones antes descritas no se satisface meramente por la sola manifestación hecha por la correspondiente autoridad nacional de que se está acatando lo dispuesto en el ordenamiento comunitario, como cuando se inserta la disposición comunitaria en el Registro o Diario Oficial, por ejemplo, y se alega que con ello lo ordenado por ésta está siendo

State cannot allege noncompliance by another State to justify its failure to comply with its own obligations. CAN law has direct effect in the territories of the member States and should be observed independently of the behavior of the other members; therefore, compliance with CAN law is an obligation owed by each State *vis-à-vis* the organization.[48]

The CJAC can review the performance of member States, but also that of CAN institutional bodies. In such cases, member States, as well as natural or legal persons whose rights have been affected, can appeal directly to the Court in cases where the Andean Council of Foreign Ministers, the Andean Community Commission, or the General Secretariat itself abstain from carrying out actions of which they are expressly in charge.[49] The CJAC has developed extensive jurisdiction in noncompliance actions.[50] The noncompliance procedure, according to the CJAC, is a way of ensuring the development and effectiveness of CAN law and of exercising control over member States' behavior.[51]

Decisions and resolutions comprising CAN law have addressed a variety of issues, notably environmental conservation and management, which are now a priority area for CAN. This organization has developed its own environmental agenda, the latest of which covers the years 2006–2010 focusing on biodiversity, climate change, and water resources.[52] Over the past few years, CAN has developed a body of law dealing

acatado. En sentir del Tribunal el cumplimiento, para que corresponda a lo que establece el ordenamiento jurídico aplicable, debe ser pleno y completo; debe materializarse en actos o conductas que efectivamente realicen lo que la norma comunitaria manda o determina o, en su caso, en abstenciones que eviten que se produzcan las situaciones que ella prohíbe" (PROCESO 8-AI-98, *Cartagena Agreement Official Gazette*, no. 512, November 26, 1999).

48 In the case *Acción de incumplimiento interpuesta por la Secretaría General de la Comunidad Andina, en contra de la República de Colombia*: "Es necesario recalcar que el incumplimiento que alega Colombia de sus obligaciones no debe estar vinculado al cumplimiento o no de otros Países Miembros, pues el efecto directo de las normas comunitarias no está condicionado a determinada conducta de otros miembros, sino a su compromiso contraído como Estado frente a la Comunidad Andina" (PROCESO 26-AI-2001, *Cartagena Agreement Official Gazette*, no. 736, November 27, 2001).

49 TCCJ, Article 37.

50 In terms of number of decisions, the court has mostly decided on noncompliance actions followed by interpretative, nullity, and omission actions, in that order.

51 In the case *Acción de incumplimiento interpuesta por la Secretaría General de la Comunidad Andina contra la República del Ecuador*: "El recurso de incumplimiento es una pieza clave en la construcción, desarrollo y vigencia del orden jurídico comunitario y mediante el cual se ejerce el control del comportamiento de los Estados. El sistema andino de integración presupone la existencia de un orden de derecho, un ordenamiento normativo comunitario frente al cual los Países Miembros que forman parte integrante del mismo tienen dos órdenes de obligaciones: las de hacer o sea adoptar las medidas necesarias para asegurar el cumplimiento de las normas que conforman el ordenamiento jurídico del Acuerdo de Cartagena y las de no hacer, o sea el no adoptar ni emplear medidas contrarias a su aplicación o que de algún modo las obstaculicen" (PROCESO 2-AI-97, *Cartagena Agreement Official Gazette*, no. 391, December 11, 1998).

52 CAN, General Secretariat, *Agenda Ambiental Andina*, SG/di 799/Rev. 1, September 22, 2006.

with different aspects of the subregion's environment, which is more advanced than most international environmental regulations in the region, and has also created specialized institutions dealing with environmental matters. By adopting decisions on areas such as access to genetic resources, industrial property, and agriculture and food security this organization has created comprehensive subregional regimes that set up common norms and procedures applicable in its member States.[53]

To ensure the implementation of those regulations, specific institutions are created, such as the Andean Committee on Genetic Resources, set up under Decision N. 391 on the Common Regime on Access to Genetic Resources.[54] Other environmental bodies created under CAN include the Andean Committee of Environmental Authorities,[55] comprising national environmental bodies, established in 1998 to help the General Secretariat and the Andean Community Commission formulate the community policy on the environment and ensure implementation of decisions and norms. In 2004, the Council of Ministers of the Environment and Sustainable Development[56] was created to assist the Andean Council of Foreign Ministers, the Andean Community Commission, and other bodies to define, harmonize, and coordinate community policies on the environment and sustainable development.

5.3. CARIBBEAN COMMUNITY

The Caribbean Community (Caricom), including the Caribbean Common Market, was set up by the 1973 Treaty Establishing the Caribbean Community (Treaty of Chaguaramas).[57] In 1989, Caricom's member States converted this common market

[53] Subregional regimes have been set up, for example, by Decision no. 391 on the Common Regime for Access to Genetic Resources of July 2, 1996; Decision no. 436 of June 11, 1998 (*Norma Andina para el Registro y Control de Plaguicidas Químicos de Uso Agrícola*); or Decision no. 486 of September 14, 2000 (*Régimen Común sobre Propiedad Industrial*). The Andean countries have also come up with common guidelines, particularly with regard to their rich biodiversity, through Decision no. 523 of July 7, 2002 on a Regional Biodiversity Strategy for Tropical Andean Countries, which among other objectives, intends to help these countries achieve common positions in international negotiation forums. These decisions are available at: http://www.comunidadandina.org/INGLES/biodiversity.htm

[54] *Cartagena Agreement Official Gazette*, no. 213, July 17, 1996.

[55] Decision no. 435, Creation of the Andean Committee of Environmental Authorities, *Cartagena Agreement Official Gazette*, no. 347, June 17, 1998.

[56] Decision no. 596, Creation of the Andean Community's Council of Environmental and Sustainable Development Ministers, *Cartagena Agreement Official Gazette*, no. 1092, June 16, 2004. This council is composed of the highest authorities on environment and sustainable development in each member State and is in charge of assisting the Andean Council of Ministers of Foreign Affairs and the Andean Community Commission, as well as other institutional bodies within the Andean Integration System, with regard to environmental community matters.

[57] 946 UNTS 17. This treaty entered into force on August 1, 1973.

into a single market and economy and consequently undertook a revision of the Treaty of Chaguaramas.[58] Under this review process, which took place from 1993 to 2000, an intergovernmental task force composed of all member States produced nine protocols that were later combined to form the 2001 Revised Treaty of Chaguaramas Establishing the Caribbean Community (RTC),[59] including the Caricom Single Market Economy. Guyana and Suriname are the only Amazon States participating in Caricom.[60]

Caricom's main organs are the Conference of Heads of Government, its supreme organ responsible for providing policy direction,[61] and the Community Council of Ministers that formulates the community's strategic planning and coordination in the areas of economic integration, functional cooperation, and external relations.[62] These bodies are assisted by four Ministerial Councils[63] and some other institutions.[64] The Council for Trade and Economic Development, and for Human and Social Development handle, among other things, issues such as renewable energy and sustainable development respectively.

In 2001, the Caribbean Court of Justice (CCJ) was created and began to operate in April 2005 in Port of Spain, Trinidad, and Tobago.[65] The Court has compulsory and exclusive jurisdiction to hear disputes concerning the interpretation and application of the RTC, and its judgments are binding on the parties involved in proceedings.[66] In principle, only disputes between member States can be submitted to the CCJ, but, with its special authorization, nationals of a contracting party may also have standing.[67]

58 As declared at the 1989 Grand Anse Declaration, adopted at the Tenth Meeting of Heads of Government of Caricom, http://www.caricom.org/jsp/communications/meetings_statements/grand_anse_declarati ojsp?menu=communications

59 2001 Revised Treaty of Chaguaramas Establishing the Caribbean Community, 2259 UNTS 295.

60 Antigua and Barbuda, Bahamas, Barbados, Belize, the Dominican Republic, Grenada, Guyana, Haiti, Jamaica, Montserrat, St. Kitts and Nevis, Saint Lucia, Suriname, and Trinidad and Tobago, http://www.caricom.org/jsp/community/member_states.jsp?menu=community

61 RTC, Article 12.

62 RTC, Article 13.

63 RTC, Article 10. Pursuant to this article, the four councils are the following: (1) Council for Finance and Planning, (2) Council for Trade and Economic Development, (3) Council for Foreign and Community Relations, and (4) Council for Human and Social Development.

64 RTC, Article 21. They include the following: (1) Caribbean Disaster Emergency Response Agency, (2) Caribbean Meteorological Institute, (3) Caribbean Meteorological Organization, (4) Caribbean Environmental Health Institute, (5) Caribbean Agricultural Research and Development Institute, (6) Caribbean Regional Centre for the Education and Training of Animal Health, (7) Veterinary Public Health Assistants, (9) Caribbean Centre For Developmental Administration, and (10) Caribbean Food and Nutrition Institute.

65 2001 Agreement Establishing the Caribbean Court of Justice, http://www.caricom.org/jsp/secretariat/legal_instruments.jsp?menu=secretariat

66 RTC, Article 22.

67 RTC, Article 24.

The decisions taken by Caricom's main bodies (i.e. the Conference of Heads of Government, the Community Council of Ministers, and the Ministerial Councils) are binding on the member States.[68] Recommendations of other institutional bodies are not mandatory,[69] but in cases of noncompliance, States are under an obligation to inform the Secretariat and give the reasons why they failed to comply with them. The RTC contains no specific provisions dealing with sanctions for breaching its provisions, and this was one aspect examined by the CCJ in the case presented by Trinidad Cement Limited (TCL) Guyana Incorporated in 2009.[70] As discussed in this case, a similar situation had occurred in the context of the European Community (EC), where no specific provisions existed with respect to sanctions for breaches of the EC Treaty. The European Court of Justice (ECJ) invoked the "principle of effectiveness," by stating that "the full effectiveness of Community rules would be impaired and the protection of the rights which they grant would be weakened if individuals were unable to obtain redress when their rights are infringed by a breach of Community law for which a Member State can be held responsible." The CCJ, held that a similar principle applies under the RTC and that the new single market involved the remedy of compensation. However, according to the CCJ, State liability is not automatic and a party would have to "demonstrate that the provision alleged to be breached was intended to benefit that person, that such breach is serious, that there is substantial loss and that there is a causal link between the breach by the State and the loss or damage to that person."

Under Caricom's institutional structure, there is no single organ dealing with environmental protection. Several institutions, listed in Article 21 of the RTC, such as the Caribbean Disaster Emergency Response Agency and the Caribbean Meteorological Institute, deal with different aspects related to the environment.[71] Caricom has adopted some decisions or agreements in the environmental domain, for example the Agreement Establishing the Caribbean Fisheries Mechanism[72] and the 2002 Agreement with the Secretariat of the United National Convention to Combat Desertification in Countries Experiencing Drought and/or Desertification,

68 RTC, Articles 28–29.

69 RTC, Article 27.

70 Case *Trinidad Cement Limited TCL Guyana Incorporated and the State of the Co-Operative Republic of Guyana*, August 20, 2009, Original Jurisdiction, CCJ, Application no. OA 2 of 2009, available at: http://www.caribbeancourtofjustice.org

71 Some of these institutions listed in Article 21 of the RTC include: the Caribbean Disaster Emergency Response Agency, Caribbean Meteorological Institute, Caribbean Meteorological Organization, Caribbean Environmental Health Institute, Caribbean Agricultural Research and Development Institute, Caribbean Regional Centre for the Education and Training of Animal Health and Veterinary Public Health Assistants, and Caribbean Food and Nutrition Institute.

72 2242 UNTS 273, http://untreaty.un.org/unts/144078_158780/3/5/11620.pdf

in order to fight land degradation in the subregion.[73] Environmental agreements adopted in the wider Caribbean have often focused on the protection of the marine environment.[74]

5.4. COMMON MARKET OF THE SOUTH

The Common Market of the South (Mercosur) was established by the 1991 Southern Common Market Agreement (or Treaty of Asunción),[75] adopted by Argentina, Brazil, Uruguay, and Paraguay to create a common market, involving the free movement of goods, services, and factors of production among countries, and a common external tariff and trade policy.[76] Venezuela applied for membership of Mercosur by the Protocol of Accession of the Bolivarian Republic of Venezuela, signed in Caracas, Venezuela, on July 24, 2006.[77]

Mercosur's structure is defined in the 1994 Additional Protocol to the Treaty of Asunción on the Institutional Structure of Mercosur (Protocol of Ouro Preto – POP).[78] The highest body is the Council of the Common Market, comprising

[73] Available at: http://www.caricom.org/jsp/secretariat/legal_instruments.jsp?menu=secretariat

[74] Such as the 1983 Cartagena Convention for the Protection and Development of the Marine Environment of the Wider Caribbean Region (and protocols) (22 ILM 227), the 1983 Protocol Concerning Cooperation in Combating Oil Spills in the Wider Caribbean Region (1506 UNTS 157), and the 1990 Protocol Concerning Specially Protected Areas and Wildlife to the Convention for the Protection and Development of the Marine Environment of the Wider Caribbean Region (W. Anderson, *Caribbean Instruments on International Law*, Barbados: Stone, 1994, p. 507).

[75] 30 ILM 1044.

[76] 1991 Southern Common Market Agreement, Article 1.

[77] According to Article 1 of this Protocol, Venezuela accedes to the Treaty of Asunción, the Protocol of Ouro Preto and the Olivos Protocol on Mercosur's dispute settlement mechanism, pending the changes required by each party for the application of the Protocol. (http://www.mercosur.int/msweb/SM/Noticias/pt/Protocolo%20Venezuela%20PT.pdf). Decision CMC no. 28/04 established the guidelines for negotiating agreements with Mercosur associated members (including the issue of the accession of new members). The possibility of the accession of new members is also foreseen by Article 20 of the Treaty of Asunción, according to which "this Treaty shall be open to accession, through negotiation, by other countries members of the Latin American Integration Association; their applications may be considered by the States Parties" and the "approval of applications shall require the unanimous decision of the States Parties." In 2005, Decision CMC no. 28/05, regulating the conditions for accession of new members, was adopted. On July 4, 2006 the Protocol of Accession of the Bolivarian Republic of Venezuela was approved and the conditions and timeframe for its accession were established. The Protocol enters into force with the ratification of five Mercosur States. Up to the present, Venezuela's accession has been approved by the parliaments of Venezuela, Argentina, Brazil, and Uruguay, while the approval of Paraguay is still pending (http://www.mercosur.int, accessed September 2010). As of January 2008, associate members of Mercosur included Chile, Bolivia, Ecuador, Colombia, and Peru. Venezuela is in the process of becoming a full member (J.S. Blyde, E. Fernández-Arias, and P. Giordano (eds.), *Deepening Integration in MERCOSUR: Dealing with Disparities*, Washington, DC: Inter-American Development Bank, 2008, p. 1).

[78] Additional Protocol to the Treaty of Asunción on the Institutional Structure of Mercosur (Protocol of Ouro Preto), 34 ILM 1248. This protocol defines in Article 3 Mercosur's institutional framework and confers international legal personality on this organization.

the ministers of foreign affairs and the economy (or their equivalents), expected to formulate policies and promote measures necessary to build the common market. The Common Market Group is Mercosur's executive body,[79] which is assisted by the Trade Commission in monitoring the application of common trade policy instruments.[80] All these bodies have decision-making powers, as stated in Article 2 of POP.[81] A dispute settlement mechanism, first regulated by Annex 3 of the Treaty of Asunción and later by the 1991 Protocol of Brasília,[82] now includes a Permanent Court of Appeal (Tribunal Permanente de Revisión), created by the 2002 Protocol of Olivos,[83] based in Asunción (Paraguay). This court is Mercosur's main judicial body. In 2005, a protocol created the Mercosur Parliament based in Montevideo, Uruguay (Protocolo Constitutivo del Parlamento del Mercosur, Decision CMC N. 23/05), which replaces the previous Parliamentary Commission (Comisión Parlamentaria Conjunta), foreseen under the Protocol of Ouro Preto. Mercosur Parliament represents the citizens of each member State and is expected to have its representatives elected by direct and universal vote.

Mercosur's legal sources are defined in Article 41 of POP, and include the Treaty of Asunción (protocols and complementary instruments), protocols and decisions of the Council of the Common Market, resolutions of the Common Market Group, and the directives of Mercosur's Trade Commission.[84] The decisions of the Council of the Common Market, resolutions of the Common Market Group and directives or proposals of the Mercosur Trade Commission are binding on the member States,[85] but, pursuant to Article 42 of POP, they must be incorporated in the member States' domestic legal systems, in accordance with the procedures provided for in each country's legislation. A Joint Parliamentary Commission is expected to speed up the internal procedures in the member States and ensure prompt and simultaneous entry into force of decisions taken by Mercosur's organs.[86] To ensure that decisions of Mercosur organs enter into force without delay and simultaneously, the member States must follow a procedure of "simultaneous entry into force," defined in Article 40 of POP.

79 POP, Article 10.

80 POP, Article 16.

81 POP, Article 2.

82 1991 Protocol of Brasília for the Solution of Controversies in Mercosur, 36 ILM 691.

83 2002 Olivos Protocol for the Settlement of Disputes, 42 ILM 2.

84 According to the Arbitral Award in the case *Obstáculos al ingreso de productos fitosanitarios argentinos en el mercado brasileño*, Article 19 of the Protocol of Brasília establishes the legal sources of Mercosur, on which the Arbitral Tribunal shall be based in order to decide a controversy: the Treaty of Asunción, the agreements contained within its framework, the decisions of the Common Market Council, and the resolutions of the Common Market Group, as well as the principles and dispositions of international law that are applicable to the matter (Arbitral Award, Item 8.9, http://www.mercosur.int/msweb/portal%20intermediario/pt/index.htm).

85 POP, Articles 9, 15, and 20.

86 POP, Article 25.

The duty to incorporate Mercosur law, specifically resolutions of the Common Market Group, was discussed in the case *Obstáculos al Ingreso de Productos Fitosanitarios Argentinos en el Mercado Brasileño* presented by Argentina against Brazil to Mercosur's Ad Hoc Tribunal on April 19, 2002.[87] According to this tribunal, Mercosur law is binding on member States from its adoption, and it is only its entry into force that is delayed until its effective incorporation into each State's national jurisdictions.[88] In any case, States are under an obligation to incorporate Mercosur's law into their domestic legal systems within a reasonable period of time[89] and may incur an international responsibility if they fail to do so.[90]

In the framework of Mercosur, environmental concerns were voiced particularly since the 1992 UNCED. In preparation for this conference, the member States adopted the Declaration of Canela in 1992,[91] in which they stated their willingness to take joint actions in relation to the environment. In the same year, the Common Market Group created the Reunión Especializada de Medio Ambiente (REMA) to examine the member States' environmental laws and propose recommendations to the Common Market Group. Following a series of meetings, REMA elaborated the basic guidelines of Mercosur's environmental policy, in a document entitled *Directrices Básicas en Matéria de Política Ambiental*, later approved by the Common Market Group.[92] In 1995, Mercosur's ministers of the environment met for the first time in Montevideo and adopted the Declaration of Taranco, which placed the environment as a priority issue in Mercorsur's agenda and institutional structure.[93] After this meeting, REMA was converted into Working Sub-Group No. 6 on the Environment – one of Mercosur's technical bodies.[94]

87 According to this Arbitral Award, Article 40 of the Ouro Preto Protocol, complemented by its Article 38, clearly establishes the obligation of member States to incorporate Mercosur's law into their respective domestic jurisdictions.

88 Ibid., Item 7.5.

89 Ibid., Item 8.3.

90 According to the case *Aplicación de Medidas Antidumping contra la exportación de pollos enteros, provenientes de Brasil*, the obligation to incorporate Mercosur law into the parties' respective jurisdictions is an obligation of performance, and if a State violates such an obligation, it incurs an international responsibility: "La obligación de incorporar la normativa MERCOSUR a los derechos internos de los Estados Partes constituye una obligación de hacer que compromete la responsabilidad internacional de los Estados en caso de incumplimiento" (Arbitral Award, Item 7.7, http://www.mercosur.int/msweb/portal%20intermediario/pt/index.htm).

91 MERCOSUR/GMC/RES 22/92, http://www.mercosur.int/msweb/portal%20intermediario/es/index.htm.

92 Approved by Resolution MERCOSUR\GMC\RES 10/94, http://www.mercosur.int/msweb/portal%20intermediario/es/index.htm.

93 Mercosur Secretariat, *Medio Ambiente en el Mercorsur*, Serie Tematica, document no. 3, 2002, p. 16.

94 This working subgroup was created by Resolution MERCOSUR/GMC/RES 20/95 and ratified by Decision CMC 59/00.

That Working Sub-Group has met regularly, since 1995, and provided recommendations to the Common Market Group on areas such as the environment and competitiveness, nontariff barriers, and common systems of environmental information. Since 1997, it has led negotiations on an environmental protocol for the Treaty of Asunción, which would introduce environmental concerns into the regional policies and decision-making processes within Mercosur.[95] As a result, in 2001, Mercosur's Framework Agreement on the Environment (CMC Decision N. 02/01) was adopted.[96] Several decisions of Mercosur's organs have so far addressed environmental issues, such as the management of pesticides and the transport of hazardous products, among others.[97] Following CMC Decision N. 02/01, an additional protocol was adopted in 2004 on the issue of cooperation and assistance in the event of environmental emergencies (Protocolo Adicional al Acuerdo Marco sobre Medio Ambiente del MERCOSUR en materia de cooperación y asistencia ante Emergencias Ambientales, Decision CMC N. 14/04). Mercosur has also adopted agreements with third-party States relevant to the environment, for example with the European Community and Canada.[98] Other decisions adopted more recently include Mercosur's Action Plan for Cooperation on the issue of Biofuels (Plan de Acción del Mercosur para la Cooperación en Materia de Biocombustibles, Decision N. 049/2007).[99]

95 This treaty was approved by the Common Market Council through resolution CMC/DEC 02/01.

96 The agreement's thematic areas are: (1) sustainable use of natural resources (e.g. wildlife, forests, protected areas, biological diversity, soils, and water resources); (2) quality of life and environmental planning (e.g. sewage, waste and dangerous residues management, transport, and renewable energy); (3) environmental policy instruments (e.g. legislation, education, environmental, and information systems); and (4) sustainable productive activities (e.g. ecotourism, sustainable forest management, and fishing).

97 In this case, the tribunal held that the environment is a priority issue among the nations of the world, and it could not be different to the States of Mercosur. Rules under Mercosur addressing the environment include: (1) CMC/DEC 2/94, Agreement on Transport of Dangerous Substances; (2) Decision GMC no. 10/94, Basic Guidelines of Environmental Policy (*Directrices Básicas en Política Ambiental*); (3) CMC/DEC 10/00 on a General Plan of Coordination and Cooperation for Regional Security in Matters Related to Environmental Torts (*Complementación del Plano General de Cooperación y Coordinación Recíproca para la Seguridad Regional, en materia de Ilícitos Ambientales*); (4) CMC/DEC 19/03 on the creation of a Meeting of Ministers of the Environment; and (5) CMC/DEC 14/04, which approves the Additional Protocol to Mercosur's Framework Agreement on Cooperation in Situations of Environmental Emergency.

98 In 1995 the Acuerdo Marco Interregional de Cooperación entre la Comunidad Europea y sus Miembros y el Mercado Común del Sur y sus Estados Partes and in 1998, the Entendimiento de Cooperación en materia de comercio e inversiones entre Canadá y MERCOSUR (Mercosur Secretariat, *Medio Ambiente en el Mercosur*, Montevideo, 2006, p. 6).

99 For Mercosur decisions relevant to the environment, see ibid.; or Programa de las Naciones Unidas para el Medio Ambiente, Centro Latino Americano de Ecología Social, *GEO Mercosur: Integracion, comercio y medio ambiente en el Mercosur*, Montevideo, 2008. For decisions in general, see http://www.mercosur.int/t_generic.jsp?contentid=526&site=1&channel=secretaria&seccion=5

As with other regional and subregional organizations, environmental concerns were gradually introduced within Mercosur's normative and institutional structures, but these are sometimes confronted with other objectives or interests deriving from its constitutional mandate. In particular, the potential tension between free trade and environmental protection was the object of a dispute brought to Mercosur's Ad Hoc Tribunal on October 25, 2005 by Uruguay against Argentina in the case *Prohibición de Importación de Neumáticos Remoldeados*,[100] later revised by the Permanent Court of Appeal on December 20, 2005.[101] Uruguay challenged the prohibition imposed by the Argentinean Law no. 25.626 of August 8, 2002 on imports of remolded tires in its territory grounded on environmental and health considerations. Argentina alleged that, at the end of their life cycle, remolded tires produced residues likely to cause damages to the environment and animals and pose risks to human health.[102] The Ad Hoc Tribunal acknowledged the existence of a conflict between two basic principles under Mercosur law: free trade and environmental protection.[103] According to the tribunal, Article 2 of the Council of the Common Market's Decision N. 2/01 foresaw the possibility of such a conflict.[104] This provision urged States to avoid taking arbitrary measures restricting free trade based on environmental justifications. The tribunal reasoned that if Article 2 envisaged the use of arbitrary measures based on environmental concerns, the legislator admitted, *a contrario sensu*, that some legitimate and nonarbitrary environmental protection measures could be permitted, even when affecting trade.[105] In conclusion, the tribunal took the view that free trade was not an absolute rule,[106] and that environmental protection was also

[100] Arbitral Award of October 25, 2005, http://www.mercosur.int/msweb/portal%20intermediario/pt/index.htm

[101] Arbitral Award of January 9, 2002, Uruguay v. Argentina. Uruguay has also presented a claim against Brazil in the case *Prohibición de Importación de Neumáticos Remoldeados (remolded) Procedentes de Uruguay*. This dispute related to a *Portaria* of the Secretariat of Foreign Trade of the Ministry of Development, Industry and Foreign Trade (SECEX) no. 8/00, September 25, 2000 enacted by Brazil prohibiting the granting of concessions of licenses for imports of remolded tires for consumption or as raw material, which hampered the entry of Uruguay's remolded tires into the Brazilian market. Uruguay claimed that Portaria (SECEX) no. 8/00, 25 violated Mercosur law. Mercosur's Ad Hoc Tribunal, without examining the environmental implications of the case, decided that Brazil should adapt its national laws and asserted that Mercosur's member States could not unilaterally define concepts such as second-hand goods even on the grounds of environmental protection.

[102] Arbitral Award, October 25, 2005, Uruguay v. Argentina, para. 95.

[103] Ibid. According to para. 95 of the Arbitral Award, the Tribunal should consider the application of two conflicting principles, free trade on the one hand and the protection of the environment on the other: "Ponderará, asimismo el Tribunal la aplicación de los mencionados principios en confronto, es decir, el libre comercio y la protección del medio ambiente, definiendo la prevalencia de uno sobre el otro, teniendo en cuenta los dictámenes del Derecho Internacional y, muy especialmente, las peculiaridades del caso sub examen."

[104] Ibid., para. 59.

[105] Ibid., para. 60.

[106] Ibid. In para. 94, the tribunal takes the view that free trade is not an absolute principle immune to any sort of exceptions: "Sin embargo, la libertad de comercio y su preservación como forma de

one of Mercosur's main principles[107] that could be invoked as an exception to other rules.[108] In view of the damage likely to be caused by introducing remolded tires, the tribunal held that Argentina had justifiable reasons for enacting Law no. 25.626 of August 8, 2002,[109] which was then compatible with the Treaty of Asunción and current principles of international law. In the Permanent Court of Appeal's Arbitral Award December 20, 2005, this ruling was overturned. The court ruled that free trade was the governing principle in a regional integration system and found no justification for any exceptions in Mercosur law. The court did not take international law into account, as it understood that other principles and norms outside the realm of Mercosur law should only be applied in a subsidiary way.[110] These rulings show that conflicts between environmental protection and other principles can occur in a particular legal regime, and that certain principles may take precedence in the event of conflict.

5.5. INTER-AMERICAN DEVELOPMENT BANK

In 1959, the Inter-American Development Bank (IDB) was established to contribute to the economic and social development of the developing member States.[111] Currently, with several country offices and technical staff based in the borrowing member countries, IDB is the main source of multilateral financing for socioeconomic development projects in Latin America and the Caribbean[112] in various sectors such as agriculture, industry, energy, transportation, the environment, and

estructurar el Mercosur no puede ser considerada principio absoluto y inderogable…surgido para solucionar todos los problemas de las relaciones comerciales e inmune a cualquier excepción."

[107] Ibid. In para. 99, the tribunal states that the protection of the environment is a basic principle of Mercosur: "La protección del medio ambiente es un principio básico para el Mercosur."

[108] Ibid. In para. 100, the tribunal holds that the defense of the environment, if justified, can be used as an exception to the general rules of regional integration: "el Tribunal entiende que la defensa del medio ambiente, desde que esté fundada en justas razones, puede ser usada como excepción a las normas generales de la integración regional y, particularmente, aquellas que regulan el libre comercio entre los países del Mercosur."

[109] Ibid., para. 102.

[110] This decision was object of complementary decisions by the Permanent Court of Appeal (Decisions 01/2006, 01/2007, and 01/2008). In this latter case (Decision 01/2008), the proceedings were initiated by Uruguay with a view to enquiring whether Argentina, by adopting Law 26.329 that modified the previous Law 25.626 prohibiting imports of remolded tires, contradicted the Court's Decision TPR 1/05 of December 20, 2005. The Court held that Law 26.329 – similarly to Law 25.626, the application of which was examined by the Court in that decision – imposed undue restrictions on trade. Therefore, the Court held that Law 26.329 was not in accordance with its previous ruling (Decision TPR 1/05) and ordered Argentina to either revoke or modify it. These decisions by the Permanent Court of Appeal can be found at: http://www.mercosur.int

[111] Agreement Establishing the Inter-American Bank, sec. 1, 426 UNTS 1986.

[112] The bank comprises borrowing member countries in Latin America and the Caribbean, and non-borrowing countries, including the United States, Japan, Canada, sixteen European countries, and Israel.

public health, among others. IDB lends money and provides grants. In 2008, IDB total loan approvals amounted to US$11.2 billion.[113] The Bank's loans cover only a part of the total cost of the projects being carried out by the borrowing countries that supply the majority of the balance.[114] Issues such as sustainable energy and climate change, and water management are among IDB's current priority areas.[115]

As early as 1979, IDB adopted an Environment Policy (OP-703) to ensure the quality of its operations and support environmental projects in the region. In the early 1990s, the bank assisted the countries in the region in preparing for the 1992 UNCED, by formulating a joint IDB–UN Development Program (UNDP) report entitled *Our Own Agenda*.[116] IDB's Eighth Replenishment of Capital in 1994 declared the environment a priority area for bank support and included a number of specific environmental mandates.[117] In 2003, a new Environment Strategy (GN-2208–4) was released outlining the bank's general policy in the support offered to borrower countries of Latin America and the Caribbean.[118] The strategy aims to render the support offered to countries in relation to environmental issues more effective. It attempts to link different sectors in order to improve governance and policies among the member States.[119] An advisory group, the Blue Ribbon Panel on the Environment, was created in 2004 to advise IDB on the draft of the Environment and Safeguards Compliance Policy that would put into operation the objectives stated in the Bank's Environment Strategy.[120] This panel released its Final Report Recommendations in 2005.[121] IDB's Board of Executive Directors approved the Environment and Safeguards Compliance Policy in 2006, which would replace the outdated Environment Policy (OP-703) and ensure that all bank operations were environmentally sustainable.[122] Some IDB organs now deal specifically with environmental issues, notably the Sustainable Development Department.

[113] IDB (Inter-American Development Bank), *Annual Report 2008: The Year in Review*, Washington, DC, 2009, p. 7.

[114] Ibid., p. 35.

[115] Ibid., p. 8.

[116] IDB and UNDP, *Our Own Agenda*, report of the Latin American and Caribbean Commission on Development and Environment, 1990.

[117] IDB, *Blue Ribbon Panel on the Environment, Final Report Recommendations*, February 23, 2005, p. 2.

[118] IDB, *Environment Strategy Document*, GN-2208–4, which was favorably reviewed by the IDB Board of Executive Directors on July 23, 2003.

[119] IDB, *Environment and Natural Resources Annual Report 2003*, Washington, DC, 2004, p. 1.

[120] IDB, Environment Division Sustainable Development Department, *Environment and Safeguards Compliance Policy*, GN-2208–18, Washington, DC: IDB, 2006. This policy document was favorably reviewed by the IDB Board of Executive Directors on January 19, 2006.

[121] IDB, *Blue Ribbon Panel*, op. cit., pp. 1–2.

[122] IDB, *Environment and Safeguards*, op. cit. This policy document (GN-2208–18) was favorably reviewed by IDB's Board of Executive Directors on January 19, 2006.

During the period between 1990 and 2000, total loans to environmental conservation programs amounted to US$10.6 billion, which represented almost 15 percent of total lending during that period.[123] Between 1961 and 2008, 33 percent of the Bank's lending was directed to Social Development, including environmental issues.[124] In the Amazon, IDB has carried out various projects, for example to strengthen protection systems (especially Brazil's SIVAN-Sipam system),[125] promote cleaner production processes,[126] and establish systems for land ownership rights.[127] IDB and ACTO signed a technical cooperation agreement in 2005 to initiate a project aimed at strengthening regional capacity for the sustainable use of Amazonian biodiversity, for example by creating cross-border ecological corridors, information systems on biodiversity, and an information clearinghouse unit.[128] In Brazil, IDB also finances programs such as that in the state of Amazonas, within the framework of the Social and Environmental Programme for the Igarapés of Manaus, to improve the environmental, urban, health, and housing conditions of the population in areas near a few water basins (Igarapés Educandos/Quarenta and Sao Raimundo).[129]

More generally, the Bank carries out the Sustainable Energy and Climate Change Initiative (SECCI), whose objectives are to expand the development and use of renewable energy sources, energy efficiency technologies, and carbon finance in Latin America and the Caribbean, as well as to promote and finance adaptation strategies to reduce the region's vulnerability to climate change. In 2008, IDB approved two policy-based loans for climate change totaling US$500 million, and nine loans totaling US$610 million for renewable energy, energy efficiency, and biofuel projects, along with thirty-eight grants totaling US$21.3 million for SECCI projects.[130]

123 IDB, *IDB and the Environment: 1990–2002: The Decade since Rio*, briefing report, 2002, p. 4.

124 IDB, *Annual Report 2008*, op. cit., p. 36.

125 IDB, *Program to Support the Amazon Protection System* (BR-T1014), ATN/JC-9556-BR, Brazil (status: implementation). This program aims to develop the full technical and operational capacity of the SIVAN-Sipam system. Also in Brazil, IDB is developing the project Sustainable Management of the Amazonas' Biodiversity (RG-T1151: BPR 38), ATN/OC-9251-RG, Brazil (status: implementation). A list of projects developed by IDB in the Amazon can be found at: http://www.iadb.org/projects/index.cfm?language=English

126 IDB, *Promotion of Cleaner Production Processes* (TC0103001), ATN/ME-7833-EC, Ecuador (status: completed).

127 IDB, *Land Titling and Registration* (EC0191), 1376/OC-EC, Ecuador (status: implementation). The objectives of this program are to establish a continuously updated system for land ownership rights by implementing it in rural cantons in the Ecuadorian Amazon (Baba, Colimes, Palenque, and Salitre on the coast; Pimampiro, Chillanes, Chaguarpamba, and Saraguro in the Sierra; and Joya de los Sachas) and to improve the system for allocating public land to producers.

128 The technical cooperation between ACTO and IDB was signed on July 25, 2005, Project ACTO-IDB, ATN/OC-9251-RG, http://www.otca.org.br

129 UNEP (United Nations Environment Programme), Amazon Cooperation Treaty Organization (ACTO) and Research Center of Universidad del Pacífico (CIUP), *Environment Outlook in Amazonia, Geo Amazonia*, 2009, p. 231, http://www.unep.org/pdf/GEOAMAZONIA.pdf

130 IDB, *Annual Report 2008*, op. cit., p. 12.

5.6. ANDEAN DEVELOPMENT CORPORATION

The Andean Development Corporation (CAF) is a multilateral financial institution established in 1968 to enhance subregional integration, by allocating investments in favor of less developed countries in the subregion and providing technical and financial assistance for the implementation of multilateral projects and granting loans.[131] This organization, comprising several countries in the Latin America and Caribbean regions as well as private banks, is a major source of multilateral financing in the Andean region.[132]

Since 1994, CAF has incorporated environmental concerns into its operational strategies with the creation of the Sustainable Development Coordination Office, responsible for the environmental assessment and follow up of its operations.[133] In 2000, it became the Sustainable Development Office under Vice-Presidency of Social and Environmental Development.[134] CAF has elaborated an Environmental Strategy[135] and other guidelines, such as an Environmental Management Plan and a System of Environmental Management and Social Operations, to introduce environmental considerations into credit approval processes and other operations.[136]

CAF has developed a number of environmental projects.[137] For example, the Environmental Unit carries out programs such as BioCAF, intended to promote the sustainable use of biodiversity and genetic resources, to support the member States in international negotiations, and to strengthen biodiversity markets. Another example is the Latin American Carbon, Clean and Alternative Energies Program, established in 1999 with a view to promote capacity building and education on issues related to climate change and its impacts. This program has promoted carbon trading, particularly in financing or buying carbon credits from the Clean Development Mechanism.[138]

[131] Agreement Establishing the Andean Development Corporation, 1968, in Venezuela, *Convenio Constitutivo de la CAF*, Edición Caracas, August 1996, http://www.caf.com/view/index.asp?ms=17

[132] An Amending Protocol of the Agreement Establishing CAF, signed on October 24, 2005, allowed new Latin American and Caribbean countries to join as full members (see http://www.caf.com). CAF is constituted by country shareholders and private banks in the region. Among the shareholders are several Amazon States, including Brazil, Bolivia, Colombia, Ecuador, Peru, and Venezuela.

[133] CAF, *Annual Report 2006*, p. 26. CAF annual reports can be found at: http://www.caf.com/view/index.asp?ms=17&pageMs=43796

[134] CAF, *Annual Report 2008*, p. 7.

[135] CAF, *La CAF y el medio ambiente: Estrategia ambiental de la CAF*, 2007, http://www.caf.com/attach/18/default/FINAL-small-sinblancas.pdf. See also CAF, *Environmental Strategy 2008*, Caracas: CAF Publications, 2008. This document sets out the principles that should guide CAF's environmental management.

[136] CAF, *Annual Report 2006*, op. cit., p. 26.

[137] For example, the Biodiversity Program, the Latin American Program on Clean and Alternative Energy, the Program on Disaster Risk Management, and the Program on Sustainable Development in Financial Institutions (ibid.).

[138] Ibid., p. 146.

One more example of existing efforts to strengthen regional cooperation is the Union of South American Nations (UNASUR) Treaty signed in 2008 (including the eight Amazon States).[139] UNASUR is an entity endowed with international legal personality (Article 1) composed of a few bodies, including a General Secretariat established in Quito, Ecuador (Article 10), and a South American Parliament, to be based in Cochabamba, Bolivia (Article 17).[140]

CONCLUSIONS

The regional and subregional organizations presented here have developed their normative and institutional frameworks over the past years by adopting legal instruments and by creating bodies with executive, legislative, and judicial powers, such as the Andean Parliament in 1997, the Caribbean Court of Justice in 2001, and Mercosur's Permanent Court of Appeal in 2002. Most of these organs can take legally binding decisions, some of which are directly applicable in the territories of the member States.

All those organizations now address, in a more or less comprehensive way, issues related to the environment and have created specialized bodies to this end. They have adopted environmental norms, such as the 1996 CAN Decision N. 391 on the Common Regime for Access to Genetic Resources and the 2001 Mercorsur Framework Agreement on Environment. In practice, they have developed environmental field projects, particularly in the Amazon. The regional development banks, notably IDB and CAF, have incorporated environmental concerns into their operational strategies, by introducing best practices, such as IDB's 2006 Environment and Safeguards Compliance Policy, and investing in green projects to be carried out in the borrowing countries.

This more complex institutional setting in which the ACT and other legal instruments are inserted does not necessarily mean improved environmental protection for the Amazon. In practice, the plurality of organizations working on similar issues may cause a duplication of efforts and increased competition for external funding. These organizations adopt norms that are applicable to the member States' respective Amazonian territories and carry out activities in that region, but without much cohesion.

The organizations presented here now deal with a broader range of issues and sometimes operate in overlapping geographical areas. As a result, conflicts of law within and among them are now more likely to occur, which will be further discussed

139 Ibid., p. 31.

140 The text of the 2008 Union of South American Nations (UNASUR) Treaty can be found at: http://www.comunidadandina.org/ingles/csn/treaty.htm. About UNASUR see www.pptunasur.com

in the next chapter. So far, no such conflicts have occurred between them and the ACT in particular. This is probably due to the fact that this treaty does not contain specific legal obligations, more likely to give rise to conflicts of law. In practice, the ACT Permanent Secretariat has interacted with other regional and subregional organizations by adopting common understandings or cooperation agreements that involve the exchange of information, consultations and the implementation of joint activities, for example with the OAS, CAN, and IDB.[141] In any case, in a complex institutional framework, where conflicts of law and other controversies are more likely to occur, more caution is required. One safeguard for preventing such conflicts is to enable consultations among these organizations, particularly prior to the adoption of norms that may conflict with those existing in other legal regimes.[142] With regards to norms related to the Amazon, the ACT Permanent Secretariat could probably be a forum where those consultations could occur. The following chapter presents some instruments comprising the legal system of the Amazon other than the ACT and looks at their interactions.

141 With the financial support of the GEF, on June 25, 2005, ACTO, the general secretariat of the OAS and the United Nations Development Program agreed to initiate the preparatory phase of the Integrated and Sustainable Management of Trans-boundary Water Resources in the Amazon River Basin Considering Climate Variability and Change Project. Information about this project is available at: http://www.otca.org.br

142 W. Jenks, "The conflict of law-making treaties", *British Yearbook of International Law*, vol. 30, pp. 401–53.

6

Other Legal Instruments Adopted by the Amazon States *Inter Se*

The international legal system of the Amazon, as suggested in this study, comprises three categories of legal instruments related in some way to the conservation and sustainable use of natural resources of the whole or parts of the Amazon region. These categories involve both legally binding and nonbinding instruments. In addition to treaties, some decisions adopted by the Andean countries (Bolivia, Ecuador, Colombia, and Peru, which are also Amazon States) under the aegis of the Andean Community (CAN) are also examined here. These decisions[1] are not only legally binding, but also directly applicable in CAN member States as of the date they are published in the agreement's *Official Gazette*.[2]

[1] CAN decisions, together with other instruments such as resolutions and protocols, constitute CAN law. Treaty Creating the Court of Justice of the Andean Community, 18 ILM 1203 (amended by the Cochabamba Protocol on May 28, 1996). According to Article 1, the CAN legal system comprises: "the Cartagena Agreement, its Protocols and additional instruments; this Treaty [Treaty Establishing the Andean Community Court of] and its Amending Protocols; the Decisions of the Andean Council of Foreign Ministers and of the Commission of the Andean Community; The Resolutions of the General Secretariat of the Andean Community; and the Industrial Complementarity Agreements and any such other agreements as the Member Countries may adopt among themselves within the context of the Andean sub-regional integration process."

[2] CAN decisions are taken by the Andean Council of Foreign Ministers or the Commission and Resolutions of the General Secretariat. Their binding nature and direct applicability are so defined by law in the Treaty Creating the Court of Justice of the Andean Community, Article 2, 18 ILM 1203 (1979) (entered into force March 10, 1996), and CAN's jurisprudence. On the direct applicability of CAN's decisions, see PROCESOS 1-IP-87, 6-IP-94, 10-IP-94, and 1-AI-96, published in *Cartagena Agreement Official Gazette*, no. 28 of February 15, 1988, no. 170 of January 23, 1995, no. 177 of April 20, 1995, and no. 234 of November 21, 1996, respectively. The Court of Justice of the Andean Community (CJAC) supports the view that no domestic legislative procedure is required to give effect to such decisions. It stated in the case *Acción de incumplimiento interpuesta por la Junta del Acuerdo de Cartagena contra la República de Venezuela* (PROCESO 3-AI-96, *Cartagena Agreement Official Gazette*, no. 261, April 29, 1997): "La sola suposición de que las Decisiones de la Comisión o las Resoluciones de la Junta, tuvieran que pasar por el tamiz legislativo de cada uno de los Países Miembros, antes de su aplicación interna, habría conducido a negar la existencia de un derecho comunitario andino." Norms adopted under the aegis of the community have direct effect

The three categories mentioned above can be divided as follows. Under the first category are the legal instruments adopted by two or more Amazon States *inter se* for the protection of the Amazonian natural environment. These instruments include bilateral treaties, for example, the 1979 agreement on the Conservation of Flora and Fauna of the Amazonian Territories adopted between Colombia and Peru,[3] or the 1978 Amazon Cooperation Treaty (ACT). The second category involves legal instruments adopted by two or more Amazon States with other States in the Latin American and Caribbean region,[4] which do not necessarily address the Amazon in particular, but are applicable to all or parts of the Amazon, such as the 2001 Common Market of the South (Mercosur) Framework Agreement on the Environment, applicable in the Brazilian Amazon (as well as other Mercosur legal instruments). Finally, the third category encompasses multilateral environmental agreements to which some or all Amazon States are parties, which are then applicable to their respective Amazonian territories. The first two categories provide evidence of cooperation efforts among the Amazon countries *inter se* or between them and other States in the region, whereas the third category of legal instruments relates to cooperation between the Amazon countries and the community of States at large. This third category will be examined separately in Chapter 7, along with the policies of major global institutions *vis-à-vis* the Amazon.

It should be noted that the 1978 ACT, which for a number of reasons has been examined in more detail in previous chapters, is only an element of the international legal system of the Amazon, inserted within a broader institutional and normative context. This treaty should then be seen in the context of other legal instruments adopted by the Amazon States *inter se* and with other States. By presenting such legal instruments, this chapter intends to look at how those States cooperate among themselves on matters concerning the protection of the Amazon.

A. COOPERATION AMONG THE AMAZON STATES *INTER SE*

The Amazon States have assumed international obligations in the field of environmental protection by adopting *inter se* legally binding and nonbinding instruments in various domains. The legally binding instruments include bilateral or multipartite agreements among two or more Amazon States, as well as binding decisions

in the territories of its member States, generating rights and obligations, such as norms within their domestic legal systems: "generan derechos y obligaciones para los particulares al igual que ocurre en las normas de los ordenamientos estatales, permitiendo la posibilidad de que aquellos puedan exigir directamente su observancia ante sus respectivos tribunales."

3 *Colombian Official Gazette*, Year CXVII, no. 35650, Law 30 of 1980, November 25, 1980, p. 474.

4 The Caribbean region is also considered here, as Guyana and Suriname are also Caribbean countries and members of the Caribbean Community.

adopted by the regional and subregional organizations to which they are parties. The number of bilateral agreements is greater than that of treaties of regional application. Nonbinding instruments include declarations, guidelines, plans of action adopted by those countries under regional and subregional organizations in which they participate.

6.1. FRESHWATER RESOURCES

Besides the early examples of bilateral treaties signed by the Amazon States to delimit their international boundaries and establish rules for commerce and fluvial navigation,[5] adopted mostly by the end of the nineteenth century, following independence from the European colonial powers, only a few treaties were concluded among those countries with respect to water resources. Examples of these treaties are the Agreement on Fluvial Transportation between Brazil and Peru, signed on November 5, 1976[6] and the Technical and Economic Cooperation Agreement for the Improvement of Navigation and Fluvial Transportation in the River Madre de Dios and its Tributaries between Bolivia and Peru, adopted on October 18, 1989. The scarcity of international treaties (and related institutions) regulating different uses of freshwater, such as irrigation, hydropower generation, domestic and industrial water supply, is surprising given the network of international rivers in the Amazon basin, offering many possibilities for international cooperation.

Many of the watercourses in the Amazon basin are shared among two or more countries, for example the Negro River (Colombia and Venezuela), the Putumayo River (Colombia, Peru and Ecuador), and the Madre de Dios River (Peru and Bolivia), among others. The Amazon basin can offer a total of 50,000 km of navigable rivers to boats weighing up to 1,000 tons and about 10,000 km are navigable to ships weighing 1,000 tons or more,[7] providing a complex communication network among its riparian States (Table 6.1).

[5] Such as the Treaty of Friendship, Commerce, Navigation, Boundaries, and Extradition, between Brazil and Bolivia, signed at La Paz, March 27, 1867 (UK, *British and Foreign State Papers*, 1868–1869, vol. LIX, London: William Ridgway, pp. 1161–69); the Treaty on Boundaries and Navigation between Brazil and Colombia, adopted at Bogota on April 24, 1907 (G. Cavelier, *Tratados de Colombia 1811–1910*, vol. 1, Bogota: Editorial Kelly, 1982, pp. 682–85); and the Treaty on Boundaries and Fluvial Navigation between Brazil and Colombia, adopted at Rio de Janeiro on November 15, 1928 (Cavelier, op. cit., vol. 2, pp. 509–12).

[6] DOFC PUB 24/04/1979, *Brazilian Official Gazette* (*Diário Oficial da União*), p. 005715 1.

[7] C. Dominguez, "The importance of rivers for the transportation system of the Amazon," in L. Aragon and M. Clüsener-Godt (eds.), *Issues of Local and Global Use of Water from the Amazon*, UNESCO, Centre for Advanced Amazonian Studies of the Federal University of Pará (UFPA), 2004, pp. 77–100, at p. 77, http://www.unesco.org.uy/mab/documentospdf/amazon.pdf

TABLE 6.1. *Some shared river basins in the Amazon*

Country where river rises	Country where river flows	Country sharing the basin
Colombia	Venezuela	Negro
Colombia	Peru-Ecuador	Putumayo
Ecuador	Colombia-Peru	San Miguel-Putumayo
Ecuador	Peru	Napo
Ecuador	Peru	Tigre
Ecuador	Peru	Pastaza
Ecuador	Peru	Morona
Ecuador	Peru	Santiago
Ecuador	Peru	Chinchipe
Peru	Bolivia	Acre
Peru	Bolivia	Madre de Dios

Note: GTZ, Fundeco and Instituto de Ecología, "Conservación de ecosistemas transfronterizos y protección y recuperación de especies Amenazadas," thematic document, Second Regional Conference, held in Lima Peru on March 26–28, 2001, http://www.comunidadandina.org/desarrollo/dct2.pdf.

The regulation of water use and conservation in the Amazon basin occurs almost exclusively through national law. Brazil, in particular, has developed extensive water legislation and, in 1997, created a National Water Agency through the National Water Law no. 9.433/97. On account of this, Bolivia requested Brazil's assistance when formulating its own national water law. As a result, in 2003, these two countries adopted a Complementary Adjustment to the Basic Agreement on Technical and Scientific Cooperation for the implementation of a joint project aimed at the formulation of Bolivia's national water legislation.[8]

Other water basins in South America, notably the La Plata River basin, have been object to a much larger number of treaties, which is in part due to higher levels of industrialization and commercial exchanges among those riparian States when compared to the Amazon basin. The major commercial use of the Amazonian rivers is for hydropower generation and the construction of dams poses significant ecological problems in the Amazon basin.[9] The scarcity of treaties among the Amazon States on the management of international rivers might also be due to the fact that

[8] Complementary Adjustment, Article 3.

[9] M. Bothe, "The impact of international law on the protection of the environment in Amazonia and Siberia," in M. Bothe, T. Kurzidem and C. Schmidt (eds.), *Amazonia and Siberia: Legal Aspects of the Preservation of the Environment and Development in the Last Open Spaces*, London: Graham & Trotman/Nijhoff, 1993, pp. 237–55, at p. 241.

most of the Amazon River and its tributaries are located in Brazil, where water use is probably more intense when compared to the other riparian States. This situation might change, however, if production activities in the upper stream States, particularly Ecuador and Peru, continue to grow.[10]

Early agreements on international boundaries and fluvial navigation adopted by the Amazon States laid down rules imposing some restrictions on navigation, for example prohibiting cabotage, limiting the number of warships or requiring prior notification, and subjecting the navigation of merchant vessels to fiscal and police regulations. As one could expect, these early treaties are silent on matters related to the conservation and use of water resources and they did not create river commissions or similar institutions.

Only few provisions of the 1978 ACT address the issue of water use and conservation. Article 3 provides for the complete freedom of commercial navigation of the Amazon and its tributaries among the riparian States, subject to fiscal and police regulations (excluding cabotage). According to Article 6, the parties are encouraged to take national, bilateral, or multilateral measures to make the Amazonian rivers navigable and to handle any specific problems affecting free and unimpeded navigation. Article 5 is the only provision addressing the rational use of water resources, and reads as follows:

> Taking account of the importance and multiplicity of the functions which the Amazonian rivers have in the process of economic and social development of the region, the Contracting Parties shall make efforts aimed at achieving rational utilization of water resources.

Up to the present, despite some attempts, no substantive obligations regulating the use of the Amazon and its tributaries have been adopted under the framework of the ACT. The Third Meeting of the Amazon Cooperation Council (ACC) held under the aegis of this treaty in 1988, requested the Pro Tempore Secretariat to convene technical meetings among the parties in order to negotiate general regulations of commercial navigation in the Amazon Rivers, with a view to harmonizing the systems of signalling, security, and control of navigation in the Amazon basin.[11] Discussions were initiated at that time, but no final agreement was reached on this matter. In 2004, the Eighth Meeting of Ministers of Foreign Affairs (MMFA) called

[10] Ibid.

[11] Third Meeting of the ACC, held in Brasília on March 16–18, 1988, Pro Tempore Secretariat, *Base Jurídica del Tratado de Cooperación Amazónica: Antecedentes Constitutivos de la Organización del Tratado de Cooperación Amazónica* (BJ), La Paz, Bolivia, 2002, p. 185. The proposal of a regulation for free commercial navigation in the international Amazon Rivers was contained in the document TCA/III CCA/SPT BR/W.P. 1/88.

for another technical meeting to advance the drafting of those general regulations,[12] but no agreement has been adopted so far.

The ACT parties have also considered the adoption of an agreement to prevent and control contamination of water resources in the Amazonian rivers and its harmful effects on human health. At the Fifth MMFA in 1995, the ACT Pro Tempore Secretariat was requested to formulate a draft agreement to be later negotiated by the member States.[13] This commitment was reflected in the 1995 Lima Declaration adopted on that occasion.[14] A draft framework agreement on the control and contamination of water resources and its harmful effects was presented at the Tenth Meeting of the ACC held in Caracas in 2000,[15] but no final agreement has been reached in this respect.

However, some efforts have been made under the Amazon Cooperation Treaty Organization (ACTO) to strengthen the institutional capacity in the Amazon basin for the sustainable management of land and water resources. To that end, ACTO launched a project on Integrated and Sustainable Management of Trans-boundary Water Resources in the Amazon River Basin, financed by the Global Environment Facility (GEF) through the UN Environment Programme (UNEP) and with the support of the Organization of American States (OAS).[16] This project mainly aims to design a common water management policy, harmonize national laws, and strengthen the institutional capacity among the Amazon States. It will also seek to enhance knowledge on the adaptation measures needed to mitigate the impacts of floods and droughts.[17] The preparatory stage of this project was carried out in 2006–2007 and additional funding has been requested to GEF for the implementation phase. In connection with this project, ACTO signed a letter of understanding in Brasília on August 30, 2004 with the General Secretariat of the Intergovernmental Coordinating Committee of the La Plata Basin Countries (CIC), headquartered in Buenos Aires, where the GEF had approved a similar project. The objective was

[12] RES/IX MRE-OTCA/06, adopted at the Eighth MMFA, Manaus, Brazil, September 14, 2004, <http://www.otca.org.br/en/institucional/index.php?id=1330>.

[13] RES/V MRE-TCA/7, BJ, pp. 93–94.

[14] BJ, pp. 96–100. A few years later, at the Ninth Meeting of the ACC held in Caracas on October 5–6, 1998, the ACT parties committed to speed up consultations on a treaty to prevent and control contamination of water resources (RES/IX CCA TCA/4, http://www.otca.org.br/ep/Institucional/index.php?id=1248).

[15] RES/X CCA TCA/2 adopted at the Tenth MMFA held in Caracas on April 4–5, 2000, BJ, pp. 275–76.

[16] UNEP, *Integrated and Sustainable Development of the Trans-boundary Water Resources in the Amazon River Basin,* Global Environment Facility Project document, http://www.otca.org.br/en/programs-projects/index.php?id=1372

[17] Amazon Cooperation Treaty Organization (ACTO), *Management Report 2007–2008*, Brasília: ACTO Permanent Secretariat, 2008, p. 28.

to maintain a regular exchange of information between those Secretariats through biannual meetings.

As a result of the few international treaties on water resources in the Amazon, only very few institutions have been created so far to manage international rivers in the basin. An early example is the Treaty on Boundaries and Fluvial Navigation, between the Empire of Brazil and Venezuela,[18] adopted on May 5, 1859, which created a working group intended to improve all aspects related to navigation. Other examples can be found in treaties such as the agreement to constitute a Technical Group to consider the establishment of a Navigation Aid System in the Amazon River, signed on November 5, 1976,[19] and the agreement on the establishment of a working group to negotiate a Technical Agreement on Navigation, adopted in Lima on February 28, 1984, between Brazil and Peru.[20] In any case, the functions of those institutions are usually restricted to the exchange of information and joint activities.

In 2006, the Technical Chamber of Trans-boundary Water Resource Management under the Brazilian National Water Agency set up a Working Group for the management of the Acre River, shared by Brazil, Bolivia, and Peru, located in the southwest of the Amazon basin, in an area of great diversity of species and ecosystems, with high hydroelectric potential and rich deposits of gas and petroleum.[21] The riparian States signed a recommendation on May 11, 2006, by which they committed to create mechanisms for the joint management of the Acre River. In particular, they agreed to constitute three working groups (one per country) to facilitate the future creation of a transboundary body formed by competent national institutions, and to formulate a work plan for the joint management of that river.[22] The Technical Chamber of Trans-boundary Water Resource Management created in September 2006 a working group to carry out activities in Brazil for the management of that River, which has been meeting regularly since then.[23]

Due to the few international norms and institutions regulating the uses of international rivers in the Amazon basin it will probably be more difficult to handle potential conflicts among the riparian States. As they carry out projects to develop

[18] Venezuelan Ministry of Foreign Affairs, *Colección de tratados públicos de Venezuela*, Caracas: Imprenta Nacional, 1910, pp. 144–48, 343–45.

[19] 1054 UNTS 218 (1977).

[20] 1353 UNTS 104 (1984).

[21] *Seminário estratégias para a gestão da bacia hidrográfica do rio Acre: articulação interinstitucional no Brasil, Bolívia e Peru*, Epitaciolandia, Brazil, May 9–11, 2006, http://www.cnrh-srh.gov.br

[22] *Recomendações para a implementação da Gestão compartilhada da bacia transfronteiriça do Rio Acre*, Oficina Aspectos Legais e Ações Estratégicas para a Gestão Compartilhada da Bacia Trinacional do Rio Acre, Peru, Brazil and Bolivia, Epitaciolandia, Brazil, May 11, 2006.

[23] The Working Group was created at the Thirty-fifth Session of the Technical Chamber of Trans-boundary Water Resource Management, held on September 27, 2006; see http://www.cnrh.gov.br/

different water uses, for example irrigation and electricity generation, the greater is the likelihood of water-related conflicts in the Amazon River basin.[24] For example, controversies exist between Brazil and Bolivia with regard to the management of the Madeira River. In February 2007, the Brazilian government launched a national investment plan, known as the Growth Acceleration Program (PAC), for the 2007–2010 period.[25] This plan foresees investments in transport of R$6.2 billion in Brazil's northern region (which encompasses most of the Brazilian Amazon), including the construction of major hydropower plants in the Brazilian Amazon.[26] The most significant of these investments will be the paving of a 1,568 km stretch of the BR-163 highway, of which 523 km are in the State of Mato Grosso and the remainder in the State of Pará (both in the Brazilian Amazon).[27] Under the PAC, Brazil has approved preliminary licenses for the construction of the Santo Antônio and Jirau hydropower plants on the Madeira River (in the state of Rondônia), one of the major tributaries of the Amazon River, lying approximately 190 and 84 km from the Bolivian border, respectively. These projects are expected to satisfy 8 percent of the national demand for electricity, whereas the total cost currently exceeds US$9 billion, excluding the transmission lines. This would open a 4,200 km industrial waterway, allowing transport of soy and timber to Atlantic ports.[28] Bolivia has contested such works for their potential harmful transboundary effects, such as the flooding of part of its territory, the increased likelihood of mercury contamination and malaria, and the extinction of species of fish of vital economic importance that, with the construction of the barrages, would be prevented from reaching reproduction sites further up the Madeira River.[29] Bolivia requested an integrated environmental impact study on the entire extent of the Madeira basin, including the area in the Bolivian territory. In some official statements, however, Brazil has alleged that this was a domestic matter.[30] Bolivia formally objected to the construction of the two dams in a letter sent on July 11, 2007 to the Brazilian Ministry of Foreign Affairs.[31] Despite the controversy, the Brazilian Institute of Environment and Renewable Natural Resources (Instituto Brasileiro do

[24] C. Walter Gonçalves, *Conflitos no Campo Brasil 2006*, Comissão Pastoral da Terra, 2007. According to Gonçalves, water conflicts increased from fourteen in 2002 to seventy-one in 2005.

[25] Information about the PAC can be found at http://www.brasil.gov.br/pac/

[26] Ibid.

[27] Information on BR 163 is available at the Departamento Nacional de Infra-Estrutuda de Transportes website, http://www.dnit.gov.br/

[28] P. Verweij et al., *Keeping the Amazon Forests Standing: A Matter of Values*, report commissioned by WWF Netherlands, p. 29.

[29] G. Switkes, *Governo brasileiro avança com o projeto de construir barragens*, Americas Program, Center for International Policy, June 12, 2007, http://americas.irc-online.org

[30] "Bolivia accuses Brazil of breaching agreements and Lula considers compensation for the Madeira River Complex," December 7, 2007, online clipping available at: http://www.amazonia.org.br

[31] "Brazil's attempt to bargain on Madeira dams irritates Bolivia," March 8, 2007, online clipping, available at: http://www.amazonia.org.br

Meio Ambiente e dos Recursos Naturais Renováveis, IBAMA) has approved the building up of both the Santo Atonio and Jirau. The construction of the Santo Antonio dam is already advanced, with nearly 16.86 percent of the works concluded in December 2009.[32] Likewise, works for the Jirau dam have also received IBAMA's approval and are underway.[33]

6.2. BIODIVERSITY

Over the last few years, the Amazon States have made some efforts to promote cooperation in the field of conservation and rational use of biodiversity, by adopting common strategies, criteria, and indicators. For example, the 2002 Regional Biodiversity Strategy for the Andean Tropical Countries, signed under the aegis of CAN (CAN Decision 523), defines common objectives with respect to the management of biodiversity. A guiding principle is that of "integration," according to which the Regional Biodiversity Strategy should "address matters of regional concern only." States are encouraged to create and manage biological corridors and transboundary ecosystems, devise regional actions for the conservation of the Amazon, develop biotechnology, and create green markets.[34] This strategy also envisages financial, information, and planning "tools" to achieve its objectives.[35] In particular, information tools able to systematize and disseminate information are deemed essential to improve monitoring systems for *in situ* conservation in the Andean subregion.

There is no such regional strategy for the Amazon region. The ACTO Permanent Secretariat proposed a 2008–2013 Regional Plan of Action for the Amazonian Biodiversity, intended in particular to guide the creation of transboundary protected areas and corridors, and control wildlife trafficking.[36] Other proposals presented in 2008 by ACTO in the field of biodiversity conservation include a Cooperation Mechanism for the Monitoring and Control of Wild Fauna and Flora Trafficking in the Amazon Region[37] and a Regional Program for the Sustainable Management of Amazonian Protected Areas, which are now under discussion by

[32] "Santo Antônio está com 16,86% das obras concluídas," January 23, 2010, http://www.amazonia.org.br/noticias/noticia.cfm?id=342520

[33] "CPI do Madeira: consórcio iniciou as obras de Jirau sem licença definitiva," January 13, 2010, http://www.amazonia.org.br. See also "Recomeçam as Obras da Usina de Jirau no Rio Madeira," June 4, 2009, http://oglobo.globo.com/economia/mat/2009/06/04/recomecam-obras-da-usina-de-jirau-no-rio-madeira-756197316.asp

[34] CAN, *Estrategia Regional de Biodiversidad para los Países del Trópico Andino*, Decision 523, July 7, 2002, http://www.comunidadandina.org/normativa/dec/D523.htm

[35] Those "instruments" or "tools" are grouped into five categories: (1) institutional, (2) financial, (3) information, (4) political and legislative, and (5) planning.

[36] ACTO, *op. cit.*, p. 24.

[37] Ibid., pp. 24–25.

the Amazon States.[38] Technical cooperation programs such as that between the ACTO Permanent Secretariat and the Inter-American Development Bank (IDB), initiated in 2005 to strengthen the regional capacity for the sustainable use of the Amazonian Biodiversity,[39] could facilitate the adoption of a common strategy in the Amazon. Other programs carried out by ACTO, such as the Andes-Amazon Room (Sala Andes-Amazonia) or the regional biotrade program, have focused in particular on promoting trade of products and services derived from native Amazonian (or Andean) biodiversity.[40]

Other initiatives intended to promote regional cooperation in the field of biodiversity conservation include the debate among South American countries on the coordinated management and monitoring of transboundary protected areas. A common pattern among all those countries is the limited capacity to monitor national parks and other protected areas. It is observed, from the South of Mexico to the South of Argentina, that existing personnel are unable to monitor even half of the national parks in those countries; there is usually only one agent to 107,000 hectares of protected areas.[41] In view of this, some emphasis has been placed on either the creation of transboundary protected areas among those countries (or on a joint management of existing ones) or sites of critical importance for biodiversity conservation. For example, the 1996 Declaration of Santa Cruz de la Sierra, adopted at the Summit of the Americas on Sustainable Development, urged States to create cross-border protected areas and national parks and adopt common criteria for biodiversity conservation.[42]

In Brazil, the Ministry of the Environment undertook, from 1998 to 2001, a consultation process to identify priority areas for biodiversity conservation in the

38 One of the components being considered under this program is a Regional Fund for Protected Areas. A contribution of 10 million euros has already been offered for this by the German government. A pre-feasibility study was carried out in 2007 (ACTO, op. cit., p. 25).

39 Agreement between ACTO and the IDB: Strengthening the Joint Regional Capacity for the Sustainable Use of Amazonian Biodiversity, LEGIII/RG-563338–05, July 25, 2005, http://www.otca.org.br/ep/proyetos-programas/index.php?id=1204

40 ACTO, op. cit., pp. 26–27 (on the program Sala Andes Amazonia) and p. 26 on the Regional BioTrade Program. The Andes-Amazon Room is a commercial space in which Amazonian Andean biotrade products are showcased to make the best use possible of the advantages offered by regional and international markets. It is the result of a strategic partnership between the national biotrade programs of Bolivia, Colombia, Ecuador, Peru, and the Ministry of Environment of Brazil, with the support of UNCTAD and CAN. The first version, launched in Sao Paulo in October 2007, had exhibitors from the five participating countries representing approximately fifteen thousand families involved in the process.

41 According to the report by Carlos Castaño, "Diagnosis of the current situation of protected areas in Latin America and the Caribbean," presented at the Second Latin American Congress on National Parks and Other Protected Areas, Bariloche, Argentina, September 2007, http://www.ipsnews.net/news.asp?idnews=39482

42 Summit of the Americas on Sustainable Development, Santa Cruz de la Sierra, Bolivia, December 7–8, 1996, http://www.summit-americas.org/Boliviadec.htm

five biomes representative of the Brazilian biodiversity (Amazonia, Pantanal and Cerrado, Caatinga, Atlantic Forest, and Southern Plains, and Coastal and Maritime Zones). As a result, 900 areas were identified, which were listed by the Decree 5.092 of May 21, 2004 (and Order 126 of the Ministry of the Environment of May 27, 2004) and included under the national project for the Conservation and Sustainable Use of Brazilian Biodiversity (PROBIO). Of that total, 385 areas located in the Brazilian Amazon and 68 of them are contiguous to or continuous with the borders of South American countries.[43] Likewise, the Andean countries conducted in 2001 under the aegis of CAN, an analysis of the state of conservation of ecosystems in the Andean region.[44] A total of twenty protected areas were identified in zones near their common borders, some of which protected on both sides of the border, such as the Tambopata Candamo Reserve and the Bahuaja Sonene National Park in Peru, which are adjacent to the Madidi National Park in Bolivia, as well as the Tamá National Park on both sides of the border between Colombia and Venezuela.[45]

Following that initiative taken in Brazil to identify priority biodiversity conservation areas, the Brazilian Ministry of the Environment undertook a similar consultation process in 2003, this time involving all South American countries.[46] The objective was to identify contiguous or continuous areas of critical biological importance where joint actions were required. As a result, potential areas of cooperation for biodiversity conservation were identified, reflected in a document entitled *National Biodiversity Strategies in South America: Perspectives for Regional Cooperation*, concluded in 2004. This report suggested that the creation of a network of bordering protected areas and ecological corridors could facilitate in particular a monitoring system (and other conservation activities) among those countries, which could function as a South American clearinghouse mechanism.[47]

At present, there are only a few examples of cooperation in transboundary protected areas in South America. For example, the transboundary protected areas in the Great American Chaco, a biodiversity-rich ecosystem shared by Argentina, Paraguay, and Bolivia, motivated the start of the Program of Subregional Action for

[43] Brazilian Ministry of the Environment, Biodiversity and Forests Secretariat, *Estratégias nacionais de biodiversidade na América do Sul: perspectivas para a cooperação regional*, Brasília, 2004, p. 231. The map of these areas can be viewed on the Ministry Web site, http://www.mma.gov.br?d_estrutura=14&id_conteudo=743

[44] GTZ, Fundeco and Instituto de Ecología, op. cit.

[45] Brazilian Ministry of the Environment, Biodiversity and Forests Secretariat, op. cit., p. 249.

[46] Participating countries: Argentina, Bolivia, Brazil, Chile, Colombia, Ecuador, Guyana, French Guyana, Paraguay, Peru, Suriname, Uruguay, and Venezuela.

[47] As agreed at the Meeting for the South American Cooperation on Biodiversity, Rio de Janeiro, 2003, and contained in para. 1 of the 2003 Declaration of the Meeting for South American Cooperation on Biodiversity, adopted on that occasion (Brazilian Ministry of Environment, Biodiversity and Forests Secretariat, op. cit., p. 24).

the Great Chaco,[48] intended to preserve and restore this ecosystem. In the Andean region, programs for the joint management of transboundary protected areas include, for example, binational programs in bordering protected areas in the Vilcabamba–Amboró corridor in the southwestern Amazon, between Bolivia and Peru. Under the ACT, initiatives such as that among Bolivia, Ecuador, Colombia, Peru, and Venezuela, taken within a regional program for the Planning and Management of Protected Areas in the Amazonian Region supported by United Nations Food and Agriculture Organization,[49] have also attempted to provide border security services and improved transport and communication between protected areas. The following sections address more specific areas related to biodiversity conservation and sustainable use.

6.2.1. Genetic Resources

The Amazon States regulate the use of genetic resources mostly by national law. The ACT is silent on this issue and little progress has been made in this respect. On the other hand, the Andean-Amazon countries (Bolivia, Colombia, Ecuador, and Peru) managed to create a common regime on access to genetic resources. The CAN Decision 391[50] established in 1996 a Common Regime on Access to Genetic Resources, which set forth common conditions and procedures for the use and equitable participation in benefits deriving from the access to genetic resources.[51] This decision applies to genetic resources of which the member countries are the countries of origin (and their by-products),[52] but also to the genetic resources of migratory species that, for natural reasons, are found in the territories of the member States.[53] It created the Andean Committee on Genetic Resources that comprises the competent authorities of each of the parties.[54]

Apart from genetic resources *per se*, this regime also protects the "intangible components" associated with those resources, defined in Article 1 as "all know-how, innovation or individual or collective practice, with a real or potential value associated with the genetic resource, its by-products or the biological resource that contains

48 Ibid., p. 242.
49 Ibid., p. 255.
50 Adopted in Caracas on July 2, 1996, *Cartagena Agreement Official Gazette*, no. 213, July 17, 1996.
51 Article 2. Other objectives include the consolidation and development of scientific, technological, and technical capacities at the local, national, and subregional levels as well as strengthening the negotiating capacity of the member countries.
52 As defined in Article 1, by-products are: "a molecule, a combination or mixture of natural molecules, including crude extracts of live or dead organisms of biological origin that come from the metabolism of living beings."
53 Article 3.
54 Article 51.

them." Therefore, in cases where access to genetic resources with an intangible component is requested, the access contract, which is signed by the supplier of the intangible component and the applicant for access in order to allow its use, shall provide for the equitable sharing of benefits. A failure to comply with this obligation constitutes grounds for the nullification of the access contract.[55]

The access procedure starts with the submission of an application for access to genetic resources to the competent national authority, containing the information required by Article 26. The national authority evaluates the application (which may involve on-site visits) and issues a technical and legal opinion about the application's admissibility or invalidity.[56] Based on that opinion, the national authority finally approves or denies the application.[57] If approved, an access contract is negotiated and drafted between the State, represented by the national authority, and the applicant requesting access.[58]

Anyone who performs access activities without the respective authorization is liable for punishment.[59] In such cases, the competent national authority may apply administrative sanctions, for example fines, confiscation, determine the temporary or definitive closing-down of establishments, or deny the right for applying for new access.[60] Further to those sanctions, the access may be cancelled, suspended, or nullified, accompanied by the payment of compensation for damages (Article 47). Interestingly, according to this Article, payment of compensation may also apply to damages or losses caused to the biodiversity itself.[61]

The Andean countries have also set up at the subregional level a common regime for the protection of the rights of breeders of new plant varieties,[62] under the CAN Decision 345 signed in 1993. According to this decision, those who have created plant varieties are given breeder's certificates, insofar as the varieties are "new, uniform, distinct and stable."[63] Here again a subregime is created, with common procedures

55 Article 35.

56 Article 29.

57 Article 30.

58 As stated in Article 32, these are the parties to the access contract. Once the contract has been adopted and signed, a corresponding resolution shall be issued in a joint act. This resolution shall then be published, together with an extract of the contract, in the official newspaper or a newspaper with wide national circulation. As of that moment, the access shall be considered to have been granted, according to Article 38.

59 Article 46.

60 Article 47.

61 For an evaluation of the application of Decision 391 for each country see: GTZ/FUNDECO/IE, "Acceso a Recursos Genéticos, Conocimientos Tradicionales y Distribución de Beneficios," Fourth Regional Workshop, Pampatar, Venezuela, 17–19 July, 2001.

62 *Cartagena Agreement Official Gazette*, no. 142, October 21, 1993, http://www.comunidadandina.org/ingles/normativa/D345e.htm

63 Article 4. According to this Article, "created" shall be understood as the production of a new variety by the application of scientific skills to the genetic improvement of plants.

applicable in the territories of the member States, and a Subregional Committee for the Protection of Plant Varieties, composed of two representatives of each of the member countries.[64]

All new plant varieties shall be registered at the National Register of Protected Plant Varieties in each State's territory.[65] To be registered, they must comply with conditions set forth in Article 7 and be accompanied by a description of the relevant breeding process.[66] Once the competent national authority gives a favorable technical report, the breeder's certificate is issued.[67] The granting of a certificate is then notified to the Board of the Cartagena Agreement, which in turn brings it to the notice of the other member States.[68] The Subregional Committee for the Protection of Plant Varieties[69] is in charge of compiling and updating information susceptible to registration. This body also proposes guidelines to standardize procedures, such as examinations and laboratory tests, devises technical criteria to distinguish plant species, and suggests common norms for the protection of essentially derived varieties.[70]

6.2.2. *Wildlife Species of Fauna and Flora*

Most bilateral treaties adopted by the Amazon basin States *inter se* on environmental protection focus on the conservation of species of native fauna and flora. Prior to the adoption of the ACT, two treaties on the Conservation of Fauna and Flora in the Amazonian Territories were adopted between Brazil and Colombia[71] in 1973, and between Brazil and Peru[72] in 1975. With very similar texts, both treaties provided for the *in situ* conservation of species of fauna and flora through biological reserves and nurseries. The import and transit of wildlife-related products should be prohibited in cases where they had been barred in the territory of the other party. Both treaties foresee the adoption of further agreements, which should be negotiated via bilateral technical meetings, on issues such as the (total or partial) prohibition of scientific and amateur hunting, forest conservation, fishing, and the handling of exotic species (both in Article 3).

64 Article 37.
65 Article 6.
66 On registration, see Article 16. According to Article 7, new plant varieties shall fulfill the conditions of novelty, distinctness, uniformity and stability and also have appropriate generic domination. "Novelty" is defined in Article 8, "distinct" in Article 10, "uniform" in Article 11, and "stable" in Article 12.
67 Articles 19 and 20.
68 Article 20.
69 Article 37. The board provides the Technical Secretariat of the committee.
70 Article 38.
71 *Brazilian Official Gazette*, Decree 78016, July 12, 1976, p. 9367.
72 *Brazilian Official Gazette*, Decree 78802, November 24, 1976, p. 15382.

After the adoption of the ACT in 1978, a series of treaties on the conservation of the Amazonian fauna and flora and, more generally, on Amazonian cooperation have been signed among the basin States, probably out of a spirit of cooperation motivated by that treaty. For example, in 1979, Colombia and Peru adopted an agreement on the Conservation of Flora and Fauna of the Amazonian Territories,[73] which provides for the exchange of information on policies, programs, and legislation on the conservation of animal and plant life, and in the same year, an agreement on Amazonian Cooperation,[74] with particular emphasis on wildlife protection. Similar treaties, intended to promote Amazonian cooperation, were signed in 1981 between Colombia and Brazil,[75] and in 1982 between Brazil and Guyana.[76] These treaties intended to promote information exchange and put particular emphasis on the need to conserve the fauna and flora in the parties' respective Amazonian territories.[77] They also created binational commissions to implement the treaty and enhance cooperation between the parties.[78]

After a relatively active period that followed the adoption of the ACT, it was only several years later that other treaties on wildlife conservation or on Amazonian cooperation were concluded among those countries. Examples are the 1995 complementary adjustment to the technical cooperation agreement for Cooperation on the Amazon and Orinoco Regions between Venezuela and Brazil,[79] and the 1999

[73] *Colombian Official Gazette*, Year CXVII, no. 35650, Law 30 of 1980, November 25, 1980, p. 474.

[74] Ibid., p. 473.

[75] *Brazilian Official Gazette*, May 19, 1986, p. 007123.

[76] Article 4.

[77] This treaty between Colombia and Brazil was intended to promote information sharing and joint activities in particular to monitor and repress illicit trafficking of Amazonian flora and fauna and their by-products (Article 10). It also provided for fluvial transport in the Amazon, Içá-Putumayo, and Negro rivers, and the strengthening of roads, and aerial and telecommunications networks. The treaty between Brazil and Guyana also had the objective of enhancing information exchange and bilateral activities in areas such as regional development, scientific research, health and sanitation, and the protection of species of fauna and flora (Article 4). To ensure the conservation and rational use of such resources, the parties agreed to convene regular expert meetings, study bioecological processes related to wildlife, carry out joint programs, and control illicit trafficking of Amazonian fauna and flora.

[78] As defined in Article 10 of the Treaty on Amazonian Cooperation between Colombia and Brazil, a mixed binational commission was created and national authorities – the Brazilian Ministry of the Interior and the Colombian Institute of Renewable Natural Resources and Environment – were designated to exchange information and conduct binational programs particularly with regard to the management of fauna and flora. Under Article 6, the Brazil–Guyana Subcommission on Amazonian Cooperation was created.

[79] Adopted on July 4, 1995 (entered into force at the same day), Brazilian Ministry of Foreign Affairs, Divisão de Atos Internacionais, http://www2.mre.gov.br/dai/amb.htm (in Portuguese: Ajuste Complementar ao Convênio Básico de Cooperação Técnica para a Cooperação Amazônica e do Orinoco, Divisão de Atos Internacionais, http://www2.mre.gov.br/dai/amb.htm).

complementary adjustment to the Basic Technical and Scientific Cooperation Agreement between Brazil and Peru.[80]

Compared to those other bilateral treaties, the agreement for the Preservation, Conservation and Monitoring of Natural Resources in Border Areas adopted between Brazil and Bolivia in 1990 involves more specific obligations.[81] This treaty contains some measures to prohibit and penalize the hunting, poaching, and trading of species of fauna and flora threatened with extinction,[82] for example the creation of conservation sites and monitoring systems in border areas.[83] A binational working group was expected to list species of fauna and flora requiring protection, set up joint programs, and supervise and carry out annual evaluations of the activities carried out under the treaty.[84] More recent examples of bilateral treaties on the protection of wildlife include the 2003 Complementary Adjustment of Technical Cooperation between Brazil and Bolivia,[85] by which Brazil committed itself to provide technical assistance for the establishment of a control and monitoring mechanism to combat illegal trade of wildlife, and the 2003 Complementary Agreement to the Basic Technical and Scientific Treaty between Brazil and Peru intended to control rabies in wildlife species.[86]

6.3. HEALTH, FOOD SAFETY, AND SANITATION

Several of the tropical diseases occurring in the Amazon basin, such as dengue, Chagas, and malaria, are common to all basin States and many of them have not

[80] Article 3. This treaty designated national authorities that functioned as executing agencies and were expected to present reports every six months on the results of projects and activities undertaken within the framework of the agreement. Both treaties between Brazil and Venezuela, and Brazil and Peru had the objective of promoting information exchange through joint seminars, workshops and programs, and designated national authorities that should meet regularly.

[81] 2060 UNTS 56.

[82] Article 1.

[83] Article 5.

[84] Article 7.

[85] Adopted on November 18, 2003, Brazilian Ministry of Foreign Affairs, Divisão de Atos Internacionais, http://www2.mre.gov.br/dai/b_boli_226_3747.htm (in Portuguese: Ajuste Complementar ao Acordo Básico de Cooperação Técnica, Científica e tecnológica entre o Governo da República Federativa do Brasil e o Governo da República da Bolívia para Implementação do Projeto Manejo de Fauna Silvestre).

[86] *Brazilian Official Gazette*, no. 137, July 18, 2003 (in Portuguese: Ajuste Complementar ao Acordo Básico de Cooperação Técnica e Científica entre o Governo da República Federativa do Brasil e o Governo da República do Peru para Implementação do Projeto Controle de Raiva Silvestre). This treaty was intended to prevent and control endemic disease and improve the sanitary conditions of animal populations (Article 1). Brazil was expected to provide technical assistance in the identification of zones of risk, implementation of plans for diagnostics and vaccination, training of personnel in activities related to the prevention of disease such as wildlife rabies, and the exchange of information. Both treaties between Brazil and Bolivia, and Brazil and Peru designated national authorities in charge of the treaty implementation (Article 5). National authorities designated under this treaty should prepare regular reports on the progress of related activities.

yet been controlled or eradicated. For this reason, since the 1970s, those States have adopted *inter se* bilateral treaties on sanitary cooperation to fight common endemic diseases and address problems such as precarious sanitary conditions and poor infrastructure, particularly in border areas. For example, the 1972 Treaty on Sanitary Cooperation for the Amazon Region between Brazil and Colombia[87] provided for the setting up of public campaigns, health programs and vaccination stations in border areas, as well as for the mandatory notification of certain diseases. Some bilateral treaties attempt to reconcile sanitary cooperation and environmental protection, such as the Treaty on Sanitary Cooperation between Brazil and Guyana[88] and the Agreement on Sanitation and Environmental Protection (Complementary to the 1972 Technical Cooperation Treaty between Brazil and Colombia) both adopted in 1981.[89] Those treaties acknowledged that social and ecological factors should be handled jointly and envisaged the designation of national institutions for the exchange of information, prevention, and control of epidemiological diseases.

A few treaties were also concluded among the Amazon States at the end of the 1970s in the field of animal health. For example, Brazil has signed separate treaties with Bolivia, Venezuela, and Colombia to foster cooperation in border areas,[90] and to prevent and combat animal diseases, such as *aftosa* fever.[91] The parties under

[87] *Colombian Official Gazette*, Year CXI, no. 34244.28, 1975, p. 186 (Law 3 of 1975).

[88] Adopted on June 8, 1981 (entered into force on June 20, 1988), *Brazilian Official Gazette*, July 29, 1988, p. 014241 (promulgated by Brazilian Decree 96430 of July 28, 1988). The objective of this treaty was to formulate a technical cooperation program involving both parties' sanitary administrations to enhance human capacity, enable the exchange of medical equipment and information, and encourage epidemiological and sanitary research in tropical environments. Each party should designate a coordinator and subsequently constitute working groups comprising nationals of both parties to carry out specific activities.

[89] Adopted in Brasília on September 2, 1981 (entered into force on February 27, 1986), Brazilian Ministry of Foreign Affairs, Divisão de Atos Internacionais, http://www2.mre.gov.br/dai/b_colo_00_1608.htm (in Portuguese: Convênio Complementar ao Acordo Básico de Cooperação Técnica entre o Governo da República Federativa do Brasil e o Governo da República da Colômbia sobre Cooperação Técnica, Cientifica e Tecnológica no campo do Saneamento Básico e de Proteção do Meio Ambiente). This treaty envisaged the development of nonconventional technologies and integrated sanitation systems (Article 2).

[90] They include the 1977 Treaty on Animal Health in Boundary Areas between Brazil and Bolivia, adopted in Brasília on August 17, 1977 (entered into force on March 5, 1979), *Brazilian Official Gazette*, April 5, 1979, p. 004929 (promulgated by Brazilian Decree 83309 of April 4, 1979); the 1979 Treaty between Brazil and Venezuela on Animal Health in Boundary Areas, adopted on November 7, 1979 (entered into force on September 25, 1981), *Brazilian Official Gazette*, October 20, 1981, p. 019806 (promulgated by Brazilian Decree 86483 of October 16, 1981); and the 1985 Agreement on Animal Sanitation between Brazil and Colombia, adopted on July 16, 1985 (entered into force on May 18, 1994), *Brazilian Official Gazette*, April 24, 2000, p. 000008 (promulgated by Brazilian Decree 3424 of April 20, 2000).

[91] 1977 Treaty on Animal Health in Boundary Areas between Brazil and Bolivia, Article 1; 1979 Treaty on Animal Health in Boundary Areas between Brazil and Venezuela, Article 1.

those treaties committed to harmonize vaccination controls, synchronize vaccination dates, mutually recognize official certificates, and share epizootic data and any other information for controlling diseases.[92] Those treaties also envisaged the creation of mixed binational commissions,[93] constituted of agencies within the parties' respective Ministries of Agriculture that should meet annually. In 1997, Brazil and Venezuela signed an agreement on Animal and Plant Health[94] (complementary to the 1977 Cooperation and Friendship Agreement), to set up joint programs and exchange experiences on methods for controlling pests and managing animal and plant by-products.[95] This treaty provides a list of rights and obligations (Articles 4 and 5). For example, the parties were to harmonize sanitary methods of sampling and inspecting animals, plants and their by-products, and also require zoophytosanitary certificates mutually agree upon by the parties in order to trade agricultural and livestock products. Notification should be exercised on the occurrence or likelihood of diseases, but also with regard to any changes in zoophytosanitary regulations likely to affect commercial exchanges. Each party was expected to designate national agencies that would elaborate annual reports on the progress of the treaty[96] and meet annually.[97]

In the context of the ACT, the parties agreed to coordinate health services and take measures to improve sanitary conditions in the Amazon and combating epidemics (Article 8). In 1988, a Special Commission on Health was created. Health is one of the "programmatic areas" of ACTO's Strategic Plan 2004–2012 and a few projects were carried out in this domain. For example, in 2005, the ACTO Permanent Secretariat concluded an understanding with the Pan American Health Organization/World Health Organization to develop joint activities for improving the health conditions of Amazonian populations.[98]

[92] 1977 Treaty on Animal Health in Boundary areas between Brazil and Bolivia, Article 1; 1979 Treaty on Animal Health in Boundary Areas between Brazil and Venezuela, Articles 2 and 3; 1985 Agreement on Animal Sanitation between Brazil and Colombia, Article 2.

[93] Such as the mixed Brazil-Bolivian Commission on Animal Sanitation, pursuant to Article 2 of the 1977 Treaty on Animal Health in Boundary Areas between Brazil and Bolivia and the 1979 Treaty on Animal Health in Boundary Areas between Brazil and Venezuela, as well as a mixed commission from Brazil and Venezuela, integrated by agencies within both parties' Ministries of Agriculture, pursuant to Article 6 of the 1985 Agreement on Animal Sanitation between Brazil and Colombia.

[94] Adopted in Brasília on December 8, 1997 (entered into force on January 7, 1998), Brazilian Ministry of Foreign Affairs, Divisão de Atos Internacionais, http://www2.mre.gov.br/dai/sanidade.htm (in Portuguese: Ajuste Complementar ao Convênio de Amizade e Cooperação em Matéria de Saúde Animal e Sanidade Vegetal).

[95] Article 2.

[96] Article 12.

[97] Article 13.

[98] As defined in clause 2 of this understanding, these organizations agreed to hold ordinary meetings to negotiate a work plan. Technical cooperation should be directed to develop the institutional

At the subregional level, the Andean-Amazon countries created in 1983 the Andean System on Agriculture, Food Safety, and Environmental Conservation Jose Celestino Mutis by CAN Decision 182.[99] The objective of this common system is to reconcile agricultural uses and environmental preservation, especially in areas affected or likely to be affected by the expansion of agriculture, and, at the same time, protect CAN member States against threats associated with food shortages.[100] The member States are expected to carry out individually the measures defined in Article 6, but should also take common actions in the field of animal and plant health for combating pests and diseases.[101] To ensure food security Decision 182 encourages States to take joint policies and actions aimed at the conservation of water, forests, and other natural resources.[102] Interestingly, the Decision enumerates some tools for fulfilling its objectives, including trade,[103] financing,[104] and conservation.[105] The structure of the Andean System on Agriculture, Food Safety and Environmental Conservation comprises the Meeting of Ministers of Agriculture,[106] the CAN General Secretariat,[107] and the Agricultural Livestock Council, composed of high-level representatives of the member States' Ministries of Agriculture.

capacity of the ACTO in order to avoid transmissible diseases, notably malaria, by means of an epidemiological surveillance network. The designated focal points in this understanding are the Pan American Health Organization/World Health Organization representative in Brazil and the secretary-general of ACTO. See the ACTO Web site, http://www.otca.org.br/ep/proyetos-programas/index.php?id=1153

99 *Cartagena Agreement Official Gazette*, no. 2, September 7, 1983.

100 Preamble.

101 For example, by adopting programs aimed at the control of phytosanitary and zoosanitary diseases and improving diagnostic centers for plant and animal health, or adopting harmonized procedures for issuing certificates and permits.

102 Article 2(l). It was also intended to ensure the supply of basic diet products, increase the Andean populations' purchasing power, facilitate subregional trade by improving transportation and distribution, enhance the development of indigenous technologies, and agree on common criteria and joint positions in particular on international trade.

103 The trade tools involve several measures, such as creating trade information systems, adopting legal instruments with a view to overcoming obstacles to subregional trade by simplifying and harmonizing custom procedures and sanitary regulations, or adopting complementary preferential agreements (Articles 9–11).

104 The financing tools involve efforts for catalyzing resources and leveraging funds with Andean financial institutions and international organizations (Articles 13–14).

105 The conservation tools include mechanisms for the rational use of soils, forests, fauna and flora, and the adoption of agreements between two or more member States for the protection of water basins or reforestation programs (Article 15).

106 The Meeting of Ministers of Agriculture is in charge of formulating general policies for the implementation of the Andean System of Food Safety (Article 16–20).

107 The CAN General Secretariat provides technical support, present proposals and studies, and evaluates the development of the system.

6.4. FORESTS

Forests are being threatened by uncontrolled degradation and overall conversion to other types of land uses. In view of this, Chapter 11 of Agenda 21 urged States to promote the sustainable management and conservation of all forests, and rehabilitate degraded areas.[108] In response to the need for halting deforestation, in particular, the Amazon States have adopted national forestry codes and policies. For example, Brazil sanctioned the Public Forest Management Law 11.284 of March 2, 2006, which introduces, among other things, forest concessions as an option for forest management. These concessions do not involve the transfer of the land ownership and require the sustainable use of forest products and services.

The Plan of Action for the Prevention and Control of Deforestation in the Legal Amazon (Plano de Ação para a Prevenção e Controle do Desmatamento na Amazônia Legal) launched by the Brazilian government in 2004, is a strategy for containing deforestation in the Legal Amazon.[109] The main guidelines include the valorization of forests; incentives for the use of already degraded areas; territorial zoning (combating illegal land grabbing or *grilagem*); better instruments for monitoring and controlling deforestation. Among the priority areas under the Plan of Action are the land use planning, monitoring, the promotion of sustainable infrastructure, institutional coordination, and sustainable economic activities. In particular, the monitoring and control of illegal activities, such as biopiracy, slave labor, drug trafficking, etc, are seen as major concerns, given the situation of "absence of the State" in many parts of the Amazon – a factor leading to deforestation.[110] The Plan of Action proposes sustainable economic activities, particularly in areas already deforested or abandoned, as a way to contain agricultural expansion in forest areas. The Legal Amazon has approximately 63 million hectares of deforested areas (of which 16.5 million hectares are abandoned), where such economic exploitation could preferably take place.[111]

In addition to that Plan of Action, the Brazilian Government also launched the Sustainable Amazon Plan in 2008 (Plano Amazônia Sustentável: Direstrizes para o Desenvolvimento Sustentável da Amazônia Brasileira, PAS), whose main objective is to introduce a new model for exploiting the Amazon based on the sustainable use of natural resources.[112] The specific objectives of PAS are similar to those

[108] Agenda 21, chap. 11, "Combating deforestation," para. 11.10, UN doc. A/CONF.151/26/Rev.1, 1992.

[109] Brazilian Ministry of Foreign Affairs, Ministry of Science and Technology, Ministry of the Environment, Ministry of Mines and Energy, and Ministry of Development, Industry and Foreign Trade, *Brazil's Contribution to Prevent Climate Change*, White Paper, Brasília, 2007, p. 37.

[110] Ibid., p. 23.

[111] Ibid., p. 25.

[112] Presidência da República, *Plano Amazônia Sustentável: Direstrizes para o Desenvolvimento Sustentável da Amazônia Brasileira*, Brasília, 2008, p. 53.

contained in the Plan of Action for the Prevention and Control of Deforestation in the Legal Amazon. For example, PAS also stresses the need to reinforce the presence of the State in the Amazon, promote territorial planning and zoning, in particular by improving the national system of protected areas (Sistema Nacional de Unidades de Conservação) and strengthening the Amazonian Protection System (Sistema de Proteção da Amazônia, Sipam).[113] The Brazilian national law considers forests to be part of the national wealth and a common asset for all inhabitants (1988 Brazilian Constitution, Article 225). The use of natural forest resources on private land is permitted, but there are restrictions. According to Brazil's Forestry Code (Law 4.771/1965 Article 16, I) 80 percent of forests (or other types of native vegetation) in the Legal Amazon must be kept as a "Legal Reserve" (Reserva Legal).[114]

Only very few treaties have been adopted by the Amazon States on the protection of forests. In general, these are bilateral treaties intended to devise early warning and monitoring systems to assess and measure changes in forest areas, as well as mechanisms to prevent forest fires and illegal activities. Among the Amazon States, Brazil has probably developed the most advanced satellite technology and remote sensoring data to monitor its forests. The first estimates of deforestation in the Brazilian Amazon in the early 1970s were provided by airborne side-looking radar data acquired as part of the RADAM project (Radar na Amazônia). Since 1988, the Brazilian National Institute for Space Research (Instituto Nacional de Pesquisas Espaciais, INPE) began to produce annual estimates of gross deforestation for the Legal Amazon using satellite technology,[115] which is now one of the most advanced systems in the world. Other projects motivated rather by national security concerns, such as the System for the Vigilance of the Amazon (Sistema de Vigilância da Amazônia, SIVAM),[116] were also instrumental in monitoring the Brazilian Amazon.

[113] The Amazonian Protection System (Sipam) is an entity linked to the Brazilian Presidency, managed by the Operations and Management Center of the Amazonian Protection System (CENSIPAM). Sipam delivers technical information about the Legal Amazon to assist the government in planning and implementing activities in the Amazon. For information about Sipam, see http://www.sipam.gov.br/

[114] For a definition of a "legal reserve," see P. Affonso Leme Machado, *Direito Ambiental Brasileiro*, Sao Paulo: Malheiros Editores, 2004, p. 717.

[115] Information available on the Brazilian Ministry of Science and Technology Web site, http://www.obt.inpe.br/prodes

[116] The origins of this project dates back before 1990, when the then Brazilian president (Fernando Collor) charged the national intelligence service with "formulating and implementing a national system for coordination, seeking the integrated work of government agencies in the suppression of illegal activities in Amazônia" (J.R. Martins Filho and D. Zirker, "Nationalism, national security, and Amazônia: Military perceptions and attitudes in contemporary Brazil," *Armed Forces Society*, vol. 27, 2000, pp. 106–29). The institutions in charge are the Secretariat for Strategic Matters of the Presidency of the Republic, the Ministry of Justice, and the Ministry of Aviation, and the main objective of this project is to collect and process information on the Amazônia Legal (information available at: http://www.sivam.gov.br/).

In particular, the Brazilian DETER/PRODES program is a real-time forest monitoring system developed by INPE to monitor deforestation of the Legal Amazon. By virtue of Brazil's expertise in monitoring deforestation, this country has committed to provide technical assistance to other basin States. For example, under the 2001 complementary adjustment to the Basic Agreements on Technical and Scientific Cooperation between Brazil and Bolivia,[117] Brazil agreed to provide assistance in the evaluation of environmental impacts and setting up of monitoring systems. In 2003, Brazil and Peru adopted a protocol for the implementation of a system of Prevention and Control of Forest Fires.[118] Under this agreement, Brazil should provide technical assistance to Peru in setting up a warning and monitoring systems to control forest fires. In particular, the Brazilian DETER/PRODES program is a real-time forest monitoring system developed by the (INPE) to monitor deforestation of the Legal Amazon.[119] Under ACTO, the Amazon States have discussed in 2007–2008 the use of this system and the possibility of creating a database with enough structure to receive information for the eight member States.[120]

A few other initiatives were taken for combating illegal activities in border areas by using joint monitoring systems. For example, the 1990 Treaty on the Preservation, Conservation and Monitoring of Natural Resources in Border Areas[121] between Brazil and Bolivia urged the parties to prohibit hunting of and trading in endangered wildlife species and also called on the parties to protect forests in border areas.[122] The parties agreed to create contiguous protected areas[123] and a system for monitoring and controlling natural resources illegally exploited in boundary areas.[124] A working group, comprising the parties' Ministries of Foreign Affairs and environmental agencies, should be created to identify wildlife species requiring protection, coordinate joint programs, and annually evaluate the activities carried out under the treaty.[125]

117 Adopted on November 15, 2001, Brazilian Ministry of Foreign Affairs, Divisão de Atos Internacionais, http://www2.mre.gov.br/dai/bibolgeamb.htm (in Portuguese: Ajuste Complementar ao Acordo Básico de Cooperação Técnica e Científica entre o Governo da República Federativa do brasil e o Governo da República da Bolívia para Implementação do Projeto Melhoramento de Procedimentos para a Gestão Ambiental na Bolívia).

118 Adopted on August 25, 2003 (entered into force on February 16, 2004), Brazilian Ministry of Foreign Affairs, Divisão de Atos Internacionais, http://www2.mre.gov.br/dai/amb.htm (in Portuguese: Protocolo para Implementação de um Sistema de Prevenção e Controle de Incêndios Florestais).

119 Information about PRODES/DETER can be found on the Web site of the Brazilian National Institute for Space Research, http://www.inpe.br/ingles/index.php

120 ACTO, op. cit., p. 22.

121 *Brazilian Official Gazette*, April 14, 1999, p. 1 (promulgated by Brazilian Decree 3026 of April 13, 1999).

122 Article 1.

123 Article 3.

124 Article 5.

125 Article 8.

The Amazon States have also adopted nonbinding instruments to combat illegal activities, such as drug trafficking or logging in border areas. For example, Brazil and Guyana signed in 2002 a memorandum of understanding to enable cooperation between their federal polices,[126] through the exchange of information, mutual assistance in training police agents, and monitoring border areas.[127] A joint committee, composed of representatives of the federal polices of Brazil and Guyana, should meet annually to evaluate the progress of joint activities and provide recommendations.[128] Brazil and Peru also signed a memorandum of understanding in 2003 with a view to creating a vigilance system in the Peruvian Amazon with the capacity to exchange information with the Brazilian Sipam/SIVAM.[129] The Amazonian Protection System, under the Executive Office of the Brazilian Presidency (Casa Civil da Presidência da República), involves a number of methodologies and technologies that have been used for the control and monitoring of the Legal Amazon. Sipam provides technical services for a number of government agencies (at the federal, state, and municipal levels).

In addition to those few bilateral treaties, there is no regional treaty on the protection of Amazonian forests. The ACT contains no specific provisions on this matter. Under this treaty States have discussed regional criteria and indicators for the management of Amazonian forests. In a meeting held in held in Tarapoto (Peru) in 1995, they agreed on twelve criteria for the management of forests, which was later adopted in 2001 also in Tarapoto.[130] This second meeting launched the so-called Tarapoto Process, under which the ACT parties were supposed to come up with an approved set of criteria.[131] Another initiative in the Amazon is that of an Application of Forest Legislation in the Amazon (ALFA). ALFA is a regional exchange and discussion forum created with a view to facilitate cooperation and coordinate policies and actions to protect biodiversity, combat illegal logging, and control other illicit practices in the Amazon.[132]

[126] Adopted on April 18, 2002, 2184 UNTS 454.
[127] Article 2.
[128] Article 1.
[129] Articles 1 and 2.
[130] ACT Pro Tempore Secretariat, *Proceso de Tarapoto sobre Criterios e Indicadores de Sostentabilidad del Bosque Amazónico*, meeting minutes, Second Regional Meeting held in Tarapoto, Peru, 2001, SPT-TCA-BOL-72-ESP 2001.
[131] Resolution RES/VII MRE – TCA/7, adopted at the Eighth MMFA, held in Santa Cruz de la Sierra, Bolivia on November 22, 2002 formally initiated the "Tarapoto Process" and approved the regional project "Validación de 15 Indicadores Priorizados de Sostenibilidad del Bosque Amazónico" (BJ, pp. 161–62).
[132] ACTO, op. cit., p. 22.

6.5. HAZARDOUS SUBSTANCES AND ACTIVITIES

There exists no treaty of regional scope among the Amazon States on the handling of hazardous substances or the conduct of activities potentially dangerous to the environment. This matter has not been addressed in the context of the ACT either. It is only at the Andean subregional level that a common mechanism for the registration, control, and management of chemical pesticides for agricultural use was established in 1998, through CAN Decision 436 of June 11, 1998. The objective of this decision is to prevent damage to health and the environment and to facilitate trade in the Andean subregion.[133] It creates a mechanism to control the production, use, and trade of such substances by designating national authorities in charge of monitoring those activities. To this end manufacturers, importers, exporters, canners, and distributors of agricultural pesticides (either natural or legal persons) are under an obligation to register with the competent national authority.[134] Except for cases of officially declared phytosanitary emergency, no import, production, or use of chemical pesticides for agricultural use is permitted in the territories of CAN member States if not previously registered.[135] Interestingly, Decision 436 also applies to chemical agricultural pesticides not originating from the Andean subregion.[136]

In addition, Decision 436 created an Andean System of Information Exchange on Chemical Pesticides of Agricultural Use, coordinated by the CAN General Secretariat. This system is expected to support the registry and control of pesticides and follow up activities related to the harmonization of norms and procedures at the subregional level.[137] The competent national authority, usually composed of each State's respective Ministries of Agriculture, ensures the implementation of Decision 436.[138] The CAN General Secretariat is responsible for the registry of chemical pesticides for agricultural use at the subregional level.[139]

6.6. DISASTER PREVENTION AND MANAGEMENT

Little progress has been made in the Amazon with respect to disaster prevention and management. Again, efforts in that direction can be observed in the Andean subregion with the adoption of CAN Decision 529 of July 27, 2002. This decision creates the Andean Committee for Disaster Prevention and Management to help reduce

[133] Article 1.
[134] Article 10.
[135] Article 33.
[136] Article 3.
[137] Article 61.
[138] Article 4.
[139] Article 7.

risks and impacts related to (natural or anthropogenic) disasters in the member States' territories. In order to do so, this organ is expected to propose and coordinate policies, strategies, and plans in the areas of disaster prevention and management, mitigation, rehabilitation, and reconstruction.[140]

The Andean Committee for Disaster Prevention and Management is constituted of the member States' designated national authorities, including representatives of civil defense organizations, Planning Ministries (or institutions performing this function), and Ministries of Foreign Affairs. This body provides recommendations to the Andean Council of Foreign Ministers and the CAN General Secretariat, at their request and on its own initiative. It should also encourage research, information exchange, the harmonization of norms, and it should create monitoring systems.

6.7. INDIGENOUS POPULATIONS

The protection of indigenous populations was virtually overlooked under the ACT. Only Article 13 makes an allusion to this matter by asserting that tourism activities should occur "without prejudice to national regulations for the protection of indigenous cultures and natural resources." A few years after the adoption of this treaty, the parties agreed to create a Special Commission for Indigenous Affairs in 1989. The issue of human settlements and indigenous peoples' affairs is now one of ACTO's "programmatic areas." ACTO has a Coordinating Office of Indigenous Affairs that has been working on the formulation of a Regional Indigenous Agenda.[141] With the establishment of the ACTO Permanent Secretariat, a few projects and initiatives have been undertaken in that domain. For example, in 2004 ACTO signed a memorandum of understanding with the Coordinating Body for the Indigenous Organizations of the Amazon Basin to consolidate their institutional relationship. This agreement set up a working group to implement activities and ensure the exchange of information.

In the context of the Andean Community, a Working Group on the Rights of Indigenous Peoples was created in 2002 through Decision 524[142] as a consultative entity within the Andean Integration System. This body should promote the participation of indigenous peoples in the economic, social, cultural, and political spheres of subregional integration.[143] It provides recommendations to the Andean Council of Foreign Ministers, the Commission, and the General Secretariat on its own

[140] Article 1.

[141] ACTO, op. cit., p. 36.

[142] Adopted on July 7, 2002, http://www.comunidadandina.org/normativa/dec/D524.htm

[143] Article 1.

initiative and at the foreign ministers' request, disseminates information and monitors the implementation of decisions within the Andean Integration System.[144]

Much of the progress in this field is owed to developments within the Inter-American system, in particular the jurisprudence of the Inter-American Human Rights Court, which has clarified the scope and application of indigenous peoples' rights in the Americas. The basic instruments related to indigenous populations in the American continent are the 1969 American Convention on Human Rights (or San José Pact), the 1948 American Declaration of the Rights and Duties of Man, and the 1948 Inter-American Charter on Social Guarantees. Despite some attempts under the framework of OAS, there is no binding regional agreement on this matter. What is now being negotiated in the context of OAS is an American Declaration on the Rights of Indigenous Peoples.[145] According to Article 18 of the draft declaration, indigenous communities have a right to a healthy environment, from which other rights derive, for example, the right to be informed and participate in the formulation, planning, and management of activities likely to have an impact on indigenous lands.[146]

Under the Inter-American system, the Inter-American Commission on Human Rights (IACHR) and the Inter-American Court of Human Rights are the organs responsible for the protection of human rights, including indigenous communities' rights.[147] Apart from processing individual complaints, the IACHR also monitors the general situation of human rights in the member States' territories.[148] From the 1980s, it has addressed the issue of indigenous peoples' rights in its special reports[149] and case

[144] The Working Group is composed of members appointed by the member States, delegates elected by indigenous communities' organizations, representatives designated by national competent authorities, and one delegate indicated from each State's Office of the Ombudsman, in addition to consultative members. It holds regular annual meetings (Article 5), and the CAN General Secretariat functions as the Working Group's Technical Secretariat (Article 6).

[145] The OAS General Assembly through AG/RES. 1022 (XIX-O/89) requested the Inter-American Commission on Human Rights "to prepare a juridical instrument relative to the rights of the Indian peoples." In 1999, by resolution AG/RES. 1610 (XXIX-O/99), the General Assembly established the Working Group of the Permanent Council to consider the "Proposed American Declaration on the Rights of Indigenous Peoples," Draft Resolution OEA/Ser.G, CP/CAJP-2757/09 adopted by the OAS Permanent Council on May 15, 2009 renewed the mandate of the Working Group to continue holding its meetings in the quest for points of consensus.

[146] OAS Permanent Council, OEA/Ser.K/XVI, GT/DADIN/doc.301/07, April 27, 2007, adopted at the Tenth Meeting of Negotiations to Prepare the Draft American Declaration on the Rights of Indigenous Peoples, held in La Paz, Bolivia, 2007, http://www.oas.org/OASpage/Events/default_ENG.asp?eve_code=11

[147] The IACHR's functions are defined in Article 41 of the American Convention. Also within the OAS framework is the Inter-American Indian Institute, created in 1940 and headquartered in Mexico City, composed of seventeen countries of the Americas.

[148] IACHR, 2005 *Annual Report of the Inter-American Commission on Human Rights*, OEA/Ser.L/V/II.124, doc. 5, February 27, 2006, para. 118.

[149] The IACHR has devoted special reports that addressed the issues of indigenous populations, for example, *Authorities and Precedents in International and Domestic Law for the Proposed American*

system.[150] In 1990, the IACHR created the position of *special rapporteur* on the rights of indigenous peoples.[151]

In addition to those instruments of regional application, the 1989 International Labor Organization Convention on Indigenous and Tribal States (no. 169) applies to all tribal peoples whose cultural and economic conditions distinguish them from other sectors of the national community. It has been ratified by Brazil, Colombia, Peru, and Venezuela. As stated in Article 1(3), the term "peoples" does not have "any implications as regards the rights which may attach to the term under international law."[152]

Most indigenous rights recognized at the international level, such as the right to usufruct natural resources, to a legal personality, and to political participation,[153] are guaranteed in the Amazon States' national constitutions. For example, the 2008 Federal Constitution of Ecuador defines the country as intercultural and plurinational (Article 1), in recognition of the various ethnic populations and cultures within its territory. Spanish is the official language of Ecuador, but indigenous languages such as the Kichwa and Shuar are also recognized (Article 2) and the State will respect and encourage their use.[154] Brazil's 1988 Federal Constitution recognizes the "original rights" of indigenous communities over their traditional lands[155] and confers standing to these communities or their members to defend their rights (the public prosecutor intervenes in all procedural acts).[156] National constitutions often

Declaration on the Rights of Indigenous Peoples, OEA/Ser.L/V/II.110, doc. 22, March 1, 2001; and *The Human Rights Situation of the Indigenous People in the Americas*, OEA/Ser.L/V/II.108, doc. 62, October 20, 2000. For example, in relation to Ecuador, the IACHR examined particularly the human rights situation in the Oriente, in response to claims that oil exploitation activities were contaminating the water, air, and soil, thereby causing the people of the region to become sick and to have a greatly increased risk of serious illness (IACHR, *Report on the Situation of Human Rights in Ecuador*, Inter-Am.C.H.R., OEA/Ser.L/V/II.96, doc. 10 rev. 1, 1997).

150 IACHR, 2005 *Annual Report*, op. cit., para. 38.

151 IACHR, *The Human Rights Situations of the Indigenous Peoples in the Americas*, op. cit., chap. 1(2).

152 For information about the status of the ratification process, see http://www.ilo.org/ilolex/cgi-lex/ratifce.pl?C169

153 This was also recognized in the previous 1998 Constitution of Ecuador (Article 1). Indigenous peoples are also often deprived of political rights. On this issue, see *Yatama vs. Nicaragua*, Judgment of June 23, 2005, I/A Court H.R. Series C, no. 127, which dealt with the political participation of indigenous communities in local elections. In this case, the court ruled that any requirement for political participation that cannot be met by groups with differentiated internal organization, such as indigenous communities, violates Articles 23 and 24 of the American Convention: "el alcance pleno de los derechos políticos se convierte en un impedimento para que los ciudadanos participen efectivamente en la dirección de los asuntos públicos. Los requisitos para ejercitar el derecho a ser elegido deben observar los parámetros establecidos en los párrafos 204, 206 y 207 de esta Sentencia" (para. 20).

154 IACHR, OEA/Ser.L/V/II.96, doc. 10 rev. 1, April 24, 1997, Ecuador; see in particular chap. IX, "Human rights issues of special relevance to the indigenous inhabitants of the country."

155 Article 231.

156 Article 232.

limit indigenous communities' land ownership by allowing the economic exploitation of resources in the subsoil.[157]

To ensure protection of indigenous populations' rights, States create by law indigenous lands or reserves, as well as mechanisms to allow those communities to claim lands they have traditionally occupied. However, despite those formal guarantees, the physical and cultural integrity of indigenous communities is often threatened in all Latin American States. This is due to the lack of land ownership titles or to ineffective mechanisms for their delimitation, demarcation, and titling, which deprive those communities of their traditional lands. Some other factors, such as the introduction of infrastructure, the creation of municipalities in indigenous areas, and the occupation by settlers using forged titles, also hamper the titling or the existence of recognized indigenous lands.[158]

The Inter-American Court of Human Rights has particularly examined the issue of ineffectiveness of land titling procedures set forth in domestic law in the case of the *Mayagna (Sumo) Awas Tingni Community v. Nicaragua*[159] and the case of the *Yakye Axa Indigenous Community v. Paraguay*.[160] In that first case, the Court examined for the first time a claim based on the lack of effective procedures for the demarcation of indigenous lands. In its conclusion, it held that the lack of effective procedures to process territorial claims by indigenous communities and excessive delays constituted *per se* violations of judicial guarantees.[161] As a result, the government of Nicaragua was ordered to establish an effective mechanism for the delimitation, demarcation, and titling of indigenous communities' properties.

Once legally recognized, indigenous peoples' possession of their lands is often hindered by different actors, such as death squads working for large landowners, guerrilla groups, small landowners, and seasonal farm workers exploiting natural resources in indigenous lands.[162] Those threats can be observed, for example, in the case *Pueblo Indígena de Kankuamo v. Colombia*,[163] in which illegal military groups

[157] IACHR, *Report on the Situation of Human Rights of Indigenous*, op. cit.; see in particular chap. IV, "The situation of human rights of indigenous peoples in certain countries of the Americas."

[158] IACHR, OEA/Ser.L/V/II.97 doc. 29, Brazil, rev.1, September 29, 1997, paras. 41, 44.

[159] Case *Mayagna (Sumo) Awas Tingni Community v. Nicaragua*, Preliminary Objections, Judgment of February 1, 2000, I/A Court H.R. Series C, no. 66.

[160] Case *Yakye Axa Indigenous Community v. Paraguay*, Merits, Reparations and Costs, Judgment of June 17, 2005, I/A Court H.R. Series C, no. 125.

[161] Ibid., para. 86. The State of Paraguay had recognized the legal personality of indigenous communities, but from the start of the administrative land claim procedures until the issuing of the judgment by the court, eleven years, eight months, and twelve days had elapsed without a definite solution to the claim of the Yakye Axa indigenous community. The Court then declared that such a long delay constituted *per se* a violation of judicial guarantees case (para. 69).

[162] IACHR, *Second Report on the Situation of Human Rights in Colombia*, OEA/Ser.L/V/II.84 doc. 39 rev., October 14, 1993.

[163] Case *Pueblo Indígena de Kankuamo v. Colombia*, Judgment of July 2, 2004, I/A Court H.R. Series C, no. 107.

operating in the Resguardo Indígena Kankuamo, an area legally recognized and protected by Colombia since 2003, carried out murders, induced food shortages, and forcibly recruited young Indians.

Large-scale infrastructure and development projects undertaken by governments and private companies in indigenous lands also pose risks to the integrity of indigenous communities' lands and cultural survival.[164] For example, from the 1960s, oil exploitation in the Oriente (Ecuadorian Amazon) opened lands traditionally occupied by indigenous populations through the construction of roads and production sites and the introduction of workers and equipment.[165] Similarly, activities of lumber and oil companies in Peru have been reported as threatening the survival of indigenous populations and their natural environment.[166]

The legality of logging and oil concessions of private companies in indigenous lands has been examined in several cases presented to the Inter-American Court of Human Rights.[167] For instance, in the case *Pueblo indígena de Sarayaku v. Ecuador*,[168] this community objected to a contract signed between Ecuador and the Argentinean Compañía General de Combustible, without its prior consent, for the exploitation of petroleum in the province of Pastaza (Ecuador), mostly located within that community's traditional lands. The Court ordered provisional measures to prevent harassments inflicted by the Ecuadorian Army and the security personnel of the Compañía General de Combustible on the Sarayaku community. On that occasion, in a separate opinion, Judge Cançado-Trindade asserted that States were under an *erga omnes* obligation to protect all persons under their jurisdiction, not only *vis-à-vis* state power, but also in relation to private persons, including illicit armed forces of any kind.[169]

In general, the jurisprudence of the Inter-American Court of Human Rights has supported the view that States must grant special protection to indigenous populations due to their particular socioeconomic conditions, practices, and

[164] IACHR, OEA/Ser.L/V/II.108 Doc. 62, October 20, 2000; see in particular Chapter IV, "The situation of human rights of indigenous peoples in certain countries of the Americas: Report on the situation of human rights of indigenous peoples in Colombia," para. 33.

[165] IACHR, OEA/Ser.L/V/II.96, Ecuador, doc. 10 rev. 1, April 24, 1997; see in particular chap. IX, "The human rights issues of special relevance to the indigenous inhabitants of the country."

[166] IACHR, *Second Report on the Situation of Human Rights in Peru*, OEA/Ser.L/V/II.106, doc. 59 rev., June 2, 2000, para. 27.

[167] For example, in the case *Mayagna (Sumo) Awas Tingni Community v. Nicaragua*, a logging concession was granted to a company called Solcarsa (Merits, Reparations and Costs; Judgment of August 31, 2001, I/A Court H.R. Series C, no. 79); and in case *Yakye Axa Indigenous Community v. Paraguay* the community proposed an action against the enterprise Torocay S.A. Agropecuaria y Forestal Estancia Loma Verde (case *Yakye Axa Indigenous Community v. Paraguay*, op. cit.).

[168] Case *Pueblo Indígena de Sarayaku vs. Ecuador*, Provisional Measures, Order of the Inter-American Court of Human Rights of June 17, 2005, I/A Court H.R. Series C, no. 125.

[169] Case *Pueblo Indígena de Sarayaku v. Ecuador*, Resolution of the IACHR, Provisional Measures of July 6, 2004, para. 10.

customs.[170] The Court has affirmed in different rulings, based on Article 1 (1) of the American Convention on Human Rights, that members of indigenous and tribal communities require "special measures" in order to guarantee the full exercise of their rights, particularly in relation to their enjoyment of property rights, essential to safeguarding their physical and cultural survival.[171] These special measures are manifested, for example, in the way that property rights are conceived and should be granted to those communities.

According to the Court, by virtue of the indigenous populations' close ties with their lands – essential for the expression of their culture, spiritual life, and economic survival – the notion of property has a different connotation.[172] In contrast to modern societies, those communities exercise a communal form of "collective property" over their traditional lands, as ownership is not centered on individuals, but on the whole community.[173] As a result, in order to enjoy basic rights, such as that of a communal property, indigenous communities must be granted a legal personality.[174] According to the Court, the mere possession of lands by indigenous populations suffices for obtaining official recognition of land ownership and consequent registration.[175] However, possession is not a prerequisite for restoring indigenous lands,[176] as they could be claimed even when possession has ceased, provided that the unique relationship between a community and its traditional lands subsists.[177] Effective recourse and due process for claiming indigenous lands are basic guarantees of the exercise of the right to communal property.[178]

[170] Case *Yakye Axa Indigenous Community v. Paraguay*, op. cit., para. 54.

[171] Case *Mayagna (Sumo) Awas Tingni Community v. Nicaragua*, op. cit., paras. 148–49, and 151; Case *Indigenous Community Sawhoyamaxa v. Paraguay*, Merits, Reparations and Costs, Judgment of March 29, 2006, I/A Court H.R. Series C, no. 146, paras. 118–21 & 131; and Case *Indigenous Community Yakye Axa v. Paraguay*, op. cit., paras. 124, 131, 135–37, & 154.

[172] Case *Mayagna (Sumo) Awas Tingni Community v. Nicaragua*, op. cit., para. 149. In the case *Yakye Axa Indigenous Community v. Paraguay*, op. cit., the court affirmed: "los Estados deben tomar en consideración las características propias que diferencian a los miembros de los pueblos indígenas de la población en general y que conforman su identidad cultural."

[173] Case *Mayagna (Sumo) Awas Tingni Community v. Nicaragua*, op. cit., para. 149.

[174] Case *Yakye Axa Indigenous Community v. Paraguay*, op. cit., para. 83.

[175] Case *Mayagna (Sumo) Awas Tingni Community v. Nicaragua*, op. cit., para. 151. In the case *Moiwana Community v. Suriname*, the IACHR held that "possession of land should suffice when it comes to obtaining official recognition of the property and the consequential registration." Members of the Moiwana community should be considered "legitimate owners" of their traditional territories, of which they have been deprived as a result of the massacre of 1986 and the subsequent failure by the State to adequately investigate the events (Preliminary Objections, Merits, Reparations and Costs, Judgment of June 15, 2005, I/A Court H.R. Series C, no. 124, paras. 131 & 134).

[176] Case *Yakye Axa Indigenous Community v. Paraguay*, op. cit., para. 128.

[177] Ibid., para. 131.

[178] Ibid., para. 66.

In the case *Saramaka People v. Suriname,*[179] the Court examined in particular the limits to property rights exercised by indigenous peoples over their traditional lands. This case had its origins in Suriname's alleged failure to legally recognize and secure the Saramaka people's rights to own and control their traditional lands, including natural resources, by granting logging and mining concessions in areas within these lands. The Saramaka are descendants of African slaves who fought to free themselves from slavery in the seventeenth century and established autonomous communities in the rainforest organized into matrilineal clans (*lös*). Although not native to the region, the Court considered them tribal people and applied largely the same rights to those pertaining to indigenous peoples.[180] Their culture was found to be similar to that of tribal peoples, as the Saramaka people also maintain a strong spiritual relationship with the ancestral territory[181] they have traditionally used and occupied.[182]

One issue of contention in that case was whether the right of the Saramaka people to exercise property over their traditional lands also extended to the use and enjoyment of natural resources that lie on and within the land (including subsoil natural resources). As asserted in previous rulings,[183] the Court found that the cultural and economic survival of indigenous and tribal peoples depends on their access to and use of the natural resources in their territory, and that Article 21 of the American Convention on Human Rights protects their right to such natural resources.[184] According to the Court's jurisprudence, indigenous communities have the right to own the natural resources they have traditionally used within their territory for the same reasons that they have a right

179 Case *Saramaka People v. Suriname*, Preliminary Objections, Merits, Reparations, and Costs, Judgment of November 28, 2007, I/A Court H.R. Series C, no. 172.

180 Forest Peoples Programme, "Indigenous peoples' rights and reduced emissions from reduced deforestation and forest degradation: The case of the *Saramaka People v. Suriname*," 2009, p. 2, http://www.rightsandresources.org/documents/files/doc_992.pdf

181 According to the IACHR, "by using the term 'territory' the Court is referring to the sum of traditionally used lands and resources. In this sense, the Saramaka territory belongs collectively to the members of the Saramaka people, whereas the lands within that territory are divided among and vested in the twelve Saramaka clans" (note 66 to para. 82, of the case *Saramaka People v. Suriname*, op. cit.).

182 Ibid., para. 79.

183 Case *Yakye Axa Indigenous Community v. Paraguay*, op. cit., note 75, para. 137; case *Indigenous Community Sawhoyamaxa v. Paraguay*, op. cit., note 75, para. 118.

184 In the case *Saramaka People v. Suriname*, Interpretation of the Judgment of Preliminary Objections, Merits, Reparations and Costs, Judgment of August 12, 2008, I/A Court H.R. Series Series C, no. 185, para. 37, the IACHR clarified the term "survival" by stating that "the phrase 'survival as a tribal people' must be understood as the ability of the Saramaka to 'preserve, protect and guarantee the special relationship that [they] have with their territory' so that 'they may continue living their traditional way of life, and that their distinct cultural identity, social structure, economic system, customs, beliefs and traditions are respected, guaranteed and protected'.... That is, the term 'survival' in this context signifies much more than physical survival."

to own the land they have traditionally used and occupied for centuries. Without them, the very physical and cultural survival of such peoples is at stake.[185] In this regard, in the case *Sawhoyamaxa Indigenous Community v. Paraguay*,[186] the Court reaffirmed that the term "property" as used in Article 21 included material things that can be possessed and any right that may be part of a person's patrimony, which included "all movable and immovable, corporeal and incorporeal elements and any other intangible object capable of having value."[187]

Although the State, in the case *Saramaka People v. Suriname*, recognized the fact that the Saramakas have traditionally occupied certain lands, it also questioned their right to use and enjoy natural resources within those lands, especially in view of logging and mining concessions previously granted. Suriname claimed that all rights to land, including its subsoil natural resources, are vested in the State, which can freely dispose of these resources through concessions to third parties.[188] This brought the Court to the issue of whether and to what extent the State may grant concessions for the exploration and extraction of natural resources found within Saramaka territory. The Court found that, although it is true that all exploration and extraction activity in the Saramaka territory could affect the use and enjoyment of some natural resources traditionally used for the subsistence of the Saramakas, Article 21 of the American Convention on Human Rights should not be interpreted in a way that prevents the State from granting any type of concession for the exploration and extraction of natural resources. According to the Court, the right to the use and enjoyment of indigenous lands, like many other rights recognized in that Convention, is subject to certain limitations and restrictions.[189] In this sense, Article 21 states that the "law may subordinate [the] use and enjoyment [of property] to the interest of society." Nevertheless, the Court indicated three safeguards that must be observed by the State to ensure that the issuance of concessions within indigenous peoples' traditional lands does not amount to a denial of their survival. In particular, Suriname was ordered, firstly, to ensure the effective participation the Saramaka people regarding any development, investment, exploration or extraction plan within their territory; secondly, to guarantee that the Saramakas will receive a reasonable benefit from any such plan within their territory; and, thirdly, to ensure that concessions within Saramaka territory are issued by independent and technically capable entities.[190]

[185] Case *Yakye Axa Indigenous Community v. Paraguay*, op. cit., para. 137; case *Indigenous Community Sawhoyamaxa v. Paraguay*, op. cit., para. 118.

[186] Case *Indigenous Community Sawhoyamaxa v. Paraguay*, op. cit.

[187] Case *Yakye Axa Indigenous Community v. Paraguay*, op. cit, para 137; case *Mayagna (Sumo) Awas Tingni Community v. Nicaragua*, op. cit., para. 144; case *Indigenous Community Sawhoyamaxa vs. Paraguay*, op. cit., para. 121.

[188] Case *Saramaka People v. Suriname*, Judgment of November 28, 2007, op. cit., para. 119.

[189] Ibid., para. 125.

[190] Ibid., para. 129.

Subsequently, Suriname requested interpretation as to the "meaning and scope" of several issues in the ruling of the *Saramaka People* case issued in November 2007. In particular, interpretation on the following: a) with whom must the State consult to establish the mechanism that will guarantee the "effective participation" of the Saramaka people? b) to whom shall a "just compensation" be given (i.e., whether it must be given to the individuals directly affected or to the Saramaka people as a whole); and c) to whom and for which development and investment activities affecting the Saramaka territory may the State grant concessions.[191] Regarding the establishment of a consultation mechanism and the determination of beneficiaries of a "just compensation" in relation to development and investment projects in Saramaka territory, the Court held that it is the Saramaka people, not the State unilaterally, who must decide which person or group of persons will represent the Saramaka people in each consultation process ordered by the tribunal or indicate the beneficiaries of a just compensation.[192] In relation to the acceptable levels of damage or impact that a development or investment project may have on the lands occupied by the Saramakas, the Court reasserted the need to undertake prior environmental and social impact assessments to measure possible impacts (including environmental and health risks) and to ensure that members of the Saramaka community are aware of possible risks, which will allow their effective participation in the process of granting concessions.[193]

The special protection afforded to indigenous populations previously mentioned is also manifested in the methods of reparation applied by the Inter-American Court of Human Rights, which have gone far beyond mere pecuniary compensation. For example, in the case of *Aloeboetoe et al. v. Suriname*,[194] the court held that reparation was due not only to the victims, but also to their tribe.[195] Reparations in the case of *Yatama v. Nicaragua*[196] consisted also of giving publicity via the community's radio to the Court's judgment in memory of the victims and as a means of preventing similar events from happening again.[197] In the case of the *Moiwana Community v. Suriname*,[198] Suriname was requested to ensure the security of all members of the Moiwana community, which included the creation of a fund on

[191] Case *Saramaka People v. Suriname*, Judgment of August 12, 2008, op. cit.

[192] Ibid., paras. 18 & 25, respectively.

[193] Ibid., para. 41.

[194] Case *Aloeboetoe et al. v. Suriname*, Merits, Judgment of December 4, 1991, I/A Court H.R. Series C, no. 11.

[195] In this case, the Maroon community is composed of African decendents.

[196] Case *Yatama v. Nicaragua*, Preliminary Objections, Merits, Reparations and Costs, Judgment of June 23, 2005, I/A Court H.R. Series C, no. 127.

[197] Ibid.

[198] Case *Moiwana Community v. Suriname*, Interpretation of the Judgment of Merits, Reparations and Costs, Judgment of February 8, 2006, I/A Court H.R. Series C, no. 145.

behalf of this community and the building of a monument in memory of the victims.[199] In this case, in a separate opinion, Judge Cançado Trindade elaborated the concept of "spiritual damage" and "damages to the project of after-life."[200]

The progressive jurisdiction of the Inter-American Court of Human Rights has provided additional protection to indigenous lands. For example, in the *Matter of Pueblo indígena de Kankuamo regarding Colombia*, the Court's provisional measures – as asserted by the representative of the victims – have positively increased the presence of the Colombian government in the Kankuamo lands after years of absence, for example, via programs of social assistance and education, among others.[201] As Judge Cançado Trindade asserted, provisional measures in cases involving indigenous communities have enabled the monitoring (*monitoreo continuo*) and the gradual formation of a right to humanitarian assistance, as States are requested to adopt specific measures to protect life and the physical integrity of community members and must keep the Court informed about actions taken to that effect.[202] In the case *Matter of Pueblo Indígena de Kankuamo*, for instance, Colombia was requested to inform the court every two months about measures that had been taken to protect the Kankuamos Indians.[203]

B. COOPERATION AMONG THE AMAZON STATES AND OTHER COUNTRIES IN LATIN AMERICA AND THE CARIBBEAN

Some regional agreements adopted in Latin America and the Caribbean, to which some or all Amazon States participate, can be also applicable in these countries' respective Amazonian territories, even if they do not address the Amazon in

[199] Ibid.

[200] Separate opinion of Judge A.A. Cançado Trindade, case *Moiwana Community v. Suriname*, Interpretation of the June 15, 2005 Judgment on the Preliminary Objections, Merits and Reparations, Judgment of 8 February 2006, para. 18: "In my Separate Opinion (which I originally wrote in English) to that Court's Judgment in the present case of the Moiwana Community versus Suriname, I recalled that the members of the Moiwana Community, at the public hearing before the Court of 9/9/2004, indicated that the massacre of 1986, planned by the State, had 'destroyed the cultural tradition ... of the Maroon communities in Moiwana' (para. 80). Beyond moral damages, in my Opinion I referred to the occurrence of a truly spiritual damage (paras. 71–81) and, beyond damages to the project of life, I dared to elaborate conceptually on the damages to the project of after-life (paras. 67–70 and the following)."

[201] Case *Pueblo Indígena Kankuamo v. Colombia*, Provisional Measures of January 30, 2007, para. 8(a): "antes de la adopción de las medidas provisionales, el territorio kankuamo había vivido un período de más de diez años en ausencia del Estado. A partir de la adopción de las medidas provisionales se ha incrementado la presencia estatal a través de programas de inversión social, tales como la salud, la niñez, la producción, la educación y la estructura vial, entre otros.

[202] According to Judge Cançado Trindade in his concurrent vote in the case *Pueblo Indígena de Sarayaku v. Ecuador*, Resolution of the IACHR, Provisional Measures, June 17, 2005.

[203] Case *Pueblo Indígena Kankuamo v. Colombia*, Provisional Measures of July 5, 2004, I/A Court H.R. Series C, no. 109.

particular. An early example is the 1940 Convention on Nature Protection and Wildlife Preservation in the Western Hemisphere, which was opened for signature by the American states at the Pan American Union on October 12, 1940, and entered into force on May 1, 1942. The objectives of this treaty were to promote *in situ* conservation of native flora and fauna, including migratory birds, and to preserve "regions and natural objects of aesthetic, historic, or scientific value"[204] through national parks and reserves, nature monuments, and wilderness reserves. The creation of such areas should be reported to the Pan American Union.[205]

That convention devised some mechanisms to protect native wildlife and "scenery of extraordinary beauty, unusual and striking geologic formations,"[206] some of which would be later used under the Convention on International Trade in Endangered Species of Wild Fauna and Flora (CITES). The parties should create protected areas[207] and provide a list of species to be annexed to the convention. The hunting, killing, capturing, or taking of those listed species would be allowed only with the permission of government authority under special circumstances.[208] Exports and transit of protected fauna and flora (or parts thereof) should only be authorized by the issuing of certificates,[209] whereas imports of species of fauna and flora (or parts thereof) protected by the country of origin required a certificate of lawful exportation.[210] Trade within national parks and strict wilderness was prohibited.[211] This convention could have been useful for the protection of wildlife and natural sites within the parties, and the Amazon States in particular, but it has never been well implemented and is recalled as an example of a "sleeping treaty."[212]

At the subregional level, Mercosur's member States adopted in 2001 a Framework Agreement on the Environment with a view to promoting environmental protection of all biomes within each of the parties (including the Brazilian Amazon). At present, this agreement contains only general obligations, for example, providing for the exchange of information and encouraging the parties to harmonize national laws and formulate policies to prevent, control, and mitigate adverse environmental

[204] 161 UNTS 193.

[205] Article 2(3). The Conservation Section was established on July 1, 1943 at the Division of Agricultural Cooperation of the Pan American Union to oversee this convention ("Conservation Section in the Pan American Union," *Bulletin of the Pan American Union*, vol. 77, 1943, pp. 714–15, at p. 714).

[206] Article 1.

[207] Article 2.

[208] Article 8.

[209] Article 9(1).

[210] Article 9(2).

[211] Article 3.

[212] P. Birnie and A. Boyle, *International Law and the Environment*, Oxford: Oxford University Press, 2002, p. 202.

impacts. If more precise obligations arise from that treaty (or other legal instruments under Mercosur) they might have implications for the preservation of the Brazilian Amazon in particular.

A few other treaties concluded by two or more Amazon States in the wider Caribbean region may provide useful examples of regulatory techniques, but are not applicable to the Amazon basin, such as the 1990 Protocol Concerning Specially Protected Areas and Wildlife, the 2000 Framework Agreement for the Conservation of Living Marine Resources on the High Seas of the South Pacific (Galapagos Agreement); or, within the inter-American system, the 1996 Inter-American Convention for the Protection and Conservation of Sea Turtles, and the 1979 Convention for the Conservation and Management of the Vicuna.

CONCLUSIONS

The overview of the international instruments presented here indicates how the Amazon States have been interacting on matters related to the protection of the Amazon. In terms of form, the Amazon States have opted for adopting bilateral treaties, instead of multipartite agreements. The 1978 Amazon Cooperation Treaty is the only regional agreement to involve all Amazon States, with the specific objective of promoting the conservation and rational use of the Amazon. The other few regional treaties to which those countries are parties, such as the 2001 Mercosur Framework Agreement on the Environment, are only applicable to parts of the Amazon (in this case, the Brazilian Amazon). Likewise, norms under more comprehensive subregional regimes, such as those adopted by the Andean Community are only applicable in a few Amazon countries (Bolivia, Colombia, Ecuador, and Peru). The Permanent Secretariat of ACTO is currently the only organ that represents exclusively and all Amazon States, able to formulate norms and policies applicable basin-wide.

As far as legal obligations are concerned, the treaties adopted by the Amazon States contain a general duty to cooperate, which is translated into obligations to carry out joint activities, harmonize national laws, and coordinate activities, among others. More specific rights and obligations are normally found in CAN decisions, for example Decision 391 on a Common Regime on Access to Genetic Resources. In general, the Amazon States are under an obligation to exchange information and notify other States in the event of emergencies or epidemic diseases, such as foreseen for example by the 1972 Treaty on Sanitary Cooperation for the Amazon Region between Brazil and Colombia.

With regards to procedural obligations and mechanisms devised under those treaties (and nonbinding legal instruments), they consist basically of the exchange of information by means of joint activities, regular meetings, and in some cases, notification. In general, designated national authorities (or binational commissions) should carry out planned activities, oversee the implementation of the treaty and facilitate information exchange.

Due to a combination of factors – few specific legal obligations, mechanisms or institutions charged with little more than the exchange of information – the Amazon States have not created effective avenues of cooperation in that region. With the exception of some decisions adopted under the Andean Community applicable in the member States' respective Amazonian territories, the international legal instruments adopted among the Amazon countries have usually had little practical impact. The ACTO Permanent Secretariat could potentially play a role in facilitating the creation of norms and mechanisms to tackle common problems in the Amazon.

7

Multilateral Treaties and Global Actors in the Amazon

The previous chapters examined the Amazon from a regional perspective, focusing in particular on how the Amazon States have interacted among themselves to protect that region. This analysis was based on bilateral and regional treaties adopted among them and on decisions taken in the framework of regional and subregional organizations. This overview of legal instruments adopted by the Amazon States gives an indication of how they have been cooperating *inter se* in matters related to the conservation and sustainable use of the Amazon.

This chapter and the subsequent ones focus on the interaction between the Amazon States and the international community as a whole. This analysis intends to assess how those States interact with the international community, represented by States, international organizations, and nongovernmental organizations (NGOs), and what role these actors may have in the protection of the Amazon. In view of the weaknesses observed within the regional legal framework of the Amazon, and the 1978 Amazon Cooperation Treaty (ACT) in particular, this and the subsequent chapters enquire whether the international community plays a role in ensuring the adoption or enforcement of legal obligations aimed at the environmental protection of the Amazon.

In particular, the present chapter looks at actual policies and activities carried out by some global actors, including States, international organizations, and NGOs, *vis-à-vis* the Amazon region, and to what extent the international community may help in its protection. The analysis focuses on the interactions between the Amazon countries and the community of States that occur under certain multilateral environmental agreements, and also between the Amazon States and international organizations and NGOs focusing on this region.

A. MULTILATERAL ENVIRONMENTAL AGREEMENTS

Apart from the international instruments adopted at the bilateral and regional levels examined in previous chapters, the Amazon is also protected by multilateral environmental agreements (MEAs) to which the basin States are parties. For example, species of wild fauna and flora occurring in their respective Amazonian territories are covered by instruments such as the 1973 Convention on International Trade in Endangered Species of Wild Fauna and Flora (CITES),[1] the 1979 Convention on the Conservation of Migratory Species of Wild Animals,[2] and, more generally, by the 1992 Convention on Biological Diversity (CBD).[3] The protection of forests in particular is the object of the 2006 International Tropical Timber Agreement,[4] the 1992 Forest Principles[5], and the 2007 Non-Legally Binding Instrument on All Types of Forests.[6]

Wildlife species of flora and fauna in the Amazon, such as the golden lion tamarin,[7] jaguar,[8] and piracucu,[9] are listed in the appendices of CITES. Similarly, tropical timber species, such as the merbau (*Intsia bijuga* and *Intsia palembanica*), ramin (*Gonystylus spp.*), and American mahogany (*Swietenia spp.*) are protected under CITES, which has contributed to the conservation of timber commercially extracted and subject to the effects of deforestation.[10] For example, the listing of mahogany (*Swietenia macrophylla*) in CITES Appendix III[11] has helped reduce illegal trade in Brazil, as CITES has furthered inspection and other control mechanisms implemented in ports of export for this species. The mahogany, which occurs from Mexico along a vast arc through Venezuela, Colombia, Ecuador, Peru, Bolivia, and Brazil,[12] has been one of the main forest products exported from Brazil and was also used domestically. Brazil requested its inclusion in CITES Appendix III in

1 All Amazon States are parties to this treaty; 993 UNTS 243.

2 Only Bolivia, Peru, and Ecuador are parties to this treaty; 9 I.L.M. 15 (1980).

3 All Amazon States are parties to this treaty; 31 ILM 818, 1992.

4 33 ILM 1016, 1994.

5 Non-Legally Binding Authoritative Statement of Principles for a Global Consensus on the Management, Conservation and Sustainable Development of All Types of Forests, A/CONF.151/26, vol. III.

6 A/C.2/62/L.5, October 22, 2007.

7 CITES listing, Appendix I, July 1, 1975.

8 Ibid.

9 Ibid.

10 B.D. Rodan, A.C. Newton, and A. Verissimo, "Mahogany conservation: Status and policy initiatives," *Environmental Conservation*, vol. 19, no. 4, 1992, pp. 331–38.

11 No documented cases of illegal trade in Brazil have been reported by the Brazilian Institute of Environment and Renewable Natural Resources. However, Brazilian timber producers have been seen working near Loreto and Madre de Dios regions in Peru, indicating that illegal trade may be taking place. According to the Peruvian authorities, the restrictive export measures taken by Brazil in recent years have caused increased harvest pressure near the two countries' borders, especially on the Peruvian side (TRAFFIC, *CITES Appendix III Implementation for Big Leafed Mahogany* Swietenia macrophylla, TRAFFIC Online Report Series, no. 1, 2002, p. 18, http://www.traffic.org).

12 J. Grogan, P. Barreto, and A. Veríssimo, *Mogno na Amazônia Brasileira: Ecologia e Perspectivas de Manejo*, Belém: Imazon, 2002, p. 16.

May 1998 and the listing became effective in July 1998.[13] Certain migratory species, such as reptiles occurring in the Upper Amazon and their habitats, are also protected according to their level of endangerment under the 1979 Convention on the Conservation of Migratory Species of Wild Animals, such as the Amazon River Dolphin (*Boto-cor-de-rosa*).[14]

With regards to forests, the 2006 International Tropical Timber Agreement (ITTA)[15] addresses the issue of international trade in tropical timber. This treaty was negotiated under the auspices of the United Nations Conference on Trade and Development, first adopted in 1983 and subject to later revisions (1994 and 2006). The major objectives of the ITTA are to promote the expansion and diversification of international trade in tropical timber and ensure the sustainable management of tropical timber-producing forests.[16]

In order to administer and supervise the operation of the ITTA, the International Tropical Timber Organization (ITTO) was created under this treaty[17] (and became operational in 1987). The ITTO has contributed to the conservation of tropical forests by generating knowledge and monitoring and funding field projects. For example, useful information produced by the ITTO include the Tropical Forest Update, the Annual Review of World Timber, and guidelines such as the ITTO Guidelines for Sustainable Use and Conservation of Biodiversity in Tropical Timber Production Forests as well as the 2005 Criteria and Indicators for the Sustainable Management of Tropical Forests. The ITTO also carries out diagnostic missions in countries such as Bolivia (1996), Guyana (2003), and Peru (2003) among others,[18] finances projects in producer countries[19] and conducts national studies at their request, as occurred, for example, in Brazil and Peru.[20] The ITTO is also committed to strengthening

[13] TRAFFIC, op. cit., p. 13.

[14] Information is available at: http://www.cms.int/pdf/en/CMS1_Species_5lng.pdf

[15] 33 I.L.M. 1014 (1994). Most Amazon States are parties, except Guyana and Suriname (United Nations Environment Programme (UNEP), Amazon Cooperation Treaty Organization (ACTO) and Research Center of Universidad del Pacífico (CIUP)), 2009, *Environment Outlook in Amazonia, Geo Amazonia*, p. 229, http://www.unep.org/pdf/GEOAMAZONIA.pdf

[16] Article 1.

[17] Article 3.

[18] P.D. Hardcastle et al., *Capability and Cost assessment of the Major Forest Nations to Measure and Monitor Their Forest Carbon for Office of Climate Change: Final Report*, Edinburgh: LTS international, April 7, 2008, pp. 31, 44, 50.

[19] For example, ITTO, *Information and Technical Assistance for Production and Trade of Tropical Timber (Brazil)*, ex-post evaluation report, Project PD 7/94, rev.3 (M,I), Thirty-eighth Session, 2006, Mérida, Yucatán, Mexico. Project PD 7/94 aimed at the development of the timber industry in the Amazon region.

[20] Under Decision 6(XXXI) of November 3, 2001, on Forest Law Enforcement in the Context of Sustainable Timber Production and Trade, the ITTO's executive director was authorized to respond to requests from producer member countries to conduct studies and provide assistance on ways to enhance forest law enforcement. Studies were completed for Brazil, Ecuador, Guyana, Honduras, Malaysia, and Peru.

collaboration with other regional institutions, such as the Amazon Cooperation Treaty Organization (ACTO).[21]

In addition to the ITTA, which is more a commodity agreement than a conservation treaty, only nonlegally binding instruments were adopted globally in relation to forests. Both the 1992 Forest Principles and the 2007 Non-Legally Binding Instrument on All Types of Forests provide general guidelines on how States should manage their forests, included the Amazon. For example, according to these instruments, States are required to sustainably manage their forests,[22] adopt national forest policies,[23] make efforts to increase forest cover, and prevent their loss and degradation,[24] as well as to produce information and knowledge on forests, among other things.[25]

Natural sites of international importance for their ecological or aesthetic value are protected under the 1971 Convention on Wetlands of International Importance especially as Waterfowl Habitat (the Ramsar Convention).[26] Ramsar sites can also include wetlands extending over the territories of more than one party (Article 5), referred to as "Transboundary Ramsar Sites."[27] Each party designates wetlands of international significance in its territories[28] for inclusion in the List of Wetlands of International Importance (the Ramsar List) maintained by the International Union for Conservation of Nature (IUCN).[29] For example, the Mamirauá Reserve, located in the Brazilian state of Amazonas, is a Ramsar site (included on October 4, 1993).[30]

Similarly, the 1972 UN Educational, Scientific and Cultural Organization (UNESCO) World Heritage Convention[31] protects natural and cultural sites of outstanding universal value from the point of view of science, conservation, and natural beauty. The properties forming part of the cultural and natural heritage are included in the World Heritage List maintained by the World Heritage Committee. In 2007,

21 Decision 7(XXX), on the Role of ITTO in International and Regional Organizations and Fora, Thirtieth Session, May 28 to June 2, 2001, Yaoundé, Cameroon, ITTC (XXX)/19, June 2, 2001.

22 1992 Forest Principles Article 2(b), A/CONF.151/26 (vol. III), June 14, 1992; 2007 Non-Legally Binding Instrument on All Types of Forests, A/C.2/62/L.5, October 22, 2007, Article 1(a).

23 1992 Forest Principles, Article 5(a); 2007 Non-Legally Binding Instrument on All Types of Forests, Article 6.

24 1992 Forest Principles Article 2(c); 2007 Non-Legally Binding Instrument on All Types of Forests, Article 5, Global Objectives on Forests (Objective 1).

25 1992 Forest Principles Article 2(c); 2007 Non-Legally Binding Instrument on All Types of Forests, Article 6(r), (v).

26 All Amazon States are parties; 11 I.L.M. 963 (1972).

27 Resolution VII.19 (1999) on international cooperation, http://www.ramsar.org

28 Ramsar Convention, Article 2(2).

29 Ibid., Article 2.

30 The List of Wetlands of International Importance is available at: http://www.ramsar.org

31 11 ILM 1358, 1972.

at the Thirty-First Meeting of the World Heritage Committee, some rainforests, for example, Atsinanana (Madagascar), the ecosystem and relict cultural landscape of Lopé-Okanda (Gabon), and the primeval beech forest of the Carpathians (Ukraine and Slovakia), were included in the World Heritage List, and since then many others have been. Of the 179 World Heritage sites included for their natural heritage values, thirty-eight (22 percent) are located in Latin America and the Caribbean. Of these, half are located in Argentina (four), Brazil (seven), Mexico (four), and Peru (four).[32] There are only very few natural heritage sites in the Amazon; they include, for example, the Noel Kempff Mercado National Park (Velasco Province, Bolivia), the Central Amazon Conservation Complex (State of Amazonas, Brazil) and the Canaima National Park (Bolivar State, Venezuela).[33]

Other multilateral agreements covering cross-cutting issues, which often require actions involving multiple fields and sectors, in particular the 1992 UN Framework Convention on Climate Change (UNFCC)[34] and the CBD, are also useful instruments for the conservation of the Amazon. These treaties contain mechanisms that can be used in relation to natural resources in the Amazon. Examples include the Clean Development Mechanism under the Kyoto Protocol, the monitoring of wildlife species, and ecosystems carried out by the United Nations Environmental Programme (UNEP) World Conservation Monitoring Centre (WCMC),[35] which provides support to the CBD Secretariat.[36]

To say that natural resources or sites in the Amazon are protected internationally through MEAs and nonbinding instruments means that the basin States enjoy the rights and should observe the principles and obligations defined under these instruments. It also involves a duty on the part of the international community to provide assistance to these countries where appropriate.

States have a customary law right to use their natural resources according to their own environmental policies (Stockholm Declaration, Article 21; Rio Declaration, Article 2). By the same token, they have sovereign rights over their forests (1992 Forest

[32] E. Guerrero and S. Sguerra (eds.), *Protected Areas and Development in Latin America: From Santa Marta 1997 to Bariloche 2007 and Perspectives for a New Decade*, Bogotá: IUCN Colombian Committee, Parques Nacionales Naturales Colombia & Fundación Natura, Colombia, 2009, p. 48.

[33] Information is available at: http://whc.unesco.org/en/list

[34] All Amazon States are parties (UNEP, ACTO, & CIUP, op. cit., p. 229).

[35] http://www.unep-wcmc.org

[36] For example, to synthesize available scientific information on ocean fertilization (through the study by CBD, UNEP, and WCMC, *Scientific Synthesis of the Impacts of Ocean Fertilization on Marine Biodiversity*, CBD Technical Series, no. 45, 2009, http://www.cbd.int/doc/publications/cbd-ts-42-en.pdf); to analyze the observed and predicted impacts of ocean acidification on marine biodiversity (through the study by CBD, UNEP and WCMC, *Scientific Synthesis of the Impacts of Ocean Acidification on Marine Biodiversity*, CBD Technical Series, no. 46, 2009, http://www.cbd.int/doc/publications/cbd-ts-42-en.pdf); or to review existing scientific literature to feed the deliberations of the CBD's Second Ad Hoc Technical Expert Group on Biodiversity and Climate Change.

Principles, Article 1), biological resources (CBD, Article 3), cultural and natural heritage (UNESCO World Heritage Convention, Article 6), wetlands (Ramsar Convention, Article 2 (3)) and over any other natural resources in their territories. Other rights derive from that exclusive sovereign right, for example, the authority of national governments to determine access to genetic resources (CBD, Article 15) and to request international assistance for property forming part of the cultural or natural heritage (UNESCO World Heritage Convention, Article 19).

The Amazon States, like other countries, are subject to customary rules and principles, most of which are included in MEAs and other instruments, which should guide national activities *vis-à-vis* the Amazon and their mutual relations.[37] For example, they should take a precautionary approach when carrying out economic or other activities in the Amazon (Rio Declaration, Principle 15) and should observe the principles of cooperation (Rio Declaration, Principle 7) and equitable utilization when handling shared natural resources, such as international rivers or migratory species. Based on a precautionary approach, States agreed, for example, to control global emissions of substances that deplete the ozone layer, in the context of the 1987 Montreal Protocol on Substances that Deplete the Ozone Layer (Preamble). Under the 1997 Convention on the Law of the Non-navigational Uses of International Watercourses, States shall use international watercourses in their respective territories in an equitable and reasonable manner (Article 5) and have a general duty to cooperate in order to attain an optimal utilization and protection of international watercourses (Article 8). States are also encouraged to cooperate for the conservation and sustainable use of biodiversity under the CBD (Article 5). Those principles (of precaution, equitable use, cooperation, etc.) are reflected in many other instruments. For example, according to the 2002 ILA International Law Association New Delhi Declaration of Principles of International Law Relating to Sustainable Development, the precautionary approach commits States, organizations, civil society, and the scientific and business communities to avoid activities that may cause significant harm to human health and the environment, even in the light of scientific uncertainty (Article 4.1).[38] The 1978 UNEP Environmental Law Guidelines and Principles on Shared Natural Resources urge States to cooperate and achieve an equitable utilization of shared natural resources, in order to control, prevent, and eliminate adverse environmental

[37] Documents containing key environmental principles include: the 1972 Stockholm Declaration, the 1978 UNEP Draft Principles, the 1982 World Charter for Nature, the 1986 World Commission on Environment and Development (WCED) Legal Principles, the 1987 WCED *Our Common Future*, the 1972 Stockholm Declaration, the 1992 Rio Declaration, the 2001 International Law Commission Draft Articles on Prevention of Trans-boundary Harm from Hazardous Activities, and the 2002 International Law Association New Delhi Declaration of Principles of International Law Relating to Sustainable Development.

[38] A/57/329, August 31, 2002.

effects that may result from the use of such resources (Principle 1). This responsibility to ensure that activities within a State's jurisdiction do not cause damage to the environment of other States is a cornerstone principle of international environmental law (Stockholm Declaration, Principle 21; Rio Declaration, Principle 2). The American States in particular are called to foster hemispheric and international cooperation in order to fight poverty and strengthen democratic governance under the 2005 La Plata Plan of Action adopted at the Fourth Summit of the Americas.

In general, States have a basic duty to preserve and ensure the sustainable use of natural resources. This duty is one of the pillars of current MEAs and other environmental instruments, even if not expressly formulated as such, and is at the basis of other forms of obligations. It is clearly stated, for example, in the 1982 UN Convention on the Law of the Sea (UNCLOS): "States have the obligation to protect and preserve the marine environment" (Article 192). The duty to ensure sustainable use of natural resources within a State's own territory or jurisdiction is a well-established principle of international law, as stated in the 2002 ILA New Delhi Declaration of Principles of International Law relating to Sustainable Development (Article 1).[39] This duty is reflected, for example, in the 1992 Forest Principles, according to which "forest resources and forest lands should be sustainably managed to meet the social, cultural and spiritual needs of present and future generations" (Principle 2b).

The duty to preserve and ensure the sustainable use of natural resources is manifested in different ways. For example, according to the UNESCO World Heritage Convention, the parties are expected to conserve and transmit to future generations their natural and cultural heritage (Article 4). The properties forming part of the cultural and natural heritage are included in the World Heritage List maintained by the World Heritage Committee. In cases where a State is unable to secure protection, conservation, or rehabilitation of property forming part of the cultural or natural heritage of outstanding universal value situated within its territory, it can request assistance from this committee.[40] To that end, the World Heritage Fund, consisting of compulsory and voluntary contributions by the parties (and other sources),[41] provides required assistance, for example, through the training of personnel, the protection and rehabilitation of cultural and natural sites, the supply of equipment, or the granting of loans and, in some cases, the provision of nonrepayable subsidies. Therefore, the listing of natural heritage sites within the Amazon in particular involves the duty of the State where the natural heritage is situated to ensure its conservation and also a right to request assistance if appropriate. Likewise, once a site

[39] New Delhi Declaration of Principles of International Law Relating to Sustainable Development, 2002, A/57/329.

[40] Article 19.

[41] Article 15.

is included in the Ramsar List, the party in whose territory it is located must ensure the maintenance of the site's "ecological character"[42] and promote its conservation and wise use.[43] Each State party is also to "consider its international responsibilities for the conservation, management and wise use of migratory stocks of waterfowl" (Article 2 (6)). These provisions are applicable to Ramsar sites within the Amazon such as the Mamirauá Reserve in the Brazilian state of Amazonas. A State party may request the inclusion of a Ramsar site in the "Montreux Record"[44] when facing problems in maintaining it and technical assistance to help solve the problem can be provided. In addition, eligible parties can apply to the Ramsar Small Grants Fund or the Wetlands for the Future Fund for financial assistance to implement wetland conservation projects.[45]

It can be noted that the international community bears a responsibility to assist developing countries parties in fulfilling their international obligations under MEAs or other legal instruments. As stated in the UNESCO World Heritage Convention, listed cultural or natural heritage "constitutes a world heritage for whose protection it is the duty of the international community as a whole to cooperate"[46] (Article 4). Under this convention, international protection of the world heritage means the "establishment of a system of international cooperation and assistance designed to support State Parties in their efforts to conserved and identify that heritage" (Article 7). Therefore, it is the duty of the international community as a whole to cooperate in the protection of world heritage sites.[47] This obligation to cooperate and assist can also be found in relation to other natural resources. For example, according to the 1992 Forest Principles, the effort of developing countries to conserve and sustainably use their forest resources should be supported by the international community (Article 9 a). This should occur through new and additional financial resources (Article 10) and by ensuring the access to and transfer of environmentally sound technologies (Article 11). It is recognized that international cooperation (including for example financial support, technology transfer, capacity building, and education) plays a crucial role in supporting developing countries as well as countries with economies in transition, to achieve sustainable forest management (2007 Non-Legally Binding Instrument on All Types of Forests, Article 2f).

In order to meet the objectives defined in MEAs, State parties create secretariats and other bodies under these treaties. The Secretariats are generally in charge

42 Ibid., Article 2 (3).
43 Ibid., Article 3.
44 See Guidelines for Operation of the Montreux Record, adopted by the Conference of the Contracting Parties, Resolution VI.1, Brisbane, 1996.
45 Information on these funds is available at: http://www.ramsar.org
46 Article 6.
47 Guerrero and Sguerra, op. cit., p. 48.

of servicing meetings of conferences of the parties, producing and transmitting reports and other information, and entering into formal arrangements with other international bodies as required for the discharge of their functions (CBD, Article 24, CITES, Article 9, UNFCCC, Article 8, etc.). Examples of other bodies include the Subsidiary Body for Scientific and Technological Advice under the United Nations Framework Convention on Climate Change (UNFCCC) (Article 9), which provides assessments of the state of scientific knowledge relating to climate change, and the Scientific and Technical Review Panel of the Ramsar Convention, created in 1993 to provide technical and scientific guidance.[48] National designated agencies located in the territories of each State party can be also created, such as CITES Management and Scientific Authorities (CITES, Article 9), or Designated National Authorities, under the 1997 Kyoto Protocol's Clean Development Mechanism.[49] Apart from these bodies created under MEAs, other global actors, including international organizations and NGOs, also conduct activities in each of the Amazon States in particular, some of which are presented here.

B. GLOBAL ACTORS

7.1. INTERNATIONAL ORGANIZATIONS

7.1.1. *The United Nations Food and Agriculture Organization*

The functions of the United Nations Food and Agriculture Organization (FAO) are to collect, analyze, interpret, and disseminate information relating to nutrition, food, and agriculture. The term "agriculture" and its derivatives include fisheries, marine products, forestry, and primary forestry products (Constitution, Article 1).[50] FAO is headquartered in Rome and works in partnership with governments, international organizations, and agencies, nongovernmental groups, the private sector, communities, and individuals through a network of offices (at the country, subregional, and regional level). This organization promotes agricultural practices and makes recommendations on various issues, related, for example, to the conservation of natural resources or methods of agricultural production (Constitution, Article 1 (2)). Among its powers is that of submitting conventions and agreements to its member States on matters related to food and agriculture. These instruments are proposed by the "Conference," composed of

[48] Resolution 5.5 (1993).

[49] CDM Decision 3/CMP.1, FCCC/KP/CMP/2005/8/Add.1, March 30, 2006: "Parties participating in the CDM shall designate a national authority for the CDM."

[50] Food and Agriculture Organization of the United Nations (FAO), Constitution, *Basic Texts of FAO*, the Functions of the Organization, Article 1 A, vol. I, Rome: FAO, 2006.

all member States (and associate members), each represented by one delegate (Constitution, Article 3). A regional office for Latin America and the Caribbean is based in Santiago (Chile) and FAO Representations are established in each of the eight Amazon countries.

In terms of environmental policy, FAO has a leading role particularly in forestry. Under the Office of the Director-General are the different FAO departments, including a Forestry Department, the objective of which is to help the member States rationally manage their forests.[51] FAO also has a Committee on Forestry (COFO), which is the highest forestry statutory body. COFO biennial sessions held at FAO headquarters in Rome bring together heads of forestry services and other senior government officials to identify policy needs and technical issues, and to advise FAO on appropriate action. Apart from the member States, other international organizations and NGOs participate in COFO.

FAO has a Strategic Plan for Forestry, which is the result of a process of consultation with member countries, institutions, and individuals, which began in 1996 and was concluded with its approval by COFO on Forestry in 1999.[52] The Strategic Plan orients FAO in forestry and related matters. According to this policy, FAO's mission is to "enhance human well-being through support to member countries in the sustainable management of the world's forests."[53] The core of FAO's work in forestry involves the technical support offered to countries and the forum it provides for the member States to discuss forest-related issues.[54] Furthermore, FAO provides advice and disseminates information and knowledge on forests, of which the *State of the World's Forests* report is an example.[55] In particular, FAO has been carrying out periodic Global Forest Resources Assessments, which are the most important global data source of country-specific forest data.[56]

At the regional level, FAO has established Regional Forestry Commissions, between 1947 and 1959 in each major region of the world, including the Latin American and Caribbean Forestry Commission,[57] to consider both policy and technical issues of interest to a particular region and gather together the heads of forestry services of each member State every two years.[58] A Regional Office for Latin America and the Caribbean was also established in Santiago (Chile)

[51] http://www.fao.org/forestry/49435/en/

[52] Information is available at: http://www.fao.org/forestry/strategic/1962/en/

[53] *FAO Strategic Plan for Forestry*, Rome: FAO, 1999, p. 9.

[54] http://www.fao.org/forestry/dev/1680/en/nga/

[55] http://www.fao.org/forestry/sofo/en/

[56] L. Westholm et al., *Assessment of Existing Global Financial Initiatives and Monitoring Aspects of Carbon Sinks in Forest Ecosystems: The Issue of REDD*, Focali report, Gothenburg: Forest, Climate and Livelihood Research Network, 2009, p. 30.

[57] http://www.rlc.fao.org/es/comisiones/coflac/

[58] http://www.rlc.fao.org/en/comisiones/

in 1955[59] to provide technical assistance to member countries in agricultural and rural development activities.[60]

Under its Field Program activities, FAO has carried out a number of projects in each of the Amazon States and their respective Amazon regions.[61] Some of these projects are intended to strengthen institutional capacity, forestry policy and regulations, and develop methodology and management techniques. Apart from forestry, field activities address issues such as fishing and aquatic resources, water, and natural resources management. In countries such as Guyana and Suriname, national and regional programs have focused on food security and related issues. In particular, FAO has provided support to the formulation of criteria and indicators for the sustainable management of the Amazonian forests negotiated since 1995 in the framework of the ACT, under the so-called "Tarapoto Process." Those criteria and indicators have been defined and should be implemented with the support of the ACTO Permanent Secretariat.[62]

7.1.2. *United Nations Environment Programme*

UNEP is the designated entity within the UN system for addressing environmental issues.[63] UNEP's responsibilities include promoting international cooperation in the field of the environment and recommending appropriate policies, monitoring the status of the global environment and disseminating environmental information, facilitating the coordination of UN activities on matters concerned with the environment, among others.[64] UNEP's five priority areas are: environmental assessment and early warning; development of policy instruments; coordination with environmental conventions; technology transfer; and support to Africa.[65] Headquartered in Nairobi (Kenya),[66] UNEP is represented worldwide by six regional offices, including one for

59 http://www.rlc.fao.org/en/quienes/

60 http://www.rlc.fao.org/en/tecnica/

61 A list of field projects can be found at: http://www.fao.org/tc/tcom

62 ACTO, *Management Report 2007–2008*, Brasília: ACTO Permanent Secretariat, 2008, p. 21. Information is also available at: http://www.fao.org/docrep/007/y5841s/y5841s12.htm

63 UNEP, "UNEP Organization Profile," p. 3, http://www.unep.og. The mandate and objectives of UNEP emanate from UN General Assembly (UNGA) Resolution 2997 (XXVII) of December 15, 1972; Agenda 21, adopted at the UN Conference on Environment and Development (UNCED) in 1992; the Nairobi Declaration on the Role and Mandate of UNEP, adopted by the UNEP Governing Council in 1997; the Malmö Ministerial Declaration and the UN Millennium Declaration, adopted in 2000; and recommendations related to international environmental governance approved by the 2002 World Summit on Sustainable Development and the 2005 World Summit.

64 UNEP, op. cit., p. 18.

65 Ibid., p. 19.

66 Ibid., p. 20.

Latin America and the Caribbean located in Mexico City.[67] It also has a number of liaison offices in different parts of the world, for instance, in South America it is located in Brasília (Brazil).[68]

UNEP is one of three implementing agencies of the Global Environment Facility (GEF), alongside the World Bank and the UN Development Programme (UNDP). UNEP's Division of Global Environment Facility Coordination provides support to GEF's Scientific and Technical Advisory Panel and to the development and implementation of GEF-funded projects. As a GEF implementing agency, UNEP works with governments, NGOs, and research institutions to develop particular projects and enhance their managerial, administrative, and technical capacity.[69] In particular, UNEP provides support to national governments, especially developing countries, to participate in international negotiations and meet their obligations under international agreements.[70] Some other initiatives carried out by UNEP include campaigns, such as the Plant for the Planet: Billion Tree Campaign launched in 1996, which resulted in more than 7.4 billion planted trees by the end of 2009 around the world.[71] UNEP has also been instrumental in monitoring the world's biodiversity and ecosystems and providing policy advice through mechanisms such as the WCMC.

In the Amazon, UNEP conducted in partnership with other organizations, including the ACTO, an integrated environmental assessment known as Global Environmental Outlook Amazonia (GEO-Amazonia), concluded in 2009, with the participation of researchers and national stakeholders from the eight Amazonian countries. GEO-Amazonia consists of an integral assessment of the state of the environment in the Amazon Region using the Global Environment Outlook, a methodology developed by UNEP.[72] Brazil is one of the countries in Latin America and the Caribbean that has made progress in undertaking environmental assessments using this methodology.[73] One of the merits of GEO-Amazonia was the presentation of a common vision of the Amazon from the perspective of the eight basin States and with the participation of Amazonian stakeholders.[74]

67 Ibid.

68 http://www.pnuma.org.br/interna.php?id=44

69 UNEP, op. cit., p. 31.

70 Ibid.

71 UNEP, *Annual Report 2009*, Nairobi: UNEP, 2009, p. 35.

72 ACTO, op. cit., p. 20.

73 UNEP Brazil Office, *Annual Report 2008*, Brasília: UNEP, 2008. Brazilian experts and institutions, including NGOs, are increasingly engaging into regional and global GEO processes and over fifty experts and institutions are now part of the GEO network in Brazil.

74 UNEP, ACTO, and CIUP, op. cit.

7.1.3. *United Nations Development Programme*

Apart from UNEP, other UN agencies are dealing with environmental issues, notably climate change mitigation and adaptation, which is the case of the United Nations Development Program. In 2008 UNDP endorsed a new climate change strategy to support the capacity of developing countries to make informed policy and investment decisions to help reduce poverty, in line with the 2000 UN Millennium Declaration[75] Goals (MDG), and limit greenhouse gas (GHG) emissions.[76]

In terms of mitigation, UNDP's efforts are directed to energy efficiency, the increased utilization of renewable energies, energy access for the poor, policy reform, and capacity development. On the adaptation side, UNDP assists countries to integrate climate risks into national development policy and plans, develops strategies to meet national adaptation costs, and shares adaptation knowledge and experiences.[77] UNDP's advantage in addressing climate change, when compared to other organizations, lies in its presence on the ground, which permits the implementation of projects at the country level.[78] UNDP has mobilized funds to promote energy access and security, energy efficiency, and clean energy development in various developing countries, mainly through the GEF cofinancing.[79] In particular, it has established the MDG Carbon Facility to assist developing countries in leveraging carbon finance for clean energy development and sustainable land use practices.[80] UNDP has a large ongoing portfolio of projects, especially on climate change, and has also carried out activities in the Amazon.[81]

Climate change and the issue of emissions from deforestation and forest degradation has been a point of convergence for UN agencies and other global actors. In 2008, UNDP, FAO, and UNEP gathered efforts, in partnership with the government of Norway, to finance the first phase of the UN Program on Reducing Emissions from Deforestation and Forest Degradation (REDD) with a view to support the dialogue on the inclusion of REDD provisions in a post-2012 climate regime.[82] This initiative aims to support demonstration REDD activities, by assisting developing countries to design national REDD strategies, monitor forest cover and carbon stocks, and implement pilot projects on forest management that maintain ecosystems and

75 A/RES/55/2, September 18, 2000.
76 *UNDP Annual Report 2009*, New York: UNDP, 2009, p. 26.
77 Ibid.
78 Ibid.
79 Ibid., p. 27.
80 *Climate Change at UNDP: Scaling up to Meet the Challenges*, New York: UNDP, 2008, p. 13. Also, UNDP, *MDG Carbon Facility: Leveraging Carbon Finance for Sustainable Development*, UNDP, New York: UNDP, 2007, p. 2.
81 For projects in the Brazilian Amazon, see http://www.pnud.org.br
82 Westholm et al., op. cit., p. 18.

maximize carbon stocks (readiness activities).[83] A few countries have been selected to pilot the REDD Programme, including one Amazon country: Bolivia.[84]

International donors and governments are currently carrying the bulk of the REDD capacity-building burden. Apart from multilateral initiatives such as the REDD Programme and the World Bank's Forest Carbon Partnership Facility (FCPF), individual countries are also supporting REDD through bilateral development assistance. For example, in 2009 Norway was the largest bilateral donor via its International Climate and Forest Initiative, which allocated funding to developing countries. The German government has promised up to 1.34 billion euros for forest conservation for 2008–2012, and more beyond,[85] while Australia has stepped up with a pledge of AUS$200 million, to be jointly administered by the Australian Department of Climate Change and AusAID.[86] At present, most international climate change funding instruments are classified as official development assistance (ODA),[87] which is official financing from general budgetary expenditure given by national governments to developing countries to promote and implement development.[88]

7.1.4. *The World Bank and the Global Environment Facility*

The World Bank was established in 1944. Headquartered in Washington, DC, it now operates out of several offices worldwide.[89] The World Bank is made up of two development institutions: the International Bank for Reconstruction and Development (IBRD) and the International Development Association. Their work is complemented by that of the International Finance Corporation (IFC), Multilateral Investment Guarantee Agency (MIGA), and the International Centre for the Settlement of Investment Disputes.[90] The World Bank is a development institution and, as such, a source of financial and technical assistance to developing countries.

83 UNDP, *Annual Report 2009*, op. cit., p. 27.

84 The other countries are Democratic Republic of Congo, Indonesia, Panama, Papua New Guinea, Paraguay, Tanzania, Vietnam, and Zambia (UN-REDD, 2009). The initial phase of the UN-REDD Program will be funded by Norway (Westholm et al., op. cit., p. 18).

85 K. Hamilton, U. Chokkalingam, and M. Bendana, *State of the Forest Carbon Markets 2009: Taking Root and Branching Out*, Washington, DC: Ecosystem Marketplace, 2010, p. 124, http://www.forest-carbonportal.com/resource/state-carbon-market-report

86 Ibid., p. 63.

87 Parker, C. et al., *The Little Climate Finance Book: A Guide to Financing Options for Forests and Climate Change*, Oxford: Global Canopy Foundation 2009, p. 23. The Organisation for Economic Co-operation and Development (OECD) estimates that the total ODA and official aid to forestry by OECD countries and multilateral agencies was an annual average commitment of US$564 million between 1996 and 2004 (Johan Eliasch, 'Climate change: financing global forests, *Eliasch Review*, London, 2008, p. 95).

88 Parker et al., op. cit., p. 23.

89 World Bank, *Annual Report 2009*, Washington, DC: World Bank, 2009, p. 28.

90 http://web.worldbank.org

In terms of environmental policy, climate change is an area of priority within the World Bank. This is so particularly given the fact that member countries have identified climate change as a development priority over the last years and have demanded its inclusion in the bank's country partnership and assistance strategies.[91] As a result, the World Bank has incorporated climate action into its projects and programs, focusing on climate risks; lower emission growth opportunities, involving energy efficiency and renewable energy; and access to climate finance. Assistance strategies supported by the bank using this approach are being prepared for countries such as Bolivia, Brazil, and Guyana, among others.[92] For example, the assistance strategy for Guyana – which is a coastal country with about half of its population living below sea level – focuses on climate change adaptation and ecoservices related, for example, to avoided deforestation (this country was also selected in 2008 for the Bank's Forest Carbon Partnership Facility).[93] Since 2003, financing for renewable energy and energy efficiency has increased at a rate of more than 20 percent per year. [94]

In 2008, the World Bank approved a Strategic Framework on Climate Change and Development, drafted in consultation with and endorsed by its member countries. This Strategic Framework guides the World Bank Group, including institutions such as IFC and MIGA,[95] in their response to development challenges posed by climate change.[96] Among its action areas, it aims to mobilize additional concessional finance and leverage private sector resources. In order to do so, the Bank has created a number of carbon funds.[97] For example, in 2008 the World Bank's board of executive directors approved the creation of the Climate Investment Funds (CIFs) to be jointly implemented with the multilateral development banks.[98] The CIFs are governed by a representation of donors and recipient countries, and comprise the Clean Technology Fund[99] and the

[91] World Bank, *Development and Climate Change: The World Bank at Work*, Washington, DC: World Bank, 2009, p. 7.

[92] Ibid.

[93] Ibid., p. 11.

[94] Ibid., p. 14.

[95] The 2008 Strategic Framework on Climate Change and Development contains six action areas, aligned with the Bali Action Plan, which aim to (1) support climate action in country-led development processes; (2) mobilize additional concessional finance; (3) facilitate the development of market-based financing mechanisms; (4) leverage private sector resources; (5) support development and deployment of new technologies; and (6) step up policy research, knowledge, and capacity building (ibid., p. 2).

[96] Ibid., p. 2.

[97] Ibid., p. 16.

[98] Ibid., p. 5. The five multilateral development banks are the African Development Bank, the Asian Development Bank, the European Bank for Reconstruction and Development, the Inter-American Development Bank, and the World Bank Group.

[99] The Clean Technology Fund aims to promote financing for the demonstration, deployment and transfer of low-carbon technologies with a significant potential for reducing GHG emissions. The

Strategic Climate Fund.[100] These funds are designed as an interim measure, pending the establishment of a post-2012 international regime. For now, they aim to assist developing countries to undertake investments to achieve their development goals toward a climate-resilient economy. At the Thirteenth Conference of Parties to the UNFCCC, the World Bank has been also appointed to be the financial trustee of the Adaptation Fund.[101]

In particular, the World Bank has given significant attention to enabling investment finance for REDD activities. In 2007, at the Thirteenth Conference of Parties to the UNFCCC held in Bali, the Bank announced the FCPF, which aims to build the capacity of developing countries in tropical and subtropical regions on REDD.[102] A number of Latin American countries were selected to participate in the FCPF, including two Amazon States: Bolivia and Colombia.[103] The World Bank acts as the convener and trustee, provides secretariat services and chairs the Participants Committee, which includes observers from forest-dependent people, NGOs, and international organizations.[104] The Forest Investment Program[105] is under the World Bank's Strategic Climate Fund and should be Complimentary to the FCPF.[106]

The FCPF was criticized during its design phase and public launch in Bali for having failed to consult properly with forest peoples and to ensure proper rights

fund finances renewable energy and efficient technologies to reduce carbon intensity; achieve greater efficiency and modal shifts in the transport sector; and promote energy efficiency in buildings, industry, agriculture, and other areas where significant GHG emission reductions can be obtained (World Bank, *Annual Report 2009*, op. cit., p. 20).

[100] The Strategic Climate Fund will provide financing to pilot new approaches or activities aimed at a specific climate change challenge. Two programs were designed to this end; firstly, a Pilot Program for Climate Resilience (PPCR) to support climate resilience in national development planning. The PPCR helps the most vulnerable developing countries by providing grants to explore practical ways to increase climate resilience in development planning and budgeting. Nine countries were selected on advice by an independent expert group to be part of the PPCR (among which was Bolivia). Secondly, there is the Forest Investment Program, which was designed to help catalyze policies and measures that reduce deforestation and forest degradation and promote the sustainable management of forests in developing countries (World Bank, *Annual Report 2009*, op. cit., p. 20).

[101] World Bank, *Development and Climate Change*, op. cit., p. 13. On the creation of the Adaptation Fund, see Decision 1/CMP.3, FCCC/KP/CMP/2007/9/Add.1 of 2007.

[102] Prompted by a range of developing and developed country stakeholders, the World Bank proposed the creation of the FCPC in 2006, World Bank, "Forest Carbon Partnership Facility information," memorandum, 2008, p. 1.

[103] On REDD country participants, see http://www.forestcarbonpartnership.org/fcp/node/14

[104] World Bank, *Development and Climate Change*, op. cit., p. 14.

[105] The Forest Investment Program (FIP) is one of the programs under the World Bank's Strategic Climate Fund. The FIP finances investments in developing countries to reduce greenhouse gas emissions from deforestation and forest degradation (ibid., p. 20). The FIP aims to coordinate its efforts with existing initiatives such as the FCPF and the UN-REDD.

[106] World Bank, *Development and Climate Change*, op. cit., p. 13.

and accountability mechanisms within its governance structure.[107] As a result, the final FCPF charter included standards and safeguards that should ensure the participation of forest-dependent indigenous peoples and forest dwellers in decisions that may affect them (Operating Principles, Section 3.1 d).[108] However, "REDD country participants" have still criticized the fact that rights holders in the tropical countries concerned have not been much involved in the FCPF decision-making structure.[109] Similar criticisms have been made with regard to the Forest Investment Program.[110] Consequently, the First Meeting of the FCPF Participants Committee agreed in 2008 to establish a participation fund for indigenous peoples and forest-dependent communities.[111] More generally, the role of the World Bank in managing global funds to fight climate change is criticized. The existing portfolio and investments (especially IFC private sector loans) in support of the fossil fuel, mining, industrial, and agribusiness sectors is said to be causing increased GHG emissions.[112] Therefore, critics say that the World Bank should only be entrusted with the role of overseeing forests and climate funds if it undergoes major reforms to improve its social and environmental due diligence and lending practices.[113]

The World Bank has increased support to Latin America and the Caribbean, especially in designing climate change strategies. Brazil, Mexico, and Argentina, respectively, were the largest borrowers in 2009, whereas sectors such as health and social services, public administration, water, and sanitation led the region in funding.[114] In this region, GHG emissions originate mainly from deforestation and the expansion of agricultural land, although energy generation and transport are other major contributors.[115] Climate change may result in the melting of glaciers in the Andean mountain chain and more frequent and intense hurricanes in the Caribbean and

[107] T. Griffiths with Francesco Martone, *Seeing 'REDD'? Forests, Climate Change Mitigation and the Rights of Indigenous Peoples and Local Communities*, Forest Peoples Programme, 2009, p. 50.

[108] International Bank for Reconstruction and Development (IBRD), Charter Establishing the Forest Carbon Partnership Facility, http://www.forestcarbonpartnership.org/fcp/node/43

[109] Griffiths, op. cit. p. 50.

[110] Ibid., p. 12. Critics argue that unless the World Bank commits itself to a rights-based and people-centered approach to forests, which includes timely consultation with forest peoples and civil society about the design, governance, and operation of the FIP; then there are real dangers that FIP funds will be channeled to conventional large-scale plantation and logging operations.

[111] Griffiths, op. cit., p. 50.

[112] Ibid., 2009, p. 51.

[113] Ibid., p. 50.

[114] World Bank, *Annual Report 2009*, op. cit., p. 46. Brazil, Mexico, and Argentina were the largest borrowers, whereas sectors such as health and social services, public administration and law, and water and sanitation led the region in funding. Support to Latin America and the Caribbean during 2009 represented 42 percent of IBRD lending.

[115] http://beta.worldbank.org/content/latin-america-caribbean

Central America, and will pose major changes in the Amazon. In view of this, the World Bank has supported some countries (Bolivia, Ecuador, and Peru) in addressing the consequences of glacier melt on power generation and the agricultural sector, particularly. Initiatives are also under way to mitigate GHG emissions in Bogota (Colombia).[116]

The World Bank has undertaken studies on low-carbon growth strategies in a few countries, including Brazil and Mexico,[117] and in 2008 released its annual flagship report, entitled *Low Carbon, High Growth: Latin American Responses to Climate Change*. The report notes that Latin America produces only 6 percent of global emissions from energy sources and has a cleaner energy mix than other parts of the world, but is already experiencing the negative impacts of climate change.[118] In conclusion, the report suggests that Latin American countries are well placed to take a leadership role in the developing world's response to climate change[119] and argues that the inclusion of REDD in a future international climate change regime is the most important issue for the region.[120]

Latin American and Caribbean countries were among the first to partner with the World Bank in addressing climate change. The region's climate change portfolio includes activities totaling US$2.5 billion, including regional studies, country assessments, IBRD lending, GEF grants, and carbon finance, among others.[121] Out of thirty-seven participating countries within the FCPF, fifteen are from that region.[122] In the Amazon, the World Bank supports projects such as the Integrated Management of Aquatic Resources (AquaBio) aimed at the conservation and sustainable use of freshwater biodiversity in Brazil's Amazon River basin.[123] It also finances programs at the regional level, for example, the Inter-American Biodiversity Information Network (IABIN)[124] and, in the Andean region, the project on Adaptation to the Impact of Rapid Glacier Retreat in the Tropical Andes.[125]

[116] Ibid.

[117] http://www.worldbank.org/climatechange

[118] World Bank, *Annual Report 2009*, op. cit., p. 48.

[119] A. de la Torre, P. Fajnzylber, and J. Nash, *Low Carbon High Growth Latin America Responses to Climate Change: An Overview*, World Bank Latin America and Caribbean Studies, Washington, DC: World Bank, 2009, p. 60.

[120] Ibid., p. 50.

[121] World Bank, *Development and Climate Change*, op. cit., p. 27.

[122] Ibid., pp. 26–27.

[123] Approval date: June 13, 2006, closing date: August 31, 2012, http://web.worldbank.org/external/projects/main?pagePK=64283627&piPK=73230&theSitePK=40941&menuPK=228424&Projectid=P066535

[124] Approval date: June 29, 2004, closing date: June 30, 2010. The IABIN was created at the Summit of the Americas on Sustainable Development in 1996. To support the development of IABIN, the GEF project seeks to develop an Internet-based, decentralized network that provides wider access to scientifically credible, biodiversity information.

[125] Approval date: May 27, 2008, closing date: September 30, 2012, http://web.worldbank.org/external/projects/main?pagePK=64283627&piPK=73230&theSitePK=40941&menuPK=228424&Projectid=P09

Within the structure of the World Bank, the GEF was created as an independent financial mechanism in 1991 to provide financing to developing countries for projects benefiting the global environment. UNDP, UNEP, and the World Bank were the three initial partners implementing GEF projects. In 1992, at the Rio Earth Summit, the GEF was restructured and moved out of the World Bank system to become a permanent and separate institution. Since then, the World Bank has served as a trustee of the GEF Trust Fund[126] and has been a partner in raising cofinancing.[127] The GEF is the financial mechanism for some multilateral environmental agreements, such as the 1992 Convention on Biodiversity, the 1992 UNFCCC, and the 2001 Stockholm Convention on Persistent Organic Pollutants. GEF projects are implemented by one or more of the designated GEF agencies, which include UNDP, UNEP, and FAO, among others.[128]

The GEF has financed projects in the Amazon, in particular. For example, it has approved a pilot project in the Amazon basin in areas experiencing droughts or floods as a result of the El Niño phenomenon.[129] National GEF projects in the Amazon region include, for example, the Protected Areas Program (Brazil), the Sustainable Use of Biodiversity by the Ashaninka (Peru), and a project on renewable energy systems (Peru).[130] Regional GEF projects[131] include one on Integrated and Sustainable Management of Transboundary Water Resources in the Amazon River Basin Considering Climate Variability and Change[132] and on a Regional Strategy for the Conservation and Sustainable Use of Natural Resources in the Amazon[133] – both involving all Amazon countries and carried out in partnership with ACTO. Also, under the GEF Tropical Forest Account, launched in 2007 at the Thirteenth Conference of the Parties to the UNFCCC in Bali, large tropical forests, including the Amazon, should receive funding for conservation purposes.[134]

As noted earlier, World Bank operations have been subject to criticism, including in relation to the Amazon. In particular, a loan granted by IFC in 2007 to one of Brazil's largest beef, dairy, and leather producers was put under the spotlight by Greenpeace when it released a report in 2009 entitled *Slaughtering the Amazon*.

8248. The objective of the Adaptation to the Impact of Rapid Glacier Retreat in the Tropical Andes Project for Andean Countries is to contribute to strengthening the resilience of local ecosystems and economies to the impacts of glacier retreat in the tropical Andes through the implementation of specific pilot adaptation activities that illustrate the costs and benefits of adaptation.

126 http://www.gefweb.org

127 World Bank, *Development and Climate Change*, op. cit., p. 8.

128 About GEFF investments, see http://www.gefweb.org

129 UNEP, ACTO, & CIUP, op. cit., p. 81.

130 http://projectdatabase.thegef.org/projectListSQL.cfm, accessed February 8, 2010.

131 Ibid.

132 Global Environmental Facility (GEF), *2008 Annual Report*, Washington, DC, GEF, 2008, p. 89.

133 http://projectdatabase.thegef.org/projectListSQL.cfm

134 Westholm et al., op. cit., p. 24.

The grant was made to facilitate the expansion of slaughterhouses on the agricultural frontier of the Brazilian Amazon.[135] Another influential study by Amigos da Terra-Amazonia Brasileira entitled *The Cattle Realm*, released in 2008, estimated that the internationally funded expansion of Brazil's beef industry was responsible for up to 12 billion tons of CO_2 emissions over the past decade.[136] IFC and foreign private banks were also criticized for loans granted to large soya producers. In particular, a report by Greenpeace-Brazil entitled *Eating up the Amazon*, released in 2006, denounced the involvement of global corporations with deforestation in the Brazilian Amazon caused by soy production. The report suggested that international lenders play an important role in the expansion of large-scale soya cultivation, providing capital for infrastructure development and agrochemical inputs for soy producers. In short, the World Bank has been criticized for making efforts to reduce tropical deforestation through initiatives such as the FCPF and, at the same time, spending millions more on the industries that are drivers of deforestation in the Amazon.[137]

7.1.5. *The Pilot Program to Conserve the Brazilian Rain Forest*

The Pilot Program to Conserve the Brazilian Rain Forest (PPG7) is a multilateral cooperation program started in 1994 through an agreement signed at the 1992 United Nations Conference on Environment and Development (UNCED) between the Brazilian government and developed countries within the Group of Seven (G7).[138] The program was proposed by the German Chancellor (Helmut Kohl)[139] at the Sixteenth G7 Summit held in Houston in 1990 in the wake of growing international concerns about rapid deforestation in the Brazilian Amazon.[140] In the following year, the G7 countries and the European Commission formally approved the program,[141] which was launched at the 1992 UNCED. The first projects under PPG7

135 Greenpeace International, *Slaughtering the Amazon*, Amsterdam: Greenpeace International, 2009, p. 5.

136 R. Smeraldi and P. May, *The Cattle Realm: A New Phase in the Livestock Colonization of the Brazilian Amazon*, São Paulo: Amigos da Terra, 2008, p. 2.

137 Daniel Howden, "The World Bank pledges to save trees, then helps cut down Amazon; a month ago it vowed to fight deforestation. Now research reveals it funds the rainforest's biggest threat," *The Independent*, January 17, 2008.

138 http://www.mma.gov.br/ppg7/

139 U.J. Lele, *Brazil: Forests in the Balance: Challenges of Conservation with Development*, Washington, DC: World Bank, Operations Evaluation Department, 2000, p. 115.

140 World Bank Brazil Rainforest Unit, *Lessons from the Rain Forest: Experiences of the Pilot Program to Conserve the Amazon and Atlantic Forests of Brazil, Ministry of the Environment, Secretariat for Coordination of the Amazon, Pilot Program to Conserve the Brazilian Rain Forest Monitoring and Analysis Project*, Washington, DC: World Bank Brazil Rainforest Unit, 2002, p. 9.

141 Ibid., p. 9.

were approved in 1994 and started operation in 1995.[142] The major PPG7 objective was to develop strategies for the protection and sustainable use of the Amazon and Atlantic Forests.[143] This program is an example of an international partnership involving a national government, civil society, and the international community. The Large Scale Biosphere-Atmosphere Experiment (LBA) in the Amazon, a multinational, multidisciplinary research effort, is another good example of an international partnership set up to investigate the biological, biogeochemical, and physical functioning of the Amazon and the effects of land-cover and use, as well as climate change on such functioning.[144]

The PPG7 was initially funded by the G7 countries, the European Union (EU), and the Netherlands, and also received financial contributions from the Brazilian government and civil society. The most significant sources of financial and technical support among donor countries have been Germany, the United Kingdom, and the EU. The World Bank was called to administer the Rainforest Trust Fund, a mechanism created to channel donor contribution and facilitate the financial management of funds.[145]

The Brazilian Ministry of the Environment is in charge of the program's overall planning and execution, in cooperation with federal and state agencies, municipal governments, environmental organizations, and the private sector.[146] The PPG7 strategic decisions are made at a biannual Participants Meeting and on a more regular basis through the Brazilian Coordination Committee and the Joint Steering Committee; both are chaired by the Brazilian Ministry of the Environment and include civil society representatives. The international partners also gather under a Donor Coordination Committee. At the executive level, a coordinating unit headed by the Executive Secretariat of the Brazilian Coordination Committee provides support for the program. An International Advisory Group, composed of Brazilian and international experts, carries out independent biannual assessments of the program and provides strategic advice.[147]

[142] Ibid.

[143] http://web.worldbank.org

[144] Teja Tscharntke et al. (eds.), *Stability of tropical rainforest margins: Linking ecological, economic and Social Constraints of Land Use and Conservation*, Heidelberg: Springer, 2007, pp. 503–44. According to this study, discussions within LBA Scientific Committee played a crucial role in establishing unprecedented level of communication and cooperation among disciplines studying the Amazon. LBA is also a very interesting experience of partnership among scientists from different countries (ibid., p. 504). In 2007, the LBA became a official program of the Brazilian government, under the National Institute for Amazonian Research, http://150.163.158.28/lba/site/

[145] World Bank Brazil Rainforest Unit, op. cit., p. 9.

[146] http://www.mma.gov.br/ppg7/

[147] World Bank Brazil Rainforest Unit, op. cit., p. 10

PPG7 developed a large portfolio of projects.[148] Success stories include, for example, the Integrated Project for the Protection of Indigenous Peoples and Land in the Amazon (PPTAL) and the Extractive Reserves Project (RESEX), both formulated in 1994. In collaboration with the Brazilian National Indian Foundation, the PPTAL supported the demarcation of indigenous lands covering an area of about 45 million hectares.[149] Apart from the effort to regularize those lands, the PPTAL was also a pioneering effort to increase the participation of indigenous peoples, particularly in the demarcation and management of their traditional lands. RESEX assisted the Brazilian Institute of Environment and Renewable Natural Resources (IBAMA) in resolving land tenure problems and creating four extractive reserves involving the protection of some 2.1 million hectares of rainforest to the benefit of rubber tappers and local communities.[150] In order to create the reserves, forest dwellers needed to be trained and organized in associations, which involved significant capacity-building activities. Environmental monitoring and enforcement tasks were also carried out by IBAMA together with local communities, which were trained to monitor the reserves and prevent illegal encroachments related to land speculation, poaching, commercial fishing, etc. Apart from these examples, the PPG7 has carried out other interesting projects, for example, to support environmental protection and resource management,[151] community mobilization for fire prevention and control,[152] and the elaboration of Positive Agendas (Agendas Positivas) in the States of the Brazilian Amazon[153] intended to provide policy guidance to activities carried out in that region.

The evaluation of the first phase of the PPG7 (1992–2000) is generally positive. The PPG7 is praised for introducing new principles of cooperation and participation in Brazil's public administration, creating innovative mechanisms for international cooperation, and strengthening institutional capacity (especially of the Ministry of Environment) for the protection of the Amazon.[154] Measurable results include the demarcation and establishment of million of hectares of indigenous lands and community-managed extractive reserves.[155] However, the PPG7 has

[148] Ibid., pp. 10, 11.
[149] World Bank Brazil Rainforest Unit, op. cit., p. 12.
[150] Ibid., p. 16.
[151] Ibid, p. 19.
[152] Ibid., p. 22.
[153] Ibid., p. 26.
[154] I. Scholz, *O Desenvolvimento de Instituições de Política Ambiental na Amazônia a partir de 1992: a contribuição do PPG7*, Brasília: Ministério do Meio Ambiente, http://www.mma.gov.br/ppg7/; Lele, op. cit., p. 116; T. Tscharntke et al., *The Stability of Tropical Rainforest Margins, Linking Ecological, Economic and Social Constraints of Land Use and Conservation: An Introduction*, Berlin & Heidelberg: Springer, 2007, p. 502.
[155] http://web.worldbank.org

been criticized in formal evaluations for a number of things, for example, weak program management, complex project design, and financing plans, which have led to costly and time-consuming project processing.[156] Other criticisms involve the lack of a long-term strategic vision, and the fact that the program has been driven mostly "by donor policies and priorities" without "sufficient trust by the donor agencies in the capacity, transparency and accountability of the Brazilian ministries and institutions."[157]

The second phase of the PPG7 started in 2003 and focuses on consolidating innovative strategies and mainstreaming lessons learned to influence public policies for the sustainable development of the Amazon and Atlantic rainforests.[158] The current main sources of funding come from governments (Germany, the Netherlands, Italy, France, Japan, Canada, the United Kingdom, and the United States), and from the European Commission and Brazilian government.[159]

7.2. INTERNATIONAL NONGOVERNMENTAL ORGANIZATIONS

As one could expect, international organizations and NGOs have a strong presence in the Amazon region. As in other parts of the world, they have been instrumental in monitoring the state of the environment, generating knowledge, and campaigning for or carrying out field projects.

For example, since the 1960s the IUCN has gathered information on wildlife plant and animal species, which is now compiled in the IUCN Red List.[160] This is a compendium of conservation-related information that classifies species according to their levels of endangerment.[161] It not only provides information on the threats to the species, but also on their ecological requirements, habitats, and conservation actions that can be used to reduce or prevent extinctions.[162] In order to collect, manage, process, and report data to the point of publication on Red List, IUCN has developed an information management system, the Species Information Service, which provides up-to-date information on species, their distribution, and conservation status.[163] A number of Amazonian species have been assessed and are classified under the IUCN Red List,[164]

[156] Lele, op. cit., p. 116.
[157] Assessment presented at the PPG7 Mid-Term Review (Tscharntke et al., op. cit., p. 504).
[158] World Bank Brazil Rainforest Unit, op. cit., p. 40.
[159] http://web.worldbank.org
[160] IUCN has a Regional Office for South America headquartered in Ecuador.
[161] J.-C. Vié, C. Hilton-Taylor and S.N. Stuart, *Wildlife in a Changing World, An Analysis of the 2008 IUCN Red List of Threatened Species*, Gland: IUCN, 2008, p. 3.
[162] IUCN, The IUCN Red List of Threatened Species: A Key Conservation Tool, Gland: IUCN, 2008.
[163] Vié, Hilton-Taylor and Stuart, op. cit., p. 5.
[164] http://www.iucnredlist.org

such as the Brazilian or South American tapir,[165] the jaguar,[166] and the giant otter.[167] Another example of a species-monitoring network is TRAFFIC, which provides information on wildlife trade issues, especially to assist decision-making processes under CITES. TRAFFIC has a regional office in South America based in Ecuador. The organization has been working with local governments, civil society organizations, and the private sector in the Amazonian regions of Ecuador, Peru, and Bolivia in order to devise ways to improve forest governance.[168] It has generated information on illegal trade in wild animals and particular Amazonian species.[169]

Global campaigning organizations, such as Greenpeace, have also played a role, especially in raising awareness of deforestation in the Amazon. For example, Greenpeace-Brazil has supported the campaign Zero Deforestation in the Amazon[170] and has released information that has attracted worldwide attention. In 2006 Greenpeace-Brazil published the report *Eating up the Amazon* that links the global demand for soy produced in the Amazon to illegal deforestation, land grabbing, slave labor, and other forms of violence. It denounces the role of Brazil's major soy producers in the deforestation of the Legal Amazon and points to the indirect responsibility of well-known global corporations. In response, the two Brazilian associations in the sector (Associação Brasileira das Indústrias de Óleos Vegetais, ABIOVE and Associação Nacional dos Exportadores de Cereais, ANEC) and large soy producers, such as Cargill, Bunge, ADM, Dreyfus, and the Amaggi Group, announced in July 2006 a two-year moratorium (*moratória da soja*) to prevent the purchase of soy from newly deforested areas in the Amazon and from farms using slave labor.[171] To monitor the crops and detect farms operating illegally a working group was created (Grupo de Trabalho da Soja, or GTS). In 2009, the soy moratorium was extended up to July 2010 by a formal agreement

165 http://www.iucnredlist.org/apps/redlist/details/21474/0. The Tapirus terrestris (Common Names: Lowland Tapir, Brazilian Tapir, OR South American Tapir), classified as vulnerable, is found in lowland regions of northern and central South America, among other areas in the Amazonian or Orinoco forests.

166 The jaguar (*Panthera onca*), classified as near threatened, is threatened by habitat loss and persecution. Much of the jaguar's remaining habitat is the rainforest of the Amazon basin, available at: http://www.iucnredlist.org/apps/redlist/details/15953/0

167 http://www.iucn.org/sur. The giant otter (*Pteronura brasiliensis*), classified as endangered (also listed in CITES, Appendix I), is considered to be so due to an inferred future population decline resulting from habitat loss and exploitation. The majority of the animals are found in the Brazilian Amazon and the regions immediately bordering this area.

168 http://www.traffic.org/timber-trade/

169 See the TRAFFIC, *Illegal Trade in Wild Animals in Brazil (Transito de comércio Ilegal de animais silvestres no Brasil)*, 1993, and the 2001 producing briefings on Big-leaf Mahogany Trade, reviewing legislation and harvest and trade controls in Bolivia, Brazil, and Peru, http://www.traffic.org/south-america/

170 http://www.greenpeace.org/brasil/amazonia/gado/a-solu-o

171 http://www.greenpeace.org/brasil/amazonia/moratoria-da-soja/entenda-a-morat-ria-da-soja

(Termo de Compromisso da Moratória da Soja 2009–10), signed in July 2009 by ABIOVE, ANEC, GTS, Greenpeace, and Brazil's Ministry of the Environment,[172] and extended once more until 2011.[173]

Another report released by Greenpeace-Brazil in 2009, *Slaughtering the Amazon*, has also denounced the role of public institutions, Brazil's beef industry, and global brands and retailers in the expansion of stock farming in the Amazon.[174] Other influencial reports, *The Cattle Realm* and A *hora da conta*, released by Amigos da Terra-Amazonia Brasileira in 2008 and 2009 respectively, suggest that most of the slaughterhouses operating in the Brazilian Amazon (including the industrial segment) should be considered illegal.[175] As of January 2007, 62 percent of enterprises on the "dirty list" of slave labor, classified as such by the Brazilian Labor Ministry, were ranches in the Amazon.[176] Following the release of those reports, a number of civil actions were initiated by the General Attorney Office (Ministério Público) particularly in the State of Pará, which led to the signature of an agreement (Termos de Ajuste de Conduta), and a federal law (12.097/2009), which regulates traceability control mechanisms in the supply chain for beef and buffalo meat.[177] Also, the Brazilian Association representing the network of major retailers in Brazil (Associação Brasileira de Supermercados) proposed a certification program (Programa de Certificação de Produção Responsável na Cadeia Bovina) to track the origins of the beef sold in supermarkets around the country. The idea was welcomed by large retailers such as Walmart, Pão de Açúcar, and Carrefour, and beef producers such as Marfig, Bertin, and JBS.[178]

Friends of the Earth is another well-established campaigning network in Latin America.[179] For example, it has also monitored the negotiation of free-trade agreements in Peru and Colombia, to counter pervasive problems such as illegal logging.[180] In Brazil, Amigos da Terra Amazônia Brasileira[181] has been actively monitoring

[172] http://www.greenpeace.org/brasil/amazonia/moratoria-da-soja

[173] "Moratória da Soja, Terceiro ano do mapeamento e monitoramento do plantio de soja no bioma Amazônia," July 2010, p. 22, http://www.bunge.com.br/sustentabilidade/2010/Bunge2010.pdf

[174] Greenpeace International, op. cit., p. vi.

[175] May, op. cit., p. 2.

[176] Ibid.

[177] "Origem da carne ainda não é totalmente identificada," November 3, 2010, http://www.estadao.com.br/estadaodehoje/20100807/not_imp591606,0.php. On the initiative "Carne Legal" by the Ministério Público Federal, see: http://www.carnelegal.mpf.gov.br/historico/ajuste-de-conduta

[178] F. Munhoz and F. Bonanome, "Queremos a certificação bovina pronta em 60 ou 90 dias," August 10, 2009, http://www.amazonia.org.br/noticias/noticia.cfm?id=322715. The "Programa de Certificação de Produção Responsável na Cadeia Bovina" was launched in 2009, http://www.abrasnet.com.br/superhiper/superhiper/direto-da-redacao/?materia=346

[179] http://www.foei.org/

[180] FoE (Friends of the Earth), *Annual Report 2008*, 2008, p. 11, http://www.foe.org/annual-reports

[181] Amigos da Terra Amazônia Brasileira, founded in 1989, is an independent public interest civil society organization, recognised by the Brazilian Ministry of Justice.

public policies related to the Brazilian Amazon, on areas such as forests, infrastructure, agrarian reform, indigenous populations, and others. The Brazilian Amazon is its priority focus, although it is not the only one. Amigos da Terra provides up-to-date information about the Amazon through a useful Web site that collects information about the Amazon in the Brazilian press,[182] and also publishes scientific papers.

In addition to those institutions, the World Wide Fund for Nature (WWF) has also carried out activities in the Amazon. WWF is a global organization with a network of local offices in various countries, which vary in their degree of autonomy. It focuses on conserving areas and regions of critical importance. Under the WWF 2020 Biodiversity Goal, some terrestrial, freshwater, and marine ecoregions were identified as global priority places, including remaining rainforests, among which the Amazon.[183] WWF has offices in most of the Amazon countries (Bolivia, Brazil, Colombia, Peru, Guyana-Suriname, and Venezuela).[184] WWF was a partner in the Brazilian Amazon Region Protected Areas Programme[185] and has supported enterprises in Acre (one of the states of the Brazilian Amazon), in collaboration with state and national institutions (Conselho Brasileiro de Manejo Florestal and Associação das Indústrias de Madeira de Manejo do Estado do Acre), to obtain certification according to Forest Stewardship Council principles.[186] WWF has also supported the creation of protected areas in the Bolivian Amazon,[187] and research on the impacts of the Jirau and Santo Antonio dams in the upper watershed of the Madeira River. In the Peruvian Amazon, WWF has also carried out projects, for example, on the management of tropical forests[188] and on freshwater conservation.[189]

182 The Web site is http://www.amazonia.org.br. This site is part of a project by Friends of the Earth Amazonia Program, carried out with the financial support from the Small Grants Program of the (IUCN Holland, and the Tropical Forests Foundation of Sweden, http://www.amazonia.org.br/english/site_info/sobreosite.cfm

183 WWF (World ildlife Fund for Nature), *A Road Map for a Living Planet: WWF Annual Review 2008*, p. 5, http://www.worldwildlife.org/who/financialinfo/2008AR/

184 http://www.panda.org

185 WWF, op. cit., p. 8.

186 http://www.wwf.org.br/natureza_brasileira/areas_prioritarias/amazonia1/nossas_solucoes_na_amazonia/areas_protegidas_na_amazonia/

187 For example, the Departmental Park and Natural Integrated Management Area Iténez, in the department of Beni; the Northern Tiquipaya Wildlife Municipal Reserve in Cochabamba; and the Bruno Racua Departmental Wildlife Reserve in Pando; see http://www.panda.org

188 Managing tropical forests in the Peruvian Amazon, start date: January 1, 2006, end date: June 30, 2009; managing office: WWF Peru Programme Office. This project seeks to contribute toward the conservation of the tropical forest ecosystem in indigenous communities of the Peruvian Amazon through a sustainable management of their forest resources; see http://www.panda.org

189 Freshwater Conservation in the Peruvian Amazon, start date: July 1, 1998, end date: June 30, 2010; managing office: WWF Peru Programme Office. The main objective is to improve livelihood opportunities in priority communities in the Abanico del Pastaza and Pacaya Samiria National Reserve and freshwater ecosystem recovery; see http://www.panda.org

C. SYNERGIES AND CONFLICTS BETWEEN NORMS AND INSTITUTIONS

As recommended by UNEP, States should avoid conflicts between environmental agreements and seek cooperation among treaties' secretariats.[190] This is particularly the case given the congestion of environmental treaties (there are now more than 500 MEAs in existence) and institutions handling environmental portfolios with limited and dispersed funding sources.[191] According to UNEP, from 1992 to 2007 the parties of eighteen major MEAs were called for 540 meetings, at which 5,084 decisions were taken.[192] In this context, institutions such as UNEP, tasked to set the normative agenda on the environment within the UN system, play a role in improving the current international environmental governance system. In view of this, in 2009 the UNEP Governing Council established a Consultative Group of Ministers or High-level Representatives to present a set of options for improving international environmental governance.[193] After a series of meetings, the Consultative Group proposed a number of changes to the current international environmental governance system.[194] In this regard, some governments have suggested the creation of a centralized global institution in charge of environmental issues. For example, the French and Brazilian presidents proposed the setting up of an "international organization devoted to the environment and sustainable development, which would give coherence to the efforts of the international community in these areas."[195]

In practice, the global actors here examined seek cooperation in different ways. For example, FAO opted for defining guidelines to clarify matters that should be considered in its relationship with intergovernmental organizations[196] and also with NGOs.[197] Similarly, the 2008 UNDP Climate Change Strategy proposes ways for

[190] UNEP, *UNEP Manual on Compliance with and Enforcement of Multilateral Environmental Agreements*, Nairobi: UNEP, 2006.

[191] UNEP, *Annual Report 2009*, op. cit., p. 52.

[192] Ibid.

[193] Ibid.

[194] The report of the Consultative Group was presented with a view of informing the UNGA (ibid.).

[195] To be established at the Rio+20 Conference in Rio de Janeiro in 2012 (Joint statement of the Brazilian president (Luis Inácio Lula da Silva) and the French president (Nicolas Sarkozy), in UNEP, *Annual Report 2009*, op. cit., p. 54).

[196] Information available in FAO, Guiding Lines regarding Relationship Agreements between FAO and Intergovernmental Organizations, Resolution no. 69/59, FAO, *Basic Texts of FAO*, op. cit., vol. II.

[197] FAO Policy concerning Relations with International Organizations (Conference Resolution 37/53). However, it is understood that cooperation between FAO and NGOs had developed and would continue to develop, which did not necessitate either consultative or specialized consultative status (Cooperation with International Non-Governmental Organizations, FAO, *Basic Texts of FAO*, vol. II, op. cit.).

UNDP to work with other UN agencies, GEF, civil society, and the private sector.[198] As noted earlier, UNDP has carried out mitigation and adaptation activities in partnership with various institutions, such as the GEF, World Bank, and regional development banks, among others.

In practice, global institutions cooperate by establishing joint programmes and partnerships. They have joined efforts, particularly in the field of climate change, to achieve a coherent strategic approach, for example, within the collective framework of Acting on Climate Change: The UN Delivering as One or under the UN REDD initiative set up by UNDP, FAO, and UNEP.[199] The World Bank is also working to broaden its partnerships with other institutions, governments, the private sector, and civil society in all aspects of climate action.[200] It acknowledges that climate change, because of its nature and the magnitude of potential impacts, requires concerted responses by the international community,[201] and no one institution or group of stakeholders can address the climate challenge alone.[202] As a co-convening agency with UNDP on climate finance, the World Bank set up climate finance vehicles such as the Climate Investment Funds or the FCPF.[203] It also supports, in partnership with IUCN, FAO, and the International Institute for Environment and Development, the initiative of Growing Forest Partnerships, which provides a platform to the international dialogue on forests and improve the connections between forests and other sectors.[204]

In the context of the Amazon, international organizations and NGOs have played a role in carrying out field projects and providing financial and technical assistance.[205] They have also established partnerships with regional organizations to which the Amazon States are parties. For example, since 2004, UNEP-Brazil is providing technical assistance to the Mercosur Sub-working Group on the Environment.[206] Despite the presence of various institutions at all levels, actions for an integrated management of natural resources in the Amazon are still limited. For example, little cooperation is observed toward an integrated management of Amazonian water resources, despite the fact that the availability of freshwater in the Amazon basin depends on equitable use and concerted actions by each basin State. In this regard, ACTO, together with UNEP, GEF, and the Organizations of

[198] UNDP, *Annual Report 2009*, op. cit., p. 26.

[199] United Nations System Chief Executives Board for Coordination (CEB), *Acting on Climate Change: The UN System Delivering as One*, New York: CEB, 2008.

[200] World Bank, *Development and Climate Change*, op. cit., p. 3.

[201] Ibid.

[202] Ibid.

[203] Ibid., p. 4.

[204] Ibid., p. 14. Information about the initiative is available at: http://www.growingforestpartnerships.org

[205] UNEP, ACTO, & CIUP, op. cit., p. 25.

[206] UNEP Brazil Office, op. cit., p. 210.

American States (OAS), has attempted to design a regional management program for the Amazon basin.[207] Information on the Amazon is still fragmented and varies in quality among Amazon States.

In a complex institutional framework, involving actors working at different levels and on similar issues, the risk of duplicating efforts or facing situations of conflicts of law is high. Two provisions of the ACT provide for the relationship between this treaty and other international agreements. According to Article 18, ACT parties can conclude bilateral or multilateral agreements, as long as they are compatible with its aims, as follows:

> Nothing contained in this Treaty shall in any way limit the rights of the Contracting Parties to conclude bilateral or multilateral agreements on specific or generic matters, provided that these are not contrary to the achievement of the common aims for cooperation in the Amazonian region stated in this instrument.

Article 19 provides for the relationship between the ACT and international treaties adopted before the signing of this treaty by stating that "neither the signing of this Treaty nor its execution shall have any effect on any other international treaties in force between the Parties." This provision was introduced especially to ensure that the ACT would not adversely affect the then incipient common market created under the 1969 Andean Subregional Integration Agreement.

Up to the present, no conflicts of law have occurred between the ACT and other regional or multilateral agreements. This is in part due to the fact that this treaty contains only general obligations and guidelines that are unlikely to conflict with other norms. In practice, the ACTO Permanent Secretariat cooperates with other regional or global institutions by adopting memoranda of understanding and setting up joint projects, with the technical and financial support of institutions such as OAS, FAO, and GEF, among others. For example, in 2004 ACTO signed an agreement with the Andean Community (CAN),[208] the Andean Development Corporation,[209] and the General Secretariat of the OAS in order to maintain close coordination in relation to issues of common interest.

Conflicts of law have occurred so far between national law and norms adopted under subregional and regional organizations; and between decisions and treaties under these institutions and international agreements. For example, disputes were observed in the framework of CAN with regard to the relationship between domestic law and CAN law, and within Mercosur in relation to allegedly conflicting principles

[207] UNEP, ACTO, and CIUP, op. cit., p. 25.

[208] Article 2. In 2004, ACTO and CAN signed a memorandum of understanding in order to maintain coordination with respect to issues of common interest within ACTO's Strategic Plan, the Regional Strategy of Biodiversity of the Tropical Andean Countries, and CAN's Follow-up Plan to the Johannesburg Summit.

[209] http://www.otca.org.br/ep/proyetos-programas/index.php?id=1151

under Mercosur law. Conflicts have also occurred between norms adopted under CAN, Mercosur, and international agreements. In both forums the prevalent understanding is that international agreements and decisions adopted respectively under CAN or Mercosur prevail over other international agreements, which are only applicable in a subsidiary manner.

The jurisprudence of the Court of Justice of the Andean Community (CJAC) supports the position that CAN law prevails over its member States' national laws and over international agreements outside the framework of CAN,[210] due to the binding character of CAN law.[211] According to the CJAC, CAN member States have the basic duty to comply with legal obligations under CAN law.[212] Therefore, CAN

[210] The supremacy of CAN law over domestic law (and over international law) is asserted in the case *Acción de Incumplimiento interpuesta por la Secretaría General de la Comunidad Andina contra la República de Colombia*: "Es decir, el hecho de que los Países Miembros de la Comunidad Andina pertenezcan a su vez a la Organización Mundial de Comercio no los exime de obedecer las normas comunitarias andinas so pretexto de que se está cumpliendo con las de dicha organización o que se pretende cumplir con los compromisos adquiridos con ella. Ello sería ni más ni menos que negar la supremacía del ordenamiento comunitario andino que como se ha dicho es preponderante no sólo respecto de los ordenamientos jurídicos internos de los Países Miembros sino de los otros ordenamientos jurídicos internacionales a que éstos pertenezcan" (Judgment of April 14, 2005, PROCESO 118-AI-2003, *Cartagena Agreement Official Gazette*, no. 1206, June 13, 2005). In the case *Interpretación Prejudicial de oficio de los artículos 81, 93 y 95 de la Decisión 344 de la Comisión del Acuerdo de Cartagena*, the CJAC supported the prevalence of CAN law: "Haciendo un análisis de la posición o jerarquía del Ordenamiento Jurídico Andino, ha manifestado que dicho ordenamiento goza de prevalecía respecto de los ordenamientos jurídicos de los Países Miembros y respecto de las normas de derecho internacional, dentro de las cuales se encuentran las de la Convención General Interamericana de Protección Marcaria y Comercial de Washington de 1929. En este marco ha establecido que en caso de presentarse antinomias entre el derecho comunitario andino y el derecho interno de los Países Miembros, prevalece el primero, al igual que al presentarse antinomias ente el derecho comunitario y las normas de derecho internacional, también prevalece el derecho comunitario andino" (PROCESO 192-IP-2006, *Cartagena Agreement Official Gazette*, no. 1478, March 21, 2007). Also on the supremacy of CAN law, see PROCESO 02-N-86, *Cartagena Agreement Official Gazette*, no. 21, July 15, 1987.

[211] As affirmed in the case *Acción de incumplimiento interpuesta por la Junta del Acuerdo de Cartagena contra la República del Ecuador*, CAN law prevails over domestic law: "La preeminencia conlleva la virtud que tiene el ordenamiento comunitario de ser imperativo de modo que cuando se trate de aplicar normas legales a actuaciones jurídicas contempladas en el derecho de integración, deberá acudirse al ordenamiento comunitario con preferencia sobre el derecho interno," and that supremacy is due to the mandatory nature of CAN law: "La preeminencia conlleva la virtud que tiene el ordenamiento comunitario de ser imperativo de modo que cuando se trate de aplicar normas legales a actuaciones jurídicas contempladas en el derecho de integración, deberá acudirse al ordenamiento comunitario con preferencia sobre el derecho interno" (PROCESO 1-AI-96, *Cartagena Agreement Official Gazette*, no. 234, November 21, 1996).

[212] In the case *Acción de incumplimiento interpuesta por la Junta del Acuerdo de Cartagena contra la República del Ecuador*: "Como fundamentos de Derecho, la Junta en primer término pone de manifiesto el artículo 5 del Tratado de Creación del Tribunal de Justicia del Acuerdo de Cartagena y en su ilustración retoma un señalamiento del mismo Tribunal, donde se resalta que los Países Miembros frente a la norma comunitaria tienen primordialmente dos obligaciones, una de asegurar y mantener este ordenamiento y la segunda de no contrariarlo" (ibid.).

member States are prevented from taking any (legislative, judicial, or administrative) measures that may conflict with CAN law,[213] and neither can they restrict in any way, through national laws or international treaties, the exercise of rights foreseen under CAN law.[214] Conversely, in the context of the Caribbean Community, the Caribbean Court is of the opinion that if domestic legislation conflicts with international treaties not yet incorporated into domestic jurisdictions, national courts may disregard the international treaty and apply national law.[215]

The relationship between CAN law and international law has been much discussed in the case *Acción de Incumplimiento interpuesta por la Secretaría General de la Comunidad Andina contra la República Bolivariana de Venezuela*.[216] The CAN General Secretariat alleged that Venezuela failed to give effect to CAN's Decision 344 on Industrial Property[217] when using the World Trade Organization's (WTO) 1995 Agreement on Trade-Related Aspects of Intellectual Property Rights (TRIPS) as a justification for granting patent to a therapeutic product prohibited under CAN Decision 344. In response, Venezuela claimed that TRIPS, adopted prior to the creation of CAN as a legal entity, was binding on all member States and that Article 143 of CAN's Decision 344 allowed its member States to adopt

[213] As asserted in the case *Acción de Incumplimiento interpuesta por la Secretaría General de la Comunidad Andina contra la República de Colombia*: "se le imponen a los Países Miembros dos obligaciones básicas con respecto al mismo: una, de hacer, dirigida a la adopción de medidas que aseguren el cumplimiento de dicho Ordenamiento dentro de su ámbito territorial; y, otra, de no hacer, conducente a que no se adopten medidas o se asuman conductas o se expidan actos, sean de naturaleza legislativa, judicial, o administrativa, que contraríen u obstaculicen la aplicación del derecho comunitario" (PROCESO 118-AI-2003, op. cit.). Also in this sense, PROCESOS 2-IP-88, 5-IP-89 and 6-IP-93, *Cartagena Agreement Official Gazette*, no. 33, July 26, 1988; no. 50, November 17, 1989; and no. 150, March 25, 1994, respectively.

[214] In the case *Interpretación Prejudicial de oficio de los artículos 81, 93 y 95 de la Decisión 344 de la Comisión del Acuerdo de Cartagena*: "el Tribunal ratifica que la potestad de las autoridades nacionales de los Países Miembros, de regular a través de normas internas o mediante la celebración de tratados internacionales, los asuntos sobre Propiedad Industrial no comprendidos en la norma comunitaria andina, no podrá ser ejercida de modo tal que signifique la introducción de restricciones adicionales al ejercicio de los derechos y facultades consagrados por la norma comunitaria. En todo caso, la norma interna o internacional que se aplique deberá ser compatible con la comunitaria" (PROCESO 192-IP-2006, op. cit.).

[215] In the Appeal from the Court of Appeal of Barbados, the Caribbean Court of Justice held that "the common law has over the centuries developed rules about the relationship between domestic and international law. The classic view is that, even if ratified by the Executive, international treaties form no part of domestic law unless they have been specifically incorporated by the legislature. In order to be binding in municipal law, the terms of a treaty must be enacted by the local Parliament" (CCJ APPEAL no. CV 2 of 2005, para. 55).

[216] In the case *Acción de Incumplimiento interpuesta por la Secretaría General de la Comunidad Andina contra la República Bolivariana de Venezuela*, PROCESO 01-AI-2001, Judgment of June 27, 2002, *Cartagena Agreement Official Gazette*, no. 818, July 23, 2002.

[217] *Cartagena Agreement Official* Gazette, no. 142, October 29, 1993.

treaties on intellectual property law. In addition, Venezuela argued that, according to CAN's Decision 486, TRIPS could be used as a source of law in the framework of CAN.

In the above case, the CJAC asserted the supremacy of CAN law over treaties adopted elsewhere by CAN member States.[218] It held that States within a regional integration system consented to limit the exercise of their sovereignty by granting legislative powers to community organs. According to the CJAC, the constituent instruments of a regional integration system (primary law) are subsequently complemented with a body of law produced by the organs endowed with legislative powers (secondary law), such as the Andean Council of Foreign Ministers, the Andean Community Commission, and the CAN General Secretariat. Decisions adopted by those organs, such as Decisions 344 and 486, were directly applicable in the member States' territories.[219] Therefore, in cases where community bodies regulated specific issues, CAN member States could use a treaty on that same matter as a source of law, but this did not mean that CAN was subordinated to it. According to the court, States had limited *motu proprio* the exercise of their legislative powers by being CAN members and, therefore should not accept international binding commitments that disregarded or contradicted CAN law.[220] The court concluded that treaties adopted by CAN member States, such as TRIPS, do not produce direct effects in the territories of the member States,[221] and that WTO membership did not

[218] According to the CJAC, CAN member States have accepted commitments when adopting the Cartagena Agreement that go beyond classic international law: "Al suscribir el Acuerdo de Cartagena, por medio del cual se puso en marcha el proceso de integración inicialmente conocido como el Grupo Andino, los países firmantes adquirieron un compromiso de carácter internacional que trasciende los principios y características del derecho internacional tradicional, puesto que a más de respetar y cumplir las cláusulas establecidas en el referido Acuerdo se obligaron, dentro del marco de los principios y características propios del derecho comunitario, a cumplir con el ordenamiento jurídico que se establecía a partir del Tratado Constitutivo y que se estructuraba además, sobre la base de las normas jurídicas supranacionales que se iban generando por los distintos organismos comunitarios" (PROCESO 7-AI-98, *CAN Gazette*, no. 490, October 4, 1999).

[219] The direct applicability of CAN law is asserted in PROCESOS 1-IP-87, 6-IP-94, 10-IP-94 and 1-AI-96, published in *Cartagena Agreement Official Gazette*, no. 28, February 15, 1988; no. 170, January 23, 1995; no. 177, April 20, 1995; and no. 234, November 21, 1996, respectively.

[220] In the case *Acción de Incumplimiento interpuesta por la Secretaría General de la Comunidad Andina contra la República Bolivariana de Venezuela*: "A causa de la citada transferencia, y en los ámbitos en que ella se produzca, los Países Miembros quedan limitados motu proprio en el ejercicio de su potestad normativa y, en consecuencia, quedan impedidos de dictar providencias o contraer compromisos, aun de alcance internacional, que menoscaben o contraríen el ordenamiento comunitario" (PROCESO 01-AI-2001, op. cit.).

[221] Ibid.: "En consecuencia, los tratados internacionales que celebren los Países Miembros por propia iniciativa, como el del Acuerdo sobre los Aspectos de los Derechos de Propiedad Intelectual Relacionados con el Comercio, no vinculan a la Comunidad, ni surten efecto directo en ella, sin perjuicio de la fuerza vinculante que tales instrumentos posean en las relaciones entre los citados Países Miembros y terceros países u organizaciones internacionales."

exempt them from observing CAN law,[222] otherwise its supremacy over other existing laws would be denied. According to the CJAC's decision, Venezuela breached both the Treaty Creating the Court of Justice of the Andean Community (Article 4) and CAN Decision 344 (Article 16), and was ordered to cease the violation by suspending the effect of the patent previously granted.

In general, the CJAC supports the position that CAN law comprises an autonomous and independent legal system.[223] Autonomy is viewed as indispensable to qualify a coherent set of structural principles and norms within an effective legal system.[224] Therefore, as the CJAC asserted, CAN law is not dependent in any way on rules within other international legal systems, neither should it necessarily be compatible with them.[225] According to the CJAC, as expressed in the case *Acción de Incumplimiento interpuesta por la Secretaría General de la Comunidad Andina contra la República Bolivariana de Venezuela*, granting supremacy to TRIPS over CAN's Decision 344 would suggest the existence of two different international

[222] Ibid.: "La circunstancia de que los Países Miembros de la Comunidad Andina pertenezcan a su vez a la Organización Mundial de Comercio no los exime de obedecer las normas comunitarias andinas so pretexto de que se está cumpliendo con las de dicha organización o que se pretende cumplir con los compromisos adquiridos con ella. Ello sería ni más ni menos que negar la supremacía del ordenamiento comunitario andino que como se ha dicho es preponderante no sólo respecto de los ordenamientos jurídicos internos de los Países Miembros sino de los otros ordenamientos jurídicos internacionales a que éstos pertenezcan." The CJAC also held that WTO membership did not exempt CAN member States from respecting CAN law in *Acción de Incumplimiento interpuesta por la Secretaría General de la Comunidad Andina contra la República de Colombia* (PROCESO 118-AI-2003, op. cit.).

[223] "El ordenamiento jurídico andino es autónomo y la aplicación de las normas comunitarias que lo conforman no depende de las de otros ordenamientos internacionales, ni debe sujetarse a que guarden compatibilidad o conformidad con ellas.... Además, el derecho comunitario andino, fuera de constituir un ordenamiento jurídico autónomo, independiente, con su propio sistema de producción, ejecución y aplicación normativa, posee los atributos, derivados de su propia naturaleza, conocidos como de aplicabilidad inmediata, efecto directo y primacía," (PROCESO 89-AI-2000, *Cartagena Agreement Official Gazette*, no. 722, October 12, 2001).

[224] In the case *Interpretación Prejudicial de oficio de los artículos 81, 93 y 95 de la Decisión 344 de la Comisión del Acuerdo de Cartagena*: "Un punto fundamental al analizar el principio de Supremacía del Derecho Comunitario Andino, como se dijo anteriormente, es el principio de "Autonomía del Ordenamiento Jurídico Andino", que no es otra cosa que el desarrollo del principio de supremacía y que consagra como un verdadero sistema jurídico al Ordenamiento Jurídico Comunitario, es decir, unido a que dicho sistema jurídico se presenta como un todo coherente y dotado de unidad, contiene un conjunto de principios y reglas estructurales que derivan de él mismo, sin derivarlas de ningún otro ordenamiento jurídico" (PROCESO 118-AI-2003, op. cit.).

[225] "El ordenamiento jurídico andino es autónomo y la aplicación de las normas comunitarias que lo conforman no depende de las de otros ordenamientos internacionales, ni debe sujetarse a que guarden compatibilidad o conformidad con ellas.... Además, el derecho comunitario andino, fuera de constituir un ordenamiento jurídico autónomo, independiente, con su propio sistema de producción, ejecución y aplicación normativa, posee los atributos, derivados de su propia naturaleza, conocidos como de aplicabilidad inmediata, efecto directo y primacía" (PROCESO 89-AI-2000, op. cit.).

legal systems. This solution is discarded, as it would grant States, depending on their own interests, discretion to choose the legal system to which they would be subject.[226]

Under the aegis of Mercosur, the Permanent Court of Appeal in the case *Prohibición de importación de neumáticos remoldeados procedentes del Uruguay* supported the view that Mercosur's "law of integration" is autonomous *vis-à-vis* other "branches of law" (including international law), disregard of which would negatively affect Mercosur's institutional and normative development. This court took the position, similarly to that of the CJAC, that international law should only be applied in a subsidiary way and never to the detriment of Mercosur law.[227]

Conflicts of norms such as the above are now more likely to occur among the Amazon countries, given the increased number of regional and global organizations involving those States, under which legal instruments addressing similar issues are adopted. Different treaties with common parties dealing with the same or a related subject do not necessarily amount to a conflict of law, as the same subject matter can be addressed from different perspectives or apply differently depending on the circumstances. However, a conflict of law in the strict sense of direct incompatibility may arise in cases where a State participating in two treaties cannot simultaneously comply with its obligations under both treaties.[228]

[226] "… supone, sin razón valedera, la coexistencia de dos ordenamientos jurídicos diferentes de carácter internacional que permitirían que los Países Miembros justificaran sus actuaciones a su elección, sujetándose al que encontraran más conveniente y dejando de cumplir el que les resultara desfavorable o inconveniente y que, en esa medida, uno fuera subalterno o dependiente del otro; y, en segundo lugar, porque pretende la convalidación, a contrapelo de lo que expresan las normas comunitarias, de la facultad de emitir normas de derecho interno con la finalidad de ajustar o hacer compatibles las normas comunitarias a los otros ordenamientos internacionales a los que también estén sujetos, de manera individual o conjuntamente con los otros países andinos, cuando, a su juicio, se presente entre tales ordenamientos alguna incongruencia o discordancia" (ibid.). In the case *Acción de Incumplimiento interpuesta por la Secretaría General de la Comunidad Andina contra la República de Colombia* (PROCESO 118-AI-2003, op. cit.), the court denied the coexistence of two different international legal systems: "es para el Tribunal inaceptable suponer que la coexistencia de dos ordenamientos jurídicos diferentes de carácter internacional permita que los Países Miembros justifiquen sus actuaciones a su elección, sujetándose al que encontraran más conveniente y dejando de cumplir el que les resultara desfavorable."

[227] According to the court, Mercosur law should have autonomy over the other branches of law: "de manera directa y primera, como desde luego corresponde en un derecho de integración (que ya lo tiene el Mercosur) y en un derecho comunitario al cual se aspira (que todavía no lo tiene el Mercosur) por la ausencia de la tan anhelada supranacionalidad. En suma, el derecho de integración tiene y debe tener suficiente autonomía de las otras ramas del Derecho. El no considerar ello contribuirá siempre de manera negativa al desarrollo de la institucionalidad y normativa mercosureña", case Prohibición de importación de neumáticos remoldeados procedentes del Uruguay (Mercosur Arbitral Award 1/2005, para. 9).

[228] W. Jenks, "The conflict of law-making treaties," *British Yearbook of International Law*, vol. 30, 1953, pp. 401–53, at p. 426.

The decisions of the CJAC and Mercosur's Permanent Court of Appeal seem to suggest that the norms adopted under their respective legal frameworks are not compatible with obligations under other treaties. They tend to frame both CAN and Mercosur legal regimes as "self-contained,"[229] the integrity of which is ensured by giving supremacy to norms they create over national laws and other international treaties.

In situations of conflict between norms, traditional techniques of interpretation, and conflict resolution can be used, such as the principle *lex specialis derogat legi generali*. This principle suggests that in cases of conflict of norms dealing with the same subject matter within a single treaty, within two or more treaties or between a treaty and a nonbinding instrument, priority should be given to the norm that is more specific.[230] According to the International Law Commission, the reason why a special law should be given priority over a general law lies in the fact that a *lex specialis*, by being more concrete, often takes better account of the particular context in which it is to be applied than any general law. Conferring priority on a special law does not mean, however, that the relevant general law shall be extinguished. The latter remains valid and applicable, providing direction for the interpretation and application of the relevant special law, and continues to be applicable in situations not provided for by the special law.[231] The basis for the *lex specialis* can be found in the law of treaties and the 2001 Articles on State Responsibility.[232] On various occasions,

[229] As J. Pauwelyn explains, references to the notion of self-contained regimes normally relate to "certain international legal regimes (in particular, those of diplomatic immunities, the European Community, and human rights treaties) that, in terms of their compliance mechanism or secondary rules, may somehow be self-contained, without any or only limited 'fallback' on general international law" (J. Pauwelyn, "The role of public international law in the WTO: How far can we go?" *American Journal of International Law*, vol. 95, pp. 535–78, at p. 537). It is claimed that multilateral environmental agreements, for example, are "autonomous sectoral legal systems," as they involve international institutions comprising an accepted body of normative prescriptions and an organized process for the making and application of these prescriptions (see T. Gehring, "International environmental regimes: Dynamic sectoral legal systems," *Yearbook of International Environmental Law*, vol. 1, 1990, pp. 35–56, at p. 56). The question of whether certain international norms have supremacy over other customary or convention rules is related to the debated question of the hierarchy of norms in international law. In principle, since the introduction of the concept of *jus cogens* in Article 53 of the Vienna Convention on the Law of the Treaties and Article 103 of the UN Charter, *jus cogens* rules and obligations are enshrined in the UN Charter, and it could therefore be said that such a hierarchy exists, and these norms would always prevail over more ordinary commitments. However, as Weiler and Paulus note, problems can be discerned in this apparently easy and undisputable positivist confirmation of a hierarchy of norms in international law (J. Weiler and A.L. Paulus, "The structure of change in international law, or is there a hierarchy of norms in international law?" *European Journal of International Law*, vol. 8, 1997, pp. 545–65, at p. 558).

[230] International Law Commission, *Report on the Work of Its 58th Session*, General Assembly, 61st Session, Supplement no. 10, A/61/10, 2006, para. 5, p. 410.

[231] Ibid., paras. 7–9, p. 409.

[232] Pauwelyn, op. cit., p. 392.

provisions of the 1969 Vienna Convention on the Law of Treaties are explicitly made subject to *lex specialis* (e.g. in Articles 19(a), 20(1), 28, 29, 30(2), 40(1), etc.) and the Articles on State Responsibility are expressly not applicable in cases where an "internationally wrongful act or the content or implementation of the international responsibility of a State are governed by special rules of international law" (Article 55).

In the event of conflicts between CAN or Mercosur norms with other treaties or national laws, those norms may prevail not due to an intrinsic supremacy, but because in most cases they may be considered *lex specialis* in relation to national laws or international treaties. As noted by the International Law Commission, the norms and principles under a special legal regime tend to prevail because they function as *lex specialis*.[233] A norm may be *lex specialis* on two grounds: subject matter or membership.[234] A norm is more special than another when its material scope is more precise or limited and it covers an issue more directly than a general law. Also (although not necessarily), a treaty with fewer parties tends to be more specific than one involving a greater number of States. By the same token, CAN and Mercosur norms tend to prevail over general law for those two reasons: They are more likely to address a subject matter with more precision than a general rule dealing with the same issue; and they involve a smaller number of parties when compared to a global treaty. However, the *lex specialis* rule does not apply in all cases. For example, when a *lex specialis* conflicts with a rule of *jus cogens*, the latter prevails.[235] A treaty can also contract out of general international law[236] by stating that, in cases of conflict, the treaty norm prevails (except over a *jus cogens* rule, which always has a superior hierarchical ranking).

In addition to those rules traditionally used in the event of conflicts of law, other principles can be also invoked in such situations, particularly in the context of the Amazon. Principles, such as that of cooperation and good neighborliness, can be used to prevent or even settle conflicts of law involving regional or global institutions. These principles are made concrete, as stated, for example, in UNCLOS (Article 197) or in Agenda 21 (Chapter 39.10), through procedures involving information exchange, notification, and consultation. States participating in different legal regimes could prevent conflicts of law by exchanging information and consulting

[233] International Law Commission, op. cit., para. 3, p. 410.

[234] Pauwelyn, op. cit., p. 389.

[235] M. Akehurst, "The hierarchy of the sources of international law," *British Yearbook of International Law*, vol. 47, 1974–75, pp. 173–285, at p. 228.

[236] J. Pauwelyn, *Conflict of Norms in Public International Law: How WTO Law Relates to Other Rules of International Law*, London & New York: Cambridge University Press, 2003, p. 394. According to Pauwelyn, the possibility of "contracting out" from general international law (other than *jus cogens*) is due to the lack of any inherent hierarchy between general international law and treaties and between any two rules of international law. As these norms derive from the same source (essentially state consent), they must in principle be equal in value.

other States prior to the adoption of new norms. This is encouraged, for example, by the 2001 Mercosur Framework Agreement on the Environment, which requires States to cooperate by exchanging information on environmental laws, policies, and practices. Procedures involving information exchange and regular consultations would benefit cooperation among regional and global institutions in general.

CONCLUSIONS

This chapter looked at a some MEAs and some international organizations and NGOs working on the Amazon, in order to address the questions: How do the Amazon States interact with the international community; and what is the role of this interaction in ensuring that international obligations assumed by those countries are met, especially the duty to preserve the environment, and the Amazon in particular.

The Amazon States interact with the international community – represented by States, international organizations, and NGOs – in relation to environmental matters under the framework of MEAs (and other legal instruments) and by creating partnerships with global actors. In the context of MEAs, States agree upon a set of rules to handle specific environmental issues, such as climate change, wildlife trade, natural and cultural heritage, among others. The Amazon States, like other countries, must observe norms and principles defined under MEAs in which they participate, as well as customary international law. These commitments are applicable in their respective territories, including the Amazon. Therefore, MEAs, as well as other nonlegally binding instruments, such as the 1992 Forestry Principles or the 1978 UNEP Environmental Law Guidelines and Principles on Shared Natural Resources, should in particular guide activities carried out in these countries' respective Amazonian territories. This is the case, for example, when one of these States is developing infrastructure, building dams on international rivers, or managing natural heritage sites in their respective portion of the Amazon.

International commitments assumed under MEAs (and other instruments) should guide the Amazon States' actions not only in their own territories, but also in their interactions with the other basin States and the wider community of States, which may be represented by the contracting parties to a particular MEA. The legal obligations (or guidelines) contained in international legal instruments are usually coupled with a set of rights. Therefore, while State parties to the UNESCO World Heritage Convention, for example, have a duty to conserve their natural heritage (Article 4), they also have the right to request assistance to maintain natural or cultural heritage situated in their territories, if appropriate (Article 19). The role of providing assistance under a MEA is to be exercised by developed country parties or multilateral financial institutions, such as the World Bank and GEF. International assistance is a common component of most MEAs and other international legal instruments. However, financial assistance provided by the international community may come from sources other than those foreseen under MEAs. For example, through ODA, States provide official financing from general budgetary expenditure to help developing countries promote and implement development.[237] At present, most international climate change funding, in particular, is classified as ODA.[238] By providing such assistance, the international community can facilitate the fulfilment of international obligations under regional and multilateral treaties. Therefore, although the international community may have a legitimate interest in the protection of the world's natural and cultural sites, for instance, it also has a duty to assist developing countries parties to meet their international obligations. Likewise, the relationship between the Amazon States and the international community with regards to the protection of the environment, or to the Amazon in particular, involves a set of rights and duties for both sides.

The Amazon States also interact with the international community, represented by global institutions, such as international organizations and NGOs, by establishing partnerships and hosting field projects. The assistance provided by those institutions has been important in many ways, for example in creating protected areas, such as seen in the context of the PPG7, monitoring wildlife species undertaken by IUCN and other institutions, or in campaigning against deforestation, as seen in the case of Greenpeace-Brazil. However, as can be observed at the regional level, the presence of these many global actors in the Amazon does not imply better environmental protection. International organizations and NGOs normally compete for international funding and work in similar areas or issues (climate change mitigation and adaptation in particular), which may lead to a duplication of efforts. Moreover, field projects conducted by those institutions vary in effectiveness. They may achieve useful outcomes, but they can also be costly, time-consuming, benefit very few people, and have only few tangible results. In some cases, international assistance, provided for example by multilateral financial institutions, for developing infrastructure or industries, even if well-intended, can be detrimental to the environment, especially in sensitive areas such as the Amazon.

In general, the (financial and technical) assistance provided by the international community in the framework of MEAs or of partnerships established with international organizations and NGOs is beneficial and necessary to ensure that legal obligations under multilateral treaties are met and that project goals achieved. However, is this assistance sufficient to help improve the current situation in the Amazon and address serious problems such as deforestation?

[237] Parker et al., op. cit., p. 23.

[238] Ibid.

Although international cooperation, especially in the form of technical and financial assistance, has occurred over years, the Amazon biome is still threatened. Problems such as deforestation is likely to continue in the absence of more effective solutions.[239] It is suggested that international assistance, notably ODA, will not deliver the scale of finance required in particular for adaptation and mitigation activities.[240] It is estimated that the average annual cost of mitigation for halving deforestation in particular requires between US$17 billion and US$33 billion per year, whereas according to the Organization for Economic Co-operation and Development (OECD), the total ODA or official aid to forestry by and multilateral agencies was an annual average commitment of US$564 million (between 1996 and 2004). On the other hand, there seems to be consensus that, in order to handle problems such as deforestation and stabilize atmospheric GHG emissions, strong and urgent action is needed.[241] In view of the magnitude of funding required and the complexity of most environmental problems, the international community is currently discussing alternatives, such as a REDD mechanism to be introduced in the post-Kyoto regime. This and other mechanisms intended to deal with current environmental issues will be examined in the subsequent chapter.

[239] It is predicted that deforestation in the tropics is substantial and projected to continue over the next half century (Eliasch, op. cit., p. 33).

[240] Parker et al., op. cit., p. 58.

[241] Eliasch, op. cit., p. 33.

8

Positive Incentives for Protecting the Amazon

It is often suggested that to achieve cooperation and handle environmental problems, treaties should use mechanisms such as sticks (negative incentives) and carrots (positive incentives), so that States find it attractive to contribute to the greater good.[1] In this regard, Barret proposes a general theory of international cooperation to provide guidance on how to negotiate more effective agreements, which involves restructuring incentives, by balancing positive (carrots) and negative (sticks) incentives.[2] Whereas positive incentives that reward law-abiding behavior of States and prevent violations of international obligations function as a "carrot," measures such as penalties or sanctions addressing noncompliance situations work as a "stick."[3] The 1987 Montreal Protocol is usually cited for having successfully combined those two elements by financially compensating developing countries for the incremental costs of complying with the Protocol and, at the same time, by using the threat of trade restrictions to enforce obligations.

Positive incentives can be divided into two major categories: one that involves market-based mechanisms, for example, carbon trading, and one that makes use of nonmarket-based financial resources, such as official development assistance, voluntary contributions from governments, the private sector, and funds under international treaties. These latter are often created under environmental treaties

1 Decision adopted at the Fourteenth Meeting of the Conference of the Parties (COP) to the 1973 Convention on International Trade in Endangered Species of Wild Fauna and Flora (CITES), held in The Hague, June 3–15, 2007, COP-14 Doc. 32, on "Incentives for Implementation of the Convention," paragraph 12, http://www.cites.org/eng/cop/14/doc/index.shtmlhttp://www.cites.org/eng/cop/14/doc/index.shtml

2 S. Barret, *Environment and Statecraft: The Strategy of Environmental Treaty-making*, Oxford & New York: Oxford University Press, 2003, p. 32. Oran Young also proposes that a mix of incentives to compete and cooperate in interactive decision-making processes is an important determinant of success of international regimes (O.R. Young, "Global environmental change and international governance," *Millennium Journal of International Studies*, vol. 19, 1990, pp. 337–46, at p. 337).

3 This definition is found in the decision CITES COP-14 Doc. 32, op. cit., para. 8.

(e.g. the Wetlands Fund, the World Heritage Fund, and the Montreal Protocol Multilateral Fund) to help developing country parties meet additional costs related to their participation in a treaty, for example scientific costs in monitoring species and habitats, reporting, covering staff expenses, operating field projects, etc. Environmental treaties can also include market-based mechanisms, such as those introduced by the 1997 Kyoto Protocol.[4]

The value of positive incentives as conservation tools has been recognized under current environmental treaties. For example, CITES Resolution Conf. 11.3 (Rev. COP-13) calls on parties to "promote incentives to secure the support and cooperation of local and rural communities in managing wildlife resources and thereby combating illegal trade."[5] The 1992 Convention on Biological Diversity (CBD)[6] requires States to "adopt economically and socially sound measures that act as incentives for the conservation and sustainable use of components of biological diversity."[7] Likewise, the 1971 Convention on Wetlands of International Importance Especially as Waterfowl Habitat (Ramsar Convention)[8] recommends the use of "incentive measures that encourage the application of the wise use principle."[9] Pursuant to the 2007 Non-legally Binding Instrument on All Types of Forests,[10] States should develop positive incentives, particularly to help developing countries and economies in transition to reduce deforestation and promote the rehabilitation of degraded forests (para. 6, d). The 2009 Declaration of Heads of States of the Amazon Cooperation Treaty Organization (ACTO)[11] calls on the parties to use mechanisms able to support and create funding required to conserve and protect their forests (para. 3). Other

[4] 1997 Kyoto Protocol to the Framework Convention on Climate, 37 ILM 22 (1998) (Kyoto Protocol). The Kyoto Protocol creates an opportunity for the use of market-based mechanisms to assist Annex I parties meet their emission reduction targets. The three Kyoto mechanisms are: Emissions Trading (Article 17), Joint Implementation between Annex I parties (Article 6), and the Clean Development Mechanism (Article 12) (CDM).

[5] CITES Resolution Conf. 11.3 (Rev. COP-13), amended at CITES COP-13, held in Bangkok, Thailand, October 2–14, 2004.

[6] 31 ILM 818, 1992.

[7] According to Article 11 of the 1992 Convention on Biological Diversity (CBD), States shall adopt "economically and sound measures that act as incentives for the conservation and use of components of biological diversity." Decisions adopted at CBD meetings of the parties have also recommended using incentive measures, such as Decision VI/15, on Incentive Measures, adopted at COP-6, held in The Hague, April 7–19, 2002, http://www.cbd.int/decisions/cop-06.shtml?m=cop-06

[8] 11 ILM 963, 1972.

[9] See also Resolution VIII:23, of the 1971 Convention on Wetlands of International Importance, especially as Waterfowl Habitat (Ramsar Convention), on Incentive Measures as Tools for Achieving the Wise use of Wetlands, which requires the State parties to develop supportive legal and policy frameworks for the design and implementation of incentive measures. Adopted at COP-8, held in Valencia, Spain, November 18–26, 2002; http://www.ramsar.org/res/key_res_viii_index_e.htm

[10] A/C.2/62/L.5, October 22, 2007.

[11] Signed on November 26, 2009, http://www.otca.org.br

legal instruments attempt to channel existing "market forces" by proposing the use of market-based instruments. For example, the 2009 Copenhagen Accord, signed at the Fifteenth Conference of the Parties (COP-15) to the 1992 United Nations Framework Convention on Climate Change (UNFCCC)[12] encourages States to "pursue various approaches, including opportunities to use markets, to enhance the cost-effectiveness of, and to promote mitigation actions" (para. 7). Other statements, such as the Pacto pela Valorização da Floresta e pelo Fim do Desmatamento na Amazônia), drafted in 2007 by a group of Brazilian nongovernmental organizations (NGOs, proposes a zero-deforestation target to be achieved by a combination of public) policies and market strategies.[13]

There is no single standardized definition of "market-based instruments." For example, they have been defined as an attempt to align private and social costs in order to reduce externalities.[14] Some propose a distinction between "traditional market-based instruments," typically applied by governments to markets, and a more recent type of "markets for environmental services" that involve developing commodities to cover the costs of maintaining certain environmental services.[15] Examples of traditional market-based instruments include income and export taxes, fines, export quotas, export taxation, price control, etc.; whereas markets for environmental services involve biodiversity conservation (with commodities such as bio-prospecting rights, conservation concessions, debt-for-nature swaps), carbon sequestration (with commodities such as certified emission reductions, carbon offsets or credits, and conservation easements), watershed protection (with commodities such as watershed management contracts, water rights, and stream flow reduction licenses), and also landscape beauty (with commodities such as access permits, package tourism services, and ecotourism concessions).[16]

In general, market-based instruments are used to alter markets facing the private sector in order to make sustainable practices more profitable and attractive than unsustainable ones. For example, the mechanism known as a "debt-for-nature swap"

[12] 31 ILM 849 (1992).

[13] This document was drawn up by a group of nongovernment organizations in support of the debate between the strategic social and economic players in the Amazon region, including Instituto Socioambiental, Greenpeace, Instituto Centro de Vida, Instituto de Pesquisa Ambiental da Amazônia, he Nature Conservancy, Conservação Internacional, Amigos da Terra-Amazônia Brasileira, IMAZON and WWF-Brasil, signed on October 3, 2007, http://www.greenpeace.org/brasil/documentos/amazonia/pacto-pela-valoriza-o-da-flor

[14] R.M. Huber, J. Ruitenbeek, and R. Seroa da Motta, "Market-based instruments for environmental policy making in Latin America and the Caribbean: Lessons from eleven countries," World Bank discussion paper, no. 381, Washington DC: World Bank, 1998, p. 1.

[15] K. Sann and K. Thornber, "Impact of market-based instruments and initiatives on the trade in forest products and sustainable forest management," LTS International, p. 2, available at: http://www.fao.org/forestry/foris/data/trade/pdf/lts.pdf

[16] Ibid., p. 10.

can be seen as a form of market-based instrument according to which an indebted developing country agrees to invest in conservation projects or environmentally friendly actions in exchange for the cancellation of part of its foreign debt.[17] Debt swaps have been used to finance environmental conservation and health projects in some developing countries. For example, under debt-for-nature swap schemes, the United States has approved some loans for conservation purposes that have benefited certain areas in the Amazon. In 1988, the US government enacted a Tropical Forest Conservation Act, by which developing countries owning tropical forests of global or regional significance could be eligible for debt forgiveness in exchange for conservation efforts. In this case, debt-for-nature swap agreements to fund local conservation projects, authorized under the Tropical Forest Conservation Act, were signed with Peru in 2002[18] and Colombia in 2004.[19]

In the forestry sector, market-based instruments may provide commercial incentives to capture the economic value of forests and promote sustainable management. They involve "traditional instruments" (e.g. levies, fines, concession conditions, and trade controls), "third-party verification schemes" (e.g. certification and labeling), and industry and government initiatives.[20] Forest certification is an instrument that is currently widely used. It emerged in the 1990s, supported by the lobbying of some NGOs against widespread logging, and as a reaction to national and multilateral actions that had failed to curb deforestation.[21] The Forest Stewardship Council (FSC), an alliance of NGOs and businesses founded in 1993, was the first widely agreed international scheme designed to trace wood products and allow verification of forest management standards.[22]

Certification can take different forms, from "self-certification" and labeling to independent audits.[23] In general, forest certification is a process by which an independent nongovernmental third party inspects a product and provides written assurance that it originates from forests managed in compliance with predefined social and environmental standards. Certification is often accompanied by a mark visible

[17] C. Parker et al., *The Little Climate Finance Book: A Guide to Financing Options for Forests and Climate Change*, Oxford: Global Canopy Foundation, 2009, p. 74: "under debt swap programmes contributing countries agree to cancel a portion of the (non-performing) debt obligation of a developing country in exchange for an investment in beneficial development projects in that developing country."

[18] Information available at: http://www.usaid.gov/our_work/environment/forestry/tfca_descs.html#Peru

[19] Information available at: http://www.usaid.gov/our_work/environment/forestry/intro_tfca.html

[20] Five categories are suggested: (1) traditional instruments (e.g. levies, fines, concession conditions, and trade controls), (2) third-party verification schemes (e.g. certification and labeling), (3) abstinence (e.g. campaigns, bans, and boycotts), (4) markets for environmental services, and (5) industry and government initiatives (Sann and Thornber, op. cit., p. 10).

[21] Ibid., p. 6.

[22] Ibid., p. 3.

[23] Ibid., p. 1.

to consumers, which assures that a process or service conforms to certain requirements.[24] Examples of international certification schemes include the FSC,[25] the Rain Forest Alliance, International Federation of Organic Agriculture Movements, Sustainable Agriculture Network, Fairtrade Labeling Organizations International, among many others. In addition, national certification schemes are also being developed, such as the Sello Ambiental Colombiano,[26] Sistema Brasileiro de Certificação (Brazil),[27] and the Programa Nacional de Certificación de Ecoturismo in Ecuador.[28] By choosing certified products or services, consumers exert an influence on demand and to some degree limit the market for those products not harvested in a sustainable way.[29]

In the Amazon, the State Forest Antimary, located in Acre (one of the States within the Brazilian Amazon) and administered by the Technical Foundation of Acre (Fundação de Tecnologia do Estado do Acre), has been certified by SmartWood, a program of the Rainforest Alliance accredited by FSC.[30] At the national level

[24] ISO/IEC Guide 65:1996, *General Requirements for Bodies Operating Product Certification, International Organization for Standardization*. Certification is defined as a "procedure by which a third party gives a written assurance that a product, process or service conforms to specified requirements." Certification can be applied voluntary, for example by producers of biodiversity products, such as "dolphin-friendly" tuna and nontimber forest products, under various certification schemes created, for example, under the Forest Stewardship Council, the Marine Stewardship Council, and ISO. Certification can also be mandated by law (known as "regulatory certification"), in which case the national government often holds the certification authority. The CITES system is an example of a regulatory certification, in which a designated national management and a scientific authority acts as "certifier," ensuring that products and goods are produced in a way nondetrimental to the survival of listed species and have been legally obtained (CITES Decision COP-14 Doc. 32, op. cit).

[25] M. Brouwer, *Amazon Your Business: Opportunities and Solutions in the Rainforest*, Bunnik: Meindert Brouwer Partner in Communicatie, 2007, pp. 60–66.

[26] Regulated by the Colombian Ministerio de Ambiente, Vivienda y Desarrollo Territorial, through Resolution no. 1555, of October 20, 2005. Later, Resolution no. 1295 of June 30, 2006 created the Comité Interno del Sello Ambiental Colombiano and its regulation. Information available at: Comité Interno del Sello Ambiental Colombiano. For more information see A.L. Arcos, *Facilitación de financiamiento para negocios basados en la biodiversidad y apoyo a actividades de desarrollo de mercados en la región andina*, Bogotá, 2005; Z. Guzmán Castro and L.C. Hillón, *Liniamentos de etiquetado ambiental para Colombia*, Bogotá: Ministerio del Medio Ambiente, 2002.

[27] Information available at: http://www.inmetro.gov.br/qualidade/comites/sbc.asp

[28] This program was initiated in 1999, supported by International Union for Conservation of Nature (IUCN) and Ecuador's Ministry of Tourism. The entreprises that meet certain standards are recognized as "empresa de Ecoturismo," recognized under the national 2002 Reglamento de Ecoturismo y Sostenabilidad. Examples of certification projects in the Ecuadorian Amazon include the Manatee Amazon Explorer, Kapawi and Yachana Lodge. For more information, see Corporación de Conservación y Desarrollo, Programa de Certificación de Turismo Sostenible para Operaciones Turísticas de Tierra, 2006, http://www.ccd.org.ec

[29] H. GreGersen and A. Contreras, *Rethinking Forest Regulations from Simple Rules to Systems to Promote Best practices and Compliance*, Washington, DC: Rights and Resources Initiative, 2010, p. 9.

[30] ITTO (International Tropical Timber Organization), "Brazil gets new forest law," *Tropical Forest Update*, 2005, http://www.itto.int

Brazil has developed a Forest Certification System (Programa Brasileiro de Certificação Florestal – CERFLOR) that started in the 1990s and became operative for planted forests in 2003.[31] CERFLOR was developed under the National System of Metrology, Standardization, and Industrial Quality (Sistema Nacional de Metrologia, Normalização e Qualidade Industrial).[32] Certification has gained ground also in other Amazon countries and tends to increase.[33] For example, Bolivia has the largest area of certified tropical forests in Latin America under the FSC scheme.[34] In general, certification is a useful tool, as it ensures sustainable practices and raises awareness among consumers. However, small landowners usually find it difficult to comply with certification standards and ultimately use these schemes, as the front-end fixed costs for implementing practices to conform with defined standards are high.[35]

Positive incentives are being used in many parts of the world to achieve different purposes, such as biodiversity conservation, watershed protection, and carbon sequestration, among others. For example, Costa Rica and Mexico have relied on payments for environmental services for protection, reforestation, and forest regeneration. As noted, these instruments have also been used in the Amazon. For example, in the Brazilian Amazon, a number of state-level fiscal and other incentives are being created to encourage and compensate individuals and communities engaged in sound economic activities. In the state of Mato Grosso incentives are being used under different denominations: *compensação ambiental, compensação financeira, taxa de reposição florestal, taxa de controle e fiscalização, ICMS Ecológico* and *Imposto de Renda Ecológico*.[36] Also in the State of Amazonas, a recently approved law on Climate Change, Environmental Conservation and Sustainable Development[37] (state Law 3.135/2007) has introduced fiscal and other incentives for individuals and communities engaged in renewable energy or ecotourism activities. In this context, the present chapter enquires whether certain incentives could be used as conservation tools and handle critical problems in the Amazon, such as deforestation.

[31] ITTO, *Status of Tropical Forest Management*, 2005, 2006, p. 212, http://www.itto.int

[32] Information available at: http://www.inmetro.gov.br/qualidade/cerflor.asp

[33] In 2007, Latin America and the Caribbean had about 12 million hectares of certified forests, or about 4 percent of all certified forests in the world. Although the certified area represented only 1.2 percent of the region's forests, this was a significant increase from 0.4 percent in 2002. Almost 80 percent was certified by FSC, and the rest under national systems (FAO [Food and Agriculture Organization of the United Nations]), *State of the World's Forests*, Rome: FAO, 2009, p. 36).

[34] The total certified forest area (as of December 2005) was 2.21 million hectares (ITTO, *Status of Tropical Forest Management*, op. cit., p. 204)

[35] GreGersen and Contreras, op. cit., p. 9.

[36] A.L.M. da Riva, L.F. Laranja da Fonseca and L. Hasenclever, *Instrumentos econômicos e financeiros para a conservação ambiental no Brasil: Uma análise do estado de arte no Brasil e no Mato Grosso, desafios e perspectivas*, São Paulo: Instituto Sócio Ambiental, 2007, p. 32.

[37] Information available at: http://www.sds.am.gov.br

8.1. CARBON TRADING

Market-based mechanisms can be further subdivided into voluntary or regulated market systems.[38] This is particularly noted in relation to carbon markets that include, for example, certified emission reductions and carbon offsets or credits. The regulated carbon market (also referred to as "mandatory" or "compliance" carbon market) was created under the 1997 Kyoto Protocol. The voluntary market uses different rules and procedures for the monitoring and verification of carbon projects compared to those applied to the Kyoto Protocol's flexible mechanisms and is under no internationally agreed institutional structure of the kind that exists in the context of the Kyoto Protocol.[39]

8.1.1. The Compliance Market

The 1992 UNFCCC set up an international framework for climate change policy, the ultimate objective of which is to "stabilize greenhouse gas concentrations in the atmosphere at a level that would prevent anthropogenic interference with the climate system" (art. 2). Prior to that convention, the international community had taken steps to address the effects of human actions on the global climate, notably by creating the Intergovernmental Panel on Climate Change (IPCC) in 1988, a scientific body established by the United Nations Environment Programme (UNEP) and the World Meteorological Organization to assess the state of climate change and its environmental and socioeconomic impacts.[40]

Following the UNFCCC, the 1997 Kyoto Protocol defined quantified emission limitation and reduction commitments to industrialized countries and some transition economies listed in Annex 1. During a first commitment period (2000–2012), Annex 1 parties should reduce their overall greenhouse gas (GHG) emissions by at least 5 percent below 1990 levels (Article 3). In line with the principle of common but differentiated responsibilities (UNFCCC, Article 3), the Kyoto Protocol imposes no reduction commitments on non-Annex 1 countries during that first commitment period, given their smaller contribution to the current stock of GHGs in the atmosphere.

An important feature of the Kyoto Protocol is that it gives Annex 1 parties enough flexibility to meet their emission reduction targets by introducing three market-based mechanisms (known as "flexible mechanisms"), namely Emissions Trading (art. 17), Joint Implementation (JI) (art. 6), and the Clean Development Mechanism

[38] L. Peskett et al., *Making REDD Work for the Poor*, prepared on behalf of the Poverty Environment Partnership, 2008, p. 23, http://www.unep-wcmc.org/climate/pdf/Making%20REDD%20work%20for%20the%20poor%20FINAL%20DRAFT%200110.pdf

[39] Ibid., p. 23.

[40] Information available at: http://www.ipcc.ch/organization/organization.htm

(CDM) (art. 12). Some countries have established (or are creating) national or regional emissions trading schemes to help them meet their emission reduction targets, such as the European Union (EU) Emission Trading Scheme that involves all EU member States and allows limited trading with the three Kyoto mechanisms.[41]

Whereas two of the flexible mechanisms under the Kyoto Protocol (Emissions Trading and Joint Implementation) involve only Annex 1 parties, the CDM includes non-Annex 1 countries. Under this mechanism, Annex 1 countries may undertake project activities in non-Annex 1 countries with a view to reducing GHG emissions. Those projects generate certified emission reductions (CERs) that Annex 1 countries may use as a way to comply with their obligations under the Kyoto Protocol. CERs are only generated if they reduce carbon emissions below what would have occurred without the CDM project activity. This mechanism was designed to provide cost-effective compliance to Annex 1 countries and also to generate resource flows and facilitate access to clean technologies for developing countries. By this means, CDM opens an avenue of cooperation between Annex 1 and non-Annex 1 countries as a form of incentive that can bring in sound technology and promote sustainable practices in developing countries.

During the period 2002–2008, China accounted for 66 percent of all primary CDM markets,[42] followed by India and Brazil (with 4 percent and 3 percent market share, respectively).[43] In terms of regional distribution, Asia-Pacific is the region with the highest number of CDM projects (and CERs issued) with a total of 78.3 percent, followed by Latin America (17.3 percent) and Africa (2.4 percent).[44] Brazil occupies the third place in number of CDM project activities (with 417 projects, or 8 percent of total projects), whereas China ranks first (with 2,024 projects, or 37 percent) and India second (with 1,446 projects, or 27 percent).[45] European buyers dominate the CDM market, in particular private sector entities.[46]

[41] More specifically, EU members may trade allowances (known as EU emissions allowances) with one another, or they may buy and sell carbon credits (ERUs and CERs) generated by Joint Implementation or CDM projects (Ecosystem Marketplace, *Forests: Taking Root in the Voluntary Carbon Market*, 2009, p. 6., http://www.katoombagroup.org/documents/cds/katoomba_xiv/documents/Forests%20II/ECM%20Forests%20II%20English%20-%20final.pdf).

[42] K. Capoor and P. Ambrosi, *State and Trends of the Carbon Market* 2009, Washington, DC: World Bank, 2009, p. 34, http://wbcarbonfinance.org/docs/State___Trends_of_the_Carbon_Market_2009-FINAL_26_May09.pdf

[43] According to the latest assessment of the CDM Executive Board, registered project activities by host party are distributed as follows: Brazil (8.69 percent), Malaysia (3.48 percent), Mexico (6.27 percent), India (24.64 percent), China (35.02 percent), Philippines (2.11 percent), Chile (1.84 percent), Republic of Korea (1.79 percent), others (16.17 percent) (*Clean Development Mechanism: Executive Board Annual Report*, 2009, p. 1, http://www.uab.edu/uasomume/cdm/reports.htm).

[44] UNEP-RISO Centre, *Energy Climate and Sustainable Development*, n.d., http://cdmpipeline.org/cdm-projects-region.htm#4

[45] Brazilian Ministry of Science and Technology, *Current Status of the Project Activities under the Clean Development Mechanism (CDM) in Brazil and the World*, 2008, p. 3, http://www.unfccc.org

[46] Capoor and Ambrosi, op. cit., p. 33.

TABLE 8.1. *The state of the carbon market in the world*

	2007		2008	
	Volume ($MtCO_2e$)	Value (US$ million)	Volume ($MtCO_2e$)	Value (US$ million)
Primary CDM	552	7,433	389	6,519
JI	41	499	20	294
Voluntary market	43	263	54	397

Source: K. Capoor and P. Ambrosi, *State and Trends of the Carbon Market 2009*, Washington, DC: World Bank, 2009, p. 34, http://wbcarbonfinance.org/docs/State_Trends_of_the_Carbon_Market_2009-FINAL_26_May09.pdf

CDM projects are not much developed in the Amazon, notably in relation to reforestation and afforestation projects. In fact, this seems a general pattern: afforestation or reforestation projects account for only 0.43 percent of registered activities under the CDM Executive Board.[47] Brazil has the largest number of CDM activities in Latin America (44 percent of the total projects in that region, followed by Mexico with 20 percent), but the majority of project activities are on energy generation and swine (65 percent).[48] Only very few afforestation or reforestation projects (0.5 percent) are being developed in this country[49] and most CDM projects are located in the southeast of Brazil (in the States of São Paulo and Minas Gerais, which account for 23 percent and 16 percent of CDM activities, respectively).[50]

The CDM market is growing rapidly[51] and is expected to generate nearly 1.67 billion CERs by the end of 2012.[52] However, it is unlikely that the Kyoto Protocol (and its flexible mechanisms) will bring about the changes required to stabilize the world climate. The Kyoto Protocol does not limit GHG emissions of three of the world's five largest emitters, in other words the United States (not a party to the Protocol), China, and India.[53] Although developing countries are not bound by individual targets under the Protocol, they account for about 50 percent of energy-related

[47] *Clean Development Mechanism*: op. cit., p. 1.

[48] Brazilian Ministry of Science and Technology, op. cit., p. 8.

[49] Ibid.

[50] Ibid., p. 11.

[51] World Bank, *Global Monitoring Report: Millennium Declaration Goals and the Environment*, Washington, DC: World Bank, 2008, p. 228.

[52] *Clean Development Mechanism*: op. cit., p. 13. According to this report, host-country statistics make clear that much remains to be done to spread the benefits of the CDM. For example, the top hosting country has about 650 projects, ten times more than the fifth-highest hosting country. Some forty countries have ten projects or less, and fifteen countries have just one project.

[53] World Bank, op. cit., p. 228.

carbon emissions, and their share is expected to rise to 70 percent by 2030 in a business-as-usual scenario.[54] The emission reduction targets under the Kyoto Protocol's first commitment period are deemed insufficient to meet the required reduction in total anthropogenic emissions for a global stabilization target of CO_2e at 445–90 parts per million (ppm) required to limit global temperature rise to 2°C, in order to avoid the worst effects of climate change, according to current scientific evidence.[55] In any case, the UNFCCC (and the Kyoto Protocol) provide a forum where countries can negotiate future commitments, which has been seen, for example, in relation to GHG emissions from deforestation.

8.1.2. *The Voluntary Carbon Market*

Unlike the compliance market, the voluntary carbon markets are not tied to mandatory country targets to generate demand.[56] They may include legally binding cap-and-trade systems, such as the Chicago Climate Exchange, and also the broader nonbinding offset market that is not driven by an emissions cap (sometimes referred to as "over-the-counter" offset market, or OTC).[57] Credits generated in those markets are commonly referred to as voluntary or verified emissions reductions (VERs).

The voluntary carbon markets are dynamic and growing markets. In 2008, they were valued at US$705 million, more than twice their value in 2007 (US$335 million).[58] Asia is the most popular project location, sourcing 45 percent of transacted credits, followed by Latin America.[59] Private companies dominate the buy side of the voluntary markets (66 percent of volume) and purchase mostly for investment and resale. Sellers include project developers, wholesalers, retailers, and brokers.[60] In terms of customer location, the United States and the EU are the major source of demand for credits.[61] Currently, almost all voluntary credits are verified by an independent

54 J Eliasch, "Climate change: Financing global forests," *Eliasch Review*, London, 2008, p. 112.

55 Ibid., p. 111. The Intergovernmental Panel on Climate Change (IPCC) estimates that stabilizing the CO_2e concentration within the 445–90 ppm range would result in a global average temperature increase of 2.0–2.4°C. The current CO_2e concentration is around 433 ppm, p. 29. See IPCC, *Fourth Assessment Report: Climate Change 2007*, http://www.ipcc.ch/publications_and_data/publications_ipcc_ fourth_assessment_report_wg1_report_the_physical_science_basis.htm

56 M. Colini Cenamo et al., *Casebook of REDD Projects in Latin America*, Manaus: Nature Conservancy, Institute for Conservation and Sustainable Development of Amazonas (IDESAM), 2009, p. 18.

57 K. Hamilton et al., *Fortifying the Foundation: State of the Voluntary Carbon Markets 2009*, New York and Washington, DC: Ecosystem Marketplace and New Carbon Finance, 2009, p. 21

58 Ibid., p. 31.

59 Ibid., p. 9.

60 Ibid., p. 13. Credits sellers normally include: (1) project developers: (2) wholesalers, who sell offsets in bulk and have ownership of a portfolio of credits: (3) retailers, who sell small amounts of credits to individuals or organizations, usually online, and have ownership of a portfolio of credits: and (4) brokers, who do not own credits, but facilitate the transactions between sellers and buyers.

61 Ibid., p. 93.

third-party organization, even if this is not a compulsory requirement.[62] The majority of projects in the OTC market of 2008 were on renewable energy, while landfill came second.[63] Similarly to the compliance market, land-based carbon sequestration projects, especially from forestry, are a minority.[64] More than three-quarters of the credits coming from Latin America and the Caribbean in the OTC market originate in Brazil (56 percent of credits, followed by Mexico, 21 percent).[65]

The voluntary carbon markets account for the bulk of forest carbon transactions.[66] Most forest carbon credits transacted have been sourced from afforestation or reforestation projects (63 percent) followed by projects aimed at Reducing Emissions from Deforestation and Forest Degradation (REDD) (17 percent).[67] In terms of location, moist tropical forests are the most popular forest type.[68] Most of forest carbon offsets originated from projects involving an internal or third-party standard.[69] Buyers of forest carbon credits are driven either by "pure voluntary" purposes (i.e. they purchase credits to offset their own emissions and to retire their credits upon purchase)[70] or by "pre-compliance" purposes[71] (i.e. buyers purchase VERs with the hope of receiving early-actor credit or selling the credits higher price under a future compliance regime).[72] Sellers are the same as those buying other offset credits in the voluntary markets, mentioned above. In 2008, Latin America accounted for 13 percent of all forest carbon reductions globally.[73]

62 Ibid., p. 62. From 87 percent of voluntary credits verified by independent third-party organizations in 2007 to 96 percent in 2008. The most utilized third-party standard was the Voluntary Carbon Standard, followed by the Gold Standard, the Climate Action Reserve, and the American Carbon Registry Standard (ibid., p. 61). The Social Carbon Company developed by the Ecológica Institute in Brazil is the first methodology proposed in the Southern hemisphere. The Social Carbon Company helps to fund the standard and exclusively sells credits verified using the Social Carbon methodology. The company is a partnership between the Ecológica Institute and Cantor Co_2e, an energy and environmental commodity brokerage firm (ibid., p. 68).

63 Ibid., p. 38.

64 Ibid., p. 43.

65 Ibid., p. 51.

66 K. Hamilton, U. Chokkalingam, and M. Bendana, *State of the Forest Carbon Markets 2009: Taking Root and Branching Out*, Washington, DC: Ecosystem Marketplace, 2010, p. viii. In 2008 and the first two quarters of 2009, the voluntary carbon markets continued to account for the bulk of forest carbon transactions: 95 percent in 2008 and 72 percent in 2009.

67 Ibid., p. x.

68 Ibid., p. 15.

69 Over time, 86 percent of all OTC forest carbon offsets originated from projects involving an internal or third-party standard. Certification to third-party standards increased from 15 percent of offsets in 2002 to a 96 percent in the first half of 2009, and account for 70 percent of all OTC offsets transacted over time (ibid., p. xi).

70 Ibid., p. 5.

71 Ibid.

72 Ibid., p. 6.

73 Ibid., p. 19. This amount decreased in 2009. In the first two quarters of 2009, Latin American project developers transacted just 9,087 tons of CO_2 (at an average price of US$11.48 per ton), which is less than 1 percent of total global volume for this period.

The voluntary carbon markets are praised for offering greater flexibility and lower transaction costs to buyers and suppliers when compared to the regulated one.[74] Also, the range of products emerging from the voluntary markets can be more innovative and flexible than those of the compliance market. However, voluntary markets have been criticized particularly for lacking impartial information and transparency. This is in part due to the fact that they are based on deals negotiated on a case-by-case basis. Many of those do not require the carbon credits to undergo a uniform verification process or to be registered with any central body. As a result, many buyers are wary of those markets, as transactions often carry risks of nondelivery, and companies buying carbon credits available in such markets fear they will be criticized if the credits purchased do not meet the highest possible standards.[75] In any case, the voluntary carbon markets seem to be well established and tend to grow, as has been observed over the past three years.[76]

8.2. REDUCING EMISSIONS FROM DEFORESTATION AND FOREST DEGRADATION

It has become all too common to hear that the equivalent of thousands of football fields are lost to deforestation every day or, per year, areas the size of entire European countries.[77] Deforestation, particularly in the tropics, is indeed very high and tends to continue.[78] The most evident impacts include climate change, biodiversity loss, and reduced availability and quality of freshwater supply, among others. For example, in Brazil, the country that contains the largest single block of tropical forest in the world, the factor that most contributes to GHG emissions is deforestation of the Amazon,[79] accounting for about 75 percent of Brazil's GHG emissions.[80]

74 Katoomba Group's Ecosystem Marketplace, *Forests: Taking Root in the Voluntary Carbon Market*, Washington, DC: Katoomba Group, 2009, p. 13.

75 Ibid., p. 12.

76 The volume of credits traded increased 64 percent (between the years 2006 and 2007) to 89 percent (between 2007 and 2008) (Colini Cenamo et al., op. cit., p. 18).

77 Until the mid-twentieth century, most deforestation occurred in temperate regions. However, in recent decades, the abandonment of agricultural land in Western Europe and North America has led to some natural reforestation. Conversely, deforestation is now progressing rapidly in the tropics. Since 1980, global forest cover is estimated to have declined by 225 million hectares due to human action. In the tropics, an estimated 13 million hectares, an area the size of England, are converted to other land uses each year (Eliasch, op. cit., p. 27).

78 Ibid., p. 33.

79 The deforestation of the Brazilian Amazon makes the largest contribution from the total amount of emissions in that country (Ministry of External Relations; Ministry of Science and Technology; Ministry of the Environment; Ministry of Mines and Energy; and Ministry of Development, Industry and Foreign Trade, *Brazil's Contribution to Prevent Climate Change*, Brasília, 2007, p. 36, http://www.mct.gov.br/upd_blob/0018/18294.pdf).

80 N. Olsen and J. Bishop *The Financial Costs of REDD: Evidence from Brazil and Indonesia*, Gland: IUCN, 2009, p. 26.

While deforestation has impacts on the global climate, global warming also badly affects rainforests. For instance, Brazil's National Institute for Space Research (INPE) has made estimates of the damage likely to be caused by global warming in the Brazilian Amazon. Based upon models used by the IPCC, it is projected that at least 18 percent of the area currently covered by rainforest could become savannah or sparse vegetation (similar to *cerrado*), and 30.4 percent of the Amazon rainforest might be lost by 2090.[81] Studies released by the Brazilian Ministry of the Environment also predict that if current patterns of deforestation continue, as agricultural and logging activities expand into the rainforest, only 53 percent of the Brazilian Amazon original forest cover will remain by 2050 (a decrease from 5.3 to 3.2 million km^2), and in a worst-case scenario, temperatures in the Amazon will rise by 8°C.[82] As temperatures rise and the climate becomes warmer, evaporation rates will increase and less rain will fall. It is feared that in such conditions the capacity of the Amazon to absorb carbon dioxide (CO_2) will decline and the rainforest will no longer be able to perform its essential role as a carbon sink; instead, it will become a significant source of GHG emissions. According to the IPCC, deforestation accounts for 17 percent of global GHG emissions, which is more than the global transport sector.[83] In particular, IPCC predicts that, by mid-century, increases in temperature and associated decreases in soil water may lead to a gradual replacement of tropical forests by savannah in the eastern Amazon.[84] Forests are, therefore, a central part of any efforts to attain the goal of the stabilization of atmospheric CO_2e.[85]

There seems to be a consensus that in order to reduce or avoid tropical deforestation and achieve the stabilization target of 445–90 ppm and a maximum temperature rise of 2°C required to avoid dangerous climate change, strong and urgent action is needed from the international community as a whole.[86] However, the means to achieve this end are still highly debatable. It is said that deforestation is not a result of irrationality, ignorance, or stupidity: people get (or expect to get) real benefits from deforestation and unsustainable forest harvesting,[87] or they simply

81 J.A. Marengo, *Mudanças climáticas globais e seus efeitos sobre a biodiversidade: caracterização do clima atual e definição das alterações climáticas para o território brasileiro ao longo do século XXI*, Brasília: Brazilian Ministry of the Environment, Secretaria de Biodiversidade e Florestas, 2006, http://www.mma.gov.br/estruturas/imprensa/_arquivos/livro%20completo.pdf

82 Ibid., p. 136.

83 Eliasch, op. cit., p. 1. According to IPCC, *Fourth Assessment Report*, op. cit., "Physical science basis, 2007,", pp. 682–83, the largest anthropogenic land cover changes today involve deforestation, and during the past two decades, the CO_2 flux caused by land use changes has been dominated by tropical deforestation."

84 Ibid. p. 50.

85 Eliasch, op. cit., p. 9.

86 Ibid., p. 172.

87 V.M. Viana, *Seeing REDD in the Amazon: A Win for People, Trees and Climate*, International Institute for Environment and Development, 2009, p. 1, http://www.iied.org/pubs/pdfs/17052IIED.pdf

depend on forests and may have no better means for exploiting them. One of the reasons why deforestation is so widespread relates to the fact that the ecosystem services they provide have not been properly valued. Forests and woodlands provide many services, such as the maintenance of water supply, carbon sequestration, and biodiversity conservation just to name a few, but as they are not much traded, there are weak economic incentives for their sustainable use.[88] Therefore, it tends to be more lucrative to deforest and sell timber or agricultural products than to maintain forests, as the costs of deforestation are not reflected in the final price of these commodities.[89] The costs of deforestation are usually viewed as an "externality" and the fact that they are not factored into the price of products a "market failure."[90]

In the absence of incentives or better alternatives, people will keep deforesting. However, there seems to be a growing recognition that "forests are worth more standing than cut."[91] For example, it is estimated that indigenous lands and extractivist reserves in the Brazilian Amazon hold about 15 billion tons of carbon,[92] whereas empirical evidence (both in Brazil and Indonesia) indicates that financial returns to a number of land uses (expressed in terms of net profits per ton of CO_2e) are below the current market prices for carbon.[93] A key question in the current debate on forests is that of how to finance an overall reduction of deforestation. This is particularly the case if one considers that the agricultural frontier in the Amazon, for example, is pushed along by a multibillion dollar per year economy.[94] Therefore, it is suggested that, if the nature of the battle is mostly economic, solutions should come with sustainable financing from public, private, and nonprofit sources.[95] Various studies have attempted so far to measure the costs of deforestation, and the finance required to reduce emissions from forest clearing, such as the Eliasch[96] and the Stern Reviews.[97]

[88] I. Bond et al., *Incentives to Sustain Forest Ecosystem Services*, London: International Institute for Environment and Development (IIED), 2009, p. 5.

[89] Eliasch, op. cit., p. 63.

[90] Ibid.

[91] Viana, op. cit., p. 1

[92] IPAM (Instituto de Pesquisa da Amazonia), "Desmistificando REDD: fortalecendo a participação dos povos indígenas e tradicionais nas discussões sobre mudanças climáticas," 2009 at the workshop Povos da Floresta do Acre querem ser beneficiados por mecanismo de REDD, press release, August 12, 2009, http://www.carbonobrasil.com

[93] Olsen and Bishop, op. cit., p. 49

[94] Viana, op. cit., p. 1

[95] Ibid., p. 1.

[96] For example, the Eliasch Review (Eliasch, op. cit., p. 28) estimates that the mean damage cost of climate change impacts derived from deforestation will be around US$1 trillion a year by 2100. On the other hand, it suggests that the finance required to reduce emissions from deforestation could be, in 2030, around US$17–33 billion per year (if forest credits are included in global carbon trading) (ibid., p. 80).

[97] N. Stern, *Stern Review: The Economics of Climate Change*, Cambridge: Cambridge University Press, 2007, p. 193, http://www.hm-treasury.gov.uk/sternreview_index.htm. This report also predicts a collapse of Amazon rainforest with a 3°C increase in global temperatures, according to some models.

Despite the different estimates in terms of actual costs, current studies tend to agree that the scale of financing required for climate change mitigation and adaptation in developing countries is of the order of hundreds of billions of dollars.[98] Despite this high scale of financing, studies such as the Stern Review have showed that the costs of inaction outweigh the costs of action.

One way of financing forest conservation, debated especially in the context of the UNFCCC is REDD. The idea behind mechanisms such as this is to compensate developing countries for their efforts in reducing GHG emissions through forest conservation.[99] The debate on REDD under the UNFCCC began in 2005 and was limited to deforestation only (the acronym then was RED). It was later expanded to REDD to include forest degradation, and more recently it involves forest conservation, sustainable forest management, and enhancement of carbon stocks (with the current acronym being REDD+ or REDD plus).[100] No such mechanism exists under the Kyoto Protocol to compensate avoided deforestation by non-Annex 1 countries. At present, carbon credits from avoided deforestation are limited to voluntary markets, but REDD will most likely be part of a UNFCCC post-Kyoto regime.

Various proposals have been made so far by governments, international organizations, and NGOs as to how REDD should operate under the UNFCCC in a post-2012 regime. Similarities among existing proposals are noted in relation to certain parameters and requirements of what REDD sets out to achieve, which include in particular "additionality," "permanence," and avoidance of "leakage."[101] Additionality is achieved when it can be evidenced that long-term, measurable GHG emission reductions would not have occurred in the absence of a particular project, policy, or activity.[102] Permanence refers to how robust a project is in the face of potential changes that could reverse the carbon benefits of the project at a future date.[103] In other words, a project would fail the test of permanence, for example, if, after its conclusion, forested areas generated by it are harvested immediately.[104] Finally, leakage is the displacement that occurs when actions to reduce emissions in one geographical area cause an increase in emissions in another area through the relocation of activities.[105] To ensure that these requirements are met, and emission reductions ultimately achieved, it is proposed, among other things, to establish baselines against which to measure

[98] Parker et al., op. cit., p. 19.
[99] World Bank, op. cit., p. 228.
[100] Olsen and Bishop, op. cit., p. 3.
[101] Peskett et al., op. cit., p. 21.
[102] Olsen and Bishop, op. cit., p. 3.
[103] N. Virgilio, *Noel Kempff Mercado Climate Action Project: A Case Study in Reducing Emissions from Deforestation and Degradation*, Arlington: Nature Conservancy, 2009, p. 13.
[104] S. Wunder, "The efficiency of payments for environmental services in tropical conservation," *Conservation Biology*, vol. 21, no. 1, 2007, pp. 48–58, at p. 54.
[105] Olsen and Bishop, op. cit., p. 5.

deforestation levels (and other land-use changes) and a monitoring, reporting, and verification system to guarantee the integrity and credibility of REDD initiatives.

As the design of a future REDD mechanism under the UNFCCC is yet to be defined, a number of issues are still open to debate. One issue of contention is whether REDD should give rise to tradable credits to be used by Annex 1 countries in order to offset their GHG emissions, or whether it should not function as a market-based mechanism. While several countries (including those in Latin America and the Caribbean) favor a "market approach," others advocate a nontradable "fund" approach.[106] Brazil, in particular, has supported a fund-based approach to de-link avoided deforestation from carbon markets.[107] In this regard, it submitted a proposal[108] at the UNFCCC COP-12 in 2006 suggesting the creation of a fund supported by multilateral institutions and Annex 1 countries by which incentives (in the form of additional financial resources, transfer of technology, capacity-building, etc.) would be provided to non-Annex 1 countries that voluntarily reduced their GHG emissions from deforestation in relation to a reference emission rate. In accordance with a "non-market" approach, in 2008 Brazil created (Decree 6527/2008) the Amazon fund (Fundo Amazônia), which is expected to finance measures to prevent, monitor, and combat deforestation in the Amazon.[109] The fund is managed by the Brazilian Development Bank and already counts with an initial US$1 billion donation by the Norwegian government, to be paid by installments until 2015, tied to performance in slowing deforestation.[110] Meanwhile, some states within the Brazilian Amazon seem to be courting the carbon markets. For example, states within the legal Amazon have been discussing ways of cooperating with the State of California to implement initiatives such as REDD.[111]

[106] A. de la Torre, P. Fajnzylber and J. Nash, *Low Carbon High Growth Latin America Responses to Climate Change: An Overview*, Washington, DC: World Bank Latin America and Caribbean Studies, 2009, p. 50.

[107] Peskett et al., op. cit., p. 22.

[108] Brazil, "Positive incentives for voluntary action in developing countries to address climate change: Brazilian perspective on reducing emissions from deforestation," Dialogue Working Paper. no. 21, 2006, presented at the Second Workshop on Dialogue on Long-Term Cooperative Action to address Climate Change by Enhancing the Implementation of the Convention, at the Twelth UNFCCC Conference of the Parties, held in Nairobi, Kenya, November 15–16, 2006.

[109] For information on the Amazon Fund, see http://www.fundoamazonia.gov.br/

[110] L. Westholm et al., *Assessment of Existing Global Financial Initiatives and Monitoring Aspects of Carbon Sinks in Forest Ecosystems: The Issue of REDD*, Focali report, Gothenburg: Forest, Climate and Livelihood Research Network 2009, p. 26.

[111] T. Rose, "Picking up the REDD Tab: Who will pay to fight deforestation and how?," p. 57, February 25, 2009, http://www.ecosystemmarketplace.com/pages/dynamic/article.page.php?page_id=6525&section=news_articles&eod=1. See on this, Katoomba Group, Forest Trends and Governo do Mato Grosso, "Avoiding Deforestation in the Amazon: REDD and PES Markets," paper presented at the 2009 Katoomba Meeting, Cuiaba-Mato Grosso (Brazil), 1–2 April 2009, p. 4, http://www.

8.2.1. *The International Negotiations on REDD*

As noted earlier, the Kyoto Protocol excludes emission reductions from deforestation as an option under its flexible mechanisms. Currently, REDD only operates in the voluntary markets.[112] Annex 1 countries under the Kyoto Protocol can purchase carbon offsets from projects in the forestry sector, but only those related to afforestation or reforestation.[113] In any case, forestry and land use have had a limited role as mitigation options due to both technical and political reasons.[114] Technical difficulties exist, for example, in measuring GHG emissions from deforestation and forest degradation, in determining "additionality" or even in understanding the drivers of deforestation. Among political factors is the concern of non-Annex 1 countries that industrialized countries may use tropical forests as an easy way to tackle climate change, instead of taking on deeper cuts in their own emissions and making changes in domestic energy consumption patterns.

In 2005, the Coalition for Rainforest Nations led by Costa Rica and Papua New Guinea submitted a formal proposal to the UNFCCC COP-11 on the possibility of introducing GHG emissions from deforestation under the international climate regime.[115] The Subsidiary Body for Scientific and Technological Advice (SBSTA) was then requested to examine the issue and report it to the UNFCCC COP-13, held in Bali in 2007.[116] The issue was then revisited and ultimately included in the Bali Action Plan (Decision 1/CP.13),[117] which provides a roadmap for the negotiation of a new regulatory framework on climate change after the expiry of the Kyoto Protocol's first commitment period in 2012.[118]

The Bali Action Plan encouraged actions to mitigate climate change, including "policy approaches and positive incentives on issues relating to reducing emissions from deforestation and forest degradation in developing countries" (Decision

katoombagroup.org/documents/cds/katoomba_xiv/documents/Agenda/Brazil%20Katoomba%20Program%20ENG%2009–0325.pdf

112 Hamilton, Chokkalingam, and Bendana, op. cit., p. 44.

113 Reference to such activities can be found in Article 2 (1) of the 1997 Kyoto Protocol, according to which Annex 1 parties, in their efforts to achieve emission reduction commitments, must "(a) Implement and/or further elaborate policies and measures… such as: (ii) protection and enhancement of sinks and reservoirs of greenhouse gases…, promotion of sustainable forest management practices, afforestation and reforestation."

114 Bond et al., op. cit., p. 1.

115 V. Holloway and E. Giandomenico, *The History of REDD Policy*, Adelaide: Carbon Planet, 2009, p. 8.

116 T. Griffiths with F. Martone, *Seeing 'REDD'?: Forests, Climate Change Mitigation and the Rights of Indigenous Peoples and Local Communities*, Forest Peoples Programme, 2009, p. 4. http://www.forestpeoples.org/documents/ifi_igo/seeing_redd_update_may09_eng.pdf

117 FCCC/CP/2007/6/Add.1, p. 3.

118 Griffiths, op. cit., p. 6.

1/CP13, Para. 1, b, iii). COP-13 adopted a specific decision on the issue of "Reducing Emissions from Deforestation in Developing Countries" (Decision 2/CP.13),[119] inviting the parties to support ongoing efforts to reduce such emissions on a voluntary basis (Article 1) and, in particular, undertake "demonstration activities" (Article 3). As a result, several pilot projects were launched to develop methods for REDD[120] and various partnerships, led by developed countries and multilateral organizations, were forged to help developing countries participate in a future REDD regime.[121] Also, the mounting evidence released in 2007–2008 of drastic climate change, resulted in a proliferation of donor funds for climate change mitigation and adaptation,[122] such as the World Bank Forest Carbon Partnership Facility and Forest Investment Programme, the GEF Tropical Forest Account, the REDD Program, the Congo Basin Forest Fund, the Norwegian Forest and Climate Initiative, the International Forest Carbon Initiative (Asia-Pacific), and the Tropical Forests Facility (Prince Charles' Rainforests Project), among others.

To facilitate further negotiations under the Bali Action Plan, COP-13 set up two bodies,[123] the 1) Ad Hoc Working Group on Further Commitments for Annex 1 Parties under the Kyoto Protocol (AWG-KP), in charge of negotiating future commitments from Annex 1 countries under the Kyoto Protocol, and the 2) Ad Hoc Working Group on Long Term Cooperative Action under the Convention (AWG-LCA), aimed at developing long-term cooperation between Annex 1 and non-Annex 1 countries on issues such as mitigation, adaptation, technology transfer, and financial assistance. REDD has been discussed within both of these groups. Also, SBSTA looks at technical aspects related to REDD, notably methodological issues.[124]

The latest Conference of the Parties of the UNFCCC held in Copenhagen in 2009 (COP-15) adopted a political agreement (known as the Copenhagen Accord),[125] which set the objective of limiting global warming to a maximum of 2°C (para. 1), but did not specify any emission reduction targets.[126] It calls for the "immediate establishment of a mechanism including REDD-plus, to enable the mobilization of financial resources from developed countries" (para. 6). Other relevant decisions adopted at COP-15 include those that extend the mandates of both AWG-LCA[127]

[119] FCCC/CP/2007/6/Add.1, p. 8
[120] Westholm et al., op. cit., 17.
[121] Bond et al., op. cit., p. 1.
[122] Griffiths, op. cit., p. 8.
[123] Holloway and Giandomenico, op. cit., p. 13.
[124] Griffiths, op. cit., p. 6.
[125] FCCC/CP/2009/L.7, December 18, 2009.
[126] M. Belton, "REDD progress at Copenhagen," *Policy Quarterly*, vol. 6, no. 2, 2010, pp. 8–10, at p. 3.
[127] Decision on Outcome of the work of the AWG-LCA, FCCC/CP/2009/L.6, December 18, 2009

and AWG-KP,[128] and one SBSTA decision[129] providing methodological guidance for REDD activities. Four new bodies were created under the Copenhagen Accord: a mechanism on REDD-plus (para. 9),[130] a High-Level Panel on the implementation of financing provisions (para. 9), a Copenhagen Green Climate Fund (Paragraph 10), and a Technology Mechanism (para. 11).[131]

Although COP-15 has firmly placed REDD under the UN climate negotiations, it did not resolve any outstanding technical or political issues surrounding this mechanism. A key outcome of COP-15 is the "collective commitment" under the Copenhagen Accord by developed countries to provide new and additional financial resources to developing countries, including investments in forestry, of about US$30 billion for the period 2010 to 2012 (Paragraph 8). Part of this funding should flow through the Copenhagen Green Climate Fund, created as "an operating entity of the financial mechanism of the Convention to support projects, programme, policies and other activities in developing countries related to mitigation including REDD-plus, adaptation, capacity-building, technology development and transfer" (para. 10). In any case, it is now yet clear through which channels this money will be mobilized, which countries will be benefited, and whether

[128] Decision 1/CMP.5. This decision mandates the AWG-KP for Annex I Parties under the Kyoto Protocol to "continue its work drawing on the draft text forwarded as part of its report on its tenth session to the Conference of the Parties serving as the meeting of the Parties to the Kyoto Protocol at its fifth session" (para. 3).

[129] Methodological guidance for activities relating to reducing emissions from deforestation and forest degradation and the role of conservation, sustainable management of forests and enhancement of forest carbon stocks in developing countries, FCCC/SBSTA/2009/L.19/Add.1, December 11, 2009. This decision requests developing country Parties (Paragraph 1) "(a) To identify drivers of deforestation and forest degradation resulting in emissions and also the means to address these; (b) To identify activities within the country that result in reduced emissions and increased removals, and stabilization of forest carbon stocks; (c) To use the most recent Intergovernmental Panel on Climate Change guidance and guidelines, as adopted or encouraged by the Conference of the Parties, as appropriate, as a basis for estimating anthropogenic forest-related greenhouse gas emissions by sources and removals by sinks, forest carbon stocks and forest area changes; (d) To establish, according to national circumstances and capabilities, robust and transparent national foresti monitoring systems and, if appropriate, sub-national systems as part of national monitoring systems." The SBSTA decision recognizes the need for engagement of indigenous peoples in monitoring and reporting, but does not extend this to full involvement, such as in the design and development of programs.

[130] "We recognize the crucial role of reducing emission from deforestation and forest degradation and the need to enhance removals of greenhouse gas emission by forests and agree on the need to provide positive incentives to such actions through the immediate establishment of a mechanism including REDD-plus, to enable the mobilization of financial resources from developed countries."

[131] IISD (International Institute for Sustainable Development), "Summary of the Copenhagen climate change conference: December 7–19, 2009", *Earth Negotiations Bulletin*, vol. 12, no. 459, December 22, 2009, p. 29, http://www.iisd.ca/climate/cop15/

developed countries will keep their promises in the absence of a binding agreement.[132] With respect to a "REDD-plus mechanism" details on "who is to pay, how much, when and how" were not mapped out in the Accord and are a long way from being resolved.[133]

8.2.2. *REDD in Practice*

Probably one of the most cited examples of REDD in South America is the Noel Kempff Climate Action Project (NKCAP), located in the Department of Santa Cruz in Bolivia (Province of Velasco), in the climatic transition zone between the Amazon, the chaco, and *cerrado* eco-regions.[134] Noel Kempff Mercado is a national park, expanded by the Supreme Decree 24457/1996[135] and a world heritage site since 2000.[136] In 1996, the Nature Conservancy and Bolivia's Fundación Amigos de la Naturaleza began to work jointly with the Bolivian government to terminate logging in areas adjacent to the park. In the following year, three energy companies (American Electric Power, BP-Amoco, and Pacificorp) entered into an agreement with the Bolivian government to protect these areas. The financing arising out of this partnership was used to buy out pre-existing logging concessions in areas adjacent to the national park. The NKCAP is managed by Fundación Amigos de la Naturaleza and financed by these three energy companies, which expect to generate carbon credits to offset their emissions. The carbon measurement is done by Winrock International Institute for Agricultural Development, Fundación Amigos de la Naturaleza, whereas validation and verification is carried out by Société Générale de Surveillance.

Positive evaluations of NKCAP suggest that the project has passed the additionally and permanence tests and managed to avoid leakage. Also, NKCAP has apparently managed to generate social co-benefits in the form of community development programs and, by this means, avoided leakage. Despite this, it has also been noted that NKCAP is not without areas for improvement.[137] In the years since this project was initiated, carbon markets, forest carbon science, and conservation approaches have evolved in many ways. For example, the methods for predicting future deforestation

[132] K. Dooley, *Forest Talks at a Standstill as Copenhagen Ends without an Agreement*, EU Forest Watch Special Report, UNFCCC Climate Talks, December 7–18, 2009, http://fern.org/sites/fern.org/files/Copenhagen%20update.pdf

[133] Belton, op. cit., p. 2.

[134] This description can be found in Virgilio, op. cit., p. 3.

[135] This Decree extends the area covered by the Noel Kempff Mercado National Park, previously established by the Supreme Decreeno. 21997/1988.

[136] The World Heritage List is available at: http://whc.unesco.org/en/list

[137] Virgilio, op. cit., p. 19.

and calculating carbon benefits are now more sophisticated than in the 1990s. In addition to the technical advances that have come about, new thinking has emerged on the design of forest carbon projects and participatory planning processes involving local people and benefit sharing.

NKCAP is, however, subject to stronger criticism. For example, according to Greenpeace, polluting energy companies such as those involved in this project are using voluntary subnational REDD offsets to project a "green image." By financing those projects, they create ways to sidestep required pollution cuts and finally do little to reduce their own emissions. Greenpeace suggests that the analysis of NKCAP's documentation, combined with on-the-ground interviews, has showed that this project has not met its own claims of having achieved goals such as additionality and permanence, or having avoided leakage, while generating social co-benefits.[138]

REDD pilot projects are also under way in Brazil, of which the Juma Reserve Project is an example. The Juma Reserve is a conservation unit located in the municipality of Novo Aripuanã in the State of Amazonas.[139] It is the first project to be implemented under the State Law on Climate Change, Environmental Conservation and Sustainable Development (Law 3.135/2007) and the State Protected Areas System (Sistema Estadual de Unidades de Conservação). The project is developed by Fundação Amazonas Sustentável in partnership with the Environment and Sustainable Development Secretary of the State of Amazonas, and the technical assistance from the Instituto de Conservação e Desenvolvimento Sustentável. It received validation by the standard Climate Community and Biodiversity Alliance issued by the German audit company Tüv Süd. The project is expected to be concluded in 2050 and offset million tons of CO_2e. The Foundation Amazonas Sustentável covers its operation costs, while its future financial sustainability should be ensured by the carbon credits trading. The hotel chain Marriott International also sponsors the project (with US$2 million allocated for a four-year period) and, in return, expects to have privileges at the carbon credits purchase. Interestingly, Marriott's guests are offered the option to offset their emissions at US$1 per night, to be directed to the Juma Reserve Project. This project is part of a broader initiative in the State of Amazonas focused on payments for environmental services (known as Bolsa Floresta). Other REDD

[138] A. Densham et al., *Carbon Scam: Noel Kempff Climate Action Project and the Push for Sub-national Forest Offsets*, Amsterdam: Greenpeace International, 2009.

[139] Information about this project is available at: http://www.fas-amazonas.org/pt/secao/projeto-juma. See also V. Viana et al., *Reserva de Desenvolvimento Sustentável do Juma: o primeiro projeto REDD na Amazônia Brasileira*, Manaus: Fundação Amazônia Sustentável, 2008.

initiatives in Brazil include, for example, the Juruá Project,[140] the Surui Project,[141] among others.[142]

In Ecuador, the Forest Partners Program (Programa Socio Bosque), announced at the UNFCCC COP-14 in Poznan (Poland), is another initiative aimed at preserving natural forests and improving the quality of life of local communities. The project covers the entire national territory with a conservation goal of 4 million hectares of forested area. It consists of direct annual monetary compensation per hectare of forest given by the government to individual landowners or communities that voluntarily protect native forests.[143] The project is implemented by Ecuador's Secretariat of the Environment.

The Forest Partners Program prioritizes areas most threatened by deforestation, of crucial importance for carbon storage (and other ecosystem services) and those with the highest levels of poverty. Both individuals and communities can participate in the program and, if eligible, a formal agreement is signed, by which they commit to protect certain forest areas in exchange of an economic compensation paid by the State. Compliance is verified through the use of remote sensing and on-site inspections. In particular, the Secretariat of Environment monitors emission reductions through a GHG monitoring system. The project is financed by the government, while additional funding from countries or organizations can be directed to a trust fund created within the National Environmental Fund (Fondo Ambiental Nacional). Programa Socio Bosque is viewed as a central component for achieving the national target of reducing deforestation by 30 percent until 2013, defined under Ecuador's National Development Plan.[144]

8.2.3. *REDD: Pros and Cons*

REDD is a good alternative to protect rich forest areas, such as the Amazon, for a number of reasons. It locks up areas for conservation and, if successfully implemented,

[140] The Carbon Project Juruá is a pilot project developed by Ecologica Forestry that may be replicated in other areas of the Amazon. The total amount of absorption of carbon expected is 1,421,767.41 tons of CO_2e. Ecologica develops the baseline methodology and applies the Social Carbon standard. Information available at: http://www.ecologica.ws/pages/6

[141] The Surui are working with Forest Trends, the Katoomba Group, Amazon Conservation Team, and other partners. By supporting processes of local consultation and planning, mapping, technical carbon assessments, and links to markets, the Katoomba Incubator and partners propose to catalyze carbon finance for the protection of over 240,000 hectares of native forest. Information available at: http://www.katoombagroup.org/events/Surui_v2.pdf

[142] Other examples of REDD projects in Latin America can be found in Colini Cenamo et al., p. 18

[143] Information on Ecuador's Forest Partners Program is available at: http://www.ambiente.gov.ec

[144] Secretaría Nacional de Planificación y Desarrollo, *Plan Nacional para el Buen Vivir: Construyendo un Estado Plurinacional e Intercultural*, 2009–13, 2009, p. 215, http://documentacion.asambleanacional.gov.ec/alfresco/d/d/workspace/SpacesStore/9fdcbda0–4f5a-4280-abe3–2dd9f1e79f08/PNBV_2009–2013.pdf

it provides incentives to avoid deforestation and ultimately achieve GHG emission reductions. As noted in the examples above, Programa Socio Bosque aims to protect 4 million hectares of forestlands in Ecuador, whereas the Juma Reserve Project expects to prevent deforestation of nearly 7,799 hectares (until 2016) and offset million tons of CO_2e.[145] REDD may also help improve the monitoring capacity of tropical forest countries and the overall surveillance over forestlands. For example, in the case of Programa Socio Bosque, emission reductions will be monitored by the Secretariat of Environment through the implementation of a GHG monitoring system. In the context of NKCAP, an endowment was created to fund the protection and management of the project site, including rangers, equipment, and infrastructure.[146] REDD may then be an incentive to improve the capacity and technology in tropical forest countries, for example, to develop forest inventories and estimate carbon stocks and emissions from deforestation through remote sensing and other appropriate technology.[147]

REDD-related financing tends to be the largest financial flow into tropical forest conservation.[148] This represents an opportunity to place economic value on standing forests and promote conservation actions that would not occur otherwise. For example, it is suggested that carbon finance at US$5 per ton (or less) could tip the balance from nonsustainable toward sustainable forest management for more than 96 percent of the Brazilian Amazon.[149] As observed with the examples described above, REDD can generate additional finance required to avoid deforestation in certain areas. More generally, efforts to avoid deforestation have already prompted the mobilization of funds, of which the Amazon Fund is an example. Although a potential source of finance for forest conservation, REDD should not be viewed as the only one. The implementation of REDD activities in itself can be very costly. For example, the costs of Programa Socio Bosque are estimated to be between US$80 and US$100 million per year.[150] Other REDD case studies in South America

[145] Information available at: http://www.fas-amazonas.org/pt/secao/projeto-juma

[146] Virgilio, op. cit., p 13.

[147] De la Torre, Fajnzylber, and Nash, op. cit., p. 53. This documents suggests that countries that are interested in moving forward with a REDD strategy may wish to consider the following steps: (1) fine-tune the estimation of emissions from land-use change at the subnational level using high-resolution imagery; (2) conduct a national forest inventory to estimate carbon stocks; (3) adopt a spatially explicit modelling approach to predict future deforestation; and (4) establish a national monitoring, reporting and verification system capable of tracking changes in deforestation and forest degradation and the resulting GHG emissions. Several Latin American and Caribbean countries are already using or planning to use high-resolution remote sensing techniques to establish their baseline deforestation trends and monitor deforestation over time.

[148] P. Verweij et al., *Keeping the Amazon Forests Standing: A Matter of Values*, report commissioned by WWF Netherlands, 2009, p. 57.

[149] Ibid.

[150] Colini Cenamo et al., op. cit., pp. 42–43.

suggest that the amount of funding required to implement REDD in tropical forest countries is higher than those available in current funding sources.[151] A significant amount of funding is required to make REDD work and should involve various sources (e.g. international donors, development agencies, multilateral development banks, and the private sector).

Despite its promise, REDD remains controversial because of a number of issues, notably technical ones. For example, it is usually difficult to measure things such as additionality, permanence, and avoidance of leakage. In particular, if "non-additional" carbon credits are issued (i.e. credits that represent emission reductions that would have happened anyway, even without a REDD project), carbon markets can be "flooded" with a large number of "hot air" carbon credits. Also, REDD requires credible baselines and methodologies in order to measure emission reductions (and other environmental benefits), as well as monitoring and verification.[152] In view of this, tropical forest countries need to develop a number of capacities, for example, to define baselines and methodologies, measure carbon stocks, ensure compliance etc. Studies on the issue of national capacity of non-Annex 1 countries show, however, a capacity gap, especially in developing and maintaining national carbon measurement and monitoring systems in the long-term.[153] In particular, the technical capacity of the Amazon States to implement REDD varies. For example, whereas some of them have significant national capacity for remote sense and forest monitoring (e.g. Bolivia, Brazil, Colombia, and Peru), others have a limited one or none (e.g. Guyana and Venezuela). Some Amazon States have national capacity for forest inventory (e.g. Bolivia and Peru), whereas others have little or none (e.g. Colombia, Guyana, and Venezuela).[154] However, certain developing countries, such as India, Brazil, and Mexico, have managed to develop advanced technology to monitor deforestation using remote sensing data.[155] This is an important and potential area for cooperation among South American countries, which would be most welcomed particularly with regard to shared ecosystems, as in the case of the Amazon.

In general, REDD (and other mechanisms) require certain conditions of good governance, which is rare in those countries facing high deforestation rates.[156] For

[151] Ibid., p. 79.

[152] P.M. Fearnside, "Carbon benefits from Amazonian forest reserves: Leakage accounting and the value of time," *Mitigation and Adaptation Strategies for Global Change*, vol.14, no. 6, 2009.

[153] M. Herold, *An Assessment of National Forest Monitoring Capabilities in Tropical non-Annex I Countries: Recommendations for Capacity Building*, report commissioned for the Prince's Rainforests Project and the Government of Norway, Friedrich Schiller University, 2009, http://unfccc.int/files/methods_science/redd/country_specific_information/application/pdf/redd_nat_capacity_report_herold_july09_publ.pdf

[154] P.D. Hardcastle et al., *Capability and Cost Assessment of the Major Forest Nations to Measure and Monitor their Forest Carbon for Office of Climate Change*," Edinburgh: LTS international, 2008.

[155] Herold, op. cit., p. 21.

[156] Peskett et al., op. cit., p. 37.

example, private or public investors involved in REDD activities tend to look for areas where land ownership issues are not an obstacle.[157] Potential risks related to insecure land tenure may involve, for example, land ownership conflicts, uncertainty in delivering REDD commitments (and lack of legal protection against nondelivery) and concerns about reputational risks that might limit investments.[158] However, insecure land tenure is an all-too-common problem in tropical forest countries. In the case of the Amazon, even if protected areas occupy a high proportion of the land, the actual control over these areas is often weak, due to factors such as poor access and lack of personnel, and technical and financial capacity (observed in Bolivia, Brazil, Colombia, Guyana, Peru, and Venezuela).[159] For example, the Brazilian Institute of Environment and Renewable Natural Resources estimates that 80 percent of logging in the Brazilian Amazon is illegal, and much is carried out by land grabbers often sponsored by powerful individuals; yet, given its remoteness and size, it is very difficult to police this region without considerable resources, especially when powerful elites capitalize on the lack of enforcement.[160] Ultimately, the success of mechanisms such as REDD depends on government support. For example, effective tenure cannot be achieved without supportive policy and institutional systems, including clear forest-use rights, effective sanctions against transgression, and institutional support for sustainable forest management.[161] REDD can potentially help, or be an incentive for, tropical forest countries to improve governance, and this might be one of the greatest challenges posed to such mechanisms.

REDD also raises equity concerns among tropical forest countries. Finance tends to be directed to areas with high historic deforestation rates or high projected future rates.[162] In this case, countries with low historical deforestation rates, such as India and the Democratic Republic of Congo (DRC), may not be properly rewarded, whereas those with high rates of deforestation (and poor performance in terms of preserving forests) are likely to gain the most from REDD.[163] Therefore, without adequate safeguards, REDD may become a disincentive for countries with low deforestation rates and could even raise the risk of forest laws being altered in ways that increase deforestation.[164] In this regard, there is a growing consensus that incentives for

157 L. Cotula and J. Mayers, *Tenure in REDD: Start-point or Afterthought?*, London: IIED, 2009, p. 3.

158 Ibid., p. v.

159 Hardcastle et al., op. cit., p. 30.

160 Peskett et al., op. cit., p. 47.

161 Cotula and Mayers, op. cit., p. 5.

162 Griffiths, op. cit., p. 12. According to this document, World Bank technicians affirm that REDD incentives have low or no potential in countries with high forest cover and low deforestation rates and have "no potential" where forest cover is expanding, confirming concerns that REDD may perversely prove to be a disincentive for countries with low or zero deforestation.

163 Peskett et al., op. cit., p. 21.

164 Ibid., p. 27.

forest conservation should also be directed to those countries that have historically conserved forests, already reduced deforestation, and even increased forestlands.[165]

Equity is also an issue in relation to REDD project participants. Some studies suggest that benefits linked to REDD projects have been captured by State agencies, local governments, and conservation NGOs, rather than by indigenous peoples and local communities.[166] It is feared that REDD may lock up forests, decoupling conservation from development, and impose asymmetric power distribution, enabling powerful REDD consortia to deprive local communities of land-development aspirations – especially as the value of forestlands increases – or erode culturally rooted not-for-profit conservation values.[167] To be equitable, REDD schemes must then ensure the participation of local individuals and communities, for example, by allowing regular consultations, demanding prior informed consent when required, and promoting benefit sharing. On the issue of indigenous peoples' participation, the 2008 Manaus Declaration, signed by indigenous and local communities of Latin America, the DRC, and Indonesia, states that REDD projects shall "recognize the capability of sustainable management of forests as exercised by indigenous peoples and traditional communities, as well as the historical role of these peoples and communities in the conservation and in the equilibrium of global climate, and develop a compensation system."[168] Equity will be a contentious issue in future REDD negotiations also from a North–South perspective. Developing countries seem highly sensitized to "economic neo-colonialism" and may fear a loss of sovereignty with their forests being locked up under REDD conservation agreements, whereas the wealth generated from REDD offsets may be used elsewhere.[169]

From an environmental perspective, a major risk associated to REDD is in using this mechanism as a cost-effective way for developed countries and heavy polluters to continue business-as-usual practices. Although the interest in providing solutions to deforestation is a positive trend, it is feared that heavy polluters may use REDD for their own purposes and as low-cost substitute for their own emission reductions.[170] It is noted that controversial issues in the REDD debate seem to deviate

[165] Prince's Charities Foundation, *The Prince's Rainforests Project: An Emergency Package for Tropical Forests*, London, UK, 2009, p. 274, http://www.rainforestsos.org/pages/emergency-package/

[166] Griffiths, op. cit., p. 134.

[167] United Nations Collaborative Program on Reducing Emissions from Deforestation and Forest Degradation in Developing Countries (UN-REDD), *UN-REDD Framework Document*, 2008, p. 4., http://www.un-redd.org

[168] Signed at the Latin American workshop, Climate Change and Forest Peoples: Reducing Emissions from Deforestation and Forest Degradation (REDD) and the Rights of Indigenous and Traditional Peoples, Manaus, Brazil, 2008, http://www.edf.org/documents/8253_ManausDeclaration.pdf

[169] Belton, op. cit., p. 3.

[170] Densham et al., op. cit., p. 5.

from the genuine necessity of achieving substantial short-term emission reductions and providing long-term financial incentives to reduce emissions and enhance removals.[171] Indeed, if REDD offsets are not connected to binding and more strict emission reduction targets, the goal of stabilizing the global climate is unlikely to be achieved, and this mechanism may end up being a form of philanthropy or green marketing.[172] In this regard, there seems to be a consensus that REDD must add to, rather than detract from, deep emissions cuts in Annex I countries.[173]

8.3. PAYMENTS FOR ENVIRONMENTAL SERVICES

Payments for environmental services (PES), also known as "market" of "reward" or "compensation" for environmental services, are a modality of incentives. They are described as a 1) voluntary transaction in which 2) a well-defined environmental service (or a land use likely to secure that service) is 3) "bought" by a (minimum of one) buyer 4) from a (minimum of one) provider 5) if – and only if – the provider continuously secures the provision of the service.[174] With regard to the criteria for participation, it is generally understood that PES providers should have at minimum effective control over land, if they do not have formal rights or legal title, as well as the ability to control access to the resources and enforce the terms of the PES agreement.[175] PES differ from other sustainable use initiatives, for example ecotourism, as they are based on a user buying a service from a provider, conditional upon the provider securing the service under a contract with a price agreed upon beforehand.[176] A key feature of PES schemes is the reliance on conditionality: sellers get paid only if they can deliver the environmental service contracted for.[177] As observed in the case of REDD, PES schemes should achieve additionality, permanence, and avoid leakage.[178] The effectiveness of PES then depends on whether the

[171] D. Zarin et al., *Reducing Emissions from Deforestation and Forest Degradation (REDD): An Options Assessment Report*, prepared for the Government of Norway, 2009, p. 3, http://www.redd-oar.org/links/REDD-OAR_en.pdf

[172] S. Lovera et al., *The Hottest REDD Issues: Rights, Equity, Development, Deforestation and Governance*, Global Forest Coalition and IUCN Commission on Environmental, Economic, and Social Policies (CEESP), 2009, p. 2, http://www.wrm.org.uy/actors/CCC/cop14/Hottest_REDD.pdf

[173] Griffiths, op. cit., p. 7.

[174] S. Wunder, "Payments for environmental services: Some nuts and bolts," Center for International Forestry Research (CIFOR), Occasional Paper no. 42, 2005, p. 3, http://www.cifor.cgiar.org/publications/pdf_files/OccPapers/OP-42.pdf

[175] M. Ogonowski et al., *Utilizing Payments for Environmental Services for Reducing Emissions from Deforestation and Forest Degradation (REDD) in Developing Countries: Challenges and Policy Options*, Washington, DC: Center for Clean Air Policy, 2009, p. 11.

[176] Ibid., p. 4.

[177] Ibid., p. 10.

[178] S. Wunder, "The efficiency of payments for environmental services," op. cit., p. 54.

agreed environmental services have been provided and on the fulfillment of those requirements compared to a business-as-usual scenario.

Some PES, notably those involving carbon sequestration services, use markets as a platform for service buyers and sellers to interact. However, it is noted that the majority of existing PES are not market based.[179] Instead, they build on bilaterally negotiated agreements between individuals or groups of buyers and sellers (e.g. upstream communities and downstream beneficiaries in a watershed). Although the terms "payments" and "buyers" invoke monetary transactions, compensation for environmental services may also include payment in kind, for example, infrastructure development, training, and education, among others.

8.3.1. *Payments for Environmental Services in Practice*

PES schemes in practice rarely meet all the five elements mentioned above.[180] Some examples of environmental services secured under PES schemes include carbon sequestration (e.g. Northern electricity companies paying farmers to plant or maintain trees); biodiversity protection (e.g. service buyers paying landholders for creating biological corridors); watershed protection (e.g. downstream water users paying upstream farmers for adopting land uses that limit soil erosion or flooding risks; or protection of landscape beauty (e.g. tourism operators paying local communities not to hunt in wildlife viewing zones).[181]

In Latin America, a classic example of PES is Costa Rica's national scheme known as Payments for Environmental Services Program (Pagos por Servicios Ambientales – PSA).[182] In response to this country's high deforestation rates, observed especially since the 1950s, a new forestry law was passed in 1966 (Law 7575/1996), which for the first time explicitly recognized environmental services provided by forests. Since then, the PSA Program has been operating, covering services such as carbon sequestration, hydrological services, biodiversity protection, and the provision of scenic beauty. Under this scheme, individual farmers eligible to participate in the program provide specific services (e.g. forest protection, reforestation, agroforestry, natural forest regeneration, etc.), defined under a mutually agreed contract, in exchange for a certain payment. The participants usually receive payments for a five-year period,

[179] E. Lee and S. Mahanty, *Payments for Environmental Services and Poverty Reduction Risks and Opportunities*, Center for People and Forests, 2009, p. 3, http://www.recoftc.org/site/fileadmin/docs/publications/The_Grey_Zone/2009/Issue_PaperISBN.pdf

[180] Ibid., p. 3.

[181] S. Wunder, "The efficiency of payments for environmental services," op. cit., p. 49.

[182] S. Wertz-Kanounnikoff, M. Kongphan-Apirak and S. Wunder, *Reducing Forest Emissions in the Amazon Basin: A Review of Drivers of Land-use Change and How Payments for Environmental Services (PES) Schemes can Affect Them*, Bogor: Center for International Forestry Research (CIFOR), November 2008, p. 10, http://www.cifor.cgiar.org/Knowledge/Publications/Detail?pid=2603

but commit to manage or protect forestlands for twenty years. Payments are suspended in cases of noncompliance.[183]

The National Fund for Forest Financing (Fondo Nacional de Financiamento Forestal) disburses the payments to landholders after project approval. A percentage of the national fossil fuel tax was allocated to this fund, whereas additional resources should come from voluntary contributions. Costa Rica's private hydropower companies, resorts, domestic airlines, and national soccer teams were among the first contributors. Initially, it was hoped that international payments for carbon sequestration would become a major source of revenue, but apart from a US$2 million purchase by Norway of 200 million tons of carbon sequestration and other few philanthropic deals, no other major sale has occurred. Financial assistance coming from sources such as the World Bank, the Global Environment Facility, and development cooperation agencies have played a major role in securing Costa Rica's PES scheme.[184] Positive outcomes of this scheme include an increased number of tree plantations in the last two decades, more than 375,000 ha protected, and nearly 5,500 PSA contracts (from 1997 to 2003).[185] Since Costa Rica pioneered the first formally labeled PES scheme, many others have been documented worldwide.[186]

Albeit at an embryonic stage, some forms of PES have also developed in the Amazon. For example, such payments occurred in the Imuya area of the Cuyabeno Wildlife Reserve in the northern Amazon region of Ecuador, where a tourism operator (called *Transturi*) provided in-kind compensation for the Zancudo community for not hunting in prime tourism visitation zones in the Imuya area.[187] In Brazil, the Program for the Socio-Environmental Development of Rural Family Production (Proambiente), under the Ministry of the Environment, is an example of a PES-like scheme created to compensate small farmers in the Amazon for providing environmental services,[188] carbon sequestration, rehabilitation of ecosystems, soil conservation, and biodiversity preservation, among others. Proambiente has so far achieved modest results, as very limited payments have been made to farmers on the basis of equally limited evidence of environmental service contributions.[189]

[183] Ogonowski et al., op. cit., pp. 5–6.

[184] H. Rankine et. al., *Innovative Socio-Economic Policy for Improving Environmental Performance: Payments for Ecosystem Services*, Bangkok: United Nations Economic and Social Commission for Asia and the Pacific, 2009, p. 10., http://www.unescap.org/esd/environment/publications/PES/ESCAP_PES%20Publication%202560.pdf

[185] A. Redondo-Brenes, "Payment for environmental services in Costa Rica: Carbon sequestration estimations of native tree plantations," *Tropical Resources Bulletin*, vol. 24, 2005, pp. 20–29, at p. 20.

[186] Rankine et al., op. cit., p. 8.

[187] M. Brouwer, *Amazon Your Business: Opportunities and Solutions in the Rainforest*, Bunnik: Meindert Brouwer Partner in Communicatie, 2007, p. 5.

[188] Ogonowski et al., op. cit., p. 5.

[189] A. Hall, "Better RED than dead: Paying the people for environmental services in Amazonia," *Philosophical Transactions of the Royal Society*, 2008, 2008, pp. 1925–32, p. 1930.

A modality of PES is currently being developed in the State of Amazonas (Brazilian Amazon) known as "Bolsa Floresta."[190] This is a State-level government-financed program created by the State Law on Climate Change, Environmental Conservation and Sustainable Development (Law 3.135/2007).[191] The scheme has been carried out since 2007 by the Foundation Amazonas Sustentável and two other Brazilian institutions, the State Secretariat for the Environment and Sustainable Development and Bradesco, one of Brazil's largest private banks.[192] Funding is equally provided by the State government and Bradesco, whereas additional funds are expected to be raised from the sale of carbon credits for avoided deforestation in the voluntary markets.[193] Under this scheme, a monthly "forest grant" (Bolsa Floresta) is paid to households and communities engaged in nondestructive activities such as extraction, fishing, and tree fruit cultivation, among others, and able to preserve forest areas.

Participation is opened to families, communities, and family associations, which must register for the program and observe sustainable practices. Compliance is monitored and a penalty applies to those that deforest beyond a certain limit or engage in unsustainable practices. Payments differ according to the type of participants (whether families, communities, or family associations).[194] Despite its budgetary limitations as a State-government initiative, this scheme has benefited local communities and has been a stimulus to expand protected status for some forest areas.[195]

8.3.2. *PES: Pros and Cons*

PES schemes can be effective tools for securing the many environmental services provided by rich ecosystems such as those in the Amazon. However, there seem to be mixed opinions about this instrument. Although some claim that PES may offer new conservation funding (notably from private sector sources) and innovative

[190] Law 3.135/2007, Article 5, II.

[191] Available at the Web site of the Sustainable Development Secretariat of the State of Amazonas, http://www.sds.am.gov.br

[192] Wertz-Kanounnikoff, Kongphan-Apirak and Wunder, op. cit., p. 11.

[193] Hall, op. cit., p. 1928.

[194] These four types are: (1) Bolsa Floresta Family: rewards families with a minimum of two years of local establishment; (2) Bolsa Floresta Association: involves communities living in conservation units that are represented by family associations and eligible for grants; (3) Bolsa Floresta Income: benefits communities that are eligible to a local investment plan; and (4) Bolsa Floresta Social: Foundation Amazonas Sustentável provides small grants to social activities developed by local communities. Information available at: http://www.fas-amazonas.org/en/index.cfm?fuseaction=conteudo&id=19

[195] According to P. Fearnside, "Bolsa Floresta is fine as a demonstration of how funds can reach traditional residents in the Amazon interior," in R.A. Butler, *How to Save the Amazon Rainforest*, January 13, 2009, http://news.mongabay.com/2009/0104-saving_the_amazon.html

solutions to current environmental problems,[196] skeptics fear that PES schemes may "bring back the fences" by delinking conservation from development, which may exclude local communities and deprive them of their development aspirations.[197] In other words, it is questioned whether this mechanism can really benefit the poor and become a tool for poverty alleviation. It is argued that the landless poor, living from degrading activities (e.g. firewood collectors, charcoal makers, bush meat hunters, etc.) could lose out from such activity-restricting schemes.[198] There is more justification for concern, as initial PES experiences have showed the risk of not reaching the poorest land users.[199]

Some of the barriers that may exclude the poor from participating in PES schemes involve things such as insecure tenure, size of landholdings, high transaction and investment costs, and a lack of technical capacity and awareness.[200] In particular, insecure land tenure is a recurrent problem in rich forest countries and one that needs to be addressed, given that PES schemes involve the exchange of services that typically require enforceable property rights. Secure land tenure is ultimately a guarantee to buyers that contracted services can be delivered and that financial investments will not be lost due to land ownership problems.[201] Even if the conditions for entering in PES schemes are flexible enough to allow the participation of those with unclear land tenure, their low control over land tends to reduce the marketability of their services. The fact that the poor usually own small landholdings may be another factor of exclusion. The costs of operating PES schemes are often higher for small plots, especially transaction costs, for instance, those related to searching for and identifying such plots, obtaining information, contracting, and monitoring.[202] In addition, PES schemes involve high investment costs in land-use practices, which poor households usually cannot afford. Even if there are good market opportunities for the poor, factors such as low awareness and education, and limited access to technical knowledge may prevent their participation. In view of these obstacles, some claim that poverty alleviation should be a central element of PES,[203] whereas others take the view that conservation should be seen as a primary objective, as such schemes have been conceived more as a conditional conservation mechanism when compared to other initiatives that aim to integrate conservation and social or development goals.[204]

[196] De la Torre, Fajnzylber, and Nash, op. cit., p. 49.
[197] Ibid., pp. 51, 52.
[198] Wunder, "The efficiency of payments for environmental services," op. cit., p. 57.
[199] Wertz-Kanounnikoff, Kongphan-Apirak, and Wunder, op. cit, p. 16.
[200] Lee and Mahanty, op. cit., p. 9.
[201] Rankine, Watkins, and Kasemsawasdi, op. cit., p. 18.
[202] Lee and Mahanty, op. cit., p. 11.
[203] Ibid., p. 5.
[204] Ibid., citing Wunder, "Payments for environmental services," op. cit., pp. 279–97.

Indeed, although PES schemes were not conceived as a poverty alleviation tool, they can (and ideally should) benefit poor communities. As such schemes can be tailored to specific circumstances and adapted to particular variables and objectives, they give enough flexibility for including social and other considerations as part of their functioning. In principle, all aspects of PES schemes, from their objectives to the type of ecosystem services and method of compensation, can be freely designed and agreed upon under a contract.[205] For example, social considerations are a key aspect of Bolsa Floresta, which is reflected in the provision of technical assistance, and access to health and education services foreseen under this scheme. Other PES schemes were designed, for example, to attend the specific objective of creating biodiversity corridors between protected areas and buffer zones at their margins.[206] PES experiences in Latin America have typically had side objectives, including poverty alleviation and regional development (biodiversity conservation has also been pursued as a co-benefit).[207] PES can, in fact, benefit from the participation especially of indigenous communities, given their traditional knowledge about the environment and management practices.[208] This is something to be considered particularly in areas such as the Brazilian Amazon, where native groups enjoy permanent usage rights over large land areas. Early and regular communication with those groups should increase their participation and also help overcome challenges and misunderstandings among PES participants.[209] As PES schemes do not always require direct cash transfers, other forms of payments, involving, for example, the building of roads, schools, health facilities, irrigation systems, etc., can also deliver social benefits to poor communities.[210] In cases where such schemes ensure the participation of poor landholders, for example, by providing access to capital, insurance, and other financial services, facilitating funding to cover upfront costs or the improvement of technical capacity, PES schemes can become a valuable new source of funding for the poor and an economic alternative to unsustainable practices. In this regard, the role of donor organizations, governments and policymakers in ensuring that participation is available to a broader range of service sellers is crucial.[211]

In practice, certain PES schemes will be more effective than others in ensuring that aspects such as additionality, permanence, and leakage are addressed and that services are effectively delivered. To this end, PES schemes require a baseline

[205] Rankine, Watkins, and Kasemsawasdi, op. cit., p. 12.
[206] Ibid.
[207] Wertz-Kanounnikoff, Kongphan-Apirak, and Wunder, op. cit., p. 16.
[208] Ogonowski et al., op. cit., p. 20.
[209] Rankine, Watkins, and Kasemsawasdi, op. cit., p. 16.
[210] Ibid., p. 14.
[211] Lee and Mahanty, op. cit., p. 9.

providing a reference against which results can be measured and some form of surveillance to ensure that environmental services are delivered as planned. Certain schemes may not achieve, for instance, good levels of additionality or permanence. It is suggested that additionality in the case of Bolsa Floresta, for example, is low, as deforestation was unlikely to occur in many of the designated areas, even in the absence of payments.[212] PES schemes also run the risk of impermanence or leakage, especially in a region of the size and remoteness as the Amazon. In such cases, it may be difficult even for well-intentioned PES participants to track and prevent activities such as illegal logging.[213] Therefore, in such cases government support is crucial to ensure that some monitoring and enforcement capacity is available. In any case, as PES schemes are conditional on the delivery of specific services under a mutually agreed contract, service buyers may take an interest in enforcing the terms of the agreement, whereas sellers have an incentive to comply with the contract and deliver the service on which payment is conditional.

Despite the apparent interest in PES schemes, many developing countries may find it difficult to use such schemes, for reasons such as the high costs of collecting data, monitoring, and creating markets for those services.[214] In Latin America, some difficulties can be noted from the demand side: too few service users have been convinced to pay; and also from the supply-side: little is known about what resource-use incentives and institutional settings are needed, how to benefit the poor, etc.[215] Also, in donor-crowded areas, such as the Andean region, individuals and communities may not enter in *quid pro quo* agreements when they are used to donors and agencies offering benefits for free, with no demand of services in return.[216] Although PES schemes are being tested in Latin America, some resistance still remains, as they can be viewed as a first step toward the expropriation of resources and sovereignty loss.[217]

PES schemes seem particularly useful in cases where forests are being cleared by small landholders for local use of fuel wood or subsistence crops, which are activities that also exert significant pressure on forests. In such cases, PES schemes can represent an economic alternative to potential service sellers, and may prevent them from using unsustainable practices or from migrating elsewhere. An interesting option in these cases is to combine conservation with other measures, for instance, improved efficiency of wood-fired stoves, agricultural productivity, and agroforestry, among others.[218] However, in Latin America, land-use change and deforestation are mostly

[212] Wertz-Kanounnikoff, Kongphan-Apirak and Wunder, op. cit., p. 14.
[213] Ogonowski et al., op. cit., p. 13.
[214] FAO, op. cit., p. 87.
[215] De la Torre, Fajnzylber and Nash, op. cit., p. 50.
[216] Ibid., p. 56.
[217] Wertz-Kanounnikoff, Kongphan-Apirak, and Wunder, op. cit., p. 17.
[218] Ogonowski et al., op. cit., p. 16.

driven by economic profitability (e.g. expansion of agriculture, stock farming, and logging, etc.) and due to factors such as lenient law enforcement and overall weak governance; and only to a much lesser extent due to poverty cycles. Poor forest dwellers (e.g. indigenous communities, smallholders, and rubber tappers) are less frequently prime Amazon deforesters – although they can be important stakeholders in stabilizing Amazonian land use.[219] In this context, PES schemes may fail to target major deforesters making large profits from forest clearing.

Most deforestation in the Brazilian Amazon in particular happens through private illegal (but often tolerated) occupation and clearing of government-owned forestlands,[220] often sponsored by large landholders. One of the main causes of deforestation in the region is the illegal sales of public lands (known as "*grilagem*"). Many factors contribute to this practice, for example. the lack of supervision by public notaries, which recognize illegal land transactions without an adequate process for verifying land ownership, and political-electoral interests involving promises of future land concessions, among others.[221] Nevertheless, PES schemes (and other forms of incentives) are based on the idea that only legal activities and legitimate landholders should be compensated. Therefore, illegal logging in the Brazilian Amazon cannot be stopped through payments of the PES type, as there are no legitimate landowners to compensate for conservation efforts.[222]

Moreover, when conservation is already been achieved relatively well by other means, for example through the creation of protected areas, PES may not be an appropriate tool.[223] In such cases, even if a minority of squatters threatens parts of a national park, for example, they should not be paid to stop, as it may trigger new waves of squatters looking for similar rewards. In general, because PES presupposes *de facto* free land-use choices, it is not a good protected-area management tool, except perhaps when the conservation of protected areas is highly threatened, as in the case of "paper parks."

CONCLUSIONS

Positive incentives are often said to be suitable tools for dealing with environmental problems, especially when compared to other techniques, for example that of command-and-control. They encourage individuals, communities, businesses, and governments to prevent environmental harm and natural resources depletion from happening rather than only remediating damage that has already occurred. Positive

[219] Wertz-Kanounnikoff, Kongphan-Apirak, and Wunder, op. cit., p. 17.
[220] Ibid.
[221] Ogonowski et al., op. cit., p. 11.
[222] Wertz-Kanounnikoff, Kongphan-Apirak, and Wunder, op. cit., p. 17.
[223] De la Torre, Fajnzylber and Nash, op. cit., p. 55.

incentives can also offer additional funding and economic alternatives to stimulate new, environmentally benign technologies and more sustainable practices.[224] They can indeed be useful conservation tools, also in the context of the Amazon, especially for handling environmental problems requiring urgent solutions and significant funding, such as deforestation.

The rationale behind positive incentives, and market-based instruments in particular, is that they can provide economic alternatives especially to local communities and at the same time be used for conservation purposes. One of their strengths lies in their ability to provide rapid responses to critical environmental problems, especially in raising funding. There seems to be a consensus that deforestation and the associated climate change are among those issues requiring solutions without further delay.[225] In view of this, developed countries recognize that financing to address climate change must be "scaled up urgently and substantially."[226] This is particularly the case as rich forest areas such as the Amazon are being rapidly degraded or lost to deforestation and other unsustainable practices. As noted by the WWF President Carter Roberts, "if the Amazon were a country, it would be in the top seven emitters of greenhouse gases in the world."[227] In such a context, opportunities for conservation are also being lost together with the clearing of rainforests. According to Nepstad, the Amazon rainforest "has already entered a dieback, in which the vicious cycle between land use (cattle ranching, logging), seasonal drought, and fire are rapidly degrading enormous swathes of rainforest each year."[228] By the same token, Fearnside warns that opportunities for reserve creation in the Amazon are quickly being lost as new areas are opened to deforestation and, at the same time, he stresses the mitigation benefits of protected areas over and extended time horizon, especially given that in the future virtually no unprotected forest is likely to remain.[229]

Nevertheless, it is argued that mechanisms such as REDD only deal with the symptoms of a bigger problem, without solving the underlying causes of deforestation, including, for example, insecure land tenure and inequality, discrimination

[224] E.H. Bulte, G. Cornelis van Kooten, and T. Swanson, "Economic Incentives and Wildlife Conservation," draft paper, October 27, 2003, p. 2, http://www.cites.org/eng/prog/economics/CITES-draft6-final.pdf

[225] Zarin et al., op. cit., p. 2. According to this document, the need to achieve emission reductions across mitigation sectors is urgent and requires mechanisms such as REDD, able to provide incentives for halting deforestation without delay.

[226] The G8 Summit, held from July 8 to 10 in L'Aquila, adopted the Declaration on Responsible Leadership for a Sustainable Future, http://www.g8italia2009.it/G8/Home/G8-G8_Layout_locale-1199882116809_Atti.htm

[227] Butler, op. cit.

[228] Ibid.

[229] Fearnside, "Carbon benefits from Amazonian forest reserves: leakage accounting and the value of time," Mitigation and Adaptation Strategies for Global Change, vol. 14, n. 6, pp. 557–567.

against indigenous peoples, corruption, over-consumption, and industrialization.[230] It is indeed crucial to address such root causes in a scenario of weak governance prevailing in most tropical forest countries, but this requires time, human, and financial resources, in order to build understanding of such causes, generate the political will to address them, modify or create laws and policies, etc. Putting in place all these measures without at the same time introducing solutions to rapidly halt deforestation and avoid further environmental degradation is something that we can no longer afford.

A major challenge today is to confer value on forests by creating the right incentives and funding to maintain them, so that deforestation and forest degradation turn out to be less attractive options. This, however, requires funding and the appropriate means. For example, taking REDD into account, the Woods Hole Research Institute estimates that reducing deforestation in the Brazilian Amazon to nearly zero within a decade would cost from US$100–600 million per year, an amount lower than the forgone profits from deforestation-dependent agriculture and ranching.[231] According to the Eliasch Review, an additional of £11–19 billion per year to halve deforestation would be needed if (in 2020) demand from carbon markets could fund around 22 percent of forestry abatement.[232]

The use of positive incentives, notably market-based mechanisms, seems attractive, especially if one considers that international assistance, equivalent to hundreds of millions of dollars in donor funds that has flowed into the Amazon in particular has not been able to revert deforestation patterns in that region.[233] A growing faith in the market can be noted, seeing it as an avenue for conferring economic value on certain ecosystems and compensating those able to maintain their services, especially as profitable economic activities, such as agriculture, stock farming, logging, and others, expand into rainforests.

In this context, positive incentives, in particular the mechanisms examined here, can be a good short-term solution to tackle critical environmental problems in the Amazon. From an environmental perspective, mechanisms such as REDD or PES are beneficial as they lock up areas for conservation. If successfully implemented,

[230] Griffiths, op. cit., p. 103

[231] "How much would it cost to end Amazon deforestation?", article released on January 27, 2008, http://news.mongabay.com/2008/0128-brazil.html. See D.C. Nepstad, "Brazil: The costs and benefits of reducing carbon emissions from the Brazilian Amazon region," in J.P. Holdren, D.C. Nepstad, and K.R. Smith, *Linking Climate Policy with Development Strategy in Brazil, China, and India,* final report to the William and Flora Hewlett Foundation from the Woods Hole Research Center, 2007, p. 8, http://whrc.org/Hewlett/Hewlett-Report-FINAL.pdf

[232] Eliasch, op. cit., p. 183. According to this report, in 2020 demand from carbon markets could fund around 22 percent of forestry abatement, representing around $7 billion of credits. An additional £11–19 billion would be needed to halve forest emissions.

[233] Butler, op. cit.

they can avoid deforestation and lead to a reduction of GHG emissions, among other co-benefits. However, in order to achieve this, such mechanisms need to ensure additionality, permanence, and avoid leakage. In other words, it must be ensured that they meet certain objectives (e.g. carbon storage, biodiversity conservation, protection, watershed protection, etc.) that would not be achieved otherwise (additionality); also, that they do not cause the deterioration of ecosystems elsewhere (leakage) and should continue for a minimum length of time (permanence).

Despite some of the limitations noted here, these instruments can provide an additional source of finance and allow conservation activities that would not happen otherwise due to a lack of funding, technical capacity, or political will. Although their implementation requires government support, it also favors the participation of the private sector, which is given a more active conservation role than that of merely avoiding environmental harm by observing codes of conduct or other guidelines. If well designed and implemented, such mechanisms can achieve environmental objectives (e.g. emission reductions and biodiversity conservation), but also benefit local individuals and communities, by providing economic alternatives to unsustainable practices. Other mechanisms, such as CDM, are also useful as they bring in investments in clean technology into developing countries, improve existing technical capacity, and allow activities such as afforestation and reforestation. Such instruments can help States, and the Amazon countries in particular, to fulfill some international obligations, for example, that of achieving a stabilization of GHG concentrations in the atmosphere (UNFCCC, art. 2), promoting the conservation and sustainable use of biodiversity (CBD, art. 1), and achieving the preservation and rational utilization of natural resources in the Amazon (ACT, art. 1).

However, as noted above, the costs of halting deforestation are high, and it is predicted that, despite the new funding mechanisms for tropical forests, there may be less money available than required.[234] Although mechanisms such as REDD or PES can indeed be a new source of finance, their implementation in itself involves costs and minimum technical capacity, which requires various sources of funding at different levels. As observed in the case of REDD, most funding allocated so far has been directed to technically prepare tropical forest countries. It has been also estimated that a PES scheme in the Brazilian Amazon similar to that in Costa Rica, for example, would cost the government about US$5 billion per year (if applied to just a third of the Amazon) and US$20 billion (if applied to the entire Amazon and other biomes).[235] Funding is then needed from various sources, in other words, it must be

[234] FAO, op. cit., p. 64.

[235] Global Forest Coalition, "Effective policies to Reduce Emissions from Deforestation in Developing Countries (REDD) must address leakage and incorporate social impact criteria," submission to the UNFCCC Secretariat, 2008, p. 2, http://unfccc.int/resource/docs/2008/smsn/ngo/016.pdf

generated by market-based mechanisms, but also from more "traditional" sources, such as donations, technical cooperation projects, official development assistance (ODA), etc. In view of this, some have dropped the "market-versus-non-market" dilemma,[236] in particular the opposition to using carbon markets to help protect forests,[237] once faced with the fact that no one funding source is the solution.

In practice, positive incentives, and market-based instruments in particular, can be useful conservation tools, depending on how they are used. For example, if Annex 1 countries perceive the flexible mechanisms under the Kyoto Protocol, or other instruments agreed upon under a post-2012 climate regime, as an easy way to continue business-as-usual practices, delaying deeper cuts in GHG emissions, such mechanisms (and others) are unlikely to prevent dangerous climate change. As noted here, the inclusion of REDD-like mechanisms under the Kyoto Protocol was objected to in part because they were viewed as a way for Annex 1 countries to avoid cutting emissions domestically from industrial and other sources. The situation is no different today. If Annex 1 countries rely on existing mechanisms without achieving the 2°C target defined under the Copenhagen Accord, they will fail to achieve the goal of stabilizing GHG emissions. In the near future, major emerging economies, such as China, India, Brazil, and South Africa, should also assume emission reduction commitments, according to their levels of development and share of responsibility, in line with the principle of common but differentiated responsibility. No specific deal has been reached in December 2009 at COP-15 in terms of emission reduction targets. So far, developed countries and major emerging economies have made political statements, recognizing the broad scientific view that the increase in global average temperature should not exceed 2°C, and have committed to deeper emission cuts. For example, under the 2009 G8 Summit Declaration, developed countries committed to reducing emissions in aggregate by 80 percent or more by 2050, compared to 1990 levels (para. 65). Pursuant to the 2009 Declaration of the Leaders of the Major Economies (including countries such as Brazil, China, Indonesia, Mexico, and South Africa), developed countries will take the lead in reaching aggregate and individual reductions by assisting developing countries to reduce emissions from deforestation and forest degradation and to enhance removals of GHG emissions by forests (para. 1).

Despite the potential of the mechanisms here discussed, they are not a panacea for all problems. For example, they are probably unsuitable to address some issues, such as transboundary pollution, overfishing, and illegal wildlife trade,

[236] Rose, op. cit.

[237] Ibid. "We are going to need a lot of money," says Pendzich of the WWF, which has abandoned its opposition to using the carbon market to help protect forests. "We're going to need it from many sources. No one funding source is the solution."

among others. Moreover, they should be used in combination with and not instead of other measures, for instance, protected areas, moratoria or bans, conservation policies, and law enforcement, which have proven to be effective in particular cases. Governments still have a prominent role in controlling deforestation by making use of good policies. For example, certain measures in the Brazilian Amazon have improved environmental law enforcement in very remote areas, including, for example, mobile courts introduced in 1998, where a judge and police force fly to areas where disputes occur and resolve the matter on the spot. Also, independent institutions such as Brazil's Federal Prosecutor's Office (Ministério Público Federal), in charge of monitoring public performance,[238] have helped improve governance in the Brazilian Amazon. Other effective policies include the state program to control deforestation in Mato Grosso (between 1999 and 2001).[239] Therefore, mechanisms such as REDD should be seen as catalytic in relation to good practices and national programs rather than as an exclusive and sufficient solution.[240] Although positive incentives can be good conservation tools, they are unlikely to prosper without government support, national laws, and policies enabling their functioning. Therefore, tropical forest countries, and the Amazon States in particular, have a crucial role in protecting their forests, as they are the ones to adopt effective policies and laws, enforce (national and international) legal obligations and facilitate the operation of market-based instruments (and other incentives). As noted by a group of South American countries, under the Plataforma Climatica Latinoamericana, despite emphasis currently put on the market, it is still for the States to facilitate and ensure the operation of market-based instruments and promote national climate change policies.[241]

238 Wertz-Kanounnikoff, Kongphan-Apirak, and Wunder, op. cit., p. 17.

239 See P. M. Fearnside, "Deforestation control in Mato Grosso: A new model for slowing the loss of Brazil's Amazon Forest," *Royal Swedish Academy of Science*, vol. 32, no. 5, 2003, pp. 343–45.

240 M.C. Bustamante, C.A. Nobre, and R. Smeraldi, *Estimating Recent Greenhouse Gas Emissions from Cattle Raising in Brazil*, 2009, http://www.amazonia.org.br/english.

241 The Plataforma Climatica Latinoamericana involves Colombia, Uruguay, Argentina, Paraguay, and Peru (Plataforma Climatica Latinoamericana, *Cambio Climatico y COP15: Reflexiones desde y para America Latina*, Quito: Plataforma Climatica Latinoamericana, 2009, p. 14).

9

The Legal Status of the Amazon

Implications for International Cooperation

The Amazon is an area of incontestable ecological significance in the region and worldwide, owning the largest portion of remaining rainforest on the planet and up to one-fifth of the world's freshwater, and as a source of environmental services of global interest, such as climatic and water regulation.[1] According to the 1988 Brazilian Constitution, the Amazon forest is part of the national patrimony and its various resources shall be used as provided by law, under conditions that ensure the preservation of the environment, including the use of the area's mineral resources.[2] The fact that the Amazon is considered a national patrimony does not mean that it is the Brazilian government's property, but rather that its exploitation should be guided by the public interest.[3] Due to the ecological importance of the Amazon in maintaining the equilibrium of the global environment for present and future generations, should this region, including its natural resources, have a special international legal status?

The legal status of natural resources[4] varies, depending on whether they are located within the territory of one State, shared by a group of States, or are outside the territorial jurisdiction of States, and these categorizations have different impacts on

[1] World Bank, *A More Equitable, Sustainable and Competitive Brazil: Country Assistance. Strategy 2004–2007*, Washington, DC: World Bank, 2003, information available in Annex 8, "A Strategic Contribution to a More Sustainable Amazon," p. 197.

[2] 1988 Brazilian Constitution, Article 225(4).

[3] R. dos Santos Vieira, "Concepts of national environmental law: Brazilian environmental law relating to Amazonia," in M. Bothe, T. Kurzidem and C. Schmidt (eds), *Amazonia and Siberia: Legal Aspects of the Environment and Development in the Last Open Spaces*, London: Nijhoff, 1993, pp. 237–55, at p. 237.

[4] As defined by Barberis, resources are those elements that man can use to satisfy own needs; the characteristics of "natural" means that there is no human intervention in their production (J.A. Barberis, *Los recursos naturales compartidos entre estados y el derecho internacinal*, Madrid: Editorial Tecnos, 1979, p. 146).

the freedom of states to exploit such resources.[5] As Birnie and Boyle note, whereas reasonably comprehensive patterns of international cooperation now exist for protecting common areas such as the high seas, the deep seabed and the Antarctic, it cannot be assumed that comparable obligations apply to areas falling wholly within the boundaries of national sovereignty.[6] Natural resources and sites located within the territories of States, albeit internationally protected under multilateral environmental agreements, such as the 1974 Convention on International Trade of Endangered Species of Wild Flora and Fauna (CITES)[7] and the 1972 Convention on Wetlands of International Importance especially as Waterfowl Habitat (Ramsar Convention),[8] are under the sovereignty of a particular State, which therefore exercises great discretion as to their use. Therefore, the legal status of certain areas and their natural resources determines to a great extent how international cooperation takes place. For example, due to the particular territorial status of the Antarctic, the concerned States are allowed to take certain measures, such as the physical inspection of vessels and certain sites, which in principle would not be permitted in relation to natural resources falling within the jurisdiction of a State without its consent. Although some environmental problems, such as greenhouse gas (GHG) emissions, clearly have global impact, threats to certain resources such as freshwater and biodiversity have more attenuated transnational implications. In these latter cases, global regulation is harder to justify, and the international community is faced with the dilemma of how to effectively protect such natural resources without impinging on national sovereignty.[9]

Natural resources falling exclusively within the territory of a State, such as species of fauna and flora and certain rivers and lakes, remain the property of the State within which they are located and are subject to its permanent national sovereignty. Some natural resources, however, transcend national boundaries, extending over the limits of one single State, for example migratory species, international rivers, ground waters, and mineral deposits throughout an area encompassing several States. These shared resources can be defined as "an element of the natural environment used by men, which constitutes a bio-geophysical unit, and is located in the territory of one or more States."[10] The 1978 UN Environment Programme (UNEP) Principles of Conduct in the Field of the Environment for the Guidance of States in the Conservation

5 P. Birnie and A. Boyle, *International Law and the Environment*, New York: Oxford University Press, 2002, p. 137.

6 Ibid., p. 89.

7 993 UNTS 243, 1974.

8 11 ILM 963, 1972.

9 H.M. Schally, "Forests: Toward an international legal regime?," *Yearbook of International Environmental Law*, vol. 4, 1993, pp. 30–50, at p. 34.

10 UNEP/IG/12/2, 1978, para. 16.

and Harmonious Utilization of Natural Resources Shared by Two or More States[11] provides an exemplificative list of shared natural resources, which include, among other things: river systems, enclosed and semienclosed seas, air sheds, mountain chains, forests, conservation areas, and migratory species.[12] According to the UNEP Principles, shared (or transboundary) natural resources entail obligations of transboundary cooperation and equitable utilization. This idea had been also introduced in the 1974 Charter of Economic Rights and Duties of States,[13] which asserts that each State must cooperate and achieve an optimum use of shared resources.

The notion of shared natural resources involves a limited form of community of interests usually involving a small group of States in geographical contiguity.[14] A main feature of these resources is their partial accessibility: only a limited number of States enjoy access to them and only these co-owners can limit the access of third States.[15] Although domestic natural resources are located exclusively within the territory of a particular State, shared natural resources expand over the territory of two or more States. As noted above, States have greater discretion with regard to natural resources falling within their own territories when compared to other types of resources. In particular, shared natural resources not only involve a right of common use by a group of States, but also an obligation to achieve the optimal and concerted utilization of those resources, taking into account the other States' interests.[16] In that sense, a State may have less discretion in using natural resources shared with other countries. In the Amazon, natural resources are either domestic or shared by two or more States. Although many of those resources are vital for maintaining the global environment, they are usually confined in the territorial limits of an individual State. However, several natural resources are shared among the basin States, for example fish species (e.g. *tambaquis, matrinchãs, surubins,* and *douradas*) or international rivers (e.g. the Amazon, Negro, Putumayo, Napo, and Tigre).

Natural resources located in areas beyond the territorial jurisdiction of individual States (or in common spaces) such as the high seas and the superjacent airspace are opened for legitimate and reasonable use by all states and cannot be appropriated by the exclusive sovereignty of a particular State.[17] In other words, no State has an exclusive right over natural resources held in common or the right to prevent others

11 UNEP Governing Council Decision 6/14 (May 19, 1978), Principles of Conduct in the Field of the Environment for the Guidance of States in the Conservation and Harmonious Utilization of Natural Resources Shared by Two or More States.

12 UNEP/GC/44, 1975, para. 86.

13 UNGA Res. 3129 XXVIII, December 13, 1973, Article 3.

14 Birnie and Boyle, op. cit., p. 139.

15 E. Benvenisti, *Sharing Transboundary Resources: International and Optimal Resource Use*, Cambridge: Cambridge University Press, 2002, pp. 2, 3.

16 Barberis, op. cit., "Introduction," p. 156.

17 Ibid., p. 141.

from joining in their exploitation. The prohibition of appropriating those resources is asserted, for example, in the 1958 Convention on the High Seas,[18] the 1982 United Nations Convention on the Law of the Sea (UNCLOS),[19] and the 1967 Treaty on Principles Governing the Activities of States in the Exploration and Use of Outer Space, Including the Moon and Other Celestial Bodies[20] (Moon Treaty).

Once more, States may be less reluctant to assume international obligations with respect to resources located in common spaces than in relation to resources falling wholly under their national jurisdiction, as they have larger discretion to use these latter resources according to their own needs and interests and may be unwilling to compromise. However, this has not prevented international law from tackling natural resources lying within the boundaries of individual States, for example under the 1968 African Convention on the Conservation of Nature and Natural Resources,[21] the 1972 World Heritage Convention[22], and the 1992 Convention on Biological Diversity (CBD).[23] States seem to recognize a legal obligation to protect the environment, including natural resources located within national borders, which is reflected in these treaties and many other legal instruments, although they do not abdicate their sovereign rights over domestic natural resources.[24]

In any case, the management (or mismanagement) of certain sites or natural resources, even if confined in precise territorial limits, may have global repercussions. For example, the loss of tropical forests has a global impact, as they are one of the most precious forms of natural wealth on our planet, sheltering most of its plant and animal species, maintaining biological diversity and climatic conditions, and protecting soil and water resources.[25] Deforestation affects the local, regional, and global climate by eliminating important "sinks" – which remove carbon dioxide (CO_2) from the atmosphere by incorporating it into biomass and releasing oxygen into the atmosphere – and by burning, which releases GHG contributing to global warming.[26] Therefore, even if certain natural resources are

18 Articles 1–2, 450 UNTS 82.

19 Articles 87 and 89, 21 ILM 1261, 1983.

20 Treaty on Principles Governing the Activities of States in the Exploration and Use of Outer Space, including the Moon and other Celestial Bodies, London, Moscow and Washington, DC, January 27, 1967, reprinted in 610 UNTS 205, Article 4.

21 1001 UNTS 3.

22 1037 UNTS 151.

23 1992 Convention on Biological Diversity (CBD), 31 ILM 818, 1992.

24 F. L. Morrison and R. Wolfrum, *International Regional and National Environmental Law*, The Hague: Kluwer International Law, 2000, p. 61.

25 N. Schrijver, P. Peters and P. Waart, "Responsibility of states in respect of the exercise of permanent sovereignty over natural resources: An analysis of some principles of the Seoul Declaration (1986)," *Netherlands International Law Review*, vol. 36, 1989/3, pp. 285–313, at p. 301.

26 K.R. Miller, W.V. Reid, and C.V. Barber, "Deforestation and species loss," in J. Tuchman Mathews (ed.), *Preserving the Global Environment, the Challenge of Shared Leadership*, New York & London, Norton, 1991, p. 79.

within precise territorial limits and under the sovereignty of a particular State, their management can be a matter that concerns the international community as a whole.

The Amazon is one of those areas whose protection seems to concern the international community at large. Given its ecological importance, can it be said that the international community has a legitimate interest in the protection of the Amazon, including its natural resources? Should this region benefit from a special international legal status, like other areas of the world beyond national jurisdiction, such as the deep seabed, the Antarctic, and the moon and outer space?

For the purposes of this study, the significance of examining whether this region has a particular status lies in the implications it may have in terms of international cooperation. The previous chapters looked at how the Amazon States cooperate among themselves and with the international community. By examining the international legal status of the Amazon, this chapter extends the analysis of the relationship between the Amazon and the international community. This is because the international legal status of certain sites and natural resources (and the absence of such a status) determine to some extent how international cooperation may take place. For example, due to its special status, certain forms of cooperation can be envisaged in Antarctica, but they might not be possible in other areas or with regard to natural resources falling within national jurisdictions. Therefore, this chapter enquires whether the Amazon has (or should have) a particular international legal status and what implications this may have in the relationship between the basin States and the international community.

9.1. THE AMAZON AS A COMMON HERITAGE OF HUMANKIND?

By the end of 1963, the UN General Assembly (UNGA) declared through Resolution 1962 (XVIII)[27] that the "exploration and use of the outer space shall be carried on for the benefit and in the interests of all mankind." A few years later, in 1970, UNGA Resolution 2749 (XXV)[28] designated as the "common heritage of mankind" the area of the seabed, ocean floor, and its subsoil beyond the limits of national jurisdiction, the exploration and exploitation of which shall be carried out for the benefit of humankind. Although the modern conceptualization of the notion of common heritage of humankind has been developed only recently, its genesis can be traced back to the nineteenth century, notably in the realm of the law of the sea negotiations, where the idea that oceans are an undivided heritage of humankind was first

27 UNGA Res. 1962 (XVIII), December 13, 1963.
28 UNGA Res. 2749 (XXV), December 17, 1970.

elaborated.[29] For example, during the 1930 Hague Conference for Codification of International Law, the seas and the resources of the sea were referred to as a common patrimony, and a few years later, the president of the 1958 UN Conference of the Law of the Sea affirmed that "the sea is the common heritage of mankind" and should be preserved for the benefit of all.[30] However, the modern concept of a common heritage of humankind is regarded as having been introduced in 1967 by Maltese Ambassador Pardo, who in an address to the UNGA proposed that the area of the seabed beyond the limits of coastal States' national jurisdiction should be declared the "common heritage of mankind."

Subsequently, this notion was incorporated in two treaties: the 1982 UNCLOS and the 1979 Agreement Governing the Activities of States on the Moon and other Celestial Bodies (Moon Treaty). The "Area" (i.e. the seabed, ocean floor, and subsoil, beyond the limits of national jurisdiction)[31] and its resources are defined as the "common heritage of mankind" (Article 136, 1982 UNCLOS). Likewise, the moon and its natural resources are also so defined (Article 11, Moon Treaty). Although the nature of the common heritage concept and its precise meaning are the object of debate,[32] the literature tends to extract four main aspects of this principle based on those two treaties: 1) exclusion of sovereignty claims over the areas in question (the moon and the Area) or over their natural resources,[33] 2) use for peaceful purposes,[34] 3) international management,[35] and 4) benefit sharing[36] derived from the exploitation of natural resources, if it occurs.[37] As regards the first aspect, it

[29] M.A. Fitzmaurice, "International protection of the environment," *Recueil des Cours de l'Académie de Droit International*, The Hague/Boston/London: Martinus Nijhoff, vol. 293, 2002, pp. 150–51.

[30] Ibid.

[31] UNCLOS, Article 1 (1).

[32] N.J. Schrijver, "Permanent sovereignty over natural resources versus the common heritage of mankind," in P. de Waart, P. Peters, and E. Denters (eds.), *International Law and Development*, Dordrecht/Boston/London: Martinus Nijhoff, 1988, pp. 87–101, at p. 98. As Schrijver notes, whereas at one extreme some claim that the common heritage of humankind is a new peremptory norm, others deny its legal value and place it in the realms of politics and morality. He suggests that the truth between those views is somewhere in the middle: The principle of the common heritage of humankind has been clarified and its basic constituent elements specified. See also B. Larschan and B.C. Brennan, "The common heritage of mankind principle in international law," *Columbia Journal of Transnational Law*, vol. 21, no. 2, 1983, pp. 305–37, at p. 319.

[33] UNCLOS, Article 137; Moon Treaty, Article 11.

[34] UNCLOS, Article 141; Moon Treaty, Article 3(1).

[35] Both UNCLOS and the Moon Treaty foresee an international regime required for the management of natural resources (R. Wolfrum, "The principle of the common heritage of mankind," *Zeitschirift für ausländisches öffentliches Recht und Völkerrecht*, vol. 43, 1983, pp. 312–37, at p. 317).

[36] Moon Treaty, Article 140, Article 160(2f(i)); UNCLOS, Article 11(7d), Article 140(2).

[37] Fitzmaurice, op. cit., p. 153; A. Kiss, "La notion de patrimoine commun de l'humanité," *Recueil des Cours de l'Académie de Droit International*, The Hague/Boston/London: Martinus Nijhoff, vol. 175, 1982, pp. 99–256, at p. 135; M.N. Shaw, *International Law*, Cambridge: Cambridge University Press, 2003, p. 362. Those four elements, with the inclusions of a fifth, can be found in this definition by

should be noted that States have a right to use natural resources under a common heritage regime, but are prevented from appropriation. Pursuant to Article 137 of UNCLOS, "no State shall claim or exercise sovereignty or sovereign rights over any part of the Area or its resources, nor shall any State or natural or juridical person appropriate any part thereof," and all rights in the resources located in the Area are "vested in mankind as a whole."[38] Similarly, the moon and its natural resources are "not subject to national appropriation by claim of sovereignty by means of use or occupation or by any other means" (Article 11, Moon Treaty).[39]

Both the Moon Treaty and UNCLOS foresee the creation of an international regime "to govern the exploitation of the natural resources of the Moon as such exploitation is about to become feasible" (Article 11, 5, Moon Treaty) or to carry out the exploration and exploitation of the Area on behalf of mankind (Article 153, UNCLOS). No such a regime yet exists as regards the moon.[40] Under UNCLOS, the international system of exploration and exploitation of the Area is conducted by the International Seabed Authority (ISA),[41] composed of all States parties. The ISA was created to regulate the deep seabed common heritage area on behalf of all States and was mainly entrusted with the role of distributing economic benefits arising out of the exploitation of the Area among the parties (including to those that do not have the technical or financial means to participate in such exploitation) and of

S.J. Shackelford in "The tragedy of the common heritage of mankind," 28 Stan. Envtl. L.J., 2009, pp. 109–67, at p. 118. According to this author, despite the lack of a universally agreed definition of the common heritage of mankind principle, a working definition would likely comprise five elements. Firstly, there can be no private or public appropriation; no one legally owns common heritage spaces. Secondly, representatives from all nations must manage resources, since a commons area is considered to belong to everyone. Thirdly, all nations must actively share with each other the benefits acquired from exploitation of the resources from the commons heritage region. Fourthly, no weaponry or military installations can be established in commons areas. Fifthly, the commons should be preserved for the benefit of future generations, and to avoid a "tragedy of the commons" scenario. Similarly, according to Telesetsky, the qualifying aspects of common heritage as expressed in the Antarctic Treaty and the Outer Space Treaty include the following: (1) states are not allowed to own resources that belong to the global commons; (2) resources held in common will be internationally managed; (3) all members of the global community should share the benefits of developing a global common resource; and (4) common heritage areas are reserved for peaceful purposes (A. Teletsky, "*Graun bilong mipela na mipela no tromweim*: The viability of international conservation easements to protect Papua New Guinea's declining biodiversity," *Georgetown International Environmental Law Review*, 2001, pp. 735–78, at p. 754).

38 "In general the resources of the Area belong to humankind as a whole on whose behalf the Authority shall act, but minerals may be alienated in some circumstances" (UNCLOS, Article 137(2)).

39 The 1967 Moon Treaty does not use the term "common heritage of mankind," but rather uses the term "province of mankind," stating that the exploration and use of outer space, including the moon and other celestial bodies "shall be the province of all mankind" (Article 1).

40 C.R. Buxton, "Property in outer space: The common heritage of mankind principle vs. the 'first in time, first in right' rule of property law," *Journal of Air Law and Commerce*, vol. 69, 2004, p. 699.

41 UNCLOS, Article 156. Pursuant to Article 158, and as defined in Article 170, "the Enterprise, is the organ of the Authority that carries out activities in the Area directly," as foreseen in UNCLOS, Article 153.

encouraging the transfer of technology.[42] Any activity in the area shall be carried out in accordance with a formal written plan of work approved by the Council, which is the executive organ of the International Seabed Authority.[43] The Authority can take any measure to ensure compliance with the provisions laid down in Part XI of UNCLOS, annexes, and related provisions of UNCLOS.[44]

Debate exists as to whether the notion of common heritage of humankind is confined to the limits of UNCLOS and the Moon Treaty, or whether it also applies to other areas and natural resources, including those located within domestic jurisdictions. One may question whether natural resources or ecosystems of critical importance to humankind could be given the status of common heritage of humankind, even if located in areas under national jurisdiction. Could such a concept be applied, for example, to the Amazon – as it has already been suggested[45] – given its role in maintaining the equilibrium of the global environment on behalf of mankind?

In examining the case of the Brazilian Amazon, Kemal Baslar comes to the conclusion that this region should also be considered a common heritage of humankind, although he is aware of the difficulty of applying this concept for areas or natural resources within national boundaries.[46] This author claims that extending this notion to certain areas, such as the Brazilian Amazon, would not represent an illegitimate interference in Brazil's domestic affairs or amount to a form of "economic colonization," as some suggest. According to Baslar, the principle of permanent sovereignty over natural resources usually invoked by developing countries was crafted decades ago in a historical context of decolonization and for purposes different from today; and, therefore, should not be used to "subjugate the legitimate interests of all mankind."[47] Brazil, in particular, should not "take refuge in an anti-colonialist permanent sovereignty principle" when the survival of humankind is at stake.[48] Baslar claims that attributing common heritage status to the Amazon would not deprive Brazil from its sovereign rights, as this concept does not imply the absence of ownership over a given territory, but rather a nonexclusive use of natural resources. In any case, he suggests that Brazil should be compensated by the

[42] Shackelford, op. cit., p. 109.

[43] The Council's powers and functions are defined in UNCLOS, Article 162. The Council is in charge of supervising and coordinating the implementation of all matters within the competence of the Authority, calling attention of the Assembly of cases of noncompliance and exercises control over activities in the area in accordance with Article 153, para. 4 and rules and procedures of the Authority.

[44] UNCLOS, Article 153(5).

[45] Baslar discusses the Brazilian Amazon in particular; see K. Baslar, *The Concept of the Common Heritage of Mankind in International Law*," The Hague/Boston/London: Martinus Nijhoff, 1998, pp. 140–46, 152–58.

[46] Ibid., p. 145.

[47] Ibid., p. 142.

[48] Ibid.

international community for preserving the Amazon through financial assistance and the transfer of technology.

According to Alexandre Kiss, the sharing of material benefits deriving from the use of natural resources in common heritage areas – which could, in principle, limit its application only to areas outside national jurisdiction – is not the main aspect of this concept. Rather, it involves a form of trust whose main elements involve exclusive use for peaceful purposes, rational utilization, and the transmission of common heritage areas to future generations.[49] In this light, Kiss suggests that the notion of common heritage applies not only to the Area or the moon, but also to Antarctica and to natural and cultural world heritage sites. If one adopts a broader interpretation of the notion of common heritage, not confined to the limits of UNCLOS and the Moon Treaty, certain natural resources and areas even if within domestic jurisdiction, could indeed be regarded as the common heritage of humankind. Conversely, some suggest that this concept should be limited to the parameters of UNCLOS and the Moon Treaty.[50]

As far as we understand, the notion of common heritage of mankind refers to areas beyond national jurisdiction, in particular the Area and the moon. As a matter of fact, its application within the domestic sphere, for example in relation to biological resources, has been generally rejected by States.[51] If this concept is extended to natural resources and areas other than those foreseen under UNCLOS and the Moon Treaty it may end up devoid of any specific meaning or practical effects; or, otherwise, it may give rise to abuse on the part of States, willing to have easier access to certain natural resources within national jurisdiction.

In any case, it is hard to compare the Amazon with essentially different common heritage areas. The law applicable to the Area or the moon regulate activities in areas that fall outside national sovereignty. The Amazon and its natural resources, unlike those areas, are under the national sovereignty of eight States, and therefore could not have a similar status. Despite changes introduced over the last years conferring more flexibility to the notion of state sovereignty, particularly in the international human rights and environmental law fields, the principle of territorial sovereignty remains one of the main corollaries of international law.[52] Even those in favor of attributing common heritage status to areas such as the Amazon agree that

[49] A. Kiss and D. Shelton, *International Environmental Law*, New York & London: Transnational, 1991, p. 250. According to Fitzmaurice (op. cit., p. 157), it would be more in-line with the idea of the whole concept if it had universal territorial applicability. It would not serve its purpose if it were territorially limited.

[50] Birnie and Boyle, op. cit., p. 143.

[51] Telesetsky, op. cit., p. 754.

[52] M. Shaw, "The heritage of states: The principle of *uti possidetis juris* today," *British Yearbook of International Law*, 1996, pp. 75–154, at p. 81.

most commentators would not support the notion of common heritage as a digestible recipe for the management and equitable sharing of natural resources within national borders.[53]

The main elements that constitute the notion of common heritage, unless otherwise provided by a new treaty or emergent customary law, are those laid down in UNCLOS and the Moon Treaty. Despite of the ecological importance of a region such as the Amazon, none of those elements, in particular the exclusion of sovereign rights, nonappropriation, and equitable sharing, are applicable in the Amazon, as contrary to the customary law rule of national sovereignty over natural resources (Principle 21 of the Stockholm and Principle 2 Rio Declaration).[54] It is not without reason that developing countries rejected the introduction of the common heritage principle in the CBD on the occasion of the 1992 UNCED.[55] If the concept of common heritage is not applicable to the Amazon, could it be given a special status, such as that attributed to certain areas of global ecological importance, for example the Antarctic?

9.2. THE AMAZON AS AN AREA OF SPECIAL PROTECTION?

Since the adoption of the Antarctic Treaty[56] in 1959, sovereignty claims over this region have been suspended, and ever since, the Antarctic is to be used for peaceful purposes only. This treaty created a framework for the evolution of the Antarctic Treaty system. It was not conceived as a conservation treaty, but by addressing issues such as demilitarization and denuclearization in the context of the Cold War, and by ultimately freezing sovereignty claims, the Antarctic Treaty achieved significant environmental outcomes.[57] Environmental concerns in the Antarctic have begun

[53] Baslar, op. cit., p. 145. The Brazilian forests, unlike Antarctica, will not become world parks, even if their care becomes a world responsibility. As Sanwal notes, if tropical forests in particular are treated as common heritage, the "transfers will only be in the nature of aid: they will be limited in amount, their utilization will be determined by the donor, and the arrangements will erode sovereignty. The developing countries will not be able to determine these transactions without becoming subject to conditions which prejudice their development and will be required to enact legislation that is impossible to enforce, especially if imposed under international bureaucracy" (M. Sanwal, "The sustainable development of all forests," *Review of the European Community*, vol. 1, no. 3, 1992, pp. 289–94, at p. 294).

[54] Stockholm Declaration of the UN Conference on the Human Environment, UN doc. A/Conf. 48/14 v. 3.7.1972, reprinted in 11 ILM 1416, 1972; and Rio Declaration on Environment and Development, UN Conference on Environment and Development, UN doc. A/CONF.151/6/Rev.1, 1992, 31 I.L.M. 874, 1992.

[55] P.-F. Mercure, "Le rejet du concept de patrimoine commun de l'humanité afin d'assurer la gestion de la diversité biologique," *Canadian Yearbook of International Law*, vol. 33, 1995, pp. 281–304.

[56] 402 UNTS 71.

[57] D.R. Rothwell, "Polar environmental protection and international law: The 1991 Antarctic Protocol," *European Journal of International Law*, 2000, pp. 591–614, at p. 608.

to prevail soon after the adoption of that treaty, giving rise to a comprehensive legal regime that covers all activities in the continent likely to have an adverse effect on the environment.[58] The first environmental instrument related to the Antarctic was the 1964 Agreed Measures for the Conservation of Antarctic Fauna and Flora that qualified Antarctic as a "special conservation area," and some species of native mammals and birds as "specially protected species" (Preamble). A series of treaties on the conservation of the Antarctic adopted afterward comprise today's Antarctic legal system.

The 1991 Protocol on Environmental Protection to the Antarctic Treaty designated the Antarctic as a "natural reserve, devoted to peace and science"[59] and set up a regime for the protection of the Antarctic environment and dependent and associated ecosystems. In 1994, the UNGA welcomed the designation of the Antarctic as a "nature reserve devoted to peace and science" and commended the prohibition on mineral resources extraction activities in the region.[60] The Antarctic legal system is one of the most comprehensive and stringent environmental protection regimes.[61] In particular, the 1991 Protocol[62] introduced an "ecosystem approach," by covering all natural resources in the "Antarctic environment and dependent and associated ecosystems,"[63] and gave effect to the precautionary principle by prohibiting mining activities in Antarctica.[64] The provisions contained in the Protocol are detailed in annexes that regulate issues such as environmental impact assessment; conservation of fauna and flora; waste disposal; and marine pollution. The obligation to conduct an environmental impact assessment (Annex 1) applied for the first time to a whole continent and was framed in great detail, one of the most significant features of the Protocol.[65]

A parallel between the Amazon and the Antarctic can hardly be drawn, given the particular territorial status of the Antarctic. The Antarctic is the only continent whose territory has not been divided among nation states. The Antarctic illustrates the benefits of attributing an international character to an area not exclusively

58 Ibid., p. 591.

59 1991 Protocol on Environmental Protection to the Antarctic Treaty, Article 2 (Antarctic Protocol).

60 UNGA Res. 49/80, December 15, 1994. Debate still remains as to whether Antarctica is a part of the common heritage. According to Rothwell (op. cit., p. 608), the existence of territorial claims on the continent and maritime claims in the Southern Ocean seem to defeat such views.

61 P. Sands, *Principles of International Environmental Law*, Cambridge: Cambridge University Press, 2003, p. 712.

62 30 ILM 1455, 1991.

63 Antarctic Protocol, Article 2.

64 Article 7. According to Rothwell (op. cit., p. 594), repeated references to the "Antarctic environment and dependent and associated ecosystems" demonstrated an ecosystem approach adopted by this instrument.

65 Ibid., p. 609.

within the territory of a particular state.[66] Unlikely the Antarctic, the Amazon lays within well-defined national boundaries and could hardly be given the status of a "natural reserve." The Amazon is under the sovereignty of eight States with a population of millions of inhabitants and is under intense economic exploitation. The Antarctic, on the other hand, has only sparsely populated scientific bases and, in part due to its unique territorial status, it is under a regime that for the most part prevents any economic exploitation.[67] Although the Amazon States recognize the importance of this region to the global environment, they reject the idea of protecting the Amazon as the "world's park" on behalf of the international community. As stated in the document *Our Own Agenda*, formulated in 1990 under the aegis of the Latin America and Caribbean Commission on the Development and Environment, Inter-American Development Bank (IDB), and the UNEP, the Amazon cannot be viewed only as an "ecological treasure" and regulator of the global climate:[68]

> The Amazon cannot be considered simply an ecological treasure and an important regulator of global climate; it is a major development resource. The preservation of the tropical rainforest, and especially the Amazon forest will depend on the mobilization of research, and development, technology and financial resources for sustainable management of this area. A commitment of support from the international community and voluntary co-operation of the eight sovereign Amazonian states is needed to preserve the Amazon as an asset for the region and the world community.
>
> Preservation of the Amazon is of interest to the world community. However, it is of great interest to the Amazonian countries. The issues must therefore be debated by them on their own terms with the support of foreign scientists and enlightened international public opinion.

A fairer comparison can probably be drawn between the Amazon and other regions of the world with a few more similarities. For example, the Amazon could be more easily compared to the Arctic. The Arctic is also under the sovereignty of eight States (Canada; Denmark, including Greenland and the Faroe Islands; Finland; Iceland; Norway; the Russian Federation; Sweden; and the United States) who

66 D. Tarlock, "The Great Lakes as an environmental heritage of humankind: An international law perspective," *University of Michigan Journal of Law Reform*, no. 40, 2007, p. 1005.

67 E.T. Bloom, "Polar regions and the development of international law," commentary on Oran R. Young, *Creating Regimes: Arctic Accords and International Governance*, Ithaca & London, Cornell University Press, 1998, in Book Reviews and Notes, *American Journal of International Law*, vol. 92, 1998, pp. 593–95, at p. 482. Here the author draws a comparison between the Arctic and Antarctic.

68 IDB and UNEP (Inter-American Development Bank and United Nations Environment Program), *Our Own Agenda*, report of the Latin America and Caribbean Commission on the Development and Environment, 1990.

share a vast biome. This region has motivated increasing regional cooperation over the past years. In 1989, on the initiative of Finland, the Arctic countries started discussions on ways of jointly protecting the Arctic environment, which culminated in the adoption of the 1991 Arctic Environmental Protection Strategy (AEPS).[69] The AEPS, together with a Declaration on the Protection of the Arctic Environment,[70] identified common objectives and areas where joint action was needed. This regional strategy, unlike the 1978 Amazon Cooperation Treaty (ACT), is not legally binding.[71] As in the Amazon, similar interests, for example that of preventing transboundary environmental damages, sharing information, and providing distinctive responses to a singular biome, has led the Arctic countries to seek regional cooperation.[72]

With a similar structure to the ACT, a high-level intergovernmental forum designated the Arctic Council, was established in 1996 to coordinate and oversee the activities carried out under the AEPS.[73] The Council is supported by a secretariat that rotates among the member states and meets every two years.[74] Among other initiatives, the Arctic Council created the Sustainable Development Program, whose purpose is to propose and adopt measures to be taken by the Arctic, including opportunities to protect the environment, and the economies, cultures, and healths of indigenous communities, according to its Terms of Reference signed in 1998.[75] The Arctic Council, unlike Amazon Cooperation Treaty Organization (ACTO), is not an international organization with legal personality.[76] The Arctic is a model of soft-law-based cooperation, using informal and nonlegally binding political understandings.[77]

69 30 ILM 1624, 1991.

70 This declaration is available under the name of the 1991 Rovaniemi Declaration at http://www.arctic-council.org/rovaniem.asp

71 D.R. Rothwell, "The Arctic environmental protection strategy and international environmental cooperation in the far north," *Yearbook of International Environmental Law*, vol. 6, 1995, pp. 65–105, at p. 101.

72 Ibid.

73 Declaration on the Establishment of the Arctic Council, signed by the representatives of the Arctic States in Ottawa on September 19, 1996, Article 1(a), 35 ILM 1382.

74 Arctic Council Rules of Procedure, adopted by the Arctic Council at the First Ministerial Meeting of the Arctic Council Iqaluit, Canada, September 17–18, 1998, Article 15, available at: http://arctic-council.npolar.no

75 Also on the sustainable development program of the Arctic Council, see the Framework Document (Chapeau) for the Sustainable Development Programme, both available at: http://arctic-council.org/section/documentation

76 Bloom, op. cit., p. 482.

77 O.R. Young, *Creating Regimes: Arctic Accords and International Governance*, Ithaca & London, Cornell University Press, 1998. Sands (op. cit., p. 731) refers to the existence of a "soft law approach" in the Arctic.

Some features of the Arctic system could possibly be tested in the Amazon, especially under the ACT. For example, indigenous communities tend to have a more active role in decision-making processes, as they were given a permanent status within the Arctic Council.[78] In the context of the ACT, those communities have a limited participation in regular meetings and decision-making processes. However, as Fearnside notes, the involvement of indigenous peoples is crucial (but still very limited) for Amazonian conservation policy, particularly given the fact that they have had a much better record of maintaining natural ecosystems. Their ability to defend and maintain their forests gives them an as yet unremunerated role in providing environmental services, which could be the key to the success of conservation policies.[79] By contrast, one of the AEPS major objectives is to accommodate traditional cultural needs, values, and practices and one of its main principles is to incorporate indigenous peoples' health, social, economic, and cultural needs into development planning.

It is also worthy of note the Arctic Monitoring and Assessment Program (AMAP), created in 1991 under the AEPS, to monitor levels of, and assess the effects of anthropogenic pollutants in all components of the Arctic environment. This program should provide regular information on the status of the Arctic environment, and also scientific advice to the Arctic States. AMAP is now an international organization that implements components of the AEPS.[80] In the Amazon, a similar initiative would be welcomed, as there is no such mechanism providing basin-wide information and monitoring. Some Amazon countries have developed monitoring systems – notably Brazil, under the National Institute of Space Research (INPE) – but only over their respective Amazonian territories. Given that neither the concept of common heritage of humankind nor a particular status such as that conferred on the Antarctic are applicable to the Amazon, what international status could be given to this region, if any, and what might be possible implications of such a status?

9.3. THE AMAZON AS A COMMON CONCERN OF HUMANKIND

In 1988, UNGA Resolution 43/53 proclaimed that "climate change is a common concern of mankind, since climate is an essential condition which sustains life on earth."[81] This concept was later introduced into a number of other instruments, for example the 1992 UN Framework Convention on Climate Change

78 As defined in the 1996 Declaration on the Establishment of the Arctic Council, Article 2.

79 P.M. Fearnside, "Conservation policy in Brazilian Amazonia: Understanding the dilemmas," *World Development*, vol. 31, no. 5, 2003, pp. 757–79, at p. 774.

80 More information is available at: http://archive.arcticportal.org/30/

81 UNGA Res. 43/53, December 6, 1988, on the Protection of Global Climate for Present and Future Generations.

(UNFCC)[82] and the 1992 CBD[83] and, at the inter-American level, in resolutions of the Organization of American States (OAS) General Assembly.[84]

Earlier references to the existence of certain values involving common interests of humankind can be traced, for example, in the *travaux preparatoires* of the 1972 Stockholm Declaration, which stated that "a growing class of environmental problems, because they are regional or global in extent, will require extensive co-operation among nations and action by the international organizations in the common interests."[85] Also, certain treaties, such as the 1980 Convention on the Conservation of Antarctic Marine Living Resources,[86] enunciated that achieving an optimum level of whale stocks was a matter of "common interest."[87]

The exact content of the notion of common concern of humankind and its legal nature is still imprecise.[88] Nevertheless, this concept seems to reflect the common interest of all States in certain forms of ecological protection.[89] According to Kiss, the notion of common concern contains two basic elements: a temporal one that takes account of the long-term consequences of environmental degradation and the related duty to protect the environment for the benefit of future generations; and a spatial element, whereby the biosphere should be considered as a unity and all its elements (either within or beyond national jurisdictions) as interdependent.[90] Kiss and Dinah Shelton suggest that, because the global environment is interdependent and its protection concerns both present and future generations,

[82] 31 ILM 849/851, 1992.

[83] The Preamble of the CBD recognizes the conservation of biodiversity as a "common concern of humankind."

[84] OAS AG/RES. 1674 (XXIX-O/99), June 7, 1999, entitled Climate Change in the Americas, adopted at the First Plenary Session of the OAS General Assembly; OAS AG/RES. 1864 (XXXII-O/02), June 4, 2002, entitled The Socioeconomic and Environmental Impact of Climate Change on the Countries of the Hemisphere, adopted at the Fourth Plenary Session of the OAS General Assembly.

[85] W. Riphagen, "International concern for the environment as expressed in the concepts of 'common heritage of mankind' and of 'shared natural resources'," in M. Bothe (ed.), *Trends in Environmental Policy and Law*, Gland: IUCN, 1980, pp. 343–62; UN.doc.A/COND.48/4, 1972, quoted by Baslar, op. cit., p. 288.

[86] 19 ILM 837, 1980.

[87] 1329 UNTS 47. Also in the Preamble of the 1980 Convention on the Conservation of Antarctic Marine Living Resources, 19 ILM 837, 1980: "believing that it is in the interest of all mankind to preserve the waters surrounding the Antarctic"; and the 1949 Inter-American Tropical Tuna Convention, examples quoted by Sands, op. cit., p. 287.

[88] According to Baslar (op. cit., p. 289), although the "common heritage of mankind" is a legal term, the "common concern of mankind" is a political one. In order to protect the rainforest from further degradation, some scholars would like to extend the aspects of "common concern" and "common heritage" to the rainforest. According to Telesetsky (op. cit., p. 754), the notion of "common concern of humankind" refers to the idea that a resource is no longer "solely within the domestic jurisdiction of states due to its global importance and consequences for all."

[89] Birnie and Boyle, op. cit., p. 144.

[90] Kiss and Shelton, op. cit., p. 252.

the world's biodiversity, climate, migratory species, oceans, and the Antarctic, as well as areas within the territory of individual States, when faced with problems such as desertification and biodiversity loss, can be considered common concerns of humankind.

Instead of the idea of common heritage, both the CBD and the UNFCCC used the notion of common concern of humankind. This indicates that the international community still resists the extension of the common heritage principle beyond traditionally recognized "unowned" common resources.[91] Public international law still relies heavily on the notion of sovereignty, which sets the boundaries between the domestic and international spheres, and has, as a corollary, the duty of states to refrain from intervening in areas of exclusive jurisdiction.[92] In any case, it is suggested that sovereignty is a malleable concept as States can freely choose to restrict it, as can be observed in international environmental law at least in two respects.[93] Firstly, it is accepted that unrestricted sovereignty cannot be exercised when a state's development activities have extraterritorial environmental effects; and, secondly, environmental agreements have progressively reduced the ambit of permanent sovereignty concerning a State's own use of its natural resources. In view of this consideration, would it be right to say that notions like that of a common concern apply to natural resources or areas under national jurisdiction, such as rivers, lakes, and forests confined within national borders?

The UNFCCC and the CBD, while recognizing States' sovereign rights to address climate change (Preamble, UNFCCC) and over biological resources (Article 3), introduced the notion of common concern, which also extends to natural resources and areas under national jurisdiction. The UNFCCC asserts the importance of marine and terrestrial ecosystems and reservoirs of greenhouse gases (Preamble), which include, for example, forests. In order to achieve the ultimate objective of the stabilization of greenhouse gas concentrations in the atmosphere (Article 1), States should conserve marine and terrestrial ecosystems within their national territories,[94] notably those that function as carbon sinks. One of the commitments assumed by States under the UNFCCC is to promote sustainable management and cooperate in the conservation of sinks and reservoirs of all greenhouse gases, including biomass, forests, and oceans, as well as other terrestrial, coastal, and marine ecosystems

91 Tarlock, op. cit., p. 1003.

92 P. Cullet and A.P. Kameri-Mbote, "Activities implemented jointly in the forestry sector: Conceptual and operational fallacies," *Georgetown International Environmental Law Review*, vol. 10, 1997, pp. 97–119, at p. 113.

93 Ibid.

94 The areas under the sovereignty of or that are subject to the jurisdiction of the coastal State, namely i.e., the seabed of the internal waters, the territorial sea, the archipelagic waters, the exclusive economic zones, and the continental shelf.

(Article 4(d)). While each State has a duty to protect terrestrial and marine ecosystems, including those under domestic jurisdiction, the UNFCCC views climate change as a common concern and calls for the "widest cooperation possible by all countries and their participation in an effective and appropriate international response" (Preamble). Therefore, from the recognition of climate change as a common concern it follows that all States should cooperate in finding solutions and that the international community has a legitimate interest in taking action.[95]

The conservation of biodiversity is also considered a common concern of humankind under the CBD and viewed as essential "for the evolution and for maintaining life sustaining systems of the biosphere" (Preamble). Among other duties, the parties are requested to promote *in situ* conservation, particularly through a system of protected areas (Article 8). To this end, they should protect ecosystems and natural habitats, and maintain viable populations of species. It can be noted that the notion of common concern also applies to resources and ecosystems within the limits of national jurisdiction. Article 4 defines the jurisdictional scope of the CBD, by stating that, in the case of components of biological diversity, the "provisions of this treaty apply, in relation to each Contracting Party ... in areas within the limits of its national jurisdiction." The protection of important areas such as the Amazon could then be considered a common concern at least from two perspectives: firstly, because it is a reservoir of greenhouse gases and plays a role in the stabilization of the global climate; and, secondly, because this region is home to a great wealth of wildlife species occurring in many and varied habitats. The mismanagement of the Amazon affects the global environment and may endanger "life sustaining systems of the biosphere."

State practice has provided evidence of the fact that the preservation of natural resources, irrespective of where they locate can be considered a common concern. This is observed particularly in international debates on forests. Since 1926, when the First World Forestry Congress was convened, States have recognized the crucial role of forests in providing social, economic, and environmental goods and services to humanity.[96] In 1966, the Sixth World Forestry Congress stated that the conservation and sound management of forests was an "international concern."[97] The debate over the protection of forests was brought to the 1972 UN Conference on the Human Environment (Stockholm Conference),[98] where States reasserted the need to protect, by appropriate technical and legal means, primeval forests containing important genetic resources.[99] At the 1992 United Nations

95 Tarlock, op. cit., p. 113.

96 Twelfth World Forestry Congress, "Forests, source of life," final statement, Quebec, September 28, 2003.

97 FAO Web site: http://www.fao.org/forestry/site/18507/en/

98 UNGA, *Report of the UN Conference on the Human Environment*, Recommendation, UN doc. A/CONF.48/14/Rev.1, Stockholm, June 5–16, 1972.

99 Ibid.

Conference on Environment and Development (UNCED),[100] this issue was again discussed, as States sought to reach a global consensus on forests through a treaty of global application. This attempt has failed, and the UNCED instead produced a Non-Legally Binding Authoritative Statement for a Global Consensus on the Management, Conservation, and Sustainable Development of all Types of Forests (Forest Principles), which recognized that the protection of forests was a matter of increased worldwide attention.[101] In 2000, the UN Economic and Social Council (ECOSOC) Resolution 2000/35 set up an international arrangement on forests to provide a forum for policy development in the management, conservation, and sustainable development of all types of forests, under which the UN Forum on Forests (UNFF) was created. In response to the need for an institutional framework for enhancing cooperation on forest issues, as indicated in Chapter 11 of UNCED's Agenda 21,[102] several forest-related organizations joined efforts under a Collaborative Partnership on Forests, set forth by that ECOSOC resolution, to provide support to the UNFF.

In different forums and on different occasions, States have proclaimed that the world's forests have a vital role in ensuring protection of the global environment and have expressed concerns over the fast-growing deforestation worldwide. For example, the Plan of Implementation of the 2002 World Summit on Sustainable Development in Johannesburg[103] urges States to fight land degradation and loss of forest biodiversity by rationally managing natural and planted forests.[104] On this occasion, the UNFF conveyed a message expressing concerns in relation to the massive deforestation of the world's forests and committed to reversing this trend.[105] Under the framework of the 1992 CBD, the 2002 Hague Ministerial Declaration[106] restated the need to halt deforestation and ensure the sustainable use of timber and nontimber resources.[107] Similarly, under the UNFF Multi-year Program of

[100] UN, *Agenda 21*, UN doc. A/CONF.151/26, vol. 1, August 12, 1992.

[101] According Schally (op. cit., p. 43), during the 1992 UNCED, developing countries ensured that several principles dear to their hearts, for example, national sovereignty, the condemnation of trade restrictions, and the linkage of sustainable forest management to the provision of financial assistance, were accepted by the international community in a political declaration. In return, industrialized countries failed to get acceptance of concepts like "global heritage" or "global stewardship," which could have been interpreted as a mandate for negotiating an international instrument.

[102] UN, op. cit., Chap. 11 on, "Combating deforestation."

[103] UN doc. A/CONF.199/20, September 4, 2002.

[104] Ibid., Plan of Implementation of the 2002 World Summit on Sustainable Development, para. 45.

[105] Ministerial Declaration, UNFF, UN doc. A/CONF.199/PC/8, March 19, 2002.

[106] Published in Secretariat of the Convention on Biological Diversity, *Handbook of the Convention on Biological Diversity*, Quebec, Canada, 2005, pp. 1453–58.

[107] The 2002 Hague Ministerial Declaration, adopted at the Sixth Conference of the Parties of the 1992 Convention on Biological Diversity, printed in *Environmental Policy and Law*, vol. 32, nos. 3–4, 2002, pp. 186–88.

Work 2007–2015,[108] States committed themselves to reversing the loss of forest cover by increasing the area of protected forests. Under the 2007 Non-legally Binding Instrument on All Types of Forests,[109] States agreed to take action at all levels to sustainably manage all types of forests through international cooperation. Article 2(a) enumerates "shared global objectives" for the preservation and sustainable use of all types of forests.

The worldwide concern over forests is an indication that the protection of certain areas or natural resources can be viewed as a common concern of humankind, even those under national jurisdiction. As stated in the Hague Recommendations on International Environmental Law (Preamble):[110] "the preservation of the environment is a common concern of humankind." Evidence that natural resources within national borders are objects of a common concern can be also found in the 1972 Convention concerning the Protection of the World Cultural and Natural Heritage.[111] According to Article 4, certain natural and cultural heritage located in the parties' territories constitute "world heritage," without detriment to their sovereign rights.

There is sufficient evidence to suggest that the protection of certain natural resources and ecosystems essential for maintaining the life-sustaining systems of the biosphere, even if under national jurisdiction, can be considered a common concern of humankind.[112] This also holds true with respect to the Amazon for its role in maintaining global ecological conditions.

9.3.1. *Consequences of a Common Concern Status*

The concept of common concern of humankind involves interests and values pertaining to the international community as a whole. This notion seems to suggest that, while all States have a legitimate interest in the preservation of natural resources and ecosystems indispensable for sustaining life on Earth, they also bear a shared responsibility for their protection.[113] This duality is reflected in most international environmental law instruments. For example, under the CBD, a State

[108] UNECOSOC, *Report of the Seventh Session*, Economic and Social Council, E/CN.18/2007/8, adopted at the UNFF Seventh Session, February 24, 2006 and April 16–27, 2007.

[109] A/C.2/62/L.5, 2007.

[110] Published in the *Yearbook of International Law*, 1992.

[111] 1037 UNTS 151, 1973. The definition of "natural heritage" is contained in Article 2.

[112] G. Handl, "Environmental security and global change: The challenge to international law," *Yearbook of International Environmental Law*, 1990, pp. 3–33, at p. 31. Boyle refers to an extension of the notion of common concern into the areas under States' jurisdiction (A. Boyle, "Remedying harm to international common spaces and resources: Compensation and other approaches," in P. Wetterstein (ed.), *Harm to the Environment*, Oxford: Oxford University Press, 1997, pp. 83–100).

[113] Sands, op. cit., p. 287.

may have access to genetic resources (Article 15), but shall also endeavor to conduct research and transfer technology to the country of origin of such resources (Article 16). Likewise, the financial mechanisms set up under certain environmental treaties to help developing countries' parties meet incremental costs related to the treaty implementation (e.g. CBD, Article 21; UNFCCC, Article 11) are an example of the dual facet of the international protection of the environment and of the notion of common concern in particular. Although the international community has a legitimate interest in the protection of certain areas indispensable for maintaining the biosphere, all States should share the burdens related to their conservation.

Common interests pertaining to the international community as a whole, such as the stabilization of greenhouse gas concentrations in the atmosphere or the conservation of biodiversity, can arguably generate *erga omnes* obligations. As defined in the much-cited ruling in the case of Barcelona Traction, Light and Power Company, Ltd, these are "obligations of a State toward the international community as a whole," and in view of the importance of the rights involved, they are a "concern of all States."[114] Similarly, certain obligations foundational to current international environmental law instruments, such as the duty to protect the environment, are obligations owed by each State toward the international community. The maintenance of the global climate, the conservation of biodiversity or the protection of common heritage areas such as the deep seabed (Article 136, UNCLOS), are also matters that concern all States and to which protection is owed by the international community as a whole.

Some suggest that one possible consequence of the concept of common concern of humankind is that any State is entitled to *locus standi* to enforce environmental rules.[115] In general, only States that have suffered a direct and tangible injury to their legal rights can seek remedies for violations of international law, by litigation or via unilateral or collective countermeasures or forceful reprisals.[116] However, when it comes to *erga omnes* obligations, a State may take remedial actions even

[114] Case *Barcelona Traction, Light and Power Company, Limited (Belgium v. Spain)*, *ICJ Reports* 1970, p. 32.

[115] Baslar, op. cit. p. 232. It is sometimes suggested that a more "liberal approach" to *locus standi* should also be applied with respect to certain environmental damages, particularly massive pollution in common areas (L. Brownlie, "A survey of international customary rules of environmental protection," *Natural Resources Journal*, vol. 13, 1973, pp. 179–89, at p. 157). Others claim that possible "candidates" for obligations *erga omnes* (with its implications in terms of *locus standi*) should be identified rigorously, as obligations *erga omnes*, by their very nature, are few (M. Ragazzi, *The Concept of International Obligations Erga Omnes*, Oxford & New York: Clarendon Press, 1997).

[116] J.I. Charney, "Third state remedies for environmental damage to the world's common spaces," in F. Francioni and T. Scovazzi (eds.), *International Responsibility for Environmental Harm*, London/Dordrecht/Boston: Graham & Trotam, 1991, pp. 149–77, at pp. 158–59.

if it has sustained no particular injury. As foreseen under the International Law Commission's Articles on State Responsibility, any State other than the injured State can invoke the responsibility of other States when the obligation is owed to the international community as a whole.[117] One may question whether a State could exercise a form of *actio popularis* when obligations *erga omnes* are involved. For example, could one State bring an action on behalf of the international community alleging an *erga omnes* obligation, such as a duty to maintain the climate system and, to this end, to protect significant carbon sinks such as the Amazon?

The protection of the environment is a matter that concerns all States. International environmental law rules, by their very nature, reflect the common interests of humanity and, unlike other norms, they do not involve direct or reciprocal advantages to States.[118] The International Court of Justice, in the case concerning the Gabčíkovo-Nagymaros Project,[119] confirmed the view that the protection of the environment is an "essential interest." In view of this, the implications derived from the particular nature of international environmental law norms is debated. According to Philippe Sands, there is doctrinal support to the view that *erga omnes* obligations might extend to environmental damages in areas beyond national jurisdiction and he also suggests that there are equally compelling policy reasons to justify an *actio popularis* in respect to environmental damages occurred in the territory of a particular State, notably in relation to obligations associated with the common concern of humankind and common heritage principles.[120]

In support of the view that an *actio popularis* may be an avenue to address environmental damages, Sands makes reference to the ruling of the World Trade Organization Appellate Body in the case *Import Prohibition of Certain Shrimp and Shrimp Products* as marking a shift toward the recognition of *actio popularis*. In this case, the Appellate Body examined the legality of an import ban imposed by the United States on shrimp harvested with commercial fishing technology that could adversely affect sea turtles. This measure, justified by the United States under Article 20(g) of the 1994 General Agreement on Tariffs and Trade (GATT),[121] was refuted on the grounds that by imposing an import ban, the United States would be applying national laws extraterritorially. However, the Appellate Body did identify a nexus between the migratory and endangered marine populations and US interests

[117] 2001 Articles on the Responsibility of Sates for Internationally Wrongful Acts, adopted by the International Law Commission at its Fifty-third Session, UNGA Res. 56/83, December 12, 2001, and corrected by UN doc. A/56/49, vol. I/Corr.4.

[118] Kiss and Shelton, op. cit., p. 24.

[119] Case *Gabčíkovo-Nagymaros Project (Hungary/Slovakia)*, 1997, *ICJ Reports*, para. 53.

[120] Sands, op. cit., p. 189.

[121] 55 UNTS 194.

for purposes of Article XX(g), para. 33[122] and recognized this country's legitimate interest in protecting migratory and endangered sea turtles. Arguably, this ruling marked a shift from earlier interpretations, such as that made by the GATT Panel, in the *Restrictions on Import of Tuna* case.[123] In this case, the embargo imposed by the United States on countries harvesting yellow-fin tuna without dolphin protection standards set up by the US Marine Mammal Protection Act was considered unjustified under Article XX(g).[124] According to the Panel, an extrajurisdictional interpretation of Article XX(g) would jeopardize the rights of all contracting parties under the GATT, as it would allow a State to determine unilaterally the conservation policies to which other countries would have to be subordinate.[125]

If *actio popularis* will become a tool for protecting areas or natural resources under common concern or common heritage categories, either within or outside national jurisdictions, is presently a merely theoretical discussion. As Sands notes, emblematic examples such as the accident at the Chernobyl nuclear power plant in 1986, where no State sought to enforce compliance by the former Union of Soviet Socialist Republics with international obligations, show that it is unlikely that a State not directly injured would seek to enforce environmental obligations owed to the international community.[126] The use of third States remedies such as *actio popularis* to enforce international obligations is even less likely when it comes to areas or natural resources falling within domestic jurisdiction. Apart from the potential use of third States remedies, what are other possible consequences of a common concern status? In particular, what are the implications in terms of rights and duties in the specific case of the Amazon?

The protection of the Amazon, as suggested here, is a common concern of humankind. However, this principle – unlike that of a common heritage of humankind – does not involve any form of ownership rights by third States over the Amazon and its natural resources. The Amazon States are the only ones to enjoy sovereignty over this region and its natural resources. As stated under the ACT, the exclusive use of natural resources within those countries' respective territories is a right inherent to their sovereignty and this right shall not be subject to any restrictions other than those arising from international law (Article 4). However, the principle of sovereignty over natural resources – apart from generating exclusive use rights, among others – also involves duties. As Nico Schrijver notes, for a long time this principle

[122] Case *Import Prohibition of Certain Shrimp and Shrimp Products (United States vs. India, Malaysia, Pakistan and Thailand), Report of the Appellate Body*, WT/DS58/AB/R, October 12, 1998.

[123] *Restrictions on Imports of Tuna (United States vs. Mexico), Report of the Panel*, 30 ILM 1594, 1991. This panel report was circulated in 1991, but not adopted.

[124] Ibid., para. 5.34.

[125] Ibid., para. 3.31.

[126] Sands, op. cit., p. 191.

has been a grammar of rights, including, for example, the right to dispose freely of natural resources, to expropriate, and the right to compensation for damages to natural resources.[127] This is so because, while formulating this principle, States tended, for obvious reasons, to define rights expanding their sovereignty rather than restricting it, and therefore less attention has been paid to possible duties incumbent on States in the exercise of their permanent sovereignty over natural resources.[128] However, according to Schrijver, that principle also gives rise to a set of duties, including, for example, the duty to respect the rights of indigenous communities, to cooperate, to conserve, and to sustainably use natural resources, among others.

In the context of the Amazon, the rights associated with the principle of sovereignty over natural resources are equally coupled with a set of duties. In fact, the Amazon States seem aware that their sovereign rights over the region also involve duties. The Preamble of the ACT states that the conservation of the environment is a responsibility inherent to the sovereignty of each member State. According to the 2004 Declaration of Manaus, those States have a "sovereign responsibility" in relation to the sustainable development of the Amazon.[129] Likewise, the 2005 Declaration of Iquitos asserts that the Amazon States have sovereignty over their resources, but also a "responsibility to preserve the natural resources."[130]

As affirmed in Principle 21 of the Stockholm Declaration and Principle 2 of the Rio Declaration (and in many other legal instruments), States have a sovereign right to exploit their own natural resources. The exercise of this right is associated with other rights, for example that of permitting the access to genetic resources (Article 15, CBD) or, more generally, of seeking development, which is framed in various UNGA's Resolutions.[131] According to Resolution 2692 (XXV) of December 11, 1970,[132] the "development of permanent sovereignty over their natural resources by

[127] N.J. Schrijver, *Sovereignty over Natural Resources: Balancing Rights and Duties*, Cambridge University Press, Cambridge, 1997, p. 258.

[128] Ibid., p. 306.

[129] 2004 Declaration of Manaus, adopted at the Eighth Meeting of Foreign Affairs Ministers of the Amazon Cooperation Treaty, on September 14, 2004, Article 2, http://www.otca.org.br

[130] 2005 Declaration of Iquitos, adopted at the Ninth Meeting of Foreign Affairs Ministers of the Amazon Cooperation Treaty Organization, November 25, 2005, Article 4, http://www.otca.org.br

[131] UNGA Res. 41/128, December 4, 1986, Declaration on the Right to Development, noted "the need for improved coordination and cooperation throughout the UN system for more effective promotion and realization of the right to development." In order to realize the right to development, according to UNGA Res. 56/150, February 8, 2002, national action and international cooperation must reinforce each other and the realization of the right to development should be conducted in the spirit of a partnership. See also UNGA Res. 52/136, December 12, 1997; UNGA Res. 61/169, February 27, 2007.

[132] UNGA Res. 2692 (XXV), December 11, 1970, on the Permanent Sovereignty over Natural Resources of Developing Countries and Expansion of Domestic Sources of Accumulation for Economic Development.

developing countries is indispensable in order that they may, *inter alia*, accelerated their industrial development."[133]

In terms of duties, the Amazon States have *inter se* (and *vis-à-vis* the international community) an obligation to observe commitments under treaties to which they are parties, as well as customary law rules (*pacta sunt servanda*). For example, under certain bilateral treaties, these States assumed obligations to set up public campaigns, health programs, and vaccination stations in border areas, and to notify others about certain diseases (e.g. the 1972 Treaty on Sanitary Cooperation for the Amazon Region Between Brazil and Colombia).[134] They also have a duty to exchange information on policies, programs, and laws on the conservation of animal and plant life (e.g. the 1979 agreement on the Conservation of Flora and Fauna of the Amazonian Territories between Colombia and Peru).[135] Under the ACT, the parties are expected to achieve the preservation of the environment and the conservation and rational utilization of natural resources (Article 1). Other more specific duties include that of taking national, bilateral, and multilateral measures to improve the navigability of the Amazonian Rivers (Article 6) or to exchange information on conservation measures aimed at the rational use of wildlife species (Article 7(b)). Those States assumed other obligations under multilateral environmental agreements (MEAs), such as the CBD and the UNFCCC. They are requested under the CBD to develop national strategies for the conservation of biodiversity (Article 6) and promote *in situ* conservation through a system of protected areas (Article 8(a)), as well as *ex situ* conservation (Article 9(b)); or, under the UNFCCC, to develop national inventories of anthropogenic greenhouse gas emissions (Article 4) and provide information on national policies (Article 4(2b)). Other types of duties are set up under other MEAs to which the Amazon States are parties. As far as customary law is concerned, those countries are subject, among others, to the basic rule (enunciated in Principle 21 of the Stockholm Declaration and Principle 2 of the Rio Declaration) by which States shall ensure that activities within their jurisdiction or control do not cause damage to other States or areas beyond the limits of their national jurisdiction.

From this duty to prevent damage to the environment of other States or of areas beyond national jurisdiction, other (substantive and procedural) obligations derive. For example, States have a duty to cooperate in order to ensure that their activities do not cause adverse transboundary effects. This is particularly the case in relation to shared ecosystems, as occurs widely in the Amazon. In this regard, the ACT affirms that cooperation is needed to help the parties conserve the Amazon (Preamble)

[133] Ibid.

[134] *Colombian Official Gazette*, Year CXI, no. 34244.28, 1975, p. 186 (Law 3 of 1975).

[135] *Colombian Official Gazette*, Year CXVII, no. 35650, Law 30 of 1980, November 25, 1980, p. 474.

and requires them to maintain regular exchange of information and cooperate *inter se* (Article 15). The duty to cooperate, which is at the basis of most environmental treaties,[136] is made operative when States are requested to enter into consultation, to notify one another about natural disasters or other emergencies (Rio Declaration, Principles 18 and 19) and, in some cases, to conduct environmental impact assessments (Rio Declaration, Principle 17).

Irrespective of any harmful transboundary effects, a State has a duty to preserve its own environment. The principle of environmental protection is the very foundation of international environmental law.[137] A series of UNGA resolutions have recalled the "historical responsibility of States for the preservation of nature for present and future generations."[138] As some commentators assert, international law has evolved to a stage where it is no longer correct to think that States have unlimited authority to injure the environment in any way they like so long as the injury is restricted to their own territory.[139] At the inter-American level, the duty to protect the environment is recalled, for example, in the first Summit of the Americas'[140] 1994 Plan of Action, according to which the American States must "intensify efforts to understand, assess, and sustainably use this living resource base."[141]

These are some of the rights and duties that apply to each Amazon State and to their interactions with other basin States. If now one looks at the relation among the Amazon States, on the one hand, and the international community, on the other,

[136] O. McIntyre, "The role of customary rules and principles of international environmental law in the protection of shared international freshwater resources," *Nature Resources Journal*, vol. 46, 2006, pp. 157–210, at p. 179.

[137] Kiss and Shelton, op. cit., p. 26.

[138] UNGA Res. 35/8, October 30, 1980; UNGA Res. 36/7, October 27, 1981 on the Historical Responsibility of States for the Preservation of Nature for Present and Future Generations. An earlier example is UNGA Res. 2996 (XXVII), December 15, 1972 on the International Responsibility of States in regard to the Environment.

[139] M.J. Glennon, "Has international law failed the elephant?", *American Journal of International Law*, vol. 84, pp. 1–43, 1990, at p. 29. See also S. Atapattu, *Emerging Principles of International Environmental Law*, Ardsley: Transnational, 2006, p. 7, who says: "we have come to a state where international law is actually dictating how states should conduct their international affairs, irrespective of their impact on the global environment."

[140] The Summits of the Americas have been dealing with environmental matters. In particular, a special Summit of the Americas on Sustainable Development held in Santa Cruz, Bolivia in 1996 defined the Summit of the Americas' environmental mandate in a number of areas. The First Summit of the Americas was held in Miami in 1994. To date there have been four Summits in total and two special summits, in 1996 on Sustainable Development and in 2004 in Monterrey, Mexico. The Fifth Summit of the Americas will be held in Trinidad and Tobago. In 2002, the OAS Secretary General signed Executive Order no. 02–03 strengthening the responsibilities of the Office of Summit and changing its name to the Secretariat of the Summit Process.

[141] Para. 22, adopted at the First Summit of the Americas Plan of Action, held in Miami, Florida, December 9–11, 1994; Summit of the Americas, Declaration of Principles and Plan of Action, December 11, 1994, reprinted in 34 ILM 808, 1995.

what rights and obligations may arise from seeing the protection of the Amazon as a common concern?

From the recognition of a legitimate interest of all States to protect the environment, notably the Amazon, it follows that the international community bears a responsibility (imprecise as it may be) to assist the Amazon basin States.[142] This duty is manifested, for example, in the 2007 Non-legally Binding Instrument on All Types of Forests, by which an effective implementation of sustainable forest management is deemed dependent on adequate resources, including financing, capacity development, and the transfer of environmentally sound technologies to developing countries. According to Article 2(d), achieving sustainable forest management, particularly in developing countries, depends on increased, new, and additional financial resources. This agreement requires high-level political commitment in financially assisting these countries[143] and in facilitating access to and transfer of technology and know-how.[144] As The Hague Recommendations on International Environmental Law (Preamble) states, "it should be noted that the preservation of the environment is a common concern of humankind. Therefore, costs should be shared equitably among States, taking into account their responsibilities and capabilities."[145]

The duty to provide assistance to developing countries is recurrent in environmental treaties and UNGA resolutions.[146] In general, these resolutions stress the need to foster international cooperation and involve all States in efforts to protect the environment, given the international dimension of environmental problems and the challenges faced by developing countries in overcoming them. This assistance may take different forms. For example, according to Resolution 1803 (XVII),[147] cooperation should involve capital investment, the exchange of goods and services, technical assistance, or the exchange of scientific information to the benefit of developing countries. It may also occur through the transfer of

[142] F. Biermann, "Common heritage of humankind: The emergence of a new concept of international environmental law," *Archiv des Völkerrechts*, vol. 34, 1996, pp. 426–81, at p. 465.

[143] Article 7(a).

[144] Article 7(k).

[145] Hague Recommendations, summarized in S. Bilderbeck (ed.), *Biodiversity and International Law: The Effectiveness of International Environmental Law*, the Netherlands: IOS Press, 1992, pp. 124–56.

[146] A series of UN resolutions on international cooperation in the field of the environment include UNGA Res. 37/217, December 20, 1982; UNGA Res. 38/165, December 19, 1983, which "bear[s] in mind the involvement of all countries in the protection of the environment" and reaffirms "the need to strengthen international cooperation in the field of the environment, particularly in order to deal with the most serious environmental problems of the developing countries"; and UNGA Res. 40/200, December 17, 1985, which recognizes "the international dimension of environmental problems" and stresses the "importance of an international exchange of experience and knowledge concerning the protection of the environment."

[147] UNGA Res. 1803 (XVII), December 14, 1962, para. 6.

technology;[148] sustainable trade;[149] or, in cases of natural disasters, through the provision of humanitarian assistance.[150]

At the inter-American level, the need to provide assistance to less-developed countries is also recognized. For example, according to the 1996 Declaration of Santa Cruz de la Sierra,[151] the development and transfer of sound technology "plays an important role in ensuring sustainable development."[152] More recently, the 2007 Declaration of Santa Cruz,[153] adopted in a meeting of high-level authorities on sustainable development, urges the international community "to support national development efforts through mobilization of resources, technical assistance, institutional strengthening, and technology transfer."[154]

Developed countries bear a "historic responsibility" to the environment in relation to other countries, as asserted in several UNGA resolutions, as they have so far caused greater damages to the environment compared to developing countries, for example, by emitting pollutants into the environment including toxic and hazardous wastes, and should thus bear the "main responsibility for combating such pollution."[155] Therefore, although all States must "take effective actions for the protection and enhancement of the environment," developed States have the "main

[148] UNGA Res. 42/184, December 11, 1987, on International Co-operation in the Field of the Environment, recognizes "the importance of the international exchange of experience and knowledge and of the promotion of the transfer of technology."

[149] UNGA Res. 2158 (XXI), November 25, 1966 on Permanent Sovereignty over Natural Resources calls upon the developed countries to "make available to the developing countries, at their request, assistance including capital goods and know-how for the exploitation and marketing of their natural resources in order to accelerate their economic development."

[150] UNGA Res. 3340 (XXX), December 9, 1975, dealing with assistance in cases of natural disasters and other disaster situations. International cooperation in situations of natural disasters has also been recognized at the inter-American level by a number of OAS General Assembly resolutions, such as AG/RES. 2024 (XXXIV-O/04), June 8, 2004 on Natural Disaster Reduction, adopted at the Fourth Plenary Session, which urges member States to "strengthen existing bilateral, subregional, and multilateral actions and institutions, such as the Inter-American Committee on Natural Disaster Reduction, and, when possible, by using technology and scientific resources to prevent the occurrence of natural disasters, as well as taking adaptive measures to mitigate their effects in order to avoid or reduce damage to the environment."

[151] Adopted at the Summit of the Americas on Sustainable Development, held in Santa Cruz de la Sierra, Bolivia, December 7–8, 1996, available at: http://www.summit-americas.org/

[152] 1996 Declaration of Santa Cruz de la Sierra, Article 7(e).

[153] Also known as the Declaration of Santa Cruz + 10, it was adopted at the First Inter-American Meeting of Ministers and High Level Authorities on Sustainable Development, held in Santa Cruz de la Sierra, Bolivia on December 4–5, 2006; OEA/XLIII.1, CIDI/RIMDS/DEC.1/06 rev. 1, January 16, 2007, http://www.summit-americas.org/

[154] 2006 Declaration of Santa Cruz, para. 48.

[155] UNGA Res. 42/229, December 22, 1989 on International Co-operation in the Field of the Environment notes that "the largest part of the current emission of pollutants into the environment, including toxic and hazardous wastes, originates in developed countries" and recognizes that "those countries therefore have the main responsibility for combating such pollution."

responsibility for taking appropriate measures urgently."[156] As noted, technical and financial assistance from the international community will serve different purposes. For example, in relation to tropical forests, it may help create natural resources inventories (as data is usually insufficient), preserve certain areas by supporting better infrastructure (e.g. guard posts, light aircraft landing strips, and tourists facilities), or achieve sustainable agroforestry and harvesting methods.[157]

The obligation to provide assistance, sometimes framed as a "duty of international environmental solidarity,"[158] is also reflected in several international environmental instruments. This duty finds justification in the different contributions of States to actual environmental degradation, but also takes account of their different capabilities. As Frank Biermann notes, by adopting national policies to protect the environment, developing countries assume significant financial burdens and usually lack adequate technologies. Because of this, they should ground their claims for assistance on the duty of developed countries to financially and technologically support their environmental policies insofar as the common concerns of humankind are affected. Biermann further suggests that a customary duty of "international environmental solidarity" can be derived from recent State practice accepted as law, particularly from consistent patterns in widely ratified treaties on ozone depletion, climate change, and biodiversity. Therefore, at least these environmental matters of common concern should involve a duty on the part of developed countries to provide financial and technological assistance.

In practice, the international community has provided assistance to developing countries for the protection of the environment and the Amazon in particular. For example, international organizations and governments have financed projects carried out in the context of the ACT (e.g. the IDB, the Global Environment Facility (GEF), the UN Conference on Trade and Development (UNCTAD), and the governments of Germany and the Netherlands, among others). In the Brazilian Amazon, the Pilot Program to Conserve the Brazilian Rain Forest (PPG7), announced by the G7 countries meeting in Houston in 1990 and launched at the 1992 UNCED, illustrates how international cooperation may help achieve conservation practices in the Brazilian Amazon and *Mata Atlantica*.[159] The provision of technical and financial assistance is the most obvious duty of the international community *vis-à-vis* the Amazon States. The following section examines what possible rights the

[156] UNGA Res. 45/94, December 14, 1990 addresses the need to ensure a health environment for the well-being of individuals.

[157] Miller et al., op. cit., p. 109.

[158] Biermann, op. cit., p. 465. On the "principle of solidarity," see Schrijver et al., op. cit., p. 289.

[159] Under this program, a fund entitled Fundo Fiduciario de Florestas Tropicais, administered by the World Bank, was created to canalize the contributions of the program's donors. More information about the PPG7 is available at: http://www.mma.gov.br/ppg7/

international community may have if the protection of the Amazon is viewed as a common concern of mankind.

9.3.2. A *Right of Surveillance in the Protection of the Amazon*

As noted above, the preservation of the Amazon can be considered a common concern of humankind. According to Abi-Saab, concepts such as those of a common heritage or common concern of humankind manifest interests of a superior value, which transcend States taken individually and reflect a legitimate interest of the international community as a whole. In the context of the international law of cooperation, discussed at the introduction of this study, those interests of a superior value give rise to a right of surveillance (*droit de regard*), exercised by the community of States.[160]

The exercise of a right of surveillance is observed especially in the field of international human rights. For example, in the inter-American system a *droit de regard* is manifested in on-site visits and rapporteurship activities carried out by the Inter-American Commission on Human Rights (IACHR) since the 1960s.[161] Under the OAS, the IACHR is the organ in charge of promoting the observance and defense of human rights.[162] With the consent, or at the invitation, of a particular government the IACHR can conduct on-site observations and make recommendations on the adoption of measures in favor of human rights.[163] In particular, under a mechanism known as "special procedure" the IACHR can address specific country situations or thematic issues. A special procedure is created by resolution and provides the mandate to designated people (mandate holders) to examine, monitor, advise, and report on human rights situations in specific countries and territories (country mandates) as well as on situations of major human rights violations (thematic mandates). For this end, an individual ("special rapporteur," "special representative," or "Independent Expert") or a few people are mandated under a working group. Activities carried out under a special procedure may include addressing individual complaints, conducting studies, and providing technical advice.[164]

160 G. Abi-Saab, *Cours général de droit international public*, The Hague: Martinus Nijhoff, 1987.
161 http://www.cidh.oas.org/what.htm. List of country reports available at: http://www.cidh.oas.org/pais.eng.htm
162 Article 1, Statute of the Inter-American Commission on human rights.
163 Article 18(g), Statute of the Inter-American Commission on human rights, approved by Resolution no. 447 taken by the General Assembly of the OAS at its ninth regular session, held in La Paz, Bolivia, October 1979.
164 http://www2.ohchr.org/english/bodies/chr/special/index.htm

For example, in March 2008 the Human Rights Council appointed a special rapporteur on the situation of human rights and fundamental freedoms of indigenous peoples in the Americas, whose attributions are defined in Resolution 6/12 of the Human Rights Council of September 28, 2007. Among other functions, the Special Rapporteur is expected to "gather, request, receive and exchange information and communications from all relevant sources, including Governments, indigenous people and their communities and organizations, on alleged violations of their human rights and fundamental freedoms" (Article 1, b).[165] In practice, the Special Rapporteur investigates and reports on the human rights situation of indigenous peoples in selected countries. The outcome is a report containing recommendations to improve the human rights conditions of indigenous peoples in a specific country. The reporting process typically involves visits to the country under review, notably to places of concern, which gives an opportunity to the Special Rapporteur to interact with government representatives, indigenous communities, and civil society.[166] The current Special Rapporteur has monitored the situation of human rights and fundamental freedoms of indigenous people in countries such as Brazil,[167] Peru,[168] and Colombia.[169]

The monitoring of human rights violations in a particular country, notably on-site investigations, under special procedures can only occur with the consent and cooperation of the concerned government.[170] To assess the situation of human rights in a country, mandate holders typically send a letter to the government in question requesting to visit the country, and, if the government agrees, an invitation is extended.[171] In order to monitor the situation of human rights and the fundamental freedoms of indigenous peoples, States are requested to "give serious consideration to the possibility of inviting the Special Rapporteur to visit their countries so as to enable him/her to fulfill the mandate effectively."[172]

A *droit de regard*, at least in the international human rights field, seems to involve the ability attributed to the international community, which can be represented by a designated individual or a working group, to monitor the situation of human rights in a country, for example by conducting on-site investigations. What justifies this

[165] J. Anaya, *Report of the Special Rapporteur on the Situation of Human Rights and Fundamental Freedoms of Indigenous People: Report on the Situation of Human Rights of Indigenous Peoples in Brazil*, A/HRC/12/34/Add.2, Human Rights Council, twelfth session, August 26, 2009

[166] Ibid.

[167] Ibid.

[168] Ibid., Addendum Report on the Situación de los Pueblos Indígenas de la Amazonía y los Sucesos del 5 de Junio y Días Posteriores en las Provincias de Bagua y Utcubamba, Perú.

[169] Anaya, op. cit.

[170] Ibid.

[171] http://www2.ohchr.org/english/bodies/chr/special/visits.htm

[172] Human Rights Council Resolution 6/12 of September 28, 2007.

external monitoring is the existence of a matter that concerns the international community as a whole, such as potential human rights violations. The exercise of a *droit de regard* is, in any case, fairly limited. It is conditioned and limited to a mandate, such as those conferred under the IACHR special procedures, and depends on State consent. However, if consent is given, the international community does have the ability to monitor specific issues, for example human rights and fundamental freedoms of indigenous peoples, pertaining in principle to the national jurisdiction of a particular State. By monitoring the situation of human rights in particular countries and by making policy recommendations the international community exercises a *droit de regard*.

The exercise of a *droit de regard* may go beyond monitoring practices and involve the use of remedies such as *actio popularis*. The awareness among States that international law obligations relating to aggression, basic human rights, and the environment are of general interest may involve a *droit de regard* that would permit States to express their concern regarding actions of others that have adverse repercussions on the community of States and its system of norms.[173] This right could be the basis of an *actio popularis* enabling a State to initiate judicial action in the common interest. In practice, those legal concepts may be abused by States acting in their own interest, and therefore it may be necessary to develop procedures in the political bodies and possibly in judicial organs to deter or limit the possibility of abuse.[174]

A *droit de regard* is said to exist, for example, in relation to the protection of the human genome.[175] More generally, it is claimed that the 1948 Universal Declaration of Human Rights gave members of the United Nations a *droit de regard vis-à-vis* the way other members were dealing with human rights in their territories. The fact that the Security Council has repeatedly approved the involvement of the United Nations in helping to resolve internal conflicts is an illustration of how such a right is exercised.[176] In this context, the UN Commission on Human Rights and the Human Rights Council – created to further the commitment under the 1945 Charter of the United Nations to "promoting and encouraging respect for human rights" (Article 1) – have both exercised a "broad, Charter-based *droit de regard* over human rights abuses regardless of whether they violated the treaty obligations of any particular state."[177] One means by which the Council, in particular, has exercised this right is

[173] O. Schachter, "International law in theory and practice," *Recueil des cours*, vol. 178, 1982, pp. 200–01.

[174] Ibid.

[175] M.-J. Bernardi, "Diversité génétique humaine: Éléments d'une politique," *Revue Juridique Themis*, vol. 35, 2001, pp. 327–405, at p. 400.

[176] Gregory F. Maggio, 'Implementation, compliance and effectiveness: International law year in review," *American Society of International Law Proceedings*, vol. 91, 1997, pp. 132–57.

[177] Simma and Alston, cited in ibid., footnote 88.

by establishing special procedures in relation to a country and a particular theme. Some view the *droit de regard* that entitles the UN to respond to gross violations of human rights in a wide variety of ways as "firmly established in customary international law,"[178] whereas others refer to it as a principle. [179]

In the international environmental law field, issues such as biodiversity conservation, climate change, the management of some parts of the sea, and the preservation of natural and cultural world heritage sites also seem to concern the international community as a whole. However, in this field mechanisms of surveillance are different than those used in the field of international human rights. The ability of the international community to monitor the situation of a particular country and, by this means, exercise a *droit de regard*, is less evident when it comes to the protection of the environment. It should also be noted that a right of surveillance has not been framed as such under any MEA, regional, or bilateral treaties among the Amazon States in particular. In any case, it can be argued that a right of surveillance is being developed in international environmental law.

The international community does have common concerns over a number of issues, such as those referred to above, but some seem to be more sensitive than others. For example, the international community has expressed concern over tropical deforestation and its impacts on climate change. However, although forest-rich countries (mostly in the Southern hemisphere) tend to agree that international assistance is required for reducing deforestation, they reject – as one could expect – any form of interference in how they manage their own resources. The disagreement between developing countries and many developed States was noted for example at the 1992 UNCED. Although most developed countries pushed for a global treaty on forests, developing States stood on their sovereign right to exploit their own forests and rejected the idea of a legally binding treaty.[180] Those disparate views surfaced at the 1992 UNCED are reflected in the extracts below:

> Statement by the Vice-President of the Dominican Republic
>
> The developed world achieved economic growth by exploiting vast environmental resources that were not preserved. Now, we countries of the so-called third world are being asked to preserve what we have. But let us be frank: this request has an inevitable economic cost. Latin America, for example, will not be able to stop exploiting

[178] Ibid.

[179] A. Korkeakivi, "Consequences of 'higher' international law: Evaluating crimes of state and *erga omnes*', *Journal of International Legal Studies*, vol. 2, Summer 1996, p. 104. Korkeakivi refers to a "principle of droit de regard" in the field of human rights.

[180] At the outcome of the UNCED, States managed to adopt no more than a non-legally binding statement on forests. Non-legally Binding Authoritative Statement of Principles for a Global Consensus on the Management, Conservation and Sustainable Development of all Types of Forests, UN doc. A/CONF.151/26, vol. III, August 14, 1992.

> its natural resources in order to preserve such lungs of the planet as Amazonia if rapid alternatives are not provided for our impoverished masses. We cannot ignore the cost of this if we want to control deforestation effectively. Let no one be fooled.[181]
>
> Statement by the President of the European Commission
>
> The Community regrets the absence of a convention on forests, where these are central to the concept of sustainable development and an essential aspect of its implementation. In this connection, let me recall that the pilot program to safeguard the Amazon forest, established by the host country of this Conference with the assistance of the European Community and the World Bank, shows that sustainable forest management can be implemented without impairing the development capacities of forest countries – quite the opposite, in fact.[182]

In practice, however, the international community does exercise, somehow, a right of surveillance over natural resources and sites, even when under national jurisdiction. In any case, as the existence of a *droit de regard* is based on State consent, its exercise cannot amount to any form of undue interference in domestic affairs, which has been condemned under international law. However, a *droit de regard* seems to be exercised at least under some multilateral environmental agreements (MEAs) and regional treaties. Under MEAs, the international community, represented by the member States, exercises surveillance equally conferred to all parties. Most of those treaties foresee regular meetings or conferences of the parties (COPs) to keep the treaty implementation under review. The parties are usually requested to submit, among other information, national reports on measures taken nationally to meet the treaty objectives. This information should provide a sense of how well the parties manage natural resources or sites protected under the treaty, which may include common heritage areas such as the seabed or ocean floor, but also those under national jurisdiction, for example sites under the World Heritage List or wildlife species listed under CITES. This information is considered and shared in the context of those regular meetings and based on this COPs may adopt decisions to make a treaty work more effectively. By this means, the parties are given the ability to collectively monitor the state of natural resources or sites (including those under national jurisdiction) and to assess States' performance. Noncompliance mechanisms, created for example under the 1987 Montreal Protocol on Substances that Deplete the Ozone Layer[183] and the Kyoto Protocol[184] to supervise States'

[181] Statement by the Vice-President of the Dominican Republic, Mr. Carlos Morales Troncoso, UN doc. A/Conf-151/26, vol. 3, p. 150.

[182] UN doc. A/Conf-151/26, vol. 3, Statement by the President of the European Commission (Mr. Jacques Delor), p. 218.

[183] Annex IV, 26 ILM 1529, 1987.

[184] Decision 24/CP.7/, FCCC/CP/2001/13/Add.3, adopted at the Seventh Conference of the Parties of the UN Framework Convention on Climate Change, November 10, 2001.

performance (and, in some cases, assist them in complying with the treaty) are also a manifestation of a *droit de regard*.

The possibility of exercising a *droit de regard* can be noted for example in the framework of the 1982 UNCLOS. Under this convention a port State is given police and judicial powers to handle cases of vessel-source marine pollution.[185] Therefore, when a vessel suspected of having caused pollution is within a State's port (or at one of its offshore terminals or in its internal waters), the port State can undertake investigations and, when there is evidence, institute proceedings to punish illegal discharges.[186] In a more novel fashion, port States may also conduct investigations and initiate proceedings in relation to violations committed outside its internal waters, territorial sea, or economic exclusion zone.[187] In principle, this competence should not be exercised in the zones of jurisdiction of other States, except when the violation threatens to cause pollution to zones under the jurisdiction of the port State. In any case, port States are given, in some circumstances, the right to monitor and safeguard the marine environment, even in areas not subject to its national jurisdiction. By being allowed to do so, the port State can exercise a right of surveillance on behalf of the collectivity of States under UNCLOS. Other environmental treaties also foresee the possibility of inspection and investigation procedures. For example, under the 1991 Antarctic Protocol, designated observers are given free access to all areas of the Antarctic, including to stations, installations and equipments, and to all ships and aircrafts.[188] The 1980 Convention on the Conservation of Antarctic Marine Living Resources envisages a system of observation and inspection by designated observers,[189] which may lead to the imposition of sanctions.

As noted, a right of surveillance may be exercised not only in relation to common areas, but also with regard to natural resources and sites within the domestic jurisdiction of a particular State. Except for areas of a particular nature such as the high seas, Antarctica, and the Outer Space (or the "global commons"),[190] most ecosystems

185 The port State, by virtue of its geographical situation and of its nature, is itself a coastal State. The quality of being a port State is circumstantial and not substantial: every coastal State may come to play the role of a port State (R.-J. Dupuy and D. Vignes (eds.), *Traité du nouveau droit de la mer*, Paris: Economica, 1985, p. 1207). It should be noted that the reference to port States is not unprecedented. The 1973 Convention for the Prevention of Marine Pollution from Ships had already referred to "port States" in Articles 5(3) and 6(2).

186 UNCLOS, Article 218.

187 Ibid.

188 Article 7(1).

189 Article 24.

190 The 1980 World Conservation Strategy mentioned Antarctica and its status as a "global commons" in its efforts to ensure that economic development is coupled with sound conservation practices. It defined a "global commons" as "parts of the earth's surface beyond national jurisdictions – notably the open ocean and the living resources found there – or held in common notably the atmosphere" (J. Grob, "Antarctica's frozen territorial claims: A meltdown proposal," *Boston College International and Comparative Law Review*, vol. 30, 2007, p. 478).

are in the internal territory of a country and decisions regarding their use are within the nation's discretion as a sovereign.[191] However, some form of surveillance can still be exercised in relation to resources and areas under national jurisdiction. For example, the ecological character of wetlands protected under the 1971 Convention on Wetlands of International Importance Especially as a Waterfowl Habitat (Ramsar Convention) are regularly monitored by COPs and other meetings.[192]

The ability of States to exercise a *droit de regard*, in relation to natural resources and sites under national jurisdiction is even more evident under the 1972 Convention Concerning the Protection of the World Cultural and Natural Heritage. For example, in 2007, a delegation from the World Heritage Committee (WHC) visited the Galapagos Islands, a world heritage site since 1978, which illustrates the overseeing capacity of that organ.[193] Since the mid-1990s, the WHC had been aware, via "reactive monitoring reports,"[194] of threats to the conservation of the islands caused by an expanding tourist industry and a boom in the size of the resident human population. The Galapagos Islands were considered for inclusion in the Danger List, but the Ecuadorian government resisted this step and, in 1998, enacted a special law on the preservation and sustainable use of Galapagos. However, subsequent reactive monitoring reports indicated that the special law was proving hard to implement. Finally, a request was made for Ecuador to invite a joint International Union for Conservation of Nature (IUCN)/WHC mission to participate in a multistakeholder meeting to develop a conservation program for the inlands. The joint mission took place in April 2007. Despite Ecuador's attempt to take remedial measures, such as the suspension of some tourism permits, the Galapagos Islands were eventually included in the List of World Heritage in Danger in 2007.

Likewise, in 1999 IUCN reported that Komodo National Park in Indonesia was subject to increases in illegal dynamite and cyanide fishing, causing damage to the

[191] Tarlock, op. cit., p. 1004.

[192] Article 6.

[193] E.J. Goodwin, "The World Heritage Convention, the environment, and compliance," *Colorado Journal of International Environmental Law and Policy*, vol. 20, 2009, pp. 176–78.

[194] The implementation of the World Heritage Convention is mainly supervised by the World Heritage Committee and by the General Conference through periodic reporting and reactive monitoring. Periodic reporting means a six-year cyclical review of States Parties' policies and legislation, as well as the organization, management, and conservation of the world heritage sites situated in the prevailing territories. It should provide an assessment of the application to the Convention by the State party and also an analysis of whether the outstanding universal value of the properties inscribed on the World Heritage List is being maintained over time. Reactive monitoring takes place in reference to properties that are under threat, which means that they are inscribed, or plan to be inscribed, on the List of World Heritage in Danger, and in the procedure for the eventual deletion of properties from the World Heritage List. It is primarily a policy guidance tool, aimed at providing benchmarks, orientations, and deadlines to the actions of the States parties (D. Zacharias, "The UNESCO regime for the protection of world heritage as prototype of an autonomy-gaining international institution," *German Law Journal*, vol. 9, 2008, pp. 1833–64, at p. 1856).

coral reefs in the World Heritage Site. Consequently, WHC requested Indonesia's permission to conduct a monitoring mission to the park in order to assess the damage and review the management of the site. Although the Indonesian government had initially proposed to set up its own mission, a joint IUCN/United Nations Educational, Scientific and Cultural Organization (UNESCO) mission was ultimately given access to the park.[195] In other cases, some of the Committee's requests received quick responses. For example, the City of Dresden and Elbe Valley was listed as a world heritage site in 2004. A few years later the city municipality initiated planning to build a motorway bridge over the river and into the heart of the city. The WHC warned Germany that this could lead to the removal of the site from the World Heritage List and, as a result, the city council agreed to stop imminent construction and reviewed the project a few days later.[196]

It should be noted that in the environmental field, international, and nongovernmental organizations (NGOs) have also assisted the community of States to exercise a right of surveillance over natural resources, by monitoring the state of the environment and States' compliance under particular MEAs. A State may as well request the assistance of international organizations to help handling environmental problems in their own territories. For example, at the request of the Congolese Institute for the Conservation of Nature, the UNESCO World Heritage Centre and IUCN conducted a joint mission to the Democratic Republic of Congo (DRC) in 2007 to investigate, in cooperation with DRC authorities, the reasons for the slaughtering of mountain gorilla populations at the Virunga National Park located in the northeast of DRC near the borders of Uganda and Rwanda, a world heritage site since 1979.[197] Also, in 1999, Global Forest Watch conducted an independent monitoring in Cameroon aimed at improving logging practices and law compliance.[198] Forest-rich countries usually lack the capacity to monitor protected areas and to control illegal logging or other practices due to problems such as weak governance and lack of technical and financial capacity. In view of this, some governments have established agreements with independent monitors in particular for

195 Goodwin, op. cit., p. 178.

196 Ibid., p. 179.

197 The World Heritage Centre under the "reactive monitoring" mechanism also conducts reporting to the Committee on the state of conservation of specific World Heritage properties under threat and undertakes on-site visits. Some examples reported at the Thirty-first Session of the World Heritage Committee, Christchurch, New Zealand, June 23 to July 2, 2007 include the report on *Reactive Monitoring Mission To Fertö/Neusiedlersee Cultural Landscape* in Austria and Hungary on February 25–26, 2007, or the report on the *Reactive Monitoring Mission to the Group of Monuments at Hampi (Karnataka, India)*. The documents related to the Thirty-first Session of the World Heritage Committee are available at: http://whc.unesco.org

198 H. GreGersen and A. Contreras, *Rethinking Forest Regulations: From Simple Rules to Systems to Promote Best Practices and Compliance*, Washington, DC: Rights and Resources Initiative, 2010, p. 10.

the surveillance of harvesting operations. To this end, some institutions such as the Overseas Development Institute, Global Witness, among others, have been helping governments to create harvest verification mechanisms and operate independent forest harvest-monitoring systems.[199]

The examples above are an indication that a *droit de regard* is being developed in the international environmental field. It has been exercised by the member States under MEAs or by a State party representing a collectivity of States, as observed under the 1982 UNCLOS. This right is exercised by monitoring the state of the environment and by evaluating States' performance under MEAs or other instruments. A *droit de regard* can be exercised with respect to areas and natural resources located both within or outside national jurisdictions. Beyond the ability to monitor natural resources and sites internationally protected, the community of States may take decisions, propose actions, and possibly use remedies such as *actio popularis* in some cases.

As a common concern of humankind, the protection of the Amazon also gives rise to a *droit de regard* on the part of the international community. In practice, this right has been exercised to some extent in the context of MEAs to which the Amazon States are parties, for example the CBD and CITES. In both cases, the parties are requested to submit information on legislative, regulatory, and administrative measures taken at the national level, which is examined under regular meetings.[200] Information is made available to member States by instruments such as the CBD Clearing-House Mechanism.[201]

The exercise of a right of surveillance, however, is limited – even more so with regards to natural resources and sites under national jurisdiction – and dependent on State consent, in line with the principles of sovereign equality and non-intervention.[202] One may question whether, particularly in the case of tropical forests, a regime limited to a right of surveillance is sufficient and whether there would be any space for an international management regime,[203] for example, such as that foreseen in global commons, involving an international authority. As the

[199] Ibid., p. 26.

[200] CBD, Article 26 and 1974; Convention on International Trade of Endangered Species of Wild Flora and Fauna, Article 7(a), 993 UNTS 243, 1974.

[201] Decision I/3, adopted at the First Conference of the Parties of the 1992 Convention on Biological Diversity, Nassau, November 28 to December 9, 1994, which created the Clearing-House Mechanism for technical and scientific cooperation to implement the provisions of Article 18, para. 3 of the CBD.

[202] Chapter 1, 1974 Charter of Economic Rights and Duties of States, A/RES/29/3281, December 12, 1974; Articles 2(4) of the 1945 United Nations Charter

[203] As questioned by L. Boisson des Chazournes, "Le droit international et la promotion des intérêts collectifs: à propos de la protection de l'environnement global," *Publications de la revue Marocaine d'administration local et de développement*, vol. 48, 2004, pp. 197–219.

international community witnesses the gradual destruction of the Amazon, one wonders whether an alternative model should be used, involving, for example, a more active participation of the international community. As Fearnside notes, the rapid pace of deforestation and other forms of destruction are closing off opportunities for conservation of the Amazon and for sustainable use both inside and outside of conservation units.[204]

In view of those concerns, proposals to protect the Amazon in view of its "public good nature"[205] are sometimes debated. For example, a French president (Francois Mitterand) suggested the renunciation of sovereignty over certain areas for the sake of the world community.[206] Others suggest the creation of a common heritage authority in the Amazon to manage the region in cooperation with the Amazon States.[207] Ideas such as those feed a widespread belief, at least in Brazil, in a "permanent conspiracy" to attack Brazilian sovereignty over the Amazon.[208] As one could expect, international schemes such as that proposed by the British environment secretary (David Miliband) in 2004 of a privatization of the Amazon for the purpose of tackling climate change are rejected by the Amazon States. The Brazilian minister of foreign affairs (Celso Amorim) reacted to that proposal – at the Eighth Meeting of Ministers of Foreign Affairs, held in 2004 in the framework of the 1978 Amazon Cooperation Treaty – by stating: "Amazonia is ours."[209] At the outcome of that ministerial meeting, the 2004 Declaration of Manaus[210] reaffirmed the ACT parties' sovereignty over the Amazon,[211] but, at the same time, referred to a "sovereign responsibility" with respect to the environmental protection of the Amazon.[212] As stated by the Brazilian minister of foreign

[204] Fearnside, op. cit., p. 775.

[205] M. McClearly, "The international community's claim to rights in Brazilian Amazonia," *Political Studies*, 1991, pp. 691–707, at p. 692.

[206] J. Goldemberg and E.R. Durham, "Amazonia and national sovereignty," *International Environmental Affairs*, vol. 2/1, 1992, pp. 22–39, at p. 24.

[207] Baslar, op. cit., p. 146.

[208] Fearnside, op. cit., p. 790. According to this author, a sociological survey of the population in Brazilian Amazon revealed that 71 percent of respondents agreed with the statement "I am afraid *Amazonia* will be internationalized," and 75 percent agreed that "Foreigners are trying to take over *Amazonia*." For example, during a series of public hearings the Amazonas State Legislature in October 1999 denounced the PP-G7 ecological corridors project as a trick to internationalize the region. This study was conducted by L.C. Barbosa, "The people of the forest against international capitalism," *Sociological Perspectives*, vol. 39, no. 2, 1996, pp. 317–32, quoted in Fearnside, op. cit., p. 790.

[209] Celso Amorim, "A Amazônia é nossa," *Jornal Hora do Povo*, September 17, 2004, p. 3, http://www.horadopovo.com.br

[210] Available at: http://www.otca.org.br

[211] Para. 3.

[212] Para. 2.

affairs on that occasion, the eight Amazon States are its "sovereign guardians"[213]. More recently, when commenting on the 2008 Sustainable Development Plan for the Amazon, the Brazilian president (Luiz Inácio Lula da Silva) said with regard to the Amazon: "We think that it needs to benefit everyone. But we also have to say loudly and clearly that it is Brazil that is in charge of looking after the Amazon."[214]

Any mechanism or form of international management, involving a supranational authority with (administrative, legislative, or executive) powers to manage the Amazon is unrealistic. At the most, what exists in the Amazon at present is an incipient "regionalization" that occurs under the ACTO. This organization has international legal personality to represent all Amazon States and is competent to adopt norms and policies applicable basin-wide. As noted above, there are two major implications from regarding the Amazon as a common concern of humankind. Whereas the international community has a legitimate interest in the protection of the Amazon, it also bears a responsibility together with the Amazon States, to preserve it. However, the Amazon, unlike other areas of the world, is under the jurisdiction of eight countries, and they have the primary role in ensuring its conservation and should be the first to benefit from its sustainable use.

CONCLUSIONS

The protection of the Amazon should be considered a common concern of humankind. This means that the Amazon countries in particular have an obligation owed to the international community to preserve the Amazon, while maintaining exclusive sovereign rights over their respective Amazonian territories. The international community, on the other hand, has a legitimate interest in the protection of this region, from which derives a right of surveillance, and, at the same time, a duty to provide financial and technical assistance to the Amazon States. Therefore, the international community plays an important role in the protection of the Amazon, although a limited one. The Amazon States have the primary responsibility for ensuring that the international commitments assumed under bilateral, regional, and multilateral treaties, as well as under customary international law, are observed, while they also enjoy exclusive rights over the Amazon and its natural resources.

A right of surveillance (or *droit de regard*) exercised by the international community in relation to natural resources or areas falling within national jurisdiction

[213] Amorim, op. cit.

[214] R. Giraldi, "Lula diz que Amazônia é da humanidade e deve ser cuidada pelo Brasil," Folha Online, Brasília, May 8, 2008.

is dependent on State sovereignty and finds limitations imposed by the principle of national sovereignty over natural resources. The exercise of a right of surveillance – which can be already observed particularly in the context of MEAs – does not (and should not) amount to any form of interference in the domestic affairs of the Amazon States in particular, which is in any case prohibited under international law.

10

General Conclusions

This study has endeavored to assess how the current situation of rapid deforestation and environmental degradation of the Amazon could be handled through enhanced international cooperation among the Amazon States *inter se*, and between them and the international community. To examine what forms of cooperation exist (and would be required) among the basin States and, in order to understand what role the international community can play in the international legal protection of the Amazon, the study attempted to systematize existing legally binding and nonbinding instruments adopted by the Amazon States *inter se*, and between them and other States, applicable to the Amazon as a whole or parts of it. These legal instruments, as well as some institutions working toward the protection of the Amazon at different levels (subregional, regional, and global), in which the Amazon States participate, comprise the proposed "international legal system of the Amazon." The overview of the existing international legal instruments and institutions at different levels provided an indication of how the Amazon States have cooperated *inter se*, with other States in the region and, more broadly, with the international community. It has also revealed some of the normative and institutional weaknesses within this system, which, if addressed, could render it more robust and effective, and ultimately contribute to tackling current environmental problems.

10.1. COOPERATION AMONG THE AMAZON STATES *INTER SE*

Due to a combination of factors, the international system of the Amazon is fragile. The existing legal instruments contain a general duty to cooperate in various fields, such as the protection of wildlife species, the prevention of forest fires, and the improvement of health and sanitation, but they lack more precise obligations requiring States to act in a certain way or take specific measures. That general duty to cooperate can be made concrete through more specific commitments, for example requiring the investigation of environmental problems, consultation, and notification, among others. In terms of procedural obligations, the Amazon States are basically required to exchange information and coordinate activities, especially in border areas. The institutions created by existing legal instruments usually take the form of binational working groups or commissions constituted by designated national authorities with limited powers to oversee the implementation of joint activities and insufficient capacity to provide administrative support. Therefore, with some exceptions, for example certain decisions adopted under the aegis of the Andean Community, international treaties adopted by the Amazon States *inter se* offer limited means and little practical guidance as to how to achieve the environmental objectives defined in them.

The international legal system of the Amazon could be strengthened if the treaties adopted among the States concerned are made more effective, involving more precise commitments and stronger institutions. A number of areas can be the object of (bilateral or regional) cooperation, some of which are already covered by existing treaties, although very little seems to be done in practice, as they foresee little more than the exchange of information. Areas where (bilateral or regional) cooperation is required include, for example, the combating of forest fires and illegal trade of wild species; the management of migratory species, shared ecosystems, international rivers, and transboundary protected areas; and issues such as transboundary water pollution, among others. A potential area of South–South cooperation, much needed in the Amazon, relates to improving technology and national capacity to monitor deforestation, measure carbon stocks and emissions from deforestation, and formulate forest inventories, etc. Among the Amazon basin States, Brazil has developed the most advanced system to monitor deforestation in the Amazon through the National Institute of Spatial Research. By developing such technology and capacity, the Amazon States would be better prepared to use certain mechanisms, such as the Clean Development Mechanism (CDM), engage in initiatives for Reducing Emissions from Deforestation and Forest Degradation (REDD), and implementing payment for environmental services (PES). In this context, one may question what could make the existing treaties adopted among the Amazon States, including the 1978 Amazon Cooperation Treaty (ACT), more effective in areas where (bilateral or regional) cooperation is required.

A quick answer to this question would be that a treaty is made effective when States manage to fulfill the obligations defined in it. However, albeit to some extent related, effectiveness is no synonym for compliance.[1] A treaty is said to be effective

[1] The term "compliance" is defined under the UNEP Guidelines on Compliance with and Enforcement of Multilateral Environmental Agreements as the "fulfillment by the contracting parties of their obligations under a multilateral environmental agreement and any amendments to the multilateral environmental agreement" (UNEP [United Nations Environment Programme], *Manual on Compliance with and Enforcement of Multilateral Environmental Agreements*, Nairobi: UNEP, 2006, para. 9).

when it is able to change States' (and individuals') behavior in such a way as to eliminate or substantially ameliorate the problem that led to its creation.[2] Therefore, States can comply with a treaty that may still be ineffective in attaining its objectives, for example, in cases where it is unable to eliminate or ameliorate the problem that led to its adoption or when obligations laid down by the treaty are inadequate. Based on evidence brought by different cases studies, it is suggested that there is even an inverse relationship between effectiveness and compliance. For example, in the case of the North Sea regime, it was noted that nonbinding standards that have not enjoyed perfect compliance were more effective than earlier binding standards where formal compliance was much higher.[3]

Conventional wisdom also holds that the most effective commitments are legally binding.[4] However, this seems not to be the case, especially in an area prolific in nonlegally binding instruments such as that of international environmental law. Some authors claim that treaties do not even need to rely exclusively on norms to achieve behavioral changes, as other means are available for attaining that purpose.[5] It is indeed noted that since the 1980s States have opted for nonbinding commitments rather than international obligations defined under legally binding treaties likely to restrict their freedom.[6] It is noted that over the years international law has been subject to changes toward a less stringent law permeated by various processes

2 O.R. Young and M.A. Levy (eds.), *The Effectiveness of International Environmental Regimes: Causal Connections and Behavioral Mechanisms*, Cambridge, MA & London: MIT Press, 1999, p. 2. These authors refer to "environmental regimes," defined as "social institutions consisting of agreed upon principles, norms, rules, procedures, and programs that govern the interactions of actors in specific issue areas" (p. 1). Similarly, effectiveness is defined as the "degree to which international environmental accords lead to changes in behavior that held solve environmental problems" (D.G. Victor, Kal Raustiala, and E.B. Skolnikoff (eds.), *The Implementation and Effectiveness of International Environmental Commitments: Theory and Practice*, Cambridge, MA & London: MIT Press, 1998, p. 1).

3 Ibid., p. 7. Two treaties were established in the early 1970s to address North Sea pollution: the Convention for the Prevention of Marine Pollution by Dumping from Ships and Aircraft (Oslo Convention) and the Convention for the Prevention of Marine Pollution from Land-Based Sources (Paris Convention), together known as OSPAR.

4 Ibid., p. 685.

5 A. Chayes, A. Handler Chayes, and R.B. Mitchell, "Managing compliance: A comparative perspective," in E.B. Weiss and H.K. Jacobson (eds.), *Engaging Countries, Strengthening Compliance with Environmental Accords*, Cambridge, MA & London: MIT Press, 1998, pp. 39–62, at p. 43. Benvenisti also argues that "when designing institutional arrangements, emphasis should not be placed on rigid obligations. The greatest attention should be directed at structures and procedures for future exchanges," E. Benvenisti, *Sharing Transboundary Resources: International and Optimal Resource Use*, Cambridge: Cambridge University Press, 2002, p. 144.

6 A. Székely, "Compliance with international environmental treaties: The empirical evidence. A commentary on the softening of international environmental law," *American Society of International Law, Proceedings of the Annual Meeting*, 1997, pp. 234–41, at p. 235.

of norm creation, in response to different needs and the requirements of rapidity, flexibility, and evolution.[7]

At first, analysis of this trend may seem to reveal a fundamental weakness of international environmental law, in particular, but both empirical and theoretical evidence seem to suggest that this is not the case.[8] Some studies show that in cases where compliance with legally binding treaties was high, their influence on behavior was low.[9] Compliance may be high because governments negotiate commitments carefully and only give their consent to binding obligations with which they can easily comply, often leading to a "lower common denominator."[10] Conversely, compliance with nonbinding agreements involving more ambitious goals may be low, but they might have a greater influence on behavior. Arguably, the fear of noncompliance *ex post* may result in "shallower" substantive rules *ex ante*.[11]

The approach favoring soft law commitments is, however, criticized as a means of cooperation because it leaves enough room for states to negotiate vague instruments and do little or nothing to comply with them.[12] It should be noted, however, that the legal nature of an international obligation is distinct from its normative content. The fact that an obligation is nonlegally binding does not mean that it has unclear normative content. However, it is true that many soft law instruments contain only vague provisions that can hardly be translated into actions. Therefore, the same empirical studies mentioned above favoring a soft law approach recognize that nonbinding instruments can be more effective when applied in tandem with legally binding ones.[13]

In the context of the Amazon, a set of clear commitments, or "rules of determinate content," under existing (bilateral or regional) treaties is required, either in the form of binding or nonbinding legal instruments.[14] However, this may also be refuted by

[7] L. Boisson de Chazournes, "Gouvernance et régulation au 21éme siècle: quelques propos iconoclastes," in L. Boisson de Chazournes and R. Mehdi (eds.), *Une société internationale en mutation: quels acteurs pour une nouvelle gouvernance?*, Brussels: Bruylant, 2005, pp. 19–40.

[8] Kal Raustiala, "Form and substance in international agreements," *American Journal of International Law*," vol. 99, no. 3, 2005, pp. 581–614, at p. 613.

[9] The International Institute for Applied Systems Analysis (IIASA) sponsored a project entitled Implementation and Effectiveness of International Environmental Commitments, involving two dozen project members whose task was to conduct fourteen case studies on how governments, private actors and international organizations put environmental agreements into practice. The results of this project are contained in Victor et al., op. cit., p. 241.

[10] Ibid., p. 242.

[11] As Raustiala notes (op. cit., p. 611). Raustiala uses the term "pledges" for nonbinding instruments and "contracts" for binding instruments.

[12] Raustiala, op. cit., p. 611.

[13] Victor et al., op. cit., p. 8.

[14] According to De Sadeleer, those rules have unequivocal meaning and are able to "regulate, prohibit, or authorize types of behaviour by reducing the risk of interpretation, and thence of contention as regards their application" (N. de Sadeleer, *Environmental Principles: From Political Slogans to Legal Rules*, New York: Oxford University Press, 2002, p. 308).

those who claim that the question is more complex than the simple formula "the more precise the better."[15] It is suggested that obligations stated in vague terms are not necessarily "weaker," and that certain treaties containing general provisions, such as the 1972 Convention Concerning the Protection of the World Cultural and Natural Heritage,[16] have experienced significant compliance.[17] Others claim that vague commitments may even have the positive effect of encouraging States to seek negotiated agreements and allow a treaty to adapt to factual changes.[18] However, as far as we understand, clear commitments are required to keep States away from the danger of inaction. As they can be more easily translated into specific actions leading to a substantial change in behavior, such commitments are in principle more effective than vague formulations.[19] Empirical evidence also shows that the implementation of and compliance with international obligations are stronger with treaties that have relatively precise procedural and substantive obligations.[20]

Apart from a set of precise commitments (in the form of legally binding or nonbinding instruments), the institutions created under the treaties adopted by the Amazon States *inter se* should have a more active role than that of merely exchanging information and overseeing the implementation of field projects. These institutions could play a role, for example, in enforcing rules aimed at the control of forest fires; deforestation, especially in border areas; or in combating the illegal trade in wildlife species. For example, under the 1994 Lusaka Agreement on Cooperative Enforcement Operations Directed at Illegal Trade in Wild Species of Fauna and Flora,[21] adopted by a group of Southern African countries, a regional enforcement network was created involving local police forces to investigate and prosecute cases of illegal trade of wildlife.[22] This treaty set up the Task Force for Cooperative Enforcement Operations Directed at Illegal Trade in Wild Fauna and Flora with an international legal personality and

[15] H.K. Jacobson and E.B. Weiss, "Assessing the record and designing strategies to engage countries," in E.B. Weiss and H.K. Jacobson (eds.), *Engaging Countries: Strengthening Compliance with Environmental Accords*, Cambridge, MA & London: MIT Press, 1998, pp. 511–54, at p. 525.

[16] 1037 UNTS 151.

[17] Ibid.

[18] Benvenisti, op. cit., p. 165.

[19] Victor et al., op. cit., p. 245. Some case studies also suggest that the incorporation of treaties into national law containing detailed specific actions to be taken by States is one of the main factors inducing compliance (I.F. Shihata, "Implementation, enforcement, and compliance with international environmental agreements: Practical suggestions in light of the World Bank experiences," *Georgetown International Environmental Law Review*, vol. 9, 1996, pp. 38–39).

[20] Jacobson and Weiss, op. cit., p. 515. This is the case of the 1987 Montreal Protocol on Substances that Deplete the Ozone Layer and the 1972 Convention on the Prevention of Marine Pollution by Dumping of Wastes and Other Matter when compared to other treaties, in other words, the 1972 Convention Concerning the Protection of the World Cultural and Natural Heritage, the 1973 Convention on International Trade in Endangered Species of Wild Fauna and Flora, and the 1983 Tropical Timber Agreement.

[21] Available at: http://www.lusakaagreement.org/faqs.html

[22] Article 4.

capacity to perform its functions in the territory of each party[23] and facilitate cooperative activities among the national bureaus.[24] Likewise, cooperation among the Amazon States should target environmental problems that concern two or more States and that can be better handled at the bilateral or regional levels. Not all issues, however, may be suitable for this type of cooperation, for instance, those related to land-tenure, demarcation of indigenous peoples' lands or protected areas, and human settlements, among others, which are usually addressed at the national level.

In practice, it has been observed that cooperation among the Amazon States *inter se* is still limited. This is noted especially in relation to areas that by their very nature involve the interests of two or more States (e.g. the management of international rivers, shared ecosystems, migratory species etc.), but which so far have not led to significant cooperation.[25] In view of this, the Amazon countries have proposed some lines of action where further cooperation is needed, which include, for example, the harmonization of environmental policies; the strengthening of regional organizations, notably Amazon Cooperation Treaty Organization (ACTO); the generation and diffusion of environmental information, for example, through an Amazonian environmental information system; and the identification of environmental services provided by the Amazon, among others.[26]

10.2. THE ROLE OF THE 1978 AMAZON COOPERATION TREATY

Cooperation can be achieved among the Amazon States at different levels and in various ways. One way to enhance cooperation and ultimately improve the effectiveness of the international legal system of the Amazon is by strengthening the 1978 ACT. When compared to other international legal instruments, this treaty is in a better position to foster regional cooperation. It is the only treaty to involve all Amazon States with the specific objective of achieving the conservation and rational use of the region's natural resources. Therefore, it is the ideal forum where norms and policies applicable basin-wide can be negotiated, implemented, and enforced. A major strength and potential role of the ACT, notably of its Permanent Secretariat, is to be a factor of cohesion among dispersed instruments and institutions operating at different levels. When compared, for example, to the secretariats of multilateral

23 Article 5.

24 Article 5(9).

25 United Nations Environment Programme (UNEP), Amazon Cooperation Treaty Organization (ACTO), and Research Center of Universidad del Pacífico (CIUP), *Environment Outlook in Amazonia, Geo Amazonia*, 2009, p. 25, http://www.unep.org/publications/search/pub_details_s.asp?ID=3997. According to this report, the continued availability of surface waters in each of the Amazonian countries depends to a great extent on the use and adequate management in each of those countries, but they have not yet attained an integrated management of Amazonian water resources.

26 Ibid., p. 27.

environmental agreements (MEAs), the Permanent Secretariat, endowed with international legal personality to represent all basin States, is better placed to understand the issues at stake and provide joint solutions to environmental problems. However, little can be done if the Amazon States do not value the ACT as a key instrument for the international protection of the Amazon and if no further steps are taken to render this treaty regime more effective. Rather than adopting a new treaty altogether, or, for example, considering an instrument of global application, which would probably be rejected by the basin States, efforts should go toward improving the ACT, given its normative framework, permanent institutions, and already existing diplomatic history, and so on.

The ACT parties manifested on many occasions their willingness to make this treaty stronger. As early as 1983 (only a few years after its entry into force), they initiated debate on ways of strengthening it. At the Third Meeting of Ministers of Foreign Affairs (MMFA), the then-president of Ecuador Rodrigo Borja Cevallos stressed the "shared responsibility" of the ACT parties in relation to the Amazon and the fact that this treaty introduced a new dimension of regional cooperation.[27] Subsequent declarations restated the need to enhance regional cooperation under the ACT. The Preamble of the 1989 Declaration of San Francisco de Quito[28] called on the parties to establish common environmental policies "owed to present and future generations." Finally, in 1995, the Permanent Secretariat and financial mechanism were formally created, both established a few years later.[29] Ever since, other ministerial declarations have reiterated the need to make regional cooperation more effective[30] in the framework of the ACT, which is viewed as an ideal instrument for achieving regional cooperation[31] and applying concerted policies.[32] In 2004, at the

[27] The then president of Ecuador was Rodrigo Borja Cevallos (Meeting Minutes, adopted at the Third MMFA, held in San Francisco de Quito, Ecuador, March 6–7, 1989, Pro Tempore Secretariat, *Base Jurídica del Tratado de Cooperación Amazónica: Antecedentes Constitutivos de la Organización del Tratado de Cooperación Amazónica* (BJ), SPT-TCA-PER-37, La Paz, Bolivia, 2002, pp. 70–72: "destacando la responsabilidad que asiste a los países Amazónicos, sin ingerencias extranjeras, en la defensa de los ecosistemas de la Amazonía. El Tratado de Cooperación Amazónica va a eso, a la defensa de la ecología, va a la preservación del medio ambiente, al respeto a las etnias preexistentes").

[28] Also adopted at the Third MMFA, BJ, pp. 73–79, at p. 75.

[29] RES/V MRE-TCA/1 creates the Permanent Secretariat to the ACT with headquarters in Brasília and RES/V MRE-TCA/2 establishes its financial mechanism, both adopted at the Fifth MMFA, held in Lima, Peru, December 4–5, 1995, BJ, pp. 90–91.

[30] 2000 Declaration of Caracas, adopted at the Sixth MMFA, held in Caracas, Venezuela, April 6, 2000, BJ 117–21, at p. 117.

[31] 2002 Santa Cruz de la Sierra Declaration, adopted at the Seventh MMFA, held in Santa Cruz de la Sierra, Bolivia, November 22, 2002, BJ, pp. 163–65. On this occasion, see also the statement by the minister of Venezuela (Roy Chaderton Matos), BJ, pp. 133–35: "Surge el Tratado de Cooperación Amazónica como un instrumento idóneo para la cooperación multilateral internacional en el esfuerzo de preservar y desarrollar sustentablemente la Cuenca Amazónica...."

[32] The 2002 Declaration of Santa Cruz de la Sierra (ibid.) reiterated the importance of the ACT as the principal instrument able to put into effect concerted policies for the sustainable development of the

first MMFA held after the establishment of the ACTO, the parties reaffirmed the importance of this treaty,[33] and at the latest MMFA, the then secretary general of ACTO, Rosalía Arteaga, stated that this was the best and most suitable forum for integrating the Amazon countries.[34]

Despite its promise, the ACT, like other international instruments adopted in the region, faces a few normative and institutional weaknesses. The treaty lacks precise commitments (in the form of legally binding or nonbinding instruments) to put its objectives into effect. For example, Article 5 requires States to "make efforts aimed at achieving rational utilization of water resources." However, in the absence of further guidance, some questions arise, for example: What kind of efforts are needed? Should they involve a group of countries? What forms of cooperation are then required? In view of this, Article 1 of the ACT envisages the implementation of "operational agreements and understandings" or other "pertinent legal instruments" to clarify and put into effect the general provisions of that treaty. However, there has been only one amendment to the ACT so far, which modified its Article 22 by replacing the Pro Tempore Secretariat with the Permanent Secretariat, but no other additional instruments have been created. In terms of procedural obligations, apart from the duty to exchange information and report on measures taken in relation to wildlife species, no other mechanisms are foreseen under the treaty or have been developed after its coming into force. This latter obligation, contained in Article 7(b), has never been met by the ACT parties. Institutions, such as the Pro Tempore Secretariats and Permanent National Commissions, have either functioned on a limited scale or have not been operative for periods of time. In general, the meetings of the ACT institutional bodies have occurred in a discontinuous way, which has affected the treaty's implementation. Overall, the ACT has had little practical impact, particularly during the period 1978–2002.

The major achievement of the ACT over the few last years has been the creation of a Permanent Secretariat (and financial mechanism), which injected more vitality into the treaty. A permanent institutional framework has enabled a more regular exchange of information (particularly under the Coordination Commission of the Amazon Cooperation Council), facilitated the leverage of funds and implementation of field projects, and permitted a debate on ways of strengthening bodies such as the National Commissions. Despite the limited budget, the obligation to provide financial contributions to the Permanent Secretariat has been timely met.

Amazon. In Spanish: "Reafirman la relevancia del Tratado de Cooperación Amazónica como principal instrumento para llevar a efecto políticas concertadas para el desarrollo sostenible de lo territorios amazónicos de los países del TCA."

33 Declaration of the Eighth MMFA, held in Santa Cruz de la Sierra, Bolivia, September 14, 2004, available at: http://www.otca.org.br

34 The then secretary general was Rosalía Arteaga; see http://www.otca.org.br

Most importantly, the ACT has succeeded over the years in maintaining some cohesion among its member States, without which the establishment of a Permanent Secretariat would not have occurred. A first step has been taken by the ACT parties in institutionally strengthening the treaty, but will this ensure a more effective regime? How can this treaty achieve tangible results and, in particular, become a factor for cohesion in that region?

In terms of its normative framework, more precise substantive obligations are required in areas where further regional cooperation is needed, for example, in the prevention of forest fires; the combating of wildlife trafficking (and other illegal activities); the management of international rivers (or the Amazon Basin as a whole), migratory species and transboundary protected areas; or in the prevention of transboundary pollution, among others. As noted, additional commitments could take the form of separate "understandings" or "agreements" under the umbrella of the ACT, as foreseen in its Article 1. Irrespective of their actual form, additional legal obligations should be clear enough to indicate what is to be targeted and how to achieve expected results. Otherwise nothing specific will be requested of the ACT parties, which may lead to inaction or disinterest in the treaty.

The substantive obligations contained in environmental treaties are often accompanied by procedural obligations (or mechanisms) that put them into effect. They are devised for different purposes, for example to allow the exchange of information, to monitor the state of the environment, or to ensure compliance with treaty objectives. As noted, with regards to procedural obligations, Article 15 of the ACT requires the parties to "maintain a permanent exchange of information and cooperation among themselves." In practice, the Permanent Secretariat has played a key role in this regard, for example, by producing progress reports (among other publications), convening meetings of its institutional bodies, and enabling debate on ways to strengthen the ACT. Reporting practices, foreseen in Article 7(b) of the treaty, should be activated and ideally extended to natural resources other than wildlife species of fauna and flora. Albeit limited in some ways,[35] reporting is a useful tool to assess how well a treaty is being implemented. Due to its potential costs, a simplified form of reporting could be envisaged under the ACT, possibly with the assistance of institutions based in the region. In practice, some environmental treaties rely on information provided by nongovernmental organizations (NGOs) or other institutions. For

[35] Although an essential tool for evaluating the progress of a treaty and particularly States' performance, national self-reporting has some limitations, especially in developing countries. They usually lack the human and financial capacity to present up-to-date and comprehensive information, and tend to present information that has already been collected for other reasons, while much information remains unreported. Furthermore, the huge demands placed on treaty secretariats' limited personnel and resources also restrict their ability to analyze data available to them; see Chayes et al., op. cit., p. 46. These authors point to two principal problems related to self-reporting: failure to report and inaccuracy.

example, Convention on International Trade in Endangered Species of Wild Fauna and Flora (CITES) has the support of TRAFFIC to identify trends in wildlife trade,[36] whereas, under the 1973 International Convention for the Prevention of Pollution from Ships, the International Chamber of Shipping, a private consortium of shipping companies, provides information about reception facilities in ports.[37]

Other instruments that could be introduced under the ACT include, for example, monitoring or implementation review and enforcement mechanisms. Monitoring in particular allows the parties to assess the state of the environment and gather information required to meet specific environmental obligations.[38] As defined under the 1992 Convention for the Protection of the Marine Environment of the Northeast Atlantic,[39] monitoring involves measuring three factors: the quality of the environment, the activities or natural anthropogenic inputs that may affect the quality of the environment, and the effects of such activities.[40] Based on this information, States can take more scientifically based and informed decisions to fulfill treaty objectives. It has been noted that monitoring makes a difference in ensuring the effectiveness of treaty regimes.[41] Up to the present, no such mechanism exists that is applicable to the Amazon as a whole, and it could be devised under the ACT. In particular, the Brazilian National Institute of Space Research (INPE) has developed remote sensing technology to monitor deforestation in the Amazon. The exchange of such capacity with other Amazon States could potentially be facilitated under the ACT, to enable them to monitor deforestation and other land-use changes in their respective Amazonian territories or with regard to the entire Amazon, possibly with the assistance of other national scientific institutions or international organizations.

Another common feature of environmental treaties is a review of the parties' performance, which in some cases may involve deterring and punishing noncompliance.[42] The mechanisms of implementation review can have that dual function of assessing States' performance and enforcing international obligations. Normally, in treaties with a "weak" structure, the review of performance (and sanctions for nonperformance) is minimal or nonexistent and the parties are usually only requested to self-report, whereas "strong" review structures add value to the information provided by the parties.[43] In this latter case, this information is often

36 Ibid., p. 831.
37 Ibid., pp. 47–48.
38 P. Sands, *Principles of International Environmental Law*, Cambridge: Cambridge University Press, 2003, p. 847.
39 32 ILM 1069, 1993.
40 Article 1, Annex IV.
41 P.H. Hass, R.O. Keohane, and M.A. Levy (eds.), *Institutions for the Earth: Sources of Effective International Environmental Protection*, Cambridge, MA and London: MIT Press, 1995, p. 16.
42 Raustiala, op. cit., p. 585.
43 Ibid.

evaluated by a competent organ and measures may be taken to make a defaulting party comply. Much emphasis is now put on implementation review and it has been observed that legal instruments (notably nonbinding ones) work better when tied to a strong review structure.[44] Of the two categories above, the ACT falls under those legal regimes with a weak implementation review structure. Under this treaty, the Amazon Cooperation Council (ACC) has the role of ensuring that "the aims and objectives of the Treaty are complied with" (Article 21), but no specific procedure is foreseen detailing how the ACC (or other bodies) should ensure compliance.

Certain environmental treaties have created noncompliance mechanisms, classic examples of which are the 1987 Montreal Protocol on Substances that Deplete the Ozone Layer[45] and the Kyoto Protocol.[46] At the regional level, the Andean Community has introduced a noncompliance procedure aimed at assessing the implementation of its decisions and enforcing legal obligations, where appropriate. Under this procedure, the General Secretariat firstly notifies the defaulting party and, if it maintains the noncompliant behavior, the General Secretariat can refer the case to the Court of Justice of the Andean Community.[47] The Court may render a mandatory verdict of noncompliance that, if not executed, allows it to define the limits within which the claimant State may restrict or suspend the benefits obtained by the defaulting party.[48] A noteworthy feature of this procedure is that not only can the General Secretariat trigger it, but other member States can also do so, as well as natural and legal persons.[49] Interestingly, any of them can directly submit complaints to the Court against alleged omission or inactivity on the part of any of the Andean Community bodies,[50] including member States and individuals. In practice, this procedure has proven useful in allowing the assessment of States' performance

44 According to Victor et al. (op. cit., p. 247), a "new approach" places greater emphasis on nonbinding instruments and implementation review, in contrast "to efforts to manage international environmental problems, which have deified binding commitments."

45 26 ILM 1550, 1987.

46 Under these procedures, the parties' performance is regularly monitored and, when cases of noncompliance are identified, the treaty secretariat (or another competent body) will enquire as to the reasons for noncompliance in order to determine whether technical or financial assistance is needed to help the defaulting party meet its obligations. In cases where the noncompliant party fails to improve its performance despite the provision of assistance, punitive measures can be applied as a last resort (Sands, op. cit., pp. 205–10).

47 TCCJ, Article 23, 18 ILM 1203, amended by the Cochabamba Protocol of May 28, 1996.

48 TCCJ, Article 27. According to Article 28, in cases where the noncessation of the behavior that motivated the procedure is likely to cause irreparable damage or damage difficult to repair, before handling down its final judgment, the Court may order the temporary suspension of the alleged violating measure.

49 Natural and legal persons can refer a claim directly refer to the Court, if the General Secretariat has failed to do so in due time (TCCJ, Article 24).

50 TCCJ, Article 37.

and to ensure compliance with legal obligations. By using implementation review mechanisms, States can also better assess the adequacy of legal commitments in terms of their practical impacts.[51]

In the Amazon, as in other parts of the world, environmental rules may look good on paper, but have little or no practical impact. Very often they fail to change the behavior of individuals, communities, and States, given the poor surveillance and enforcement capacity, especially among developing countries. In areas of great size and remoteness such as the Amazon, better surveillance and enforcement remain a major challenge.[52] Given the magnitude and complexity of certain environmental problems, some States in other parts of the world have opted for collectively enforcing international legal obligations, as noted, for example, in the case of the 1994 Lusaka Agreement on Cooperative Enforcement Operations Directed at Illegal Trade in Wild Species of Fauna and Flora,[53] mentioned above. In general, rather than resorting to coercion for punishing "violators," the preferred approach in the environmental field is that of exploring the accountability of states, for instance, by rendering their performance transparent to the scrutiny of the international community through practices such as reporting.[54] By this and other means, State parties to a treaty can collectively manage situations of noncompliance. The possibility of collectively enforcing legal obligations is particularly attractive in the Amazon, given the weak enforcement capacity of the Amazon States taken individually. Such collective mechanisms could possibly be developed under the ACT, especially to enforce legal obligations in border areas.

Finally, in terms of its institutional structure, one of the main weaknesses of the ACT lies in the fact that the meetings of its institutional bodies, notably the MMFA and the ACC, have not occurred regularly. However, such meetings are essential, particularly in terms of the normative strengthening of that treaty, as they can adopt decisions and negotiate additional legal commitments. Those meetings are also a way of overseeing the treaty's implementation and assessing the parties' performance. It has also been noted that these meetings have been often limited to diplomatic circles, with little or no involvement of other actors, such as NGOs and indigenous communities. However, the latter's participation could bring greater

[51] Victor et al., op. cit., p. 16. These authors define "systems for implementation review" as comprising "rules and procedures by which the parties to international agreements (as well as interest groups, administrative bodies and the like) exchange data, share information on implementation, monitor activities, assess the adequacy of existing commitments and handle problems of poor implementation."

[52] S. Margulis, *Causes of Deforestation in the Brazilian Amazon*, Brazil: World Bank, 2003, p. 65. This study concerns only the Brazilian Amazon.

[53] Available at: http://www.lusakaagreement.org/faqs.html

[54] A. Chayes and A.H. Chayes, "Adjustments and compliance processes in international regulatory regimes," in J. Tuchman Mathews (ed.), *Preserving the Global Environment: The Challenge of Shared Leadership*, New York & London: Norton, 1991, pp. 280–308, at p. 291.

visibility to the ACT and better inform its decision-making processes. Other institutional weaknesses relate to the fact that the ACT executive organs (i.e. Permanent National Commissions and Special Commissions) in charge of carrying out the treaty provisions and the decisions taken by the MMFAs and the ACC have not been operative or functioned on a very limited scale. Also, the practice of appointing Pro Tempore Secretariats proved inadequate, as such bodies failed to ensure the continuity of activities, regular exchange of information, and a minimum level of stability. Until the creation of the Permanent Secretariat, most activities planned under the ACT failed to get beyond the formulation stage. In this sense, the establishment of a Permanent Secretariat (and financial mechanism) contributed considerably to improving that treaty's institutional framework.

By addressing some of the normative and institutional weaknesses indicated above, the ACT could probably be strengthened and eventually contribute to making the international legal system of the Amazon more effective. This might be achieved especially if the ACT functions as a factor of cohesion among existing legal instruments and institutions. To this end, the role of the Permanent Secretariat, as a forum where environmental policies and norms can be negotiated, implemented, and enforced basin-wide, should be enhanced. By doing so, ACTO might confer more coherence on the international legal system of the Amazon, as a forum where institutions working at different levels (bilateral, subregional, regional, and global) could exchange information, coordinate activities, harmonize laws and policies, and enter into consultations, especially before adopting international treaties or undertaking measures likely to affect other States. The Permanent Secretariat has played this "coordinating" role, for example, by providing a forum where the Amazon States adopted common positions in preparation for the United Nations Forum on Forests (UNFF) negotiations. This is probably one of the major strengths of ACTO, as the only regional organization involving exclusively all Amazon States.

In this regard, the ACT parties requested at the Ninth MMFA in 2005 (under RES/IX MRE-OTCA/05) that the Permanent Secretariat continues playing that role as a facilitator in international negotiation forums. When compared to the prior Pro Tempore Secretariats, the Permanent Secretariat has broader powers to represent its member States in the international sphere,[55] for example, by entering into agreements on their behalf when so mandated by the MMFA or ACC and in accordance with its work plan (ACTO Regulation, Articles 3 and 4).

[55] An organization endowed with international legal personality is "subject of international law so as itself to enjoy rights, duties or powers established in international law, and, generally, the capacity to act on the international plane" (R. Jennings and A. Watts (eds.), *Oppenheim's International Law*, vol. 1, London: Longman, 1991, p. 119).

As noted, the Amazon States have on different occasions expressed their willingness to achieve further regional cooperation and strengthen the ACT in particular, of which ACTO is a concrete manifestation. All basin States share a common interest in the conservation and sustainable use of the Amazon and view the ACT as a suitable forum, at least for enhancing regional cooperation. However, even if strengthened, this treaty could not tackle the current environmental problems of the Amazon single-handedly. In particular, it could not address all types of environmental problems in the Amazon, such as those pertaining mostly to the national jurisdiction of each basin State or requiring the mobilization of the international community as a whole, but it could play a key role in complementing efforts taken at different levels toward the protection of the Amazon.

10.3. COOPERATION BETWEEN THE AMAZON STATES AND THE INTERNATIONAL COMMUNITY

The Amazon States interact with the international community, represented by States, international organizations, and NGOs, under the framework of MEAs (and other international legal instruments) to which those countries are parties, and by creating partnerships with international organizations and NGOs. The Amazon States, like other parties to a particular MEA, must comply with the norms and observe the principles defined in it, as well as international customary law. International legal obligations set up under such instruments apply to their respective Amazonian territories and should guide their activities, for example, when building infrastructure and managing forests or world heritage sites, as well as their interactions with other basin States and the international community. In general, natural resources and sites in the Amazon already are internationally protected under MEAs, such as wildlife species of flora and fauna occurring in the Amazon listed under CITES, as well as wetlands, natural, or cultural heritage protected by the 1971 Ramsar Convention[56] and the UN Educational, Scientific and Cultural Organization (UNESCO) World Heritage Convention,[57] respectively. To say that natural resources and sites in the Amazon are internationally protected means that the basin States enjoy the rights and should observe the principles and obligations defined under a particular MEA (or other legal instruments) conferring such protection. It also involves a duty on the part of the international community to provide assistance to these countries where appropriate.

The legal obligations contained in environmental treaties are usually coupled with a set of rights. For example, while the parties to the UNESCO World Heritage

[56] 1971 Convention on Wetlands of International Importance Especially as Waterfowl Habitat, 11 I.L.M. 963 (1972).

[57] 1972 UN Educational, Scientific and Cultural Organization (UNESCO) World Heritage Convention, 11 ILM 1358, 1972.

Convention have a duty to conserve their natural heritage (Article 4), they also have a right to request assistance in order to maintain natural or cultural heritage in their territories (Article 19). Financial or other forms of assistance are often provided by developed country parties or multilateral institutions, such as the World Bank or Global Environmental Facility. Therefore, the relationship between the Amazon States and the international community in the context of such treaties involves a set of rights and duties for both sides. The Amazon States also interact with the international community, represented by global institutions, such as international organizations and NGOs, by setting up partnerships and hosting field projects. The assistance provided by such institutions has been important is many ways, for example, in creating protected areas and monitoring wildlife species, and in the campaigning against deforestation and other forms of environmental degradation.

In general, both the technical and financial assistance provided by the international community over the past years have been crucial in particular to helping the Amazon States achieve certain environmental objectives defined under international legal instruments; create protected areas, as observed in the context of the Pilot Program to Conserve the Brazilian Rain Forest (PPG7); and conduct scientific research, for example. under the Large Scale Biosphere-Atmosphere Experiment (LBA) in the Amazon.[58] Despite this, the usual patterns of deforestation and environmental degradation still persist. In fact, it is suggested that international assistance, notably in the form of official development assistance (ODA), will not deliver the scale of finance required for adaptation and mitigation activities in particular.[59] On the other hand, there seems to be a consensus that serious environmental problems such as deforestation, require urgent action by the international community. In view of the limits observed with "traditional" forms of international assistance (e.g. capacity building and ODA, among others) and due to the magnitude of funding and the complexity of certain environmental problems, such as climate change, more attention needs to be given to finding alternatives involving, for example, market-based mechanisms.

Certain mechanisms such as REDD, now at the top of the international agenda and already being tested to some extent in the Amazon and elsewhere, can provide a good short-term solution to environmental problems requiring urgent responses, in particular deforestation. The use of positive incentives, notably market-based mechanisms, seems attractive, especially if one considers that the hundreds of millions of dollars in donor funds flowed into the Amazon over the past years have not been able to reverse current deforestation patterns. In this context, a growing faith

[58] UNEP, ACTO, and CIUP, op. cit., p. 25. According to this report, international cooperation and international organisms have played an important role in facilitating financial and technological resources for carrying out activities aimed at the conservation and sustainable use of the Amazon.

[59] C. Parker et al., *The Little Climate Finance Book: A Guide to Financing Options for Forests and Climate Change*, Oxford: Global Canopy Foundation, 2009, p. 58.

in the markets is observed, based on the idea that, market-based mechanisms can be used as conservation tools, especially as the profitable agricultural frontier and other economic activities rapidly expand into rainforests.

Mechanisms such as REDD and PES can be useful as they lock up areas for conservation and, if well implemented, may avoid deforestation and reduce greenhouse gas (GHG) emissions, among other (social and environmental) co-benefits. From an environmental perspective, such instruments are commendable, provided that they achieve additionality and permanence, and avoid leakage. In other words, they must ensure that emission reductions that would not occur otherwise are achieved (additionality), without causing the deterioration of ecosystems elsewhere (leakage), and that the expected outcomes will continue for a minimum length of time (permanence). These mechanisms are an additional source of finance in developing countries, which can allow conservation activities that would not occur otherwise due to a lack of funding, technical capacity, or political will. Their implementation requires the commitment of governments, but also stimulates the participation of the private sector. Instead of only avoiding environmental harm, by observing codes of conduct or guidelines, the private sector is given a more active role in helping design and finance conservation projects. These instruments can also benefit individuals and local communities by providing economic alternatives to unsustainable practices. In particular, mechanisms such as the CDM are beneficial in that they bring investments in clean technology into developing countries, improve the existing technical capacity, and enable conservation activities such as afforestation and reforestation. Therefore, they are instrumental in helping both developed and developing countries achieve certain objectives, for example, reducing GHG emissions (UNFCCC, Article 2), promoting the conservation and sustainable use of biodiversity (CBD, Article 1), and, in particular, achieving the preservation and rational utilization of natural resources in the Amazon (ACT, Article 1).

In any case, the costs, particularly those related to halting deforestation, are high. It is estimated that, despite the new funding mechanisms for tropical forests, there may be less money than required.[60] Although mechanisms such as REDD or PES can be a source of finance, their own implementation is costly and might require various sources of funding. As noted with REDD, so far a great part of donor funding has been directed only to enable tropical forest countries to take part in the mechanism, by improving their technical capacity to carry out future project activities, which has led to the creation of various international funds. Therefore, funding will be needed from various sources, including the more "traditional" ones, for example donations, ODA, and technical cooperation,

[60] FAO (Food and Agriculture Organization), *State of the World's Forests*, Rome: FAO, 2009, p. 64.

among others. The use of market-based instruments will then go along with other types of incentives, such as voluntary contributions, as well as capacity building, and conservation programs.

In practice, market-based mechanisms, such as those under the Kyoto Protocol, can be counterproductive if used by developed States as an easy way to maintain business-as-usual practices, delaying changes in their energy-intensive economies and reliance on fossil fuels. For example, if Annex 1 countries rely on the flexible mechanisms under the Kyoto Protocol or others agreed upon in a post-2012 climate regime as a way to avoid deeper cuts in their GHG emissions, the goal of stabilizing GHG emissions in the atmosphere and avoiding dangerous climate change will not be achieved. Also, major emerging economies such as China, India, Brazil, and South Africa should have emission reduction targets imposed on them in the near future, according to their levels of development and share of responsibility, in-line with the principle of common but differentiated responsibility. No specific deal was reached in December 2009 at the United Nations Framework Convention on Climate Change (UNFCCC) COP-15 and thus individual and exact emission reduction targets for a post-2012 regime are still up for negotiation. In this context, the market-based instruments here examined provide an alternative to problems such as climate change and tropical deforestation, as deeper emission cuts and structural changes are needed both from developed and some developing countries.

Despite the potential of market-based instruments and other types of incentives to be used as conservation tools, they are not a panacea to all environmental problems. Firstly, they are probably not suited to dealing with all types of issues, such as transboundary pollution, illegal wildlife trade, and overfishing, among others. Secondly, they should be used in combination with and not instead of other measures that have proven effective in certain circumstances, for example, those involving the creation and management of protected areas, moratoriums or bans, conservation policies, and law enforcement measures. The use of positive incentives, and particularly market-based mechanisms, may help improve governance in tropical forests countries, but not without an enabling normative framework. As has been observed in practice, good governance, involving, for example, some clarity in relation to land tenure, minimum technical capacity, and supportive laws and policies, is a key factor determining the success of such mechanisms. For this reason, instruments currently debated, such as REDD, should be seen as catalytic in relation to good governance rather than an exclusive and sufficient solution.[61]

[61] M.M.C. Bustamante, C.A. Nobre, and R. Smeraldi, *Estimating Recent Greenhouse Gas Emissions from Cattle Raising in Brazil*, 2008, http://www.amazonia.org.br

10.4. INTERNATIONAL LEGAL STATUS FOR THE AMAZON AND POSSIBLE IMPLICATIONS

The inquiry into a possible international legal status for the Amazon has shown that the preservation of the Amazon can be considered a matter of common concern of humankind. State practice has provided evidence of the fact that the preservation of natural resources, irrespective of where they are located, can be a common concern. This has been observed especially in the international debate on forests, where States have recognized in different forums and on different occasions the role of forests in providing vital services to humanity. Evidence that natural resources within national borders are objects of a common concern can be found, for example, in Article 4 of the 1972 Convention Concerning the Protection of the World Cultural and Natural Heritage,[62] according to which natural and cultural heritage sites situated in the parties' territories constitute "world heritage." Other instruments go further, saying that the preservation of the environment as a whole is a "common concern of humankind," as stated under The Hague Recommendations on International Environmental Law (Preamble).[63]

The change in the Earth's climate and the conservation of biological diversity are considered a common concern of humankind (Preamble of the UNFCCC and Convention on Biological Diversity (CBD), respectively), which extends to natural resources or areas under national jurisdiction. For example, in order to achieve the goal of stabilizing GHG emissions in the atmosphere, the UNFCCC requires States to preserve marine or terrestrial ecosystems, including those located within their internal waters, and also sinks and reservoirs of all GHGs, such as biomass and forests (Article 4, d). The common concern over the global climate therefore requires the adequate management of natural resources within national jurisdiction. Like the Earth's climate, the conservation of biodiversity is deemed essential for "maintaining life sustaining systems of the biosphere," and is also viewed as a common concern of humankind (CBD, Preamble). Article 4 clearly defines the jurisdictional scope of the CBD by stating that, in the case of components of biological diversity, the "provisions of this treaty apply, in relation to each Contracting Party ... in areas within the limits of its national jurisdiction." In this case, the notion of common concern clearly applies to natural resources within national borders. In this light, the protection of areas such as the Amazon can be also considered a common concern from at least two perspectives: as a reservoir of GHG gases, the Amazon plays a vital role in stabilizing the global climate; and it is also a shelter for biodiversity occurring in its many and varied habitats. Threats to the Amazon are

[62] 1037 UNTS 151, 1973.

[63] Published in the *Yearbook of International Law*, 1992.

likely to affect the global environment and can be considered a concern of the international community. The protection of certain areas and natural resources essential for maintaining life sustaining systems, such as the Amazon, can therefore be object of a common concern of humankind, even if under national jurisdiction.

The concept of common concern of humankind involves interests and values pertaining to the international community as a whole. Albeit imprecise, this notion seems to suggest that all States may have a legitimate interest in the preservation of certain natural resources and ecosystems deemed crucial for sustaining life on Earth, as well as a shared responsibility for their protection.[64] In this sense, two main consequences derive from seeing the protection of the Amazon as a common concern of humankind. On the one hand, while the international community bears a responsibility to assist the Amazon States in preserving the environment and rationally using natural resources, they may also exercise a right of surveillance over this region. The Amazon States, on the other hand, have an obligation owed to the international community to protect the Amazon, while maintaining sovereignty over their respective Amazonian territories.

The concept of common concern – unlike that of a common heritage of humankind – does not involve elements such as the exclusion of sovereignty claims, international management, and benefit sharing. The Amazon States remain the only ones to enjoy sovereignty over this region and its natural resources. Article 4 of the ACT recalls that the exclusive use of natural resources within these countries' respective Amazonian territories is a right inherent to their sovereignty, which shall not be subject to any restrictions other than those arising from international law. First and foremost, national sovereignty over natural resources involves the right to dispose freely of such resources, apart from other rights (e.g. to expropriate and to be compensated for environmental harm), but also duties (e.g. to respect the rights of indigenous communities, to cooperate, and to preserve and sustainably use natural resources).[65]

The Amazon States seem aware that the rights associated with their exclusive sovereignty over that region and its natural resources are coupled with a set of duties. For example, according to the ACT, the conservation of the environment is a responsibility inherent to the sovereignty of each member State (Preamble). The 2004 Declaration of Manaus affirms their "sovereign responsibility" in relation to the sustainable development of the Amazon.[66] Likewise, pursuant to the 2005 Declaration of Iquitos, the Amazon States enjoy sovereignty over that region, but

[64] Sands, op. cit., p. 287.

[65] N. Schrijver, *Sovereignty over Natural Resources: Balancing Rights and Duties*, Cambridge: Cambridge University Press, 1997, p. 258.

[66] 2004 Declaration of Manaus, adopted at the Eighth MMFA of the ACTO, on September 14, 2004, Article 2, http://www.otca.org.br

also have a "responsibility to preserve the natural resources."[67] In terms of duties, the Amazon States have *inter se* (and *vis-à-vis* the international community) an obligation to comply with international treaties to which they are signatories, as well as customary law rules (*pacta sunt servanda*). Particularly under the ACT, they are committed to preserving the environment and rationally using natural resources (Article 1). Other duties derive from that obligation to protect the Amazon, for example, the duty to cooperate and ensure that activities undertaken in one State do not cause adverse transboundary effects. In general, the Amazon States must comply with legal obligations agreed upon under treaties to which they are parties, which are then applicable to their respective Amazonian territories, such as the duty to develop national strategies for the conservation of biodiversity (CBD, Article 6), to promote *in situ* conservation (CBD, Article 8, a), and to develop national inventories of anthropogenic GHG emissions (UNFCCC, Article 4).

On the other hand, the international community has a legitimate interest in the protection of the Amazon, from which rights and duties also derive. From a common concern over the Amazon flows the obligation to assist the Amazon basin States. As stated in The Hague Recommendations on International Environmental Law (Preamble), "the preservation of the environment is a common concern of humankind. Therefore, costs should be shared equitably among States, taking into account their responsibilities and capabilities." The duty to provide assistance to developing countries is recurrent in environmental treaties and UN General Assembly (UNGA) resolutions. These latter recall the "historic responsibility" of developed countries to provide assistance, as they have often caused greater damages to the environment and have the means to take remedial actions. Therefore, it is recognized that, although all States must "take effective actions for the protection and enhancement of the environment," developed States have the "main responsibility for taking appropriate measures urgently."[68]

Apart from this basic duty to provide technical and financial assistance, the international community may exercise a right of surveillance (or *droit de regard*) with respect to areas and natural resources of common concern such as the Amazon. A right of surveillance exercised by the international community in relation to natural resources and areas falling within national jurisdiction is dependent on State consent and limited by the customary law rule of national sovereignty over natural resources. However, this right has not been framed as such under MEAs or any other treaties adopted, in particular, among the Amazon States. In any case, it can be argued that such a right is still being developed under international environmental law.

67 2005 Declaration of Iquitos, adopted at the Ninth MMFA of the ACTO, November 25, 2005, Article 4, http://www.otca.org.br

68 UNGA Res. 45/94, December 14, 1990.

In practice, the international community exercises somehow a *droit de regard* over natural resources and sites, even when under national jurisdiction, for example, in the context of MEAs. In this case, the international community, represented by the member States, exercises surveillance equally conferred on all parties. Most of these treaties foresee regular meetings or conferences of the parties to keep the treaty's implementation under review. The parties are then requested to submit information on national measures taken to fulfill treaty objectives, which is made available and evaluated by all parties. By this and other means, the parties collectively monitor the state of natural resources or sites, including those under national jurisdiction, and assess each State's performance. For example, under the 1972 Convention Concerning the Protection of the World Cultural and Natural Heritage, the World Heritage Committee has conducted *in situ* missions to countries, with their consent, in order to evaluate threats to natural and cultural heritage sites. A right of surveillance can then be exercised not only in relation to common areas, such as the high seas, Antarctica, and Outer Space, but also to natural resources and sites within countries' domestic jurisdiction, within the limits imposed by the principle of national sovereignty.

Albeit important, the role of the international community in the protection of the Amazon is a limited one. The Amazon States have the primary responsibility in ensuring that agreed international commitments under (bilateral, regional, and multilateral) treaties, as well as under customary international law, are observed, while they also enjoy exclusive rights over the Amazon and its natural resources. Despite this, international cooperation at all levels (bilateral, regional, and global) remains crucial for the protection of the Amazon, as it allows these countries to handle environmental problems that require joint actions by two or more States, and to address complex issues that can only be properly addressed by the international community as a whole. Through international cooperation, in the form of technical and financial assistance, the international community has helped the Amazon States in particular to fulfill legal obligations under MEAs (and other legal instruments). The international community shall also play a key role in handling current environmental problems such as deforestation, in particular, by enabling the adoption of positive incentives in tropical forest countries such as those sharing the Amazon.

Bibliography

I. Official Documents

1. *Treaties and Other Instruments (arranged by date)*

Treaty of Tordesillas of June 7, 1494, in G.F. Martens, *Suppléments au Recueil des traités*, vol. 1, Gottingue: Dieterich, 1802–1808.

Treaty of Madrid of January 13, 1750, in A. Cantillo, *Tratados, convenios y declaraciones de paz y de comercio*, Madrid: Imprenta de Alegria y Charlain, 1843, p. 400.

Treaty San Ildefonso of October 1, 1777, in R. Aranda, *Colección de tratados del Peru*, vol. 2, Lima: Imprenta del Estado, 1843, pp. 637–52.

Boundary and Navigation Treaty between Brazil and Venezuela Signed in Caracas on May 5, 1859, in Venezuela, *Collección de tratados públicos de Venezuela*, Caracas: Imprenta Nacional, 1910, pp. 144–48, 343–45.

Treaty between Brazil and Britain Providing for the Delimitation of the Boundary Line between Brazil and the Colony of British Guyana Convened in London through Arbitration on November 6, 1901, in UK, *British and Foreign State Papers*, vol. 94, pp. 23–29.

Treaty of Petropolis Adopted by Brazil and Bolivia on November 17, 1903, in UK, *British and Foreign State Papers*, vol. 128, pp. 677–68.

Treaty between Brazil and the Netherlands Defining the Brazil–Suriname Boundary Signed at Rio de Janeiro, on May 5, 1906, in UK, *British and Foreign State Papers*, vol. 99, pp. 932–33.

Boundary and Navigation Treaty between Brazil and the Republic of Gran Colombia signed April 24, 1907, in G. Cavelier, *Tratados de Colombia 1811–1910*, vol. 1, Bogota: Editorial Kelly, 1982, pp. 682–85.

Treaty on Commerce and Navigation of April 24, 1907 between Brazil and Colombia Recognizing Ample Freedom of Terrestrial Transit and Fluvial Navigation, in G. Cavelier, *Tratados de Colombia 1811–1910*, vol. 1, Bogota: Editorial Kelly, 1982, pp. 682–85; vol. 2, 1984, pp. 509–12.

Treaty Ensuring Perpetual Freedom of Navigation of the Napo and Putumayo Rivers, in G. Cavelier, *Tratados de Colombia 1811–1910*, vol. 1, Bogota: Editorial Kelly, 1982, pp. 682–85; vol. 2, 1984, pp. 509–12.

Agreement Regulating the Navigation of the Yapurá River, on April 15, 1908, in UK, *British and Foreign State Papers*, vol. 96, pp. 383–87.
Treaty of Commerce and River Navigation between the United States of Brazil and Bolivia, 1910, United Nations Treaty Series (UNTS), vol. 888.
Boundary Treaty between Brazil and Great Britain to Demarcate the Boundary Along the Watershed between the Amazon Basin and the Basins of the Essequibo and Courantyne Rivers as far as the Point of Junction with Surinam, signed on April 22, 1926, in UK, *British and Foreign State Papers*, vol. 123, p. 468.
Treaty of Delimitation of the Brazil-Colombia Frontier signed at Rio de Janeiro on November 15, 1928 by Defining Its Southern Section, in G. Cavelier, *Tratados de Colombia 1811–1910*, vol. 1, Bogota: Editorial Kelly, vol. 2, 1984, pp. 509–12.
Boundary Treaty between Brazil and Bolivia signed in Rio de Janeiro on December 25, 1928, in UK, *British and Foreign State Papers*, vol. 128, pp. 677–68.
Convention for the Regulation of Whaling, Geneva, September 24, 1931, 155 League of Nations Treaty Series (LoNTS), 351.
Convention on the Preservation of Fauna and Flora in their Natural State, London, November 8, 1933, reprinted in 172 LNTS, 242.
Convention on Nature Protection and Wild Life Preservation in the Western Hemisphere, of October 12, 1940, UNTS, vol. 161, p. 193.
Agreement Establishing the Inter-American Ban, 426 UNTS 1986, 1959.
American Declaration of the Rights and Duties of Man, 1948, OAS AG/RES. 1591 (XXVIII-O/98)
Inter-American Charter on Social Guarantees, 1948, in E.J. Osmanczyk, *Encyclopedia of the United Nations and International Agreements*, New York, Routledge, 1990.
Convention for the Establishment of an Inter-American Tropical Tuna Commission, reprinted in 80 UNTS 3, 1949.
Preliminary Protocol Concerning permanent Navigation on the Bolivian and Brazilian Rivers of the Amazon River System, 1958, UNTS 1972.
Antarctic Treaty, Washington, DC, December 1, 1959, reprinted in 402 UNTS 71.
International Covenant on Economic, Social and Cultural Rights, New York, December 16, 1966, reprinted in 6 ILM 360, 1967.
International Covenant on Civil and Political Rights, with Two Optional Protocols, New York, December 16, 1966, reprinted in 999 UNTS 171, 302.
Treaty on Principles Governing the Activities of States in the Exploration and Use of Outer Space, including the Moon and other Celestial Bodies, London, Moscow, and Washington, DC, January 27, 1967, reprinted in 610 UNTS 205.
African Convention on the Conservation of Nature and Natural Resources, 1968, 1001 UNTS 4.
Agreement Establishing the Andean Development Corporation, 1968, in Venezuela, *Convenio Constitutivo de la CAF*, Edición Caracas, August 1996.
American Convention on Human Rights (or San José Pact), 1969, 1144UNTS 143.
Sub-regional Integration Agreement (Andean Pact or Cartagena Agreement), 8 ILM 910, 1969.
La Plata Basin Treaty, 875 UNTS 11, 1969.
Convention on the Law of Treaties, Vienna, May 23, 1969, reprinted in 8 ILM 679, 1969.
Convention on Wetlands of International Importance especially as Waterfowl Habitat, Ramsar, February 2, 1971, reprinted in 11 ILM 963, 1972.

Cooperation Agreement on Sanitation in the Amazon Region, between Brazil and Colombia, 1972, in Brazilian Ministry of Foreign Relations, *Coleção de Atos Internacionais, Divisão de Atos Internacionais*, vol. 847, 1976.

Cooperation Agreement on Sanitation between Brazil and Bolivia, 1972, in Brazilian Ministry of Foreign Relations, *Coleção de Atos Internacionais, Divisão de Atos Internacionais*, vol. 936, 1978.

Convention for the Protection of World Cultural and Natural Heritage, Paris, November 16, 1972, reprinted in 1037 UNTS 151, 1973.

Treaty between Brazil and Colombia on the Conservation of Fauna and Flora of their Respective Amazonian Territories, 1973, in Brazilian *Diário Oficial da União* of July 12, 1976 (Decree 78016), p. 9367.

Treaty Establishing the Caribbean Community (Treaty of Chaguaramas), 1973, reprinted in 12 ILM 1033, 1973.

Convention on International Trade of Endangered Species of Wild Flora and Fauna, Washington, DC, 3 March 1973, reprinted in 993 UNTS 243, 1974.

Treaty between Brazil and Peru on the Conservation of Flora and Fauna in the Amazonian Territories, 1975, in Brazilian *Diário Oficial da União* of November 24, 1976 (Decree 78802), p. 15382.

Convention on the Protection of the Archeological, Historical and Artistic Heritage of American Nations, 15 ILM 1350, 1976.

Treaty of Friendship, Cooperation, and Trade between Brazil and Surinam, 1976, in Brazilian Ministry of Foreign Affairs, Divisão de Atos Internacionais, *Coleção de Atos Internacionais*, vol. 875, 1978.

Treaty of Friendship and Cooperation between Brazil and Venezuela, 1977, in Brazilian Ministry of Foreign Relations, Divisão de Atos Internacionais, *Coleção de Atos Internacionais*, vol. 941, 1978.

Agreement on Animal Health in the Frontier Areas between the Federative Republic of Brazil and the Republic of Bolivia, 1977, UNTS 1137.

Amazon Cooperation Treaty, 1978, reprinted in 17 ILM 1045, 1978.

Bern Convention on the Conservation of European Wildlife and Natural Habitats, *European Treaty Series*, n. 104, 1979.

Treaty Constituting the Andean Parliament, 1979, http://www.congreso.gob.pe/parlamento-andino

Agreement Governing the Activities of States on the Moon and Other Celestial Bodies, New York, December 18, 1979, reprinted in 18 ILM 1434, 1979.

Convention on Migratory Species of Wild Animals, Bonn, June 23, 1979, reprinted in 19 ILM 15, 1980.

Convention on Long-Range Transboundary Air Pollution, Geneva, November 13, 1979, reprinted in 18 ILM 1442, 1980.

Convention on the Conservation of Antarctic Marine Living Resources, Canberra, May 20, 1980, reprinted in 19 ILM 837, 1980.

United Nations Convention on the Law of the Sea, Montego Bay, December 10, 1982, reprinted in 21 ILM 1261, 1983.

Cartagena Convention for the Protection and Development of the Marine Environment of the Wider Caribbean Region, 1983, reprinted in UNTS, vol. 1506, p. 158.

Protocol Concerning Co-operation in Combating Oil Spills in the Wider Caribbean Region, 1983, reprinted in UNTS, vol. 1506, p. 224.

International Tropical Timber Agreement, Geneva, 1983 (extracts), in H. Hohmann (ed.), *Basic Documents of International Environmental Law*, vol. 3, London: Nijhoff, 1992.

Protocol on Long-Term Financing of the Co-operative Program for Monitoring and Evaluation of the Long-Range Transmission of Air Pollutants in Europe, Geneva, September 28, 1984, reprinted in ILM 701, 1988.

Convention on the Protection of the Ozone Layer, Vienna, March 22, 1985, reprinted in 26 ILM 1529, 1987.

ASEAN Agreement on the Conservation of Nature and Natural Resources, 1985, www.aseansec.org.

Convention for the Protection of the Natural Resources and Environment in the South Pacific Region, 1986, reprinted in 26 ILM 38, 1987.

Protocol on Substances that Deplete the Ozone Layer, Montreal, September 16, 1987, reprinted in 26 ILM 1550, 1987.

Convention on Control of Transboundary Movements of Hazardous Wastes and their Disposal, Basel, March 22, 1989, reprinted in 28 ILM 657, 1989.

ILO Convention on Indigenous and Tribal States, 1989, 28 ILM 1382, 1989.

Port of Spain Accord on the Management and Conservation of the Caribbean Environment Adopted at the First Caricom Ministerial Conference on the Environment on May 31 to June 2, 1989 in Port of Spain, Trinidad, and Tobago.

Agreement between the Government of the Federative Republic of Brazil and the Government of the Republic of Bolivia for the Preservation, Conservation and Monitoring of Natural Resources in the Border Area, 1990, UNTS, 2060.

Convention on Environmental Impact Assessment in a Transboundary Context, Espoo, February 25, 1991, reprinted in 30 ILM 800, 1991.

Southern Common Market Agreement (or Treaty of Asunción), 1991, reprinted in ILM, vol. 30, p. 1044.

Protocol of Brasilia, 1991, http://www.mercosur.int/msweb/portal%20intermediario/pt/index.htm

Protocol on Environmental Protection to the Antarctic Treaty, Madrid, October 4, 1991, reprinted in 30 ILM 1455, 1991.

Convention on the Protection and Use of Transboundary Watercourses and Lakes, Helsinki, March 17, 1992, reprinted in 31 ILM 1312 (1992).

Convention on the Protection of the Marine Environment of the Baltic Sea, Helsinki, April 2, 1992, reprinted in 1507 UNTS 167.

United Nations Framework Convention on Climate Change, New York, May 9, 1992, reprinted in 31 ILM 849/851, 1992.

Convention on Biological Diversity, Rio de Janeiro, June 5, 1992, reprinted in 31 ILM 818, 1992.

Protocol Concerning Specially Protected Areas and Wildlife to the Convention for the Protection and Development of the Marine Environment of the Wider Caribbean Region, 1990, in *Caribbean Instruments on International Law*, 1994, p. 507.

Protocol of Ouro Preto, 1994, ILM, vol. 34, p. 1248.

United Nations Convention to Combat Desertification in Those Countries Experiencing Serious Drought and/or Desertification, Particularly in Africa, Paris, June 17, 1994, reprinted in 33 ILM 1016, 1994.

Convention on Cooperation for the Protection and Sustainable Use of the Danube River, Sofia, June 29, 1994, available at http://www.icpdr.org/

Convention on the Protection of the Marine Environment and the Coastal Region of the Mediterranean, Barcelona, June 10, 1995, reprinted in 6 *Yearbook of International Environmental Law (YBIEL)* 887, 1995.

Agreement for the Conservation and Management of Straddling Fish Stocks and Highly Migratory Fish Stocks (Big Fishes Accord), New York, December 4, 1995, reprinted in 34 ILM 1542, 1995.

Treaty Creating the Court of Justice of the Andean Community (amended by the 1996 Cochabamba Protocol), 18 ILM 1203, 1979.

Additional Protocol to the Treaty Establishing the Andean Parliament, 1997, http://www.comunidadandina.org/INGLES/normativa/ande_trie5.htm

Treaty Creating the Court of Justice of the Andean Community Amended by the Cochabamba Protocol on May 28, 1996.

Protocol of Modification of the Treaty Establishing the Court of Justice of the Andean Community, 1996, http://www.comunidadandina.org/ingles/normativa/ande_trie2.htm

Additional Protocol to the Treaty establishing the Andean Parliament, 1997, http://www.comunidadandina.org/ingles/treaties.htm

Convention on the Law Relating to the Uses of International Watercourses for Purposes Other Than Navigation, New York, May 21, 1997, reprinted in 36 ILM 719, 1997.

Protocol to the United Nations Framework Convention on Climate Change, Kyoto, December 11, 1997, reprinted in 37 ILM 22, 1998.

Protocol on Persistent Organic Pollutants, Aarhus, June 24, 1998, reprinted in 37 ILM 505, 1998.

Convention on Access to Information, Public Participation in Decision-Making and Access to Justice in Environmental Matters, Aarhus, June 25, 1998, reprinted in 38 ILM 517, 1999.

Cartagena Protocol on Biosafety, Montreal, January 29, 2000, reprinted in 39 ILM 1027, 2000.

Agreement Establishing the Caribbean Court of Justice, 2001, available at: http://www.caricom.org

Convention on Persistent Organic Pollutants, Stockholm, May 22, 2001, reprinted in 40 ILM 532, 2001.

Protocol of Olivos, 2002, reprinted in 42 ILM 2, 2003.

Non-Legally Binding Authoritative Statement of Principles for a Global Consensus on the Management, Conservation and Sustainable Development of All Types of Forests, A/CONF.151/26, 2007.

Union of South American Nations (UNASUR) Treaty, 2008, http://www.comunidadandina.org/ingles/csn/treaty.htm

United Nations Collaborative Program on Reducing Emissions from Deforestation and Forest Degradation in Developing Countries (UN-REDD), *UN-REDD Framework Document*, 2008, http://www.un-redd.org

2. *United Nations General Assembly Resolutions*

Res. 96 (I), December 11, 1946.

Res. 174 (III), September 21, 1947.

Res. 523 (VI), January 12, 1951.

Res. 626 (VII), December 21, 1952.

Res. 837 (IX), December 14, 1954.

Res. 1314 (XIII), December 12, 1958.
Res. 1515 (XV), December 15, 1960.
Res. 1803 (XVII), December 14, 1962.
Res. 1962 (XVIII), December 13, 1963.
Res. 2158 (XXI), November 25, 1966.
Res. 2625 (XXV), October 24, 1970.
Res. 2692 (XXV), December 11, 1970.
Res. 2749 (XXV), December 17, 1970.
Res. 2996 (XXVII), December 15, 1972.
Res. 2997 (XXVII) of December 15, 1972.
Res. 3129 XXVIII, December 13, 1973.
Res. 3201 (S-VI), May 1, 1974.
Res.3281 (XXIX), December 13, 1974.
Res. 3314 (XXIX), December 14, 1974.
Res. 3362 (S-VII), September 16, 1975.
Res. 3340 (XXX), December 9, 1975.
Res. 35/8, October 30, 1980.
Res. 36/7, October 27, 1981.
Res. 37/217, December 20, 1982.
Res. 38/165, December 19, 1983.
Res. 40/200, December 17, 1985.
Res. 41/128, December 4, 1986.
Res. 42/184, December 11, 1987.
Res. 42/229, December 22, 1989.
Res. 45/94, December 14, 1990.
Res. 52/136, December 12, 1997.
RES. 55/2, September 18, 2000.
Res. 56/150, February 8, 2002.
Res. 61/169, February 27, 2007.

3. *Other UN Documents*

A/5209 and Corr.1, June 26, 1962.
A/CONF. 48/10/Annex I, December 10, 1971.
A/CONF. 48/14 v. 3.7.1972.
A/CONF.151/6/Rev.1, 1992.
A/CONF.151/26, vol. III, August 14, 1992.
A/CONF.199/PC/8, March 19, 2002.
A/CONF.199/20, September 4, 2002.
A/C.2/62/L.5, 2007.

4. *United Nations Economic and Social Council Decisions and Resolutions*

ECOSOC Res. 32 (IV), 1947.
ECOSOC Res. 288 B(X), February 27, 1950.

5. *Resolutions and Decisions of Other Organizations*

Amazon Cooperation Treaty Organization (ACTO)

RES/VII MRE-TCA/7.
RES/V MRE-TCA/1.
RES/V MRE-TCA/4.
RES/VI MRE-TCA/6.
RES/VI CCA 2.
RES/VII MRE-TCA/6.
RES/IX CCA-TCA/6.
RES/VI MRE-TCA/1.
RES/VII MRE-TCA/4.
RES/VII MRE-TCA/5.
RES/IX CCA-TCA/8.
RES/V MRE-TCA/6.
RES/V MRE-TCA/10.
RES/IX MRE-OTCA/07.
RES/VI CCA/4.
RES/VI CCA/1.
RES/VI CCA 3.
RES/V MRE-TCA/3.
RES/V MRE-TCA/2.
RES/VII MRE-TCA/4.
RES/VII MRE-TCA/5.
RES/VII MRE-TCA/8.
RES/VI MRE-TCA/2.
RES/VIII CCOOR-OTCA/02/04.
RES/VI MRE-TCA/6.
RES/VIII MRE-OTCA/09.
RES/IX MRE-OTCA/05.
RES/IX MRE-OTCA/12.
RES/IX MRE-OTCA/04.
TCA/III CCA/SPT BR/W.P. 1/88.

Andean Community (CAN)

Decision 182 of July 25, 1983.
Decision 345 of June 5, 1992.
Decision 391 of July 2, 1996.
Decision 436 of June 11, 1998.
Decision 486 of September 14, 2000.
Decision 523 of July 7, 2002.
Decision 596 of July 11, 2004.
Decision 645 of September 20, 2006.

Common Market of the South (Mercosur)

MERCOSUR/GMC/RES no. 20/95.
MERCOSUR/GMC/RES no. 22/92.
MERCOSUR/GMC/RES no. 10/94.
CMC/DEC 2/94.
CMC/DEC 10/00.
CMC 59/00.
CMC/DEC 02/01.
CMC/DEC 19/03.
CMC/DEC 14/04.
CMC 28/04
CMC 28/05

Economic Commission for Latin America and the Caribbean (ECLAC)

ECLAC Resolution 611(XXX), 2004.

International Tropical Timber Organization (ITTO)

Decision 6(XXXI) of November 3, 2001, on Forest Law Enforcement in the Context of Sustainable Timber Production and Trade.

Decision 7(XXX), on the Role of ITTO in International and Regional Organizations and Fora, Thirtieth Session, May 28 to June 2, 2001, Yaoundé, Cameroon, ITTC (XXX)/19, June 2, 2001.

United Nations Framework Convention on Climate Change (UNFCCC)

Decision 3/CMP.1, FCCC/KP/CMP/2005/8/Add.1, March 30, 2006.
Decision 1/CMP.3, FCCC/KP/CMP/2007/9/Add.1, 2007.

Organization of American States (OAS)

AG/RES. 210 (VI-O/76).
AG/RES. 1022 (XIX-O/89).
AG/RES. 1440 (XXVI-O/96).
AG/RES. 1610 (XXIX-O/99).
AG/RES. 1864 (XXXII-O/02).
AG/RES. 2024 (XXXIV-O/04).
OEA/XLIII.1, CIDI/RIMDS/DEC.1/06 rev. 1.
OEA/Ser.K/XVI, GT/DADIN/doc.301/07.
OEA/Ser.G, CP/CAJP-2757/09.
OEA/Ser.G, CP/CSH-951/08 corr. 1.

United Nations Framework Convention on Climate Change (UNFCC)

Decision 24/CP.7/, FCCC/CP/2001/13/Add.3, adopted at the Seventh Conference of the Parties of the UN Framework Convention on Climate Change, November 10, 2001.

6. *International Law Commission Documents (arranged by date)*

Report by Mr. Roberto Ago, Chairman of Sub-Committee on State Responsibility (approved by the Sub-Committee), contained in UN doc. A/CN.4/152, *Yearbook of the International Law Commission*, vol. II, 1963, pp. 227–28.

Report on the Work of its Fifteenth Session, UN General Assembly, Eighteenth Session, Official Records, Supp. no. 9 (A/5509), *Yearbook of the International Law Commission*, 1966, pp. 11–12.

The International Wrongful Act of the State: Source of International Responsibility, third report on state responsibility by Mr. Roberto Ago, Special Rapporteur, contained in UN doc. A/CN.4/246 and Add. 1–3, *Yearbook of the International Law Commission*, vol. II, 1971.

Draft Article 19 (3) on State Responsibility, *Yearbook of the International Law Commission*, vol. II, part II, 1976, pp. 95–96.

Report on the Work of its Fifty-third Session, contained in doc. UN GAOR, Fifty-sixth Session, Supplement no. 10, UN doc. A/56/10, 2001.

Report of the ILC, Fifty-fifth Session, 2003, UN doc. A/58/10.

First Report on State Responsibility of International Organizations by ILC Special Rapporteur Gaja, contained in UN doc. A/CN.4/532, 2003.

Report on the Work of Its Fifty-eighth Session, General Assembly, Sixty-first Session, Supplement no. 10, UN doc. A/61/10, 2006.

7. *Declarations and Other Soft Law Instruments (arranged by date)*

Resolution XCIII on the Conference on Indian Life, Seventh International Conference of American States, Montevideo, December 3–26, 1933.

Declaration XI on the Protection of the American Indigenous Populations, Resolution XIII on the Problem of the Indigenous Woman, Conference of Experts on Indian Life in the Americas, Eighth International Conference of American States, Lima, December 9–27, 1938.

Declaration of the United Nations Conference on the Human Environment, Stockholm, June 16, 1972.

UNEP Draft Principles of Conduct in the Field of Environment for the Guidance of States in the Conservation and Harmonious Utilization of Natural Resources Shared by Two or More States, approved by the UNEP Governing Council on May 19, 1978 by Decision 6/14 (UNEP/IG. 12/2).

World Charter for Nature, A/RES/37/7, adopted at the General Assembly's Forty-eighth Plenary Meeting, October 28, 1982.

Process of Preparation of the Environmental Perspective to the Year 2000 and Beyond, A/RES/38/161, adopted at the General Assembly's 102nd Plenary Meeting, December 19, 1983.

Protection of Global Climate for Present and Future Generations of Mankind, adopted at the General Assembly's Seventieth Plenary Meeting, December 6, 1988, A/RES/43/53.

San Francisco de Quito Declaration, 1989, in Base Jurídica del Tratado de Cooperación Amazónica: Antecedentes Constitutivos de la Organización del Tratado de Cooperación Amazónica (BJ), La Paz, Bolivia, 2002, pp. 73–79.

Latin America and Caribbean Declaration Sustainable Development, Brasília, March 31, 1989, reprinted in 28 ILM 1303, 1989.

Amazon Declaration, adopted by the presidents of the States parties to the Treaty for Amazonian Cooperation in Manaus, Brazil, May 6, 1989.

Declaration of Caracas, adopted by the Ministers of Foreign Affairs of the Member Countries of the Group of 77, on the Occasion of the Twenty-fifth Anniversary of the Creation of the Group, held in Caracas, Venezuela, June 21–23, 1989.

Declaration on Sustainable Development of the United Nations European Economic Commission, Bergen, May 16, 1990.

Declaration on International Economic Cooperation, A/RES/S-18/3, adopted at the General Assembly's Eleventh Plenary Meeting, May 1, 1990.

Declaration of Canela, 1992, available on Mercosur's Web site: http://www.mercosur.int/msweb/portal%20intermediario/es/index.htm.

Rio Declaration on Environment and Development, UN Conference on Environment and Development, UN doc. A/CONF.151/6/Rev.1, 1992, 31 I.L.M. 874, 1992.

International Co-operation for Economic Growth and Development, A/RES/47/152, adopted at the General Assembly's Ninety-second Plenary Meeting, December 18, 1992.

Declaration of Taranco, 1995, in Secretaria Administrativa del Mercosur, Serie Tematica, Documento no. 3, *Medio Ambiente en el Mercorsur*, 2002, p. 16.

Lima Declaration, 1995, http://www.otca.org.br

Declaration on the Establishment of the Arctic Council, signed by the representatives of the Arctic States in Ottawa on September 19, 1996, 35 ILM 1382.

Manaus Declaration Signed by the Heads of States of the Amazon Countries on February 10–11, 1992, in, BJ, p. 42.

Declaration of Iquitos, 2005, BJ, p. 96.

Paracas Declaration, April 1, 2005, http://www.comunidadandina.org.

Declaration of Heads of States of the Amazon Cooperation Treaty Organization, November 26, 2009, http://www.otca.org.br

8. *Statements and Speeches*

ADRIAZOLA VALDA, Oscar, speech of the Bolivian minister of foreign affairs at a celebration ceremony offered by the Brazilian government to the ACT minister of foreign affairs, July 4, 1978, "Governo brasileiro condecora chanceleres dos países amazônicos," Brazilian Ministry of Foreign Affairs, *Resenha de Politica Exterior*, vol. 18, 1978, pp. 23–24.

AMAZON COOPERATION TREATY ORGANIZATION, "OTCA busca posição comum amazônica sobre florestas," SP-OTCA Comunicado no. 26, Brasília, April 1, 2005.

CARDOSO, F.H., speech of the Brazilian president at the signing ceremony of the headquarters agreement between Brazil and the Amazon Cooperation Treaty Organization, "Discurso do presidente da República, Fernando Henrique Cardoso, na cerimônia de assinatura do Acordo de Sede entre o Brasil e a Organização do Tratado de Cooperação Amazônica," http://www.radiobras.gov.br.

CONSALVI, Alberto, speech of the Venezuelan minister of foreign affairs at the ACT signing ceremony in Brasília on July 3, 1978, "A assinatura do tratado de cooperação multilateral na Amazônia," Brazilian Ministry of Foreign Affairs, *Resenha de Política Exterior do Brasil*, no. 18, 1978, pp. 11–13.

DA SILVEIRA, Azeredo, speech of the Brazilian minister of foreign affairs (Azeredo da Silveira) at the Brazilian National Parliament, "Silveira no congresso expõe bases da diplomacia brasileira," Brazilian Ministry of Foreign Affairs, *Resenha de Política Exterior*, vol. 10, 1976, pp. 57–65.

speech of the Brazilian minister of foreign affairs at the opening session of the First Preparatory Meeting of the Amazon Cooperation Treaty, "Silveira abre a primeira reunião preparatória do pacto amazônico," Brazilian Ministry of Foreign Affairs, *Resenha de Política Externa*, vol. 15, 1977, pp. 53–55.

interview of the Brazilian minister of foreign affairs with Venezuelan television on November 14, 1977, "Silveira explica política externa brasileira à tv da Venezuela," Brazilian Ministry of Foreign Affairs, *Resenha de Política Externa*, vol. 15, 1977, pp. 11–15.

speech of the Brazilian minister of foreign affairs on the occasion of the official dinner offered to the ministers of foreign affairs, "Tratado amazônico reafirma unidade fundamental dos países latino-americanos," Brazilian Ministry of Foreign Affairs, *Resenha de Política Exterior do Brasil*, vol. 18, 1978, pp. 13–15.

GEISEL, Ernesto, speech of the Brazilian president at the Amazon Cooperation Treaty signing ceremony in Brasilia on July 3, 1978, "Tratado de Cooperação Amazônica," Brazilian Ministry of Foreign Affairs, *Resenha de Política Exterior do Brasil*, Brasília, 1978, pp. 59–65.

speech of the Brazilian president on the occasion of the official dinner offered to the Venezuelan president (Carlos Andrés Perez) in Brasília, November 16, 1977, Brazilian Ministry of Foreign Affairs, "Entre Brasil e Venezuela, idêntico espírito de cooperação," *Resenha de Política Exterior do Brasil*, vol. 15, 1977, pp. 15–17.

MEIRA FILHO, Luiz Gilvan, speech in Federação das Indústrias de São Paulo – FIESP (São Paulo Industrial Federation), São Paulo, Brazil, January 20, 1999.

II. Reports and Yearbooks

AMAZON COOPERATION TREATY ORGANIZATION (ACTO), *Base Jurídica del Tratado de Cooperación Amazónica: Antecedentes Constitutivos de la Organización del Tratado de Cooperación Amazónica*, La Paz: Pro Tempore Secretariat, 2002.

A Proposal for Defining the Geographical Boundaries Amazonia, Luxembourg: European Commission, Directorate General, Joint Research Center, Office for Official Publications of the European Communities, 2005.

Integrando la Amazonía Continental, report, November 2005–October 2006, Brasília: ACTO Permanent Secretariat, 2007.

Management Report 2007–2008, Brasília: ACTO Permanent Secretariat, 2008.

ANDEAN DEVELOPMENT CORPORATION (CAF), *Annual Report*, 2005.

Annual Report, 2006, http://www.caf.com/view/index.asp?ms=17&pageMs=43796

La CAF y el medio ambiente: Estratéegia ambiental de la CAF, 2007, http://www.caf.com/attach/18/default/FINAL-small-sinblancas.pdf

Environmental Strategy 2008, Caracas: CAF Publications, 2008.

ANAYA, James, *Report of the Special Rapporteur on the Situation of Human Rights and Fundamental Freedoms of Indigenous People: Report on the Situation of Human Rights of Indigenous Peoples in Brazil*, A/HRC/12/34/Add.2, Human Rights Council, Twelfth session, August 26, 2009.

ARANIBAR QUIROGA, A., *La OTCA, algunas claves de su evolución histórica*, Brasília, July 2003.

ARANIBAR QUIROGA, A. and MARCOVITCH, J., *Lineamientos estratégicos para la Organización del Tratado de Cooperación Amazónica: contribuciones para la reflexión a 25 años de suscripción del Tratado de Cooperación Amazónica*, report prepared for the ACTO Permanent Secretariat, 2003.

BRAZILIAN INSTITUTE OF GEOGRAPHY AND STATISTICS (IBGE), *Anuário Estatístico do Brasil*, Brasília, 1980.

Censo agropecuário 2006, Rio de Janeiro: IBGE, 2006.

Report on the Agricultural Frontier in Amazônia Legal, Brasília: IBGE, 2007.

BRAZILIAN MINISTRY OF THE ENVIRONMENT, *What Brazilians Think of Environmental Issues and Sustainable Development*, National Opinion Poll (1992, 1997, and 2001).

O Estado das Águas no Brasil 2001–2002, Brasília, 2003.

Estratégias nacionais de biodiversidade na América do Sul: perspectivas para a cooperação regional, Brasília: Secretaria de Biodiversidade e Florestas, 2004.

Ministério do Meio Ambiente apresenta estudos sobre mudanças climáticas e biodiversidade, Brasília, 2007, http://www.meioambiente.gov.br

BRAZILIAN MINISTRY OF EXTERNAL RELATIONS et al., *Brazil's Contribution to Prevent Climate Change*, Brasília, 2007.

BRAZILIAN MINISTRY OF SCIENCE AND TECHNOLOGY, *Current Status of the Project Activities under the Clean Development Mechanism (CDM) in Brazil and the World*, 2008, p. 3, http://www.unfccc.org

CARRERA DE LA TORRE, L., Organización del Tratado de Cooperación Amazónica: periodo 2004–2005, report prepared for the ACTO Permanent Secretariat, Brasília, 2003.

CASA CIVIL DA PRESIDÊNCIA DA REPÚBLICA et al., *Plano Amazônia Sustentável Diretrizes para o Desenvolvimento Sustentável da Amazônia Brasileira*, Brasília, 2008.

CASTANO-URIBE, C., Diagnostico y situación actual de las áreas protegidas en America Latina y el Caribe 2007, Informe Regional, Corporación Andina de Fomento (CAF), 2008.

CATTANEO, A., *Balancing Agricultural Development and Deforestation in the Brazilian Amazon*, Washington, DC: International Food Policy Research Institute, 2002.

CDM (Clean Development Mechanism), *Clean Development Mechanism: Executive Board Annual Report*, 2009, p. 1, http://www.uab.edu/uasomume/cdm/reports.htm

CHOMITZ, K.M., *At Loggerheads? Agricultural Expansion Poverty Reduction and Environment in the Tropical Forests*, Washington, DC: World Bank, 2007.

CISNEROS, P., McBREEN, J., Superposición de territorios indígenas y áreas protegidas en América del Sur, IUCN, 2010.

COMITE INTERMINISTERIAL SOBRE MUDANÇA DO CLIMA, *Plano Nacional de Mudanca do Clima*, 2008 (Decree 6.263, 21.11.2007), Brasília, 2007.

Plano Amazônia Sustentável Diretrizes para o Desenvolvimento Sustentável da Amazônia Brasileira, Brasília, 2008.

COMMISSION ON THE ENVIRONMENT AND DEVELOPMENT FOR AMAZONIA, *Amazonia without Myths*, Inter-American Development Bank, United Nations Development Program, ACT, 1992.

DOOLEY, K., *Forest Talks at a Standstill as Copenhagen Ends without an Agreement*, EU Forest Watch Report, UNFCCC Climate Talks, December 7–18, 2009, http://fern.org/sites/fern.org/files/Copenhagen%20update.pdf

ECONOMIC COMMISSION FOR LATIN AMERICA AND THE CARIBBEAN (ECLAC), Draft Program of Work of the ECLAC System, 2006–2007, Distr. GENERAL, LC/G.2238(SES.30/6), June 19, 2004.

Anuário da Cepal, Santiago, 2007.

FAO (FOOD AND AGRICULTURE ORGANIZATION), *State of the World's Forests 1997*, Oxford: Worlds and Publications, 1997.

State of the World's Forests 2007, Oxford: Worlds and Publications, 2007.
State of the World's Forests from 1990 to 2005, report, 2007, Oxford: Worlds and Publications, 2007.
State of the World's Forests, Oxford: Worlds and Publications, 2009.
FOE (FRIENDS OF THE EARTH), *Annual Report 2008*, 2008, p. 11, http://www.foe.org/annual-reports
GASQUES, J.E. and YOKOMIZO, C., *Avaliação dos Incentivos Fiscais na Amazônia*, Brasília: IPEA, 1987.
GLOBAL ENVIRONMENTAL FACILITY (GEF), 2008 *Annual Report*, Washington, DC, GEF, 2008.
GREENPEACE INTERNATIONAL, *Eating up the Amazon*, Amsterdam: Greenpeace International, 2006.
Slaughtering the Amazon, Amsterdam: Greenpeace International, 2009.
HALTIA, O. and KEIPI, K., *Financing Forest Investments in Latin America: The Issue of Incentives*, Washington, DC: Inter-American Development Bank, 1997.
HANDL, G. "International Law and the Protection of the Atmosphere," *Proceedings of the Eighty-third American Society of International Law*.
HARDCASTLE, P.D. et al., *Capability and Cost assessment of the Major Forest Nations to Measure and Monitor Their Forest Carbon for Office of Climate Change: Final Report*, Edinburgh: LTS international, April 7, 2008.
HEROLD, M., *An Assessment of National Forest Monitoring Capabilities in Tropical Non-Annex I Countries: Recommendations for Capacity Building*, report commissioned for The Prince's Rainforests Project and The Government of Norway, Jena: Friedrich Schiller University, 2009, http://unfccc.int/files/methods_science/redd/country_specific_information/application/pdf/redd_nat_capacity_report_herold_july09_publ.pdf
INSTITUTO BRASILEIRO DE GEOGRAFIA E ESTATÍSTICA (IBGE), *Statistical Yearbook*, Brasília: IBGE, editions for 1970, 1977, 1980, and 2000.
Report on the Agricultural Frontier in Amazônia Legal, Brasília, 2007.
INTER-AMERICAN COMMISSION ON HUMAN RIGHTS (IACHR), *Second Report on the Situation of Human Rights in Colombia*, OEA/Ser.L/V/II.84 doc. 39 rev., October 14, 1993.
Report on the Situation of Human Rights in Ecuador, OEA/Ser.L/V/II.96, doc. 10 rev. 1, 1997.
Second Report on the Situation of Human Rights in Peru, OEA/Ser.L/V/II.106 doc. 59 rev., June 2, 2000.
The Human Rights Situation of the Indigenous People in the Americas, OEA/Ser.L/V/II.108, doc. 62, October 20, 2000.
Authorities and Precedents in International and Domestic Law for the Proposed American Declaration on the Rights of Indigenous Peoples, OEA/Ser.L/V/II.110, doc. 22, March 1, 2001.
2005 *Annual Report of the Inter-American Commission on Human Rights*, OEA/Ser.L/V/II.124, doc.5, February 27, 2006.
INTER-AMERICAN DEVELOPMENT BANK (IDB), *Blue Ribbon Panel on the Environment: Final Report Recommendations*, Washington, DC: IDB, February 23, 2005.
IDB and the Environment: 1990–2002: The Decade since Rio, IDB briefing report, January 2002.
Environment Strategy Document, GN-2208–4, July 23, 2003.

Environment and Natural Resources Annual Report 2003, Washington, DC: IDB, 2004.
Annual Report 2005, Washington, DC: IDB, 2006.
Environment Division Sustainable Development Department, *Environment and Safeguards Compliance Policy*, GN-2208–18, Washington, DC: IDB, 2006.
Annual Report 2008: The Year in Review, Washington DC: IDB, 2009.

INTER-AMERICAN DEVELOPMENT BANK (IDB) and UN DEVELOPMENT PROGRAM, *Our Own Agenda*, report of the Latin America and Caribbean Commission on the Development and Environment, 1990.

INTERGOVERNMENTAL PANEL ON CLIMATE CHANGE (IPCC), *Climate Change 2007: Synthesis Report*, 2007, http://www.aussmc.org/IPCC_Synthesis_Report.php
Fourth Assessment Report: Climate Change 2007, http://www.ipcc.ch/publications_and_data/publications_and_data_reports.htm

INTERNATIONAL LAW COMMISSION, *Report on the Work of Its Fifty-eighth Session*, General Assembly, Sixty-first Session, Supplement no. 10, A/61/10, 2006.

INTERNATIONAL TROPICAL TIMBER ORGANIZATION (ITTO), *Status of Tropical Forest Management*, 2005, 2006, http://www.itto.int/
Information and Technical Assistance for Production and Trade of Tropical Timber (Brazil), ex-post evaluation report, Project PD 7/94, rev.3 (M,I), Thirty-eighth Session, 2006, Mérida, Yucatán, Mexico.

KLEMM, C. and SHINE, C., "Biological diversity conservation and the law: Legal mechanisms for conserving species and ecosystems," World Conservation Union (IUCN), Environmental Policy and Law Paper, no. 29, Gland, Switzerland, 1993.

LENTINI, M., PEREIRA, D., CELENTANO, D., and PEREIRA, R., *Fatos Florestais da Amazônia*, Belém: Instituto do Homem e Meio Ambiente da Amazonia (IMAZON), 2005.

MARENGO, J.A., *Mudanças climáticas globais e seus efeitos sobre a biodiversidade brasileira*, Brasília: Brazilian Ministry of the Environment, Secretaria de Biodiversidade e Florestas, 2006.

MERCOSUR SECRETARIAT, *Medio Ambiente en el Mercosur*, Montevideo, 2006.

MINISTÉRIO DA INTEGRAÇÃO NACIONAL ET AL., *Plano de Ação para a Prevenção e Controle do Desmatamento na Amazônia Legal*, Brasília, 2004.

MINISTÉRIO DO DESENVOLVIMENTO, SECRETARIA DO COMERCIO EXTERIOR, *Balança comercial Dezembro de 2007*, Brasília: 2007.

MINISTRY OF SCIENCE AND TECHNOLOGY OF BRAZIL/GENERAL COORDINATION ON GLOBAL CLIMATE CHANGE, *Brazil's Initial National Communication to the United Nations Framework Convention on Climate Change*, Brasília: Ministério da Ciência e Tecnologia, 2004.

ORGANIZATION OF AMERICAN STATES (OAS), *Second Report on the Situation of Human Rights in Colombia*, OEA/Ser.L/V/II.84, doc. 39 rev., October 14, 1993.
Inter-Am. C.H.R., *Annual Report* 24, OEA/Ser.L/V/II.66, doc. 10, rev. 1, 1985, doc. 10 rev. 1, 1997.
Report on the Situation of Human Rights in Ecuador, Inter-Am. C.H.R., OEA/Ser.L/V/II.96, doc. 10 rev. 1, 1997.
Report on the Situation of Human Rights in Brazil, Inter-Am. C.H.R., OEA/Ser.L/V/II.97, doc. 29 rev.1 September 29, 1997.
Second Report on the Situation of Human Rights in Peru, OEA/Ser.L/V/II.106, doc. 59 rev., June 2, 2000.

The Human Rights Situation of the Indigenous People in the Americas, OEA/Ser.L/V/II.108, doc. 62, October 20, 2000.

Authorities and Precedents in International and Domestic Law for the Proposed American Declaration on the Rights of Indigenous Peoples, OEA/Ser.L/V/II.110, doc. 22, March 1, 2001.

La Plata River Basin: A Framework for the Sustainable Management of its Water Resources with Respect to the Hydrolocial Effects of Climate Variability and Change, OAS, Water Project Series, no. 6, October 2005.

Amazon River Basin: Integrated and Sustainable Management of Transboundary Water Resources in the Amazon River Basin, Office for Sustainable Development and Environment, Water Project Series, no. 8, October 2005, http://www.oas.org/dsd/Events/english/Documents/OSDE_8Amazopdf

2005 *Annual Report of the Inter-American Commission on Human Rights*, OEA/Ser.L/V/II.124, doc. 5, February 27, 2006.

ORGANIZATION FOR ECONOMIC COOPERATION AND DEVELOPMENT (OECD), *DAC Orientation for Development Cooperation in Support of Private Sector Development*, Paris: OECD, Development Assistance Committee, 1994.

PASQUIS, R. FERREIRA NUNES, B., LE TOURNEAU, F.M., MACHADO, L., and DE MELLO, N.A., *As Amazônias: um mosaico de visões*, report, Brasilia: World Bank, 2003.

PRESIDÊNCIA DA REPÚBLICA, *Plano Amazônia Sustentável: Direstrizes para o Desenvolvimento Sustentável da Amazônia Brasileira*, Brasília, 2008.

Programa de las Naciones Unidas para el Medio Ambiente, Centro Latino Americano de Ecología Social, *GEO Mercosur: Integracion, comercio y medio ambiente en el Mercosur*, Montevideo, 2008.

SALMAN, S.M. and UPRETY, K., *Conflict and Cooperation on South Asia's International Rivers: A Legal Perspective*. Washington, DC: World Bank, 2002.

SCHNEIDER, R. et al., *Amazônia Sustentável: limitantes e oportunidades para o desenvolvimento rural*, Brasília: Instituto do Homem e Meio Ambiente da Amazonia (Imazon), 2000.

TIETENBERG, T., *Private Enforcement of Environmental Regulations in Latin America and the Caribbean: An Effective Instrument for Environmental Management?*, Washington, DC: Inter-American Development Bank, 1996.

TRAFFIC, *Illegal Trade in Wild Animals in Brazil (Transito de comércio Ilegal de animais silvestres no Brasil)*, 1993, http://www.traffic.org/south-america/

CITES Appendix III Implementation for Big leafed Mahogany Swietenia macrophylla, TRAFFIC Online Report Series, no. 1, 2002, http://www.traffic.org

UNITED NATIONS COMMISSION ON SUSTAINABLE DEVELOPMENT, *Report of the Expert Group Meeting on Identification of Principles of International Law for Sustainable Development*, Geneva, Switzerland, September 26–28, 1995; prepared by the Division for Sustainable Development for the Commission on Sustainable Development, Fourth Session, April 18 to May 3, 1996, New York.

UNITED NATIONS DEVELOPMENT PROGRAM (UNDP), *UNDP Annual Report 2009*, New York: UNDP, 2009.

UNITED NATIONS ENVIRONMENTAL PROGRAM (UNEP), *Geo Yearbook 2004/5: An Overview of Our Changing Environment*, Nairobi: UNEP, 2005.

UNITED NATIONS ENVIRONMENTAL PROGRAM (UNEP)/ANA, *Report on the Status of Water Resources in Brazil*, Brasília: UNEP and ANA, GEO Brasil, Water Resources, January 2007.

UNITED NATIONS ENVIRONMENTAL PROGRAM (UNEP), *Annual Report 2009*, Nairobi, UNEP, 2009.

UNEP-RISO Centre, *Energy Climate and Sustainable Development*, n.d., http://cdmpipeline.org/cdm-projects-region.htm#4

UNITED NATIONS ENVIRONMENTAL PROGRAM Brazil Office, *Annual Report 2008*, Brasília: UNEP, 2008.

UNITED NATIONS EDUCATIONAL, SCIENTIFIC AND CULTURAL ORGANIZATION (UNESCO), *General Information on the Conference for the Establishment of the International Institute of the Hylean Amazon*, IIHA/1, Nat.Sci./42, Paris, February 12, 1948.

Provisional Programme of Action of the Hylean Amazon Project during 1948, IIHA/7, Nat. Sci./52, Paris, April 7, 1948.

Final Act of the Conference Creating the International Institute of the Hylean Amazon, Nat. Sci./IIHA/10, Paris, June 14, 1948, Annex 1.

Report on the Measures Undertaken with a View to Creating the International Institute of the Hylean Amazon, UNESCO, doc. NS/IIHA/12, October 18, 1948.

UNITED NATIONS ECONOMIC AND SOCIAL COUNCIL, *Report of the Seventh Session*, Economic and Social Council, E/CN.18/2007/8, adopted at the UNFF Seventh Session, February 24, 2006 and April 16–27, 2007.

UNITED NATIONS SECRETARIAT, *Secretary-General's Bulletin*, February 7, 2000, secs. 11 and 12, ST/SGB/2000/5.

UNITED NATIONS WORLD COMMISSION ON ENVIRONMENT AND DEVELOPMENT, *Our Common Future*, report adopted at the Ninety-sixth General Assembly's Plenary Meeting on December 11, 1987 (A/RES/42/187).

VERWEIJ, P. et al., *Keeping the Amazon Forests Standing: A Matter of Values*, report commissioned by WWF Netherlands, 2009.

VOSTI, S.A, WITCOVER, J., and CARPENTIER, C.L., *Agricultural Intensification by Smallholders in the Western Brazilian Amazon: From Deforestation to Sustainable Land Use*, Washington, DC: International Food Policy Research Institute, 2002.

WALKER, R. et al., Protecting the Amazon with Protected Areas, *Proceedings of the National Academy of Sciences of the United States of America* (PNAS), vol. 106, no. 26, 2009.

WESTHOLM, L. et al., *Assessment of Existing Global Financial Initiatives and Monitoring Aspects of Carbon Sinks in Forest Ecosystems: The Issue of REDD*, Focali report, Gothenburg: Forest, Climate and Livelihood Research Network, 2009.

WORLD BANK, *Strengthening Forest Law, Enforcement and Governance, Addressing a Systemic Constraint to Sustainable Development*, Washington, DC: World Bank, 2006.

Global Monitoring Report: Millennium Declaration Goals and the Environment, Washington, DC: World Bank, 2008.

Annual Report 2009, Washington, DC: World Bank, 2009.

Development and Climate Change: The World Bank at Work, Washington, DC: World Bank, 2009.

WORLD BUSINESS COUNCIL FOR SUSTAINABLE DEVELOPMENT, *DaimlerChrysler – POEMAtec Alliance, Partening for Mutual Success*, Case Study 2004, POEMAtec Alliance Partnering for Mutual Success, 2004.

WORLD CONSERVATION UNION (IUCN), *World Conservation Strategy*, Gland, Switzerland, 1980.

WORLD WILDLIFE FUND FOR NATURE (WWF), *A Road Map for a Living Planet: WWF Annual Review 2008*, 2008, http://www.worldwildlife.org/who/financialinfo/2008AR/

ZARIN, D. et al., *Reducing Emissions from Deforestation and Forest Degradation (REDD): An Options Assessment Report*, prepared for the Government of Norway, 2009, p. 3, http://www.redd-oar.org/links/REDD-OAR_en.pdf

III. Decisions of International and National Tribunals

1. *Judgements and Advisory Opinions of the Permanent International Court of Justice*

Lotus Case (France v. Turkey), 1927 PCIJ Series A, no. 9.

2. *Judgements and Advisory Opinions of the International Court of Justice*

Reparation for Injuries in the Service of the United Nations, Advisory Opinion, *ICJ Reports* 1949.

Corfu Channel (United Kingdom v. Albania), Judgment, *ICJ Reports* 1949.

South West Africa, Second Phase, Judgment, *ICJ Reports* 1966.

Barcelona Traction, Light and Power Company, Limited (Belgium v. Spain), Second Phase, *ICJ Reports* 1970.

Legal Consequences for States of the Continued Presence of South Africa in Namibia (South West Africa) notwithstanding Security Council Resolution 276 (1970), Advisory Opinion, *ICJ Reports* 1971.

Nuclear Tests (Australia v. France), Judgment, *ICJ Reports* 1974.

Nuclear Tests (New Zealand v. France), Judgment, *ICJ Report*, 1974.

Military and Paramilitary Activities In and Against Nicaragua (Nicaragua v. United States of America), Judgement, *ICJ Reports* 1986.

Legality of the Threat or Use of Nuclear Weapons, Advisory Opinion, *ICJ Reports*, 1996.

Gabcíkovo-Nagymaros Project (Hungary v. Slovakia), *ICJ Reports* 1997.

3. *Arbitral Awards*

1922 Swiss Federal Council Arbitral Award (Colombia–Venezuela), United Nations, Reports of International Arbitral Awards, vol. I.

Trail Smelter Case (Canada v. United States), March 11, 1941, *International Environmental Law Reports*, no. 1.

Lac Lanoux Case, Recueil des Sentences Arbitrales, XII, 1957, pp. 281–317.

Obstáculos al Ingreso de Productos Fitosanitarios Argentinos en el Mercado Brasileño, Argentina v. Brazil, Arbitral Award, April 19, 2002 – MERCOSUL.

Productos Fitosanitarios Argentinos en el Mercado Brasileño, Arbitral Award – MERCOSUL.

Aplicacion de Medidas Antidumping contra la exportacion de pollos enteros, provenientes de Brasil, Resolucion no. 574/2000 – MERCOSUL.

Prohibición de Importación de Neumáticos Remoldeados (remolded) Procedentes de Uruguay, Uruguay v. Argentina, Arbitral Award of January 9, 2002 – MERCOSUL.

Arbitral Award, October 25, 2005, Uruguay v. Argentina, *Mercosur Official Gazette*, no. 9, August 2002 – MERCOSUL.

Arbitral Award no. 1/2005 – MERCOSUL.

4. *Decisions of the Court of Justice of the Andean Community (CJAC)*

PROCESO 02-N-86, *Cartagena Agreement Official Gazette*, no. 21, July 15, 1987.
PROCESO 1-IP-87, *Cartagena Agreement Official Gazette*, no. 28, February 15, 1988.
PROCESO 2-IP-88, *Cartagena Agreement Official Gazette*, no. 33, July 26, 1988.
PROCESO 2-AI-97, *Cartagena Agreement Official Gazette*, no. 391, December 11, 1998.
PROCESO 5-IP-89, *Cartagena Agreement Official Gazette*, no. 50, November 17, 1989.
PROCESO 6-IP-93, *Cartagena Agreement Official Gazette*, no. 150, March 25, 1994.
PROCESO 6-IP-94, *Cartagena Agreement Official Gazette*, no. 170, January 23, 1995.
PROCESO 10-IP-94, *Cartagena Agreement Official Gazette*, no. 177, April 20, 1995.
PROCESO 1-AI-96, *Cartagena Agreement Official Gazette*, no. 234, November 21, 1996.
PROCESO 3-AI-96, *Cartagena Agreement Official Gazette*, no. 261, April 29, 1997.
PROCESO 8-AI-98, *Cartagena Agreement Official Gazette*, no. 512, November 26, 1999.
PROCESO 2-AI-97, *Cartagena Agreement Official Gazette*, no. 391, December 11, 1998.
PROCESO 7-AI-98, *Cartagena Agreement Official Gazette*, no. 490, October 4, 1999.
PROCESO 89-AI-2000, *Cartagena Agreement Official Gazette*, no. 722, October 12, 2001.
PROCESO 26-AI-2001, *Cartagena Agreement Official Gazette*, no. 736, November 27, 2001.
PROCESO 01-AI-2001, *Cartagena Agreement Official Gazette*, no. 818, July 23, 2002.
PROCESO 118-AI-2003, *Cartagena Agreement Official Gazette*, no. 1206, June 13, 2005.
PROCESO 192-IP-2006, *Cartagena Agreement Official Gazette*, no. 1478, March 21, 2007.

5. *Caribbean Court of Justice*

Appeal from the Court of Appeal of Barbados, CCJ APPEAL N. CV 2 of 2005.
Trinidad Cement Limited TCL Guyana Incorporated and the State of the Co-Operative Republic of Guyana, August 20, 2009, Original Jurisdiction, Application No. OA 2 of 2009.

6. *Inter-American Court of Human Rights*

Mayagna (Sumo) Awas Tingni Community v. Nicaragua, Preliminary Objections, Judgment of February 1, 2000, I/A Court H.R. Series C, no. 66.
Mayagna (Sumo) Awas Tingni Community v. Nicaragua, Merits, Reparations and Costs; Judgment of August 31, 2001, I/A Court H.R. Series C, no. 79.
Pueblo Indígena de Kankuamo v. Colombia, Judgment of July 2, 2004, I/A Court H.R. Series C, no. 107.
Pueblo indígena de Sarayaku v. Ecuador, Provisional Measures, Order of the Inter-American Court of Human Rights of June 17, 2005, I/A Court H.R. Series C, no. 125.
Yakye Axa Indigenous Community v. Paraguay, Merits, Reparations and Costs, Judgment of June 17, 2005, I/A Court H.R. Series C, no. 125.
Pueblo Indígena Kankuamo (on top, judgment of July 2, 2004), Provisional Measures of July 5, 2004, I/A Court H.R. Series C, no. 109.
Moiwana Community v. Suriname, Preliminary Objections, Merits, Reparations and Costs, Judgment of June 15, 2005, I/A Court H.R. Series C, no. 124.
Yatama v. Nicaragua, Preliminary Objections, Merits, Reparations and Costs, Judgment of June 23, 2005, I/A Court H.R. Series C, no. 127.

Moiwana Village Community v. Suriname, Interpretation of the June 15, 2005, Judgment on the Preliminary Objections, Merits and Reparations, Judgment of February 8, 2006.
Indigenous Community Sawhoyamaxa v. Paraguay, Merits, Reparations and Costs, Judgment of March 29, 2006, I/A Court H.R. Series C, no. 146.
Pueblo Indígena Kankuamo, Provisional Measures, January 30, 2007.

7. *International Tribunal for the Law of the Sea*

MOX Plant case, *Ireland v. United Kingdom*, Provisional Measures Order, December 3, 2001.

8. *Brazilian Courts*

Processo 2003.39.02.001733–3, classe 9.108, MPF Informação Técnica no. 065/00 4a CCR, July 7, 2000, dossier no. 030/2000–4a CCR.
Agravo de instrumento N.2000.01.00.019713/1/PA, Tribunal Regional Federal da 1a Região, September 29, 2003.
Ação de Atentado com pedido de liminar, Ministerio Publico Federal, Procuradoria da Republica no Municipio de Santarém, Santarém, December 12, 2003.
Processo 2003.39.02.001733–3, classe 9.108, Santarém, January 8, 2004.
Agravo de instrumento no. 2004.01.00.000377–2/PA, Processo na Origem 200339020017333, Brasília, January 14, 2004.
Processo 2000.39.02.000141–0 07100, Ação civil Publica Santarém, May 18, 2005.

9. *Mercosur Permanent Court of Appeal*

Decision TPR 1/05.
Decision 01/2006.
Decisions 01/2007.
Decision 01/2008.

IV. Books and Compilations

ABI-SAAB, G., *Cours général de droit international public*, The Hague: Martinus Nijhoff, 1987.
AMADO CERVO, L. *Relações internacionais na América Latina: velhos e novos paradigmas*. Brasília: Instituto Brasileiro de Relações Internacionais, Universidade de Brasília, 2001.
ANDERSON, W., *Caribbean Instruments on International Law*, Barbados: Stone, 1994.
ARANDA, R., *Colección de tratados del Perú*, Lima: Imprenta del Estado, 1843.
ARIMA, E., BARRETO, P., and BRITO, M., *Cattle Ranching in the Amazon: Trends and Implications for Environmental Conservation*, Belém: Instituto do Homem e Meio Ambiente da Amazônia (Imazon), 2006.
ARCHIVO DIPLOMÁTICO DEL PERÚ, *Congresos Americanos de Lima: recopilación de documentos*, vol. 1, Lima, 1938.
ARCOS, A. L., *Facilitación de financiamiento para negocios basados en la biodiversidad y apoyo a actividades de desarrollo de mercados en la región andina*, Bogotá, 2005.

ATAPATTU, S. *Emerging Principles of International Environmental Law*, Ardsley, NY: Transnational, 2006.

BASLAR, K. *The Concept of the Common Heritage of Mankind in International Law*, The Hague/Boston/London: Martinus Nijhoff, 1998.

BARBERIS, J.S., *Los recursos naturales compartidos y el derecho internacional*, Madrid: Editorial Tecnos, 1979.

BARRET, Scott, *Environment and Statecraft: The Strategy of Environmental Treaty-making*, Oxford and New York: Oxford University Press, 2003.

BARRETO, P., ARIMA, E., and VERISSIMO, A., *Florestas nacionais na Amazônia: consulta a empresários e madeireiros e atores afins à política florestal*, Brasília, 2002.

BARRETO, P. et al., *Human Pressure on the Brazilian Amazon Forests*, Belem: IMAZON, Global Forest Watch and World Resources Institute, 2006.

BECKER, B., *Geopolítica da Amazônia*, Secretaria de Assuntos Estratégicos, Documento de Arquivo, 1994.

Amazônia, São Paulo: Atica, 1998.

Amazônia: Geopolítica na virada do III milênio, Rio de Janeiro: Garamond Universitária, 2007.

BENVENISTI, E., *Sharing Transboundary Resources, International Law and Optimal Resources Use*, Cambridge: Cambridge University Press, 2002.

BENHARDT, R. (ed.), *Encyclopedia of Public International Law*, vol. 4, Amsterdam: North Holland, 2000.

BERGONZINI, J-C. and LANLY, J.-P., *Les forêts tropicales*, Montpellier: Librairie du Cirad, 2000.

BIGER, G. (ed.), *The Encyclopedia of International Boundaries*, Jerusalem: Jerusalem Publishing House, 1995.

BILDERBECK, S. (ed.), *Biodiversity and International Law: The Effectiveness of International Environmental Law*, The Netherlands: IOS Press, 1992.

BIRNIE, P. and BOYLE, A. *International Law and the Environment*, Oxford: Oxford University Press, 2002.

BISWAS, Asit K. et al. (eds.), *Management of Latin America River Basins: Amazon, Plata and Sao Francisco*, Tokyo: United Nations University Press, 1999.

BLYDE, J.S., FERNÁNDEZ-ARIAS, E. and GIORDANO, P. (eds.), *Deepening Integration in MERCOSUR: Dealing with Disparities*, Washington, DC: Inter-American Development Bank, 2008.

BOGGS, W. *International Boundaries: A Study of Boundary Functions and Problems*, New York: Columbia University Press, 1940.

BOISSON DE CHAZOURNES, L. *Les contre-mesures dans les relations internationales économiques*, Paris: A. Pedone, 2003.

BOND, I. et al., *Incentives to Sustain Forest Ecosystem Services*, London: IIED, 2009.

BOOTH, W., COLOMBE, G., and WILLIAMS, S., *The Craft of Research*, Chicago and London: University of Chicago Press, 1995.

BOTHE, M., KURZIDEM, T. and SCHMIDT, C., *Amazonia and Siberia: Legal Aspects of the Environment and Development in the Last Open Spaces*, London: Nijhoff, 1993.

BRAZILIAN INSTITUTE OF GEOGRAPHY AND STATISTICS, *Recursos Hídricos: Anuário estatístico do Brasil, Brasília*, 2000, http://www.ibge.gov.br

BRAZILIAN MINISTRY OF THE ENVIRONMENT, BIODIVERSITY AND FORESTS SECRETARIAT, *Estratégias nacionais de biodiversidade na América do Sul: Perspectivas para a cooperação regional, Brasília*, 2004.

BRAZILIAN MINISTRY OF FOREIGN AFFAIRS, DIVISÃO DE ATOS INTERNACIONAIS, *Coleção de Atos Internacionais*, vol. 875, *Brasília*, 1978.

Tratado de Cooperação Amazônica, Brasília, 1978.

BROUWER, M., *Amazon Your Business: Opportunities and Solutions in the Rainforest*, Bunnik: Meindert Brouwer Partner in Communicatie, 2007.

BROWN WEISS, E. (ed.), *Environmental Change and International Law: New Challenges and Dimensions*, Tokyo: United Nations University Press, 1992.

BROWNLIE, I. (ed.), *State Responsibility for Transboundary Air Pollution in International Law*, New York: Oxford University Press, 2000.

Principles of Public International Law, Fifth ed., Oxford: Oxford University Press, 1998.

BRUNDTLAND, Gro Harlem, *Our Common Future*, Oxford: Oxford University Press, 1987.

BUSTAMANTE, M.M.C., NOBRE, C.A., and SMERALDI, R., *Estimating Recent Greenhouse Gas Emissions from Cattle Raising in Brazil*, 2008, http://www.amazonia.org.br/english/

Estimating Recent Greenhouse Gas Emissions from Cattle Raising in Brazil, 2009, http://www.amazonia.org.br/english

CAFLISH, L. and ZACKLIN, R. (eds.), *The Legal Regime of International Rivers and Lakes*, The Hague/Boston/London: Nijhoff, 1981.

CALASANS, J.T., *Le concept de ressource naturelle partagée. Application aux ressources en eau: l'exemple de l'Amérique du Sud*, thèse de doctorat en droit, Université Paris I, Panthéon-Sorbonne, Septentrion Presses Universitaires, 1996.

CALVERT, P. (ed.), *Border and Territorial Disputes of the World*, London: John Harper, 2004.

CAMPIGLIO, Luigi *et al.* (eds.), *The Environment after Rio: International Law and Economics*, London/Dordrecht/Boston: Graham and Trotman/Martinus Nijhoff, 1994.

CANTILLO, A., *Tratados, convenios y declaraciones de paz y de comercio*, Madrid: Imprenta de Alegria y Charlain, 1843.

CAPOOR, K. and AMBROSI, P., *State and Trends of the Carbon Market 2009*, Washington, DC: World Bank, 2009, p. 34, http://wbcarbonfinance.org/docs/State_Trends_of_the_Carbon_Market_2009-FINAL_26_May09.pdf

CARDENAS, M. CORREA, D. H., and BARON, M. G. (eds.), *Derechos territoriales indigenas y ecologia en las selvas tropicales*, Bogota: Gaia Fundation, 1992.

CAVELIER, G., *Tratados de Colombia 1811–1910*, vol. 1, Bogota: Editorial Kelly.

CBD, UNEP, and WCMC, *Scientific Synthesis of the Impacts of Ocean Fertilization on Marine Biodiversity*, CBD Technical Series, no. 45, 2009, http://www.cbd.int/doc/publications/cbd-ts-42-en.pdf

CENTRE FOR STUDIES AND RESEARCH IN INTERNATIONAL LAW AND INTERNATIONAL RELATIONS, *Water Resources and International Law*, The Hague: Nijhoff, 2001.

CERVO, A.L. and BUENO, C., *História da política exterior do Brasil*, São Paulo: Atica, 1992.

COTULA L. and MAYERS, J., *Tenure in REDD: Start-point or Afterthought?*, London: IIED, 2009.

DA RIVA, A.L.M., LARANJA DA FONSECA, L.F., and HASENCLEVER, L., *Instrumentos econômicos e financeiros para a conservação ambiental no Brasil: Uma análise do estado de arte no Brasil e no Mato Grosso, desafios e perspectivas*, São Paulo: Instituto Sócio Ambiental, 2007.

DA SILVEIRA, E.D., *Direito ambiental: Tratado de Cooperação Amazônica*, Curitiba: Juruá Editora, 2005.

DE LA TORRE, A., FAJNZYLBER, P., and NASH, J., *Low Carbon High Growth Latin America Responses to Climate Change: An Overview*, Washington, DC: World Bank Latin America and Caribbean Studies, 2009.

DENSHAM, A. et al., *Carbon Scam: Noel Kempff Climate Action Project and the Push for Sub-national Forest Offsets*, Amsterdam:Greenpeace International, 2009.

DE OLIVEIRA, J.M., *Actos diplomáticos do Brasil*, vol. 1, Rio de Janeiro: Jornal do Commercio, 1912.

DE SADELEER, N., *Environmental Principles: From Political Slogans to Legal Rules*, New York: Oxford University Press, 2002.

DUPUY R.-J. and VIGNES D. (eds.), *Traité du nouveau droit de la mer*, Paris: Economica, 1985.

EBESSON, J., *Compatibility of International and National Environmental Law*, London, The Hague and Boston, Kluwer Law International, 1996.

Ecosystem Marketplace, *Forests: Taking Root in the Voluntary Carbon Market*, 2009, http://www.katoombagroup.org/documents/cds/katoomba_xiv/documents/Forests%20II/ECM%20Forests%20II%20English%20-%20final.pdf

FALK, R. (ed.), *Peaceful Uses of International Rivers: The Euphrates and Tigris Rivers Dispute*, Ardsley: Transnational, 2002.

FAO (Food and Agriculture Organization), *FAO Strategic Plan for Forestry*, Rome: FAO, 1999.

Basic Texts of FAO, Rome: FAO, 2006.

State of the World's Forests, Rome: FAO, 2007.

State of the World's Forests, Rome: FAO, 2009.

FREESTONE, D. and HEY, E. (eds.), *The Precautionary Principle and International Law: The Challenge of Implementation*, The Hague/London/Boston: Kluwer Law International, 1996.

FREGAPANI, G., *Amazônia e a grande cobiça internacional*, Brasília: Editora Thesaurus, 2000.

FRIEDMANN, Wolfgang, *The Changing Structure of International Law*, London: Stevens, 1964.

FRIENDS OF THE AMAZON, BRAZILIAN AMAZON, *Time to Pay the Bill: The Current Situation of Cattle Ranching in the Amazon*, São Paulo: Friends of the Amazon, Brazilian Amazon, 2009.

GALE, F.P, *The Tropical Timber Trade Regime*, New York: St. Martin's Press, 1998.

GASQUES, J.G., CONCEICAO, J. and BASTOS, T., *Indicatores de Estrutura Fundiária na Amazonia Legal*, Brasília: DISET/IPEA, 2000.

GASQUES, J.G. and YOKOMIZO, C., *Avaliação dos Incentivos Fiscais na Amazônia*, Brasília: Instituto de Pesquisas Econômicas Aplicadas, IPEA, 1987.

GIROT, P. (ed.), *The Americas: World Boundaries*, vol. 4, London: Routledge, 1994.

GOIDANICH, S.P ., *Amazônia Internacionalização e Cooperação*, master's dissertation, Faculdade de Direito da Universidade de São Paulo, 1997.

GONÇALVES, C. Walter, *Conflitos no Campo Brasil 2006*, Comissão Pastoral da Terra, 2007.

GOODMAN, D. and REDCLIF, M. *Environment and Development in Latin America: The Politics of Sustainability*, Manchester: Manchester University Press, 1991.

GOULDING, M., BARTHEM, R.B., and FERREIRA, E., *The Smithsonian Atlas of the Amazon*, Washington, DC: Smithsonian Books, 2003.

GREGERSEN, H. and CONTRERAS, A., *Rethinking Forest Regulations: From Simple Rules to Systems to Promote Best Practices and Compliance*, Washington, DC: Rights and Resources Initiative, 2010.

GRIFFITHS, T. with F. MARTONE, *Seeing 'REDD'? Forests, Climate Change Mitigation and the Rights of Indigenous Peoples and Local Communities*, Forest Peoples Programme, 2009, http://www.forestpeoples.org/documents/ifi_igo/seeing_redd_update_may09_eng.pdf

GROGAN, J., BARRETO, P., and VERÍSSIMO, A., *Mogno na Amazônia Brasileira: Ecologia e Perspectivas de Manejo*, Belém: Imazon, 2002.

GUERRERO, E. and SGUERRA, S. (eds.), *Protected Areas and Development in Latin America: From Santa Marta 1997 to Bariloche 2007 and Perspectives for a New Decade*, Bogotá: IUCN Colombian Committee, Parques Nacionales Naturales Colombia and Fundación Natura, Colombia, 2009.

GUZMÁN, Castro Z., and HILLÓN, L.C., *Liniamentos de etiquetado ambiental para Colombia*, Bogotá: Ministerio del Medio Ambiente, 2002.

HALL, A. *Sustaining Amazonia: Grassroots Actions for Productive Conservation*, New York: Manchester University Press, 1998.

HAMILTON, K. et al., *Fortifying the Foundation: State of the Voluntary Carbon Markets 2009*, New York and Washington, DC: Ecosystem Marketplace and New Carbon Finance, 2009.

HAMILTON, K., CHOKKALINGAM, U. and BENDANA, M., *State of the Forest Carbon Markets 2009: Taking Root and Branching Out*, Washington, DC: Ecosystem Marketplace, 2010, http://www.forestcarbonportal.com/resource/state-carbon-market-report

HASS, P., *Saving the Mediterranean: The Politics of International environmental Cooperation*, New York: Columbia University Press, 1990.

KEOHANE, R.O. and LEVY, M.A. (eds.), *Institutions for the Earth: Sources of Effective International Environmental Protection*, Cambridge, MA and London: MIT Press, 1995.

HOLLOWAY, V. and GIANDOMENICO, E., *The History of REDD Policy*, Adelaide: Carbon Planet, 2009.

HOHMANN, H. (ed.), *Basic Documents of International Environmental Law*, 3 vols., London: Nijhoff, 1992.

INSTITUTO SÓCIO AMBIENTAL (ISA), *Instrumentos econômicos e financeiros para a conservação ambiental no Brasil: uma análise do estado da arte no Brasil e no Mato Grosso, Desafios e perspectivas*, 2007, http://www.socioambiental.org

INTERNATIONAL ENERGY AGENCY, *Beyond Kyoto: Energy Dynamics and Climate Stabilisation*, Paris: OECD/IEA, 2002.

INTERNATIONAL UNION FOR THE CONSERVATION OF NATURE (IUCN), *The IUCN Red List of Threatened Species: A Key Conservation Tool*, Gland: IUCN, 2008.

IRELAND, G., *Boundaries, Possessions and Conflicts in South America*, Cambridge, MA: Harvard University Press, 1938.

JACOB WERKSMAN, J.C. and RODERICK, P., *Improving Compliance with International Environmental Law*, London: Earthscan, 1996.

JENNINGS, R. and WATTS, A. (eds.), *Oppenheim's International Law*, vol. 1, London: Longman, 1991.

JOINT RESEARCH CENTRE, *A Proposal for Defining the Geographical Boundaries Amazônia*, Luxembourg: European Commission, Directorate General, JRC, ACTO, Office for Official Publications of the European Communities, 2005.

KATOOMBA GROUP'S ECOSYSTEM MARKETPLACE, *Forests: Taking Root in the Voluntary Carbon Market*, Washington, DC: Katoomba Group, 2009.

KINDGDOM, Sir Donald, *The Laws of British Guyana*, vol. 7, London: Waterlow, 1954.

KISS, A. and SHELTON, D., *International Environmental Law*, New York and London: Transnational, 1991.

KOHEN, M.G., *Possession contestée et souveraineté territoriale*, Paris: Presses Universitaires de France, 1997.

LEAGUE OF NATIONS, *Treaty Series*, vol. CI, Geneva.

LEE, E. and MAHANTY, S., *Payments for Environmental Services and Poverty Reduction Risks and Opportunities*, Center for People and Forests, 2009, http://www.recoftc.org/site/fileadmin/docs/publications/The_Grey_Zone/2009/Issue_PaperISBN.pdf

LELE, U.J., *Brazil: Forests in the Balance: Challenges of Conservation with Development*, Washington, DC: World Bank, Operations Evaluation Department, 2000.

LENTINI, M. et al., *Fatos Florestais da Amazônia*, Belém: Instituto do Homem e Meio Ambiente da Amazônia (Imazon), 2005.

LE PRESTRE, P. (ed.), *Governing Global Biodiversity: The Evolution and Implementation of the Convention on Biological Diversity*, Hampshire and Burlington: Ashgate, 2002.

LOVERA, S. et al., *The Hottest REDD Issues: Rights, Equity, Development, Deforestation and Governance*, Global Forest Coalition and CEESP, 2009, http://www.wrm.org.uy/actors/CCC/cop14/Hottest_REDD.pdf

MACEDO SOARES, J.C., *Fronteiras do Brasil no regime colonial*, Rio de Janeiro: Livraria José Olimpio, 1939.

MAGALHÃES PIMENTA, J.F., *The Treaty of Amazonian Cooperation: An Analysis of the Brazilian Proposal in the Light of Brazil's Regional and International Constraints*, master's thesis, George Washington University, 1982.

MALLOY, W.M., *Treaties, Conventions, International Acts, Protocols and Agreements between the United States and Other Powers, 1776–1909*, vol. 2, Washington, DC: Government Printing Office, 1910.

MARGULIS, S., *Causes of Deforestation in the Brazilian Amazon*, Brazil: World Bank, July 2003.

MARTENS, G.F., *Suppléments au recueil des traités*, vol. 1, Gottingue: Dieterich, 1802–1808.

MCCAFFREY, S., *The Law of International Watercourses, Non-Navigational Uses*, Oxford: Oxford University Press, 2003.

MIRANDA, E.E., *Quando o Amazonas corria para o Pacífico: Uma história desconhecida da Amazônia*, Petrópolis: Editora Vozes, 2007.

MOORE, J.B., *Digest of International Law*, vol. 1, Washington, DC: Government Printing Office, 1906.

MOREIRA ALVES, M.H., *State and Opposition in Military Brazil*, Austin: University of Texas Press, 1985.

MORRISON F.L. and WOLFRUM, R., *International Regional and National Environmental Law*, The Hague: Kluwer International Law, 2000.

MOTTA PINTO COELHO, P., *Fronteiras na Amazônia, Um espaço integrado*, Brasilia: Fundação Alexandre Gusmão, Instituto de Pesquisa de Relações Internacionais, 1992.

MUNOZ, H., *Environment and Diplomacy in the Americas*, London: Lynne Rienner, 1992.

OGONOWSKI, M. et al., *Utilizing Payments for Environmental Services for Reducing Emissions from Deforestation and Forest Degradation (REDD) in Developing Countries: Challenges and Policy Options*, Washington, DC: Center for Clean Air Policy, 2009.

OLSEN, N. and BISHOP, J. *The Financial Costs of REDD: Evidence from Brazil and Indonesia*, Gland: IUCN, 2009.

ORGANIZATION OF AMERICAN STATES (OAS), *The Inter-American System: Treaties Conventions and Other Documents* (compilation annotated by F.V. García-Amador), vol. 1, Washington, DC, Sub-Secretary of Legal Affairs, General Secretary of the OAS, 1982.

PAEHLKE, R. (ed.), *Conservation and Environmentalism: An Encyclopedia*, New York and London: Garland, 1995.

PAN AMERICAN UNION, *Manual of Inter-American Relations: A Systematic Classification of the Treaties, Conventions, Resolutions, Declarations and Recommendations Adopted at Inter-American Conferences and Meetings of Consultation*, Washington, DC: Division of Conferences and Organizations, Department of International Law, Pan American Union, 1956.

PAQUEROT, S., *Le statut des ressources vitales en droit international: essai sur le concept de patrimoine commun de l'humanité*, Bruxelles: Bruylant, 2002.

PARK, C.C., *Tropical Rainforests*, London and New York: Routledge, 1992.

PARKER, C. et al., *The Little Climate Finance Book: A Guide to Financing Options for Forests and Climate Change*, Oxford: Global Canopy Foundation, 2009.

PARRY, C., *The Consolidated Treaty Series*, vol. 46 (1775–1778), London: Oceana, 1979.

PAUWELYN, J., *Conflict of Norms in Public International Law: How WTO Law Relates to Other Rules of International Law*, London and New York: Cambridge University Press, 2003.

PEARCE, D. and MORAN, D., *The Economic Value of Biodiversity*, London: Earthscan, 1994.

PERREZ, F.X., *Cooperative Sovereignty: From Independence to Interdependence in the Structure of International Environmental Law*, The Hague: Kluwer Law International, 2000.

PESKETT, L. et al., *Making REDD Work for the Poor*, prepared on behalf of the Poverty Environment Partnership, 2008, p. 23, http://www.unep-wcmc.org/climate/pdf/Making%20REDD%20work%20for%20the%20poor%20FINAL%20DRAFT%200110.pdf

PINTO, L.F., *Amazônia Brasileira no rastro do saque*, São Paulo: Editora Hucitec, 1980.

Plataforma Climática Latinoamericana, *Cambio Climático y COP15: Reflexiones desde y para America Latina*, Quito: Plataforma Climática Latinoamericana, 2009.

PRINCE'S CHARITIES FOUNDATION, *The Prince's Rainforests Project: An Emergency Package for Tropical Forests*, London, 2009, http://www.rainforestsos.org/pages/emergency-package/

PROCOPIO, A. *L'Amazonie et la mondialisation*, Paris: L'Harmattan, 2000.

Ragazzi, M., *The Concept of International Obligations*, Erga Omnes, Oxford and New York: Clarendon Press, 1997.

RANKINE, H. et. al., *Innovative Socio-Economic Policy for Improving Environmental Performance: Payments for Ecosystem Services*, Bangkok: United Nations Economic and Social Commission for Asia and the Pacific, 2009, p. 10., http://www.unescap.org/esd/environment/publications/PES/ESCAP_PES%20Publication%202560.pdf

REDONDO-BRENES, A., "Payment for environmental services in Costa Rica: Carbon sequestration estimations of native tree plantations," *Tropical Resources Bulletin*, vol. 24, 2005, pp. 20–29.

REIS, A.C., *A Amazonia e a cobiça internacional*, Rio de Janeiro: Editora Nova, 1965.

REY CARO, E.J., *El derecho de los cursos de agua internacionales: la navigacion*, Cordoba: Universidad Nacional, 1986.

ROMANO, C.P.R., *The Peaceful Settlement of International Environmental Disputes: A Pragmatic Approach*, The Hague/London/Boston: Kluwer Law International, 2000.

SALMAN, S.M. and UPRETY, K., *Conflict and Cooperation on South Asia's International Rivers: A Legal Perspective*, Washington, DC: World Bank, 2002, p. 193.

SALMON, J. (ed.), *Dictionnaire de droit international public*, Brussels: Universités Francophones, Bruylant, 2001.

SAND, P.H. (ed.), *The Effectiveness of International Environmental Agreements: A Survey of Existing Legal Instruments*, Cambridge: Grotius, 1992.

SANDLER, T. *Global Challenges*, Cambridge: Cambridge University Press, 1997.

SANDS, P. *Principles of International Environmental Law: Frameworks, Standards and Implementation*, Manchester: Manchester University Press, 1995.

Principles of International Environmental Law, Second ed., Cambridge: Cambridge University Press, 2003.

SANDS, P. and GALIZZI, P., *Documents in International Environmental Law*, Cambridge: Cambridge University Press, 2004.

SAUTCHUK, J., *Projeto Jari: A invasão Americana*, São Paulo: Editora Brasil Debates, 1979.

SCHERMERS, H.G. and BLOKKER, N.M., *International Institutional Law: Unity within Diversity*, Fourth revised ed., Leiden and Boston: Martinus Nijhoff, 2003.

SCHOLZ, I., *O Desenvolvimento de Instituições de Política Ambiental na Amazônia a partir de 1992: A contribuição do PPG7*, Brasília: Ministério do Meio Ambiente, http://www.mma.gov.br/ppg7/

SCHRIJVER, N. *Sovereignty Over Natural Resources: Balancing Rights and Duties*, Cambridge: Cambridge University Press, 1997.

SECRETARÍA NACIONAL DE PLANIFICACIÓN Y DESARROLLO, *Plan Nacional para el Buen Vivir: Construyendo un Estado Plurinacional e Intercultural, 2009–2013*, 2009, p. 215, http://documentacion.asambleanacional.gov.ec/alfresco/d/d/workspace/SpacesStore/9fdcbda0–4f5a-4280-abe3–2dd9f1e79f08/PNBV_2009–2013.pdf

SECRETARIAT OF THE CONVENTION ON BIOLOGICAL DIVERSITY, *Handbook of the Convention on Biological Diversity*, Quebec, Canada, 2005.

SHANY, Y. *The Competing Jurisdiction of International Courts and Tribunals*, New York: Oxford University Press, 2003.

SHAW, M. N., *International Law*, Cambridge: Cambridge University Press, 2003.

SICKING, L., *Frontières d'Outre-Mer: la France et le Pays-Bas dans le monde atlantique au XIXe siècle*, Paris: Les Indes Savantes, 2006.

SILVA OLIVEIRA, L.P., *Globalização e soberania: o Brasil e a biodiversidade Amazônica*, Brasília: Fundação Milton Campos, Conselho de Reitores de Universidades Brasileiras, 2002.

SIMMA, B. (ed.), *The Charter of the United Nations: A Commentary*, Second ed., New York: Oxford University Press, 2002.

SMERALDI, R. and MAY, P.H., *The Cattle Realm: A New Phase in the Livestock Colonization of the Brazilian Amazon*, São Paulo: Amigos da Terra, 2008.

SNYDER, F.E. and SATHIRATHAI, S., *Third World Attitudes Toward International Law*, Dordrecht/Boston/Lancaster: Martinus Nijhoff, 1987.

SOSA-RODRIGUEZ, C., *Le droit fluvial international et les fleuves de l'Amérique Latine*, Paris: Pedone, 1935.

SOUSA FRANCO FILHO, G., *O Pacto Amazônico: idéias e conceitos*. Belém: Falangola, 1979.

STERN, N., *Stern Review: The Economics of Climate Change*, Cambridge: Cambridge University Press, 2007, http://www.hm-treasury.gov.uk/sternreview_index.htm

SWANSON, T. and JOHNSTON, S., *Global Environmental Problems and International Environmental Agreements: The Economics of International Institutional Building*, Northampton, MA: Edward Elgar, 1999.

SWITKES, G., *Governo brasileiro avança com o projeto de construir barragens*, Americas Program, Center for International Policy, June 12, 2007, http://americas.irc-online.org

TSCHARNTKE, T. et al. (eds.), *The Stability of Tropical Rainforest Margins, Linking Ecological, Economic and Social Constraints of Land Use and Conservation: An Introduction*, Heidelberg: Springer, 2007.

VICTOR, D.G., RAUSTIALA, K., and SKOLNIKOFF, E.B. (eds.), *The Implementation and Effectiveness of International Environmental Commitments: Theory and Practice*, London: MIT Press, 1998.

UNITED KINGDOM (UK), *British and Foreign State Papers*, various vols., London: William Ridgway.

UNITED NATIONS (UN), *Agenda 21*, UN doc. A/CONF.151/26, August 12, 1992.

UNITED NATIONS CONFERENCE ON TRADE AND DEVELOPMENT (UNCTAD), *Developing Country Interests in Climate Change Actions and Implications for a Post-2012 Climate Change Regime*, Rome:, UNCTAD, 2009.

UNITED NATIONS DEVELOPMENT PROGRAM (UNDP), *MDG Carbon Facility: Leveraging Carbon Finance for Sustainable Development*, UNDP, New York: UNDP, 2007.

Climate Change at UNDP: Scaling Up to Meet the Challenges, New York: UNDP, 2008.

UN ECONOMIC AND SOCIAL COUNCIL (ECOSOC), *Illicit Trafficking in Protected Species of Wild Flora and Fauna and Illicit Access to Genetic Resources*, report of the secretary general, E/CN.15/2003/8, Commission on Crime Prevention and Criminal Justice, Twelfth Session, held in Vienna, May 13–22, 2003.

UNITED NATIONS ENVIRONMENTAL PROGRAM (UNEP), *Integrated and Sustainable Development of the Trans-boundary Water Resources in the Amazon River Basin*, Global Environment Facility Project document, http://www.otca.org.br/en/programs-projects/index.php?id=1372

Manual on Compliance with and Enforcement of Multilateral Environmental Agreements, Nairobi: UNEP, 2006.

UNITED NATIONS ENVIRONMENT PROGRAMME (UNEP), Amazon Cooperation Treaty Organization (ACTO) and Research Center of Universidad del Pacífico (CIUP), *Environment Outlook in Amazonia, Geo Amazonia*, 2009, http://www.unep.org/pdf/GEOAMAZONIA.pdf

UNITED NATIONS GENERAL ASSEMBLY (UNGA), *Report of the UN Conference on the Human Environment*, UN doc. A/CONF.48/14/Rev.1.

UNITED NATIONS SYSTEM CHIEF EXECUTIVES BOARD FOR COORDINATION (CEB), *Acting on Climate Change: The UN System Delivering as One*, New York: CEB, 2008.

VENEZUELAN MINISTRY OF FOREIGN AFFAIRS, *Collección de tratados públicos de Venezuela*, Caracas: Imprenta Nacional, 1910.

VIANA, V.M., *Seeing REDD in the Amazon: A Win for People, Trees and Climate*, International Institute for Environment and Development, 2009, http://www.iied.org/pubs/pdfs/17052IIED.pdf

VIANA, V. et al., *Reserva de Desenvolvimento Sustentável do Juma: o primeiro projeto REDD na Amazônia Brasileira*, Manaus: Fundação Amazônia Sustentável, 2008.

VIÉ, J.-C., HILTON-TAYLOR, C., and STUART, S.N., *Wildlife in a Changing World, An Analysis of the 2008 IUCN Red List of Threatened Species*, Gland: IUCN, 2008.

VIRGILIO, N., *Noel Kempff Mercado Climate Action Project: A Case Study in Reducing Emissions from Deforestation and Degradation*, Arlington: Nature Conservancy, 2009.

WEISS, E.B., MCCAFFREY, S., MAGRAW, D., SZASZ, P., and LUTZ, R., *International Environmental Law and Policy*, New York: Aspen Law and Business, 1998.

WERTZ-KANOUNNIKOFF, S., KONGPHAN-APIRAK, M., and WUNDER, S., *Reducing Forest Emissions in the Amazon Basin: A Review of Drivers of Land-use Change and How Payments for Environmental Services (PES) Schemes Can Affect Them*, Bogor: Center for International Forestry Research (CIFOR), November 2008.

WORLD BANK, *A More Equitable, Sustainable and Competitive Brazil: Country Assistance Strategy 2004–2007*, Washington, DC: World Bank, 2003.

WORLD BANK BRAZIL RAINFOREST UNIT, *Lessons from the Rain Forest, Experiences of the Pilot Program to Conserve the Amazon and Atlantic Forests of Brazil*, Ministry of the Environment, Secretariat for Coordination of the Amazon, Pilot Program to Conserve the Brazilian Rain Forest Monitoring and Analysis Project, Washington, DC: World Bank Brazil Rainforest Unit, 2002.

WOUTERS, P., *Rivers of the World, Fundamental Principles of the Law of International Watercourses*, thèse no. 596, Geneva: Institut Universitaire de Hautes Études Internationales, 1999.

YOUNG, H.P., *Equity in Theory and Practice*, Princeton: Princeton University Press, 1994.

Creating Regimes: Arctic Accords and International Governance, Ithaca and London: Cornell University Press, 1998.

YOUNG, H.P. and LEVY, M.A. (eds.), *The Effectiveness of International Environmental Regimes: Causal Connections and Behavioral Mechanisms*, Cambridge, MA and London: MIT Press, 1999.

V. Articles, Courses, Contributions to Collective Works

ABI-SAAB, G. "Cours générale de droit international public," *Recueil des Cours de l'Académie de Droit International*, vol. 207, The Hague: Martinus Nijhoff, 1987.

"La souveraineté permanente sur les ressources naturelles," in BEDJAOUI, M., *Droit international: bilan et perspectives*, Paris: Editions A. Pedone, 1991.

"The changing world order and the international legal order: The structural evolution of international law beyond the state-centric model," in SAKAMOTO, Y. (ed.), *Global Transformation: Challenges to the State System*, Tokyo: United Nations University Press, 1994, pp. 439–61.

"Some thoughts on the principle of non-intervention," in WELLENS, K. (ed.), *International Law: Theory and Practice: Essays in Honour of Erik Suy*, The Hague: Kluwer Law International, 1998, pp. 225–50.

AGO, R. "Le droit international dans la conception de Grotius," *Recueil des Cours de l'Académie de Droit International*, vol. 182, 1983.

"Conclusions du colloque responsabilité des états pour les dommages à l'environnement," in FRANCIONI, F. and SCOVAZZI, T. (eds.), *International Responsibility for Environmental Harm*, International Environmental Law and Policy Series, London: Graham and Trotman, 1991, pp. 493–99.

AKEHURST, M. "The hierarchy of the sources of international law," *British Yearbook of International Law*, vol. 47, 1974–75, pp. 173–285.

"Equity and general principles of law," *International and Comparative Law Quarterly*, vol. 25, no. 4, 1976, pp. 801–25.

AMARAL SOUZA NETO, E.J., "O Tratado de Cooperação Amazônica," in FRANÇA DANESE, S. (ed.), *Ensaios de historia diplomática do Brasil (1930–1986): Cadernos do Instituto de Pesquisa de Relações Internacionais*, vol. 2, Brasília: Fundação Alexandre Gusmão, 1989, pp. 165–75.

AMERICAN JOURNAL OF INTERNATIONAL LAW, "The arbitral award in the Peru-Bolivia boundary controversy," editorial comment, vol. 3, 1909.

AMORIM, Celso "A Amazônia é nossa," *Jornal Hora do Povo*, September 17, 2004, p. 3, http://www.horadopovo.com.br

ANTONOPOULOS, C. "The principle of uti possidetis iuris in contemporary international law," *Revue Hellénique de Droit International*, vol. 49, 1996.

APESTEGUY, C., MARTINIERE, G., and THERY, H., "Frontières en Amazonie: la politique du Brésil et l'intégration de l'Amérique du Sud," *Problèmes de l'Amérique Latine*, vols. 4533–34, pp. 76–98.

ARTEAGA, R., "Organização do Tratado de Cooperação Amazônica (OTCA): um desafio permanente," *Diplomacia, estratégia, política*, vol. 4, 2006, pp. 87–119.

BALDWIN, S.E., "The international congresses and conferences of the last century as forces working towards the solidarity of the world," *American Journal of International Law*, vol. 1, 1907, pp. 565–78.

BARBERIS, J.S., "Introduction," in *Los recursos naturales compartidos entre estados y el derecho internacinal*, Madrid: Editorial Tecnos, 1979.

"Les règles spécifiques du droit international en Amérique Latine," *Recueil des Cours de l'Académie de Droit International*, vol. 235, 1992, pp. 81–230.

BARBOSA, L.C., "The people of the forest against international capitalism," *Sociological Perspectives*, vol. 39, no. 2, 1996, pp. 317–32.

BECKER, B.K., "Amazonian frontiers in the beginning of the 21st century," presentation at the Open Meeting of the Global Environmental Change Research Community, Rio de Janeiro, October 6–8, 2001.

BELTON, M., "REDD progress at Copenhagen," *Policy Quarterly*, vol. 6, no. 2, 2010, pp. 8–10.

BENATTI, J.H., "Unidades de conservação e as populaces tradicionais: uma análise jurídica da realidade brasileira," *Novos Cadernos*, vol. 2, no. 2, 1999, pp. 107–25.

BERNARDI, M.-J., "Diversité génétique humaine: Éléments d'une politique," *Revue Juridique Themis*, vol. 35, 2001, pp. 327–405.

BIERMANN, F., "Common heritage of humankind: The emergence of a new concept of international environmental law," *Archiv des Völkerrechts*, vol. 34, Hamburg, 1996, pp. 426–81.

BINSWANGER, H. P. "Brazilian policies that encourage deforestation in the Amazon," *World Development*, vol. 19, no. 7, 1991, pp. 821–29.

BLOOM, E.T., "Polar regions and the development of international law," commentary on Oran R. Young, *Creating Regimes: Arctic Accords and International Governance*, Ithaca and London, Cornell University Press, 1998, in Book Reviews and Notes, *The American Journal of International Law*, vol. 92, 1998, pp. 593–95.

BODANSKY, D. "The precautionary principle: Scientific uncertainty and international environmental law," *Proceedings of the American Society of International Law*, Washington, DC, 1991, pp. 413–17.

"Prologue to the Climate Change Convention," in MINTZER, IRVING M. and LEONARD, J.A. (eds.), *Negotiating Climate Change: The Inside Story of the Rio Convention*, Cambridge/New York/Melbourne: Cambridge University Press, 1994, pp. 45–74.

"What's so bad about unilateral action to protect the environment?", *European Journal of International Law*, 2000, vol. 11, no. 2, pp. 339–47.

BOISSON DE CHAZOURNES, L. "La mise en œuvre du droit international dans le domaine de la protection de l'environnement: enjeux et défis," *Revue de Droit International Public*, 1995, pp. 37–76.

"Variations juridiques sur le thème de l'ingérence écologique," in SABELLI, F. (éd.), *Ecologie contre nature: Les Nouveaux Cahiers de L'institut Universitaire d'Études du Développement*, Paris: Presses Universitaires de France, 1995, pp. 53–59.

"La mise en oeuvre du droit international dans le domaine de la protection de l'environnement: enjeux et défis," *Revue Générale de Droit International Public*, no. 1, 1995, pp. 37–76.

"La Convention-Cadre des Nations Unies sur les Changements Climatiques," in RENS, Ivo, (ed.), *Le Droit Internationale face à l'Éthique et à la Politique de l'Environnement*, Genève: SEBES, 1996, pp. 135–41.

"La gestion de l'intérêt commun à l'épreuve des enjeux économiques: le Protocole de Kyoto sur les Changements Climatiques," *Annuaire Français de Droit International*, vol. XLIII, 1997, pp. 700–15.

"The global environmental facility galaxy: On linkages among institutions," In *Max Planck Yearbook of United Nations Law*, vol. 3, The Hague/London/Boston: Kluwer Law International, 1999, pp. 243–85.

"Unilateralism and environmental protection: Issues of perception and reality of issues," *European Journal of International Law*, vol. 11, no. 2, 2000, pp. 315–38.

"À propos de la régulation juridique de stratégies économiques dans le domaines de l'environnement," in MALJEAN-DUBOIS, S. (éd.), *L'outil économique en droit international et Européen de l'environnement*, Paris, Monde Européen et International, collection dirigée par Jacques Bourrinet, Université d'Aix-Marseille III, 2002.

"Le droit international et la promotion des intérêts collectifs: à propos de la protection de l'environnement global," *Publications de la revue Marocaine d'administration locale et de développement*, vol. 48, 2004, pp. 197–98.

"Gouvernance et régulation au 21éme siècle: quelques propos iconoclastes," in BOISSON DE CHAZOURNES, L. and MEHDI, R. (eds.), *Une société internationale en mutation: quels acteurs pour une nouvelle gouvernance?*, Bruxelles: Bruylant, 2005.

"The Global Environmental Facility (GEF): A unique and crucial institution," *Review of European Community and International Environmental Law*, vol. 14, no. 3, 2005, pp. 193–201.

"Technical and financial assistance," in BODANSKY, D. BRUNÉE, J. and HEY, E. (eds.), *The Oxford Handbook of International Law*, Oxford: Oxford University Press, 2007, pp. 948–73.

BOTHE, M., "The impact of international law on the protection of the environment in Amazonia and Siberia," in BOTHE, M., KURZIDEM, T., and SCHMIDT, C. (eds.), *Amazonia and Siberia: Legal Aspects of the Preservation of the Environment and Development in the Last Open Spaces*, London: Graham and Trotman/Martinus Nijhoff, 1993, pp. 237–55.

BOWMAN, M., "The nature, development and philosophical foundations of the biodiversity concept in international law," in BOWMAN, M. and REDGWELL, C. (eds.), *International Law and the Conservation of Biological Diversity*, The Hague: Kluwer Law International, 1996, pp. 5–31.

BOYLE, A. "Remedying harm to international common spaces and resources: Compensation and Other approaches," in WETTERSTEIN, P. (ed.), *Harm to the Environment*, Oxford: Oxford University Press, 1997, pp. 83–100.

"Some reflections on the relationship of treaties and soft law," in GOWLLAND-DEBBAS, V. (ed.), *Multilateral Treaty-Making: The Current Status of Challenges to and Reforms Needed in the International Legislative Process. Papers presented at the Forum Geneva, held in Geneva, Switzerland, May 16, 1998*, The Hague: American Society of International Law, Institut Universitaire des Hautes Études Internationales de Genève and Martinus Nijhoff, 2000, pp. 25–38.

BRANDAO, A.S.P., REZENDE, G.C., and MARQUES, R.W., "Crescimento agrícola no período 1999–2004: explosão da área plantada com soja e meio ambiente no Brasil," IPEA, Discussion Paper no. 1062, January 2005, pp. 16–17.

BRAZILIAN MINISTRY OF FOREIGN AFFAIRS, "Pacto vai evitar interferência na região amazônica," *Tratado de Cooperação Amazônica*, Brasília, 1978 (originally published in *Correio Braziliense*, Brasília, July 4, 1978).

"El tratado organizará vínculos entre los 8 países Amazônicos," *Tratado de Cooperação Amazônica*, Brasília, 1978, pp. 131–35 (originally published in *La Prensa*, Lima, July 5, 1978).

"Silveira revives ideal of Shuman in Latin America," *Tratado de Cooperação Amazônica*, Brasília, 1978, pp. 144–45 (originally published in *Sunday Chronicle*, Georgetown, July 23, 1978).

BROWNLIE, L., "A survey of international customary rules of environmental protection," *Natural Resources Journal*, vol. 13, 1973, pp. 179–89.

"Legal status of natural resources in international law," *Recueil des cours, Hague Academy of International Law*, 1979, vol. 162, pp. 245–318.

BRUNNÉE, J. "Common interest: Echoes from an empty shell? Some thoughts on the common interest and international environmental law," *Zeitschrift für ausländisches öffentliches Recht un Völkerrecht*, vol. 49, 1989, pp. 791–808.

BULTE, E.H., CORNELIS VAN KOOTEN, G., and SWANSON, T., "Economic incentives and wildlife conservation," draft paper, October 27, 2003, http://www.cites.org/eng/prog/economics/CITES-draft6-final.pdf

BUXTON, C. R., "Property in outer space: The common heritage of mankind principle v. the "first in time, first in right" rule of property law," *Journal of Air Law and Commerce*, vol. 69, 2004, pp. 689–707.

CAFLISH, L., "Règles générales du droit des cours d'eau internationaux," in *Collected Courses of the Hague Academy of International Law*, vol. VII, Dordrecht: Nijhoff, 1989, pp. 117–18.

CARVALHO, G., MOUTINHO, P., NEPSTAD, D., MATTOS, L., and SANTILLI, M., "An Amazon perspective on the forest–climate connection: Opportunity for climate mitigation, conservation and development?", *Environment, Development and Sustainability*, vol. 6, 2004, pp. 163–74.

CASSESE, A., "Ex iniura ius oritur: Are we moving towards international legitimation of forcible humanitarian countermeasures in the world community?", *European Journal of International Law*, vol. 10, 1999, p. 21.

CASTAÑO, C. "Diagnosis of the current situation of protected areas in Latin America and the Caribbean," paper presented at the 2nd Latin American Congress on National Parks and Other Protected Areas, Bariloche, Argentina, September 2007, http://www.ipsnews.net/news.asp?idnews=39482

CASTELLINO, J. and ALLEN, S., "The doctrine of uti possidetis: Crystallization of modern post-colonial identity," *German Book of International Law*, vol. 43, 2000.

CAUBET, C.G., "Le Traité de Coopération Amazonienne: régionalisation et développement de L'Amazonie," *Annuaire Français de Droit International*, vol. 30, 1984, pp. 803–18.

CERVO, A.L., "Os primeiros passos da diplomacia brasileira," *Relações Internacionais*, no. 3, 1978, pp. 43–62.

CHARNEY, J.I., "Third state remedies for environmental damage to the world's common spaces," in FRANCIONI, F. and SCOVAZZI, T. (eds.), *International Responsibility for Environmental Harm*, London/Dordrecht/Boston: Graham and Trotam, 1991, pp. 149–77.

CHAYES, A. and CHAYES, A.H., "Adjustments and compliance processes in international regulatory regimes," in TUCHMAN MATHEWS, J. (ed.), *Preserving the Global Environment: The Challenge of Shared Leadership*, New York and London: Norton, 1991.

CHAYES, A. and CHAYES, A.H. and MITCHELL, R.B., "Managing compliance: A comparative perspective," in BROWN WEISS, E. and JACOBSON, H.K. (eds.), *Engaging Countries, Strengthening Compliance with Environmental Accords*, Cambridge, MA and London, MIT Press, 1998.

COLINI CENAMO, M. et al., *Casebook of REDD Projects in Latin America*, Manaus: Nature Conservancy, Institute for Conservation and Sustainable Development of Amazonas (IDESAM), 2009.

COLINI CENAMO, M., *COLUMBIA ENCYCLOPEDIA*, "Dutch Wars," 2008.

CORDEIRO, N.V., "Environmental management issues in the Plata basin," in BISWAS, A. K., CORDEIRO, N.V., BRAGA, B.P.F. and TORTAJADA, C. (eds.), *Management of Latin America River Basins: Amazon, Plata and Sao Francisco*, Tokyo: United Nations University Press, 1998, pp. 148–73.

CORTEN, O., "The controversies over the customary prohibition on the use of force: A methodological debate," *European Journal of International Law*, vol. 16, no. 5, 2006, pp. 803–22.

CRAVEN, M. "Humanitarism and the quest for smarter sanctions," *European Journal of International Law*, vol. 13, no. 1, 2002, pp. 43–61.

CULLET P. and KAMERI-MBOTE, A.P., "Activities implemented jointly in the forestry sector: Conceptual and operational fallacies," *Georgetown International Environmental Law Review*, vol. 10, 1997, pp. 97–119.

DAVIS, S.H., "La crisis del Amazonas y el destino de los Indios," *El Universo Amazonico y la Integracion Latinoamericana*, 1981, pp. 17–44.

DÉJEANT-PONS, M., "European biodiversity: The Bern Convention of September 19,1979 on the Conservation of European Wildlife and Natural Habitats," in MACDONALD,

R. St. J. (ed.), *Essays in Honour of Wang Tieya*, Dordrecht/Boston/London: Martinus Nijhoff, 1994, pp. 211–26.

DOLBEC, J. et al., "Sequential analysis of hair mercury levels in relation to fish diet of an Amazonian population, Brazil," *The Science of the Total Environment*, vol. 271, nos. 1–3, pp. 87–97.

DOMINGUEZ, C., "The importance of rivers for the transportation system of the Amazon," in ARAGON, L. and CLÜSENER-GODT, M. (eds.), *Issues of Local and Global Use of Water from the Amazon*, UNESCO, Centre for Advanced Amazonian Studies of the Federal University of Pará (UFPA), 2004, pp. 77–100.

DOS SANTOS VIEIRA, R., "Concepts of national environmental law: Brazilian environmental law relating to Amazonia," in BOTHE, M., KURZIDEM, T. and SCHMIDT, C. (eds.), *Amazonia and Siberia: Legal Aspects of the Environment and Development in the Last Open Spaces*, London: Nijhoff, 1993.

DRUMMOND, J.A., "A extração sustentável de produtos florestais na Amazônia brasileira," *Estudos Sociedade e Agricultura*, 1996, pp. 115–37.

DUPUY, Pierre-Marie, "Due diligence in the international law of liability," in OECD, *Legal Aspects of Transfrontier Pollution*, Paris: OECD, 1977.

ELIASCH, J., "Climate change: Financing global forests," *Eliasch Review*, London, 2008.

EVANS, G., "The responsibility to protect: Moving towards a shared consensus," in KERBER, G., WEIDERUD, P., and ASFAW, S. (eds.), *The Responsibility to Protect: Ethical and Theological Reflections*, Geneva: World Council of Churches, 2005, pp. 3–9.

FARELLA, N. et al., "Deforestation modifying terrestrial organic transport in the Rio Tapajós," *Organic Geochemistry*, vol. 32, 2001, pp. 1443–58.

FEARNSIDE, P.M., "Brazil's Amazon settlement schemes: Conflicting objectives and human carrying capacity," *Habitat International*, vol. 8, no.1, 1984, pp. 45–61.

"Biodiversity as an environmental service in Brazil's Amazonian forests: Risks, value and conservation," *Environmental Conservation*, vol. 26, no. 4,1999, pp. 305–21.

"Soybean cultivation as a threat to the environment in Brazil," *Environmental Conservation*, vol. 28, no. 1, 2001, pp. 23–38.

"Conservation policy in Brazilian Amazonia: Understanding the dilemmas," *World Development*, vol. 31, no. 5, 2003.

"Deforestation control in Mato Grosso: A new model for slowing the loss of Brazil's Amazon Forest," *Royal Swedish Academy of Science*, vol. 32, no. 5, 2003, pp. 343–45.

"Carbon benefits from Amazonian forest reserves: Leakage accounting and the value of time," *Mitigation and Adaptation Strategies for Global Change*, vol.14, no. 6, 2009.

FENZL, N. and MATHIS, A., "Pollution of natural water resources in Amazonia: Sources, risks and consequences," in CLÜSENER-GODT, M. and ARAGÓN, L.E. (eds.), *Water from the Amazon: Issues of Local and Global Use*, Montevideo: Núcleo de Altos Estudos Amazônicos, Universidade Federal do Pará and UNESCO, 2004, pp. 51–75.

FERRIS, E., "The Andean Pact and the Amazon Treaty: Reflections of changing Latin American relations," *Journal of Inter-American Studies and World Affairs*, vol. 23, 1981, pp. 147–75.

FITZMAURICE, M.A., "International protection of the environment," *Collected Courses of The Hague Academy of International Law*, The Hague/Boston/London: Martinus Nijhoff, vol. 293, 2002.

FOREST PEOPLES PROGRAMME, "Indigenous peoples' rights and reduced emissions from reduced deforestation and forest degradation: The case of the *Saramaka People v. Suriname*," 2009, p. 2, http://www.rightsandresources.org/documents/files/doc_992.pdf

FREESTONE, D. and BOISSON DE CHAZOURNES, L. "Environmental law: Concepts and issues," in SERAGELDIN, I. and MARTIN-BROWN, J. (eds.), *Partnerships for Global Ecosystem Management*, Washington, DC: World Bank, 1997, pp. 182–92.

GASQUES, J.G., "Padrão de Crescimento da Amazônia e Instrumentos Econômicos," seminário Instrumentos Econômicos para o Desenvolvimento da Amzônia Brasileira, Brasília, July 24–25, 2002, Ministério do Meio Ambiente, IPEA, 2002.

GEHRING, T., "International environmental regimes: Dynamic sectoral legal systems," *Yearbook of International Environmental Law*, vol. 1, 1990, pp. 35–56.

GERMAN TECHNICAL COOPERATION (GTZ), FUNDECO and INSTITUTO DE ECOLOGÍA, "Conservación de ecosistemas transfronterizos y protección y recuperación de especies Amenazadas," thematic document, Second Regional Conference, held in Lima, Peru, March 26–28, 2001, http://www.comunidadandina.org/desarrollo/dct2.pdf

GIRALDI, R., "Lula diz que Amazônia é da humanidade e deve ser cuidada pelo Brasil," Folha Online, Brasília, May 8, 2008.

GLENNON, M.J., "Has international law failed the elephant?", *American Journal of International Law*, vol. 84, 1990, pp. 1–43.

Global Forest Coalition, "Effective policies to Reduce Emissions from Deforestation in Developing Countries (REDD) must address leakage and incorporate social impact criteria," submission to the UNFCCC Secretariat, 2008, http://unfccc.int/resource/docs/2008/smsn/ngo/016.pdf

GOLDEMBERG, J. "Global environment and the changing nature of states: The role of energy," in CHASEK, P. S. (ed.), *The Global Environment in the Twenty-First Century: Prospects for International Cooperation*, Tokyo/New York/Paris: United Nations University Press, 2000, pp. 64–79.

GOLDEMBERG, J. and DURHAM, R.E., "Amazonia and national sovereignty," *International Environmental Affairs*, no. 2/1, 1990, pp. 22–39.

GOLDIE, L.F.E., "Reconciling values of distributive equity and management efficiency in the international commons," in DUPEYED, R. (ed.), *Hague Academy Workshop on the Settlement of Disputes on the New Natural Resources*, 1983, p. 335.

GOMAA, M.M., "Non-binding agreements in international law," in BOISSON DE CHAZOURNES, L. and GOWLLAND-DEBBAS, V. (eds.), The International Legal System in Quest of Equity and Universality/L'Ordre juridique international, un système en quête d'équité et d'universalité, The Hague/London/Boston: Martinus Nijhoff, 2001, pp. 229–50.

GOODWIN, E.J., "The World Heritage Convention, the environment, and compliance," *Colorado Journal of International Environmental Law and Policy*, vol. 20, 2009, pp. 157–98.

GOWLLAND-DEBBAS, V., "The limits of unilateral enforcement of community objectives in the framework of UN peace maintenance," *European Journal of International Law*, vol. 11, no. 2, pp. 361–83.

GREÑO VELASCO, J.E., "Implicancias del Pacto Amazonico," *Revista Argentina de Relaciones Internacionales*, no. 10, 1978, pp. 56–75.

"Pacto Amazónico y Tratado de la Cuenca del Plata: Analogías y diferencia," *Revista de política internacional*, vol. 165, 1979, pp. 75–92.

GROB, J., "Antarctica's frozen territorial claims: A meltdown proposal," *Boston College International and Comparative Law Review*, vol. 30, 2007, pp. 461–84.

HAAS, P.M., "Introduction: Epistemic communities and international policy coordination," *International Organizations*, vol. 46, no. 1, 1992, pp. 1–36.
HAAS, E.B., "Words can hurt you; or, who said what to whom about regimes," in KRASNER, S.D. (ed.), *International Regimes*, Ithaca: Cornell University Press, 1983, pp. 23–59.
HALL, A., "Better RED than dead: Paying the people for environmental services in Amazonia," *Philosophical Transactions of the Royal Society*, 2008, pp. 1925–32, p. 1930.
HANDL, G., "International law and the protection of the atmosphere," *American Society of International Law, Proceedings* 62, 1989, p. 61.
"Environmental security and global change: The challenge to international law," *Yearbook of International Environmental Law*, 1990, pp. 3–33.
HILPOLD, P., "Humanitarian intervention: Is there a need for a legal reappraisal?," *European Journal of International Law*, vol. 12, no. 3, 2001, p. 444.
HOYLE, P.A., "The Guyana–Suriname maritime boundary dispute and its regional context," *Boundary and Security Bulletin*, 2001.
HUBER, R.M., RUITENBEEK, J., and SEROA DA MOTTA, R., "Market-based instruments for environmental policy making in Latin America and the Caribbean: Lessons from eleven countries," World Bank discussion paper, no. 381, Washington DC: World Bank.
HURRELL, A., "The politics of Amazonian deforestation," *Journal of Latin American Studies*, no. 23/1, 1991, pp. 197–217.
IISD (INTERNATIONAL INSTITUTE FOR SUSTAINABLE DEVELOPMENT), "Summary of the Copenhagen climate change conference: 7–19 December 2009", *Earth Negotiations Bulletin*, vol. 12, no. 459, December 22, 2009, p. 29, http://www.iisd.ca/climate/cop15/
IMPERIALI, C. (ed.), "L'effectivité du Droit International de l'Environnement, Contrôle de la Mis-en-Oeuvre des Conventions Internationales," Paris: Economica, 1998.
JACOBSON, H.K. and WEISS, E.B., "Assessing the record and designing strategies to engage countries," in BROWN WEISS, E. and JACOBSON, H.K. (eds.), *Engaging Countries: Strengthening Compliance with Environmental Accords*, Cambridge, MA and London: MIT Press, 1998.
JENKS, W., "The conflict of law-making treaties," *British Yearbook of International Law*, vol. 30, 1953, pp. 401–53.
JOHNSTON, S., "Sustainability, biodiversity and international law," in BOWMAN, M. and REDGWELL, C. (eds.), *International Law and the Conservation of Biological Diversity*, The Hague: Kluwer Law International, 1996, pp. 5–31.
KATOOMBA GROUP, FOREST TRENDS, and GOVERNO DO MATO GROSSO, "Avoiding deforestation in the Amazon: REDD and PES Markets," paper presented at the 2009 Katoomba Meeting, Cuiaba-Mato Grosso (Brazil), April 1–2, 2009, p. 4, http://www.katoombagroup.org/documents/cds/katoomba_xiv/documents/Agenda/Brazil%20Katoomba%20Program%20ENG%2009–0325.pdf
KEMPFF BACIGALUPO, R., "Proceso de Integración de la Cuenca del Plata," *Revista de Política Internacional*, no. 159, 1978, pp. 93–106.
KINTISCH, E., "Improved monitoring of rainforests helps pierce haze of deforestation," *Science*, April 27, 2007.
KISS, A. "La notion de patrimoine commun de l'humanité," *Recueil des Cours de. l'Académie de Droit International*, The Hague/Boston/London: Martinus Nijhoff, vol. 175, 1982, pp. 99–256.

"The international protection of the environment," in McDONALD, R.St.J. and JOHNSTON, D.M. (eds.), *The Structure and Process of International Law*, The Hague: Martinus Nijhoff, 1983, pp. 1069–93.

"Present limits to the enforcement of state responsibility for environmental damage," in FRANCIONI, F. and SCOVAZZI, T. (eds.), *International Responsibility for Environmental Harm*, International Environmental Law and Policy Series, London: Graham and Trotman, 1991, pp. 3–14.

"Environnement et développement ou environnement et survie?", *Journal de Droit International*, no. 2, 1991, 263–82.

"Les Traités-Cadres: une technique juridique caractéristique du droit international de l'environnement," *Annuaire Français de Droit International*, vol. 39, 1993, pp. 792–97.

"The Rio Declaration on Environment and Development," in CAMPIGLIO, L. et al. (eds.), *The Environment after Rio: International Law and Economics*, London/Dordrecht/Boston: Graham and Trotman/Martinus Nijhoff, 1994, p. 64.

"Economic globalization and the common concern of humanity," in KISS, A. SHELTON, D. and ISHIBASHI, K. (eds.), *Economic Globalization and Compliance with International Environmental Agreements*, The Hague/London/New York: Kluwer Law International, 2003, pp. 3–11.

KISS, A. and DRUMMOND, J.A., "Natureza, Povos Pobres? Questões conceituais e analíticas sobre of papel dos recursos naturais na prosperidade contemporânea," *Ambiente e Sociedade*, no. 10, 2002, pp. 45–68.

KOH, K.-L. and ROBINSON, N.A., "Regional environmental governance: Examining the Association of Southeast Asian Nations (ASEAN) Model," *Global Environmental Governance*, Yale Center for Environmental Law and Policy, 2002.

KOHEN, M.G., "La libre determinación de pueblos y su relación con el territorio," *Estudios de Derecho Internacional en Homenaje al Professor Ernesto J. Rey Caro*, Córdoba: Marcos Lerner Editora Ano, 2002.

"L'uti possidetis revisité: l'arrêt du 11 septembre 1992 dans l'affaire el Salvador/Honduras," *Revue Générale de Droit International Public*, vol. 97, no. 4, 1993, pp. 939–73.

"La contribución de América Latina al Desarrollo Progresivo del Derecho Internacional en Materia Territorial," *Anuario de Derecho Internacional*, Universidad de Navarra, vol. 16, 2001, pp. 57–77.

"Manifeste pour le droit international du XX siècle," in BOISSON DE CHAZOURNES, L. and GOWLLAND-DEBBAS, V. (eds.), *The International Legal System in Quest of Equity and Universality*, Nijhoff, 2001.

KOLK, A., "From conflict to cooperation: International policies to protect the Brazilian Amazon," *World Development*, vol. 26, 1998, pp. 1481–93.

KORKEAKIVI, A., "Consequences of 'higher' international law: Evaluating crimes of state and *erga omnes*', *Journal of International Legal Studies*, vol. 2, Summer, 1996, pp. 81–117.

KOSKENNIEMI, M., "Breach of treaty or non-compliance? Reflections on the enforcement of the Montreal Protocol," *Yearbook of International Environmental Law*, vol. 3, 1992, pp. 123–62.

"Solidarity measures: State responsibility as a new international order?", *British Yearbook of International Law*, no. 72, 2001, pp. 337–56.

KUCINSKI, B., "A Amazônia e a geopolítica do mundo," *Encontros com a civilização Brasileira*, vol. 11, 1979, pp. 12–19.

LANDAU, G. D., "Tratado de Cooperación Amazónica: nuevo ensayo de integracion," *Integración Latinoamericana*, no. 27, 1978, pp. 1–10.
"The Treaty for Amazonian Cooperation: A bold new instrument for development," *Georgia Journal of International and Comparative Law*, vol. 10, 1980, pp. 463–89.
LARSCHAN, B. and BRENNAN, B.C., "The common heritage of mankind principle in international law," *Columbia Journal of Transnational Law*, vol. 21, no. 2, 1983, pp. 305–37.
LAURANCE, W.F., COCHRANE, M.A., BERGEN, S., FEARNSIDE, P.M., DELAMÔNICA, P., BARBER, C., D'ANGELO, S., and FERNANDES, T., "The future of the Brazilian Amazon," *Science*, 2001, pp. 438–39.
LEE, T.R., "The management of shared water resources in Latin America," *Natural Resources Journal*, vol. 35, 1995, pp. 541–53.
LEME MACHADO, P.A., "Le droit comparé et le droit de l'environnement de l'Amazonie Brésilienne," *Rivista Trimestrale di Diritto Pubblico*, 1991, pp. 315–28.
Direito Ambiental Brasileiro, Sao Paulo: Malheiros Editores, 2004.
LEWIS, S.L., "Tropical forests and the changing earth system," *Philosophical Transactions of the Royal Society*, 2006, pp. 195–210.
MAGGIO, G.F., 'Implementation, compliance and effectiveness: International law year in review," *American Society of International Law Proceedings*, vol. 91, 1997, pp. 132–57.
MALHI, Y. and PHILLIPS, O.L., "Tropical forests and global atmospheric change: A synthesis," *The Royal Society*, 2004, vol. 359, pp. 549–55.
MANI, V.S. "Humanitarian intervention today," *Recueil des cours*, vol. 313, 2005, pp. 9–323.
MARTINS FILHO, J.R. and ZIRKER, D., "Nationalism, national security, and Amazônia: Military perceptions and attitudes in contemporary Brazil," *Armed Forces Society*, vol. 27, 2000.
"Pacto Amazônico precisa ser reativado," *Cadernos do terceiro mundo*, vol. 19, 2000.
MCCLEARLY, R.M., "The international community's claim to rights in Brazilian Amazonia," *Political Studies*, no. 39/4, 1991, pp. 691–707.
MCINTYRE, O., "The role of customary rules and principles of international environmental law in the protection of shared international freshwater resources," *Nature Resources Journal*, vol. 46, 2006, pp. 157–210.
MEDINA M., "Algunos aspectos políticos en el proceso de integración," *Universo Amazónico y la Integración Latino-americana*, Caracas: Universidad Simón Bolívar, 1983.
MEIRA MATTOS, A., "Pacto Amazonico: cooperacao e integracao," *Revista Brasileira de estudos politicos*, vol. 53, 1981, pp. 91–152.
"Pacto Amazônico precisa ser reativado," *Cadernos do terceiro mundo*, vol. 19, 2000, pp. 24–25.
MENON, P.K., "International boundaries: A case study of the Guyana–Surinam boundary," *International and Comparative Law Quarterly*, vol. 27, 1978, pp. 738–68.
MERCADO-JARRIN, E., "Analisis de los aspectos politicos relacionados con el Pacto Amazonico y la integracion Latinoamericana," *El universo Amazónico y L Integración Latinoamericana*, Caracas: Universidad Simón Bolivar, Instituto de Altos Estudios de América Latina, Fundación Bicentenario de Simon Bolívar, 1983, pp. 237–47.
MERCURE, P.-F., "Le rejet du concept de patrimoine commun de l'humanité afin d'assurer la gestion de la diversité biologique," *The Canadian Yearbook of International Law*, vol. 33, tome 33, 1995, pp. 281–304.
MILES, L., GRAINGER, A., and PHILLIPS, O., "The impact of global climate change on tropical forest biodiversity in Amazonia," *Global Ecology and Biogeography*, vol. 13, 2004, pp. 553–65.

MILLER, K.R., REID, W.V., and BARBER, C.V., "Deforestation and species loss," in MATHEWS, J.T. (ed.), *Preserving the Global Environment: The Challenge of Shared Leadership*, New York/London: Norton, 1991, pp. 76–111.

MONTENEGRO, M., "Política externa e cooperação amazônica: a negociação do Tratado de Cooperação Amazônica," *Sessenta anos de política externa*, São Paulo: Fundação Getulio Vargas, 2000, pp. 355–80.

MUNHOZ, F. and BONANOME, F., "Queremos a certificação bovina pronta em 60 ou 90 dias," August 10, 2009, http://www.amazonia.org.br/noticias/noticia.cfm?id=322715

MUNOZ, H., "El estudio de las políticas exteriores latinoamericanas: temas y enfoques dominantes," *Estudios Internacionales*, no. 80, 1987, pp. 406–34.

"The environment in inter-American relations," in MUNOZ, H., *Environment and Diplomacy in the Americas*, London: Lynne Rienner, 1992, pp. 88–89.

NEPSTAD, D., "Brazil: The costs and benefits of reducing carbon emissions from the Brazilian Amazon region," in J.P. HOLDREN, D.C. NEPSTAD and K.R. SMITH, *Linking Climate Policy with Development Strategy in Brazil, China, and India*, final report to the William and Flora Hewlett Foundation from the Woods Hole Research Center, 2007, http://whrc.org/Hewlett/Hewlett-Report-FINAL.pdf

NEPSTAD, D., et al., "The end of deforestation in the Brazilian Amazon," *Science*, vol. 326, no. 5958, 2009.

NOBRE, C.A. et al., "Estimating recent greenhouse gas emissions from cattle raising in Brazil," http://www.amazonia.org.br

O'CONNELL, M. E. "The role of soft law in a global order," in SHELTON, D. (ed.), *Commitment and Compliance: The Role of Non-binding Norms in the International Legal System*, Oxford and New York: Oxford University Press, 2000, pp. 100–14.

OSSA HENAO, C., "Aspectos de la implementación del Tratado de Cooperación Amazónica en las relaciones bilaterales entre Brasil y Colombia: El Plan Modelo para el desarrollo integrado de las comunidades Vecinas del Eje Tabainga-Apaporis," *Boletim da Sociedade Brasileira de Direito Internacional*, vols. 79/80, 1992, pp. 81–89.

OTERO, L.M., "El Tratado de Cooperación Amazónica," *Revista de Política Internacional*, no. 166, 1979, pp. 63–71.

PACHT, L., "Commerce international du bois tropical et protection de l'environnement: chronique d'une scission annoncée?", *Revue Générale de Droit International Public*, vol. 105, 2001, pp. 654–75.

PALMER, G., "New ways to make international law," *American Journal of International Law*, vol. 86, no. 2, pp. 259–83.

PAN AMERICAN UNION, "Conservation Section in the Pan American Union," *Bulletin of the Pan American Union*, vol. 77, 1943, pp. 714–15.

PARQUIS, R., "Conservación de la biodiversidade Amazónica: un enfoque regional," *Bois et Forêts des Tropiques*, vol. 290, 2006.

PAUWELYN, J., "The role of public international law in the WTO: How far can we go?" *American Journal of International Law*, vol. 95, pp. 535–78.

PICANCO, Garcia M.A., "La Cuenca del Amazonas y la Cuenca del Plata: zonas clave para la integración económica en America Latina," *Integración Latinoamericana*, no. 152, 1989, pp. 33–40.

POCHAT, V., "Water-resources management of the Plata basin," in BISWAS, Asit K., CORDEIRO, N.V., BRAGA, Benedito P.F. and TORTAJADA, C. (eds.), *Management of Latin America River Basins: Amazon, Plata and Sao Francisco*, Tokyo: United Nations University Press, 1998, pp. 123–47.

PRANCE, G., "The future of the Amazonian rainforest," *Futures*, 1990, pp. 891–903.

PRIEUR, M. and LAMBRECHTS, C., "Mankind and the environment: what rights for the twenty-first century?; Les Hommes et l'environnement: quels droits pour le vingt-et-unième siècle?," *Etudes en homage à Alexandre Kiss*, Paris: Editions Frison-Roche, 1998.

RAGAZZI, M., *The Concept of International Obligations Erga Omnes*, Oxford and New York: Clarendon Press, 1997.

RAUSTIALA, K. "Domestic institutions and international regulatory cooperation: Comparative responses to the Convention on Biological Diversity," *World Politics*, vol. 49, no. 4, 1997, pp. 482–509.

"Form and substance in international agreements," *American Journal of International Law*," vol. 99, no. 3, 2005, pp. 581–614.

RECONDO, G., "Comités de frontera: nuevos mecanismos para la integracion regional," *Integración Latino Americana*, vol. 132, 1988, pp. 35–47.

REVISTA VEJA, *Veja: Especial Amazônia*, ed. 2053, vol. 41, no. 12, March 26, 2008.

RIPHAGEN, W., "International concern for the environment as expressed in the concepts of 'common heritage of mankind' and of 'shared natural resources," in BOTHE, M. (ed.), *Trends in Environmental Policy and Law*, Gland: IUCN, 1980, pp. 343–62.

RICÚPERO, R., "O Tratado de Cooperação Amazônica," *Revista de Informação Legislativa*, no. 81, Senado Federal, 1984, pp. 177–96.

RODAN, B.D., NEWTON, A.C., and VERISSIMO, A., "Mahogany conservation: Status and policy initiatives," *Environmental Conservation*, vol. 19, no. 4, 1992, pp. 331–38.

ROSE, T., "Picking up the REDD Tab: Who will pay to fight deforestation and how?," p. 57, 25 February 2009, http://www.ecosystemmarketplace.com/pages/dynamic/article.page.php?page_id=6525&section=news_articles&eod=1

ROTHWELL, D.R., "The Arctic environmental protection strategy and international environmental cooperation in the far north," *Yearbook of International Environmental Law*, vol. 6, 1995, pp. 65–105.

"Polar environmental protection and international law: The 1991 Antarctic Protocol," *European Journal of International Law*, 2000, pp. 591–614.

ROULET, M. et al., "Spatio-temporal geochemistry of Hg in waters of the Tapajós and Amazon rivers, Brazil," *Limnology and Oceanography*, vol. 46, no. 5, 2001, pp. 1158–70.

ROWE, R., NARENDRA, S., and BOWDER, J., "Deforestation: Problems, causes and concerns," in SHARMA, N.P. (ed.), *Managing the Worlds Forests: Looking for Balance between Conservation and Development*, New York: World Bank, Kendall Hunt, 1992, pp. 33–45.

SALMAN, M.A. and BOISSON DE CHAZOURNES, L. (eds.), "Cours d'eaux internationaux, renforcer la cooperation et gérer les differends," *Actes du Semiaire de la Banque Mondiale: Rapport Technique de la Banque Mondiale*, no. 414, Washington, DC: World Bank, 1999.

SANCHEZ, F. and VALDERRAMA, A., "La Soberanía permanente sobre los recursos naturales: orígenes y contenidos," *Anuario de Derecho Internacional*, Universidad de Navarra, vol. 5, 1979, 1980, 1981, pp. 3–41.

SANDLER, Todd, "Global and regional public goods: A prognosis for collective action," *Fiscal Studies*, vol. 19, no. 3, 1998, pp. 221–47.

SAND, P.H., "Environmental summitry and international law," *Yearbook of International Environmental Law*, vol. 13, 2002, pp. 21–41.

SANDS, P. "The 'greening' of international law: Emerging principles and rules," *Indiana Journal of Global Legal Studies*, vol. 1, 1994, pp. 293–323.

"International law in the field of sustainable development: Emerging legal principles," in LANG, W. (ed.), *Sustainable Development in International Law*, London: Graham and Trotman, 1995, pp. 53–66.

SANN K. and THORNBER, K. "Impact of market-based instruments and initiatives on the trade in forest products and sustainable forest management," LTS International, http://www.fao.org/forestry/foris/data/trade/pdf/lts.pdf

SANWAL, M., "The sustainable development of all forests," *Review of the European Community*, vol. 1, no. 3, 1992, pp. 289–94.

SCHACHTER, O., "International law in theory and practice," *Recueil des cours*, vol. 178, 1982.

SHACKELFORD, S.J., "The tragedy of the common heritage of mankind," 28 *Stanford Environmental Law Journal*, 2009, pp. 109–67.

SCHALLY, H.M., "Forests: Toward an international legal regime?", *Yearbook of International Environmental Law*, vol. 4, 1993, pp. 30–50.

SCHRIJVER, N.J., "Permanent sovereignty over natural resources versus the common heritage of mankind," in DE WAART, P., PETERS, P. and DENTERS, E. (eds.), *International Law and Development*, Dordrecht/Boston/London: Martinus Nijhoff, 1988, pp. 87–101.

"Sovereignty and the sharing of natural resources," in WESTING, A.H. (ed.), *Transfrontier Reserves for Peace and Nature: A Contribution to Human Security*, Nairobi: UNEP, 1993, pp. 21–33.

"The changing nature of state sovereignty," *British Year Book of International Law*, vol. 70, 1999, pp. 65–98.

PETERS, P. and WAART, P., "Responsibility of states in respect of the exercise of permanent sovereignty over natural resources: An analysis of some principles of the Seoul Declaration (1986)," *Netherlands International Law Review*, vol. XXXVI, 1989/3, pp. 285–313.

SHARMA, N.P. et al., "World forests in perspective," in SHARMA, N.P. (ed.), *Managing the Worlds Forests: Looking for Balance between Conservation and Development*, New York: World Bank, Kendall Hunt, 1992, pp. 17–31.

SHAW, M.N., "The heritage of states: The principle of *uti possidetis juris* today," *British Yearbook of International Law*, 1996, pp. 75–154.

SHIHATA, I.F., "Implementation, enforcement, and compliance with international environmental agreements: Practical suggestions in light of the World Bank experiences," *Georgetown International Environmental Law Review*, vol. 9, 1996, pp. 38–39.

SIOLO, H. "Bases ecológicas de Amazonía," *El Universo Amazínico y la Integracíín Latinoamericana*, 1981, pp. 27–45.

STAHN, C., "Responsibility to protect: Political rhetoric or emerging legal norm?", *American Journal of International Law*, vol. 101, pp. 99–120.

STASETT, S.J., "Theological reflections on vulnerability, sovereignty and the responsibility to protect," in KERBER, G., WEIDERUD, P., and ASFAW, S. (eds.), *The Responsibility to Protect: Ethical and Theological Reflections*, Geneva: World Council of Churches, 2005, pp. 49–56.

STROMSETH, J.E., "Law and force after Iraq: A transitional moment," *American Journal of International Law*, vol. 97, 2003, pp. 628–42.

SUÁREZ, J.L., "El uti possidetis y los límites americanos," *Diplomacia Universitaria Americana, Argentina en el Brasil, Ciclo de Conferencias*, Buenos Aires: Escoffier, Caracciolo, 1918.

SWANSON, T. "Why is there a Biodiversity Convention? The international interest in centralized developing planning," *Royal Institute of International Affairs*, vol. 75, no. 2, 1999, pp. 307–31.

SZÉKELY, A., "Compliance with international environmental treaties: The empirical evidence. A commentary on the softening of international environmental law," *American Society of International Law, Proceedings of the Annual Meeting*, 1997.

SKUBISZEWSKI, K., "Implied powers of international organizations," in DINSTEIN, Y. (ed.), *International Law at a Time of Perplexity*, Dordrecht: Kluwer Academic, 1989, pp. 855–68.

SOUSA FRANCO FILHO, G., "Aspectos controversos do Pacto Amazônico," *Revista do Tribunal de Justiça do Estado do Pará*, vol. 21, 1980, pp. 35–53.

TARLOCK, D., "The Great Lakes as an environmental heritage of humankind: An international law perspective," *University of Michigan Journal of Law Reform*, no. 40, 2007, pp. 995–1020.

TEIXEIRA SOARES, A., "Tratado de Cooperação Amazônica," *Segurança de Desenvolvimento, Associação dos Diplomados da Escola Superior de Guerra*, vol. 188, Rio de Janeiro, 1981, pp. 6–14.

TELESETSKY, A., "*Graun bilong mipela na mipela no tromweim*: The viability of international conservation easements to protect Papua New Guinea's declining biodiversity," *Georgetown International Environmental Law Review*, 2001, pp. 735–78.

TIETENBERG, T.H., "Managing the transition: The potential role for economic policies," in TUCHMAN MATHEWS, J. (ed.), *Preserving the Global Environment: The Challenge of Shared Leadership*, New York and London, Norton, 1991, pp. 280–308.

VELASCO, J.E., "Pacto Amazónico y Tratado de la Cuenca del Plata: analogía y diferencias," *Revista de Política Internacional*, no. 165, 1979, pp. 75–92.

VICE, D., "Implementation of biodiversity treaties: Monitoring, fact-finding, and dispute resolution," *Journal of International Law and Politics*, vol. 29, no. 4, 1997, pp. 577–635.

VICTOR, D.G., "The use and effectiveness of non-binding instruments in the management of complex international environmental problems," *American Society of International Law, Proceedings of the Annual Meeting*, 1997, pp. 241–50.

VICTOR, D.G., RAUSTIALA, K. and SKOLNIKOFF, E.B. (eds.), *The Implementation and Effectiveness of International Environmental Commitments: Theory and Practice*, Cambridge MA, London: MIT Press, 1998.

VIEIRA, R. dos Santos, "Concepts of national environmental law, Brazilian environmental law relating to Amazonia," in BOTHE, M., KURZIDEM, T. and SCHMIDT, C. (eds.), *Amazonia and Siberia: Legal Aspects of the Preservation of the Environment and Development in the Last Open Spaces*, London: Graham and Trotman/Martinus Nijhoff, 1993, pp. 105–29.

VIEIRA, S.C., "A Proteção do meio ambiente e o Direito Internacional," *Revista Direito USF*, Bragança Paulista, vol. 10, no. 1, 1993, pp. 63–73.

"Environmental damage evaluation in Brazil, in BOWMAN M. and BOYLE, A. (eds.), *Environmental Damage in International and Comparative Law*, Oxford: Oxford University Press, 2002, pp. 297–303.

"The right to sustainable development in Amazonas," in JAYAGOVIND, A. (ed.) *Reflections on Emerging International Law*, Bangalore: International Law Association, Calcutta Centre, 2004, pp. 141–56.

WARE, D. "The Amazon Treaty: A turning point in Latin America cooperation?", *Texas International Law Journal*, vol. 15, 1980, pp. 118–37.
WEIL, P. "Towards relative normativity in international law?", *American Journal of International Law*, vol. 77, 1983, pp. 413–42.
WEILER, J. and PAULUS, A.L., "The structure of change in international law, or is there a hierarchy of norms in international law?", *European Journal of International Law*, vol. 8, 1997, pp. 545–65.
WOLFRUM, R., "The principle of the common heritage of mankind," *Zeitschirift für ausländisches öffentliches Recht und Völkerrecht*, vol. 43, 1983, pp. 312–37.
WOOD, M.C., "The interpretation of Security Council resolutions," *Max Planck Yearbook of United Nations Law*, vol. 2, The Hague: Kluwer Law International, 1997.
WUNDER, S. "Payments for environmental services: Some nuts and bolts," Center for International Forestry Research (CIFOR), Occasional Paper no. 42, 2005,, http://www.cifor.cgiar.org/publications/pdf_files/OccPapers/OP-42.pdf
"The efficiency of payments for environmental services in tropical conservation," *Conservation Biology*, vol. 21, no. 1, 2007, pp. 48–58.
YARZABAL, L. ESPINAL, C. and ARAGON, L.E., "Enfoque integral de la salud humana en la Amazonia," *Serie Cooperación Amazónica*, no. 10, Caracas: Universidad Central de Venezuela and UNAMAZ, 1992, pp. 1–59.
YOUNG, O.R., "Global environmental change and international governance," *Millennium Journal of International Studies*, vol. 19, 1990, pp. 337–46.
ZACHARIAS, D., "The UNESCO regime for the protection of world heritage as prototype of an autonomy-gaining international institution," *German Law Journal*, vol. 9, 2008, pp. 1833–64.
ZIMMERMAN, K. and MCGEE Jr., H., "The deforestation of the Brazilian Amazon: Law, politics and international cooperation," *Inter-American Law Review*, vol. 21, 1990, pp. 513–50.

VI. Main Electronic Documents and Internet Sites

Amazon Cooperation Treaty Organization: http://www.otca.org.br
Amigos da Terra – Amazônia Brasileira: http://www.amazonia.org.br
Andean Community: http://www.comunidadandina.org
Andean Development Corporation (CAF) http://www.caf.com
Caribbean Community: http://www.caricom.org/
Convention on Biological Diversity (CBD) www.cbd.int
Convention on International Trade in Endangered Species of Wild Fauna and Flora (CITES) http://www.cites.org
Food and Agriculture Organization of the United Nations (UNFF) http://www.fao.org/
Friends of the Earth (FoE) http://www.foe.co.uk
Global Environment Facility (GEF) http://www.thegef.org/gef/
Greenpeace http://www.greenpeace.org/international/
Instituto Brasileiro de Geografia e Estatística: http://www.ibge.gov.br
Instituto Nacional de Pesquisas Espaciais: http://www.inpe.br
Inter-American Biodiversity Information Network: http://www.iabin.info
Inter-American Commission on Human Rights: http://www.cidh.oas.org
Inter-American Court of Human Rights: http://www.corteidh.or.cr

Inter-American Development Bank: http://www.iadb.org
International Court of Justice: http://www.icj-cij.org
Mercado Comun del Sur: http://www.mercosur.int
Mercorsur: http://www.mercosur.org.uy/
Ministério das Relações Exteriores http://www.itamaraty.gov.br/
Ministério do Meio Ambiente: http://www.mma.gov.br
Ministério da Ciência e Tecnologia: http://www.mct.gov.br/clima
Organization of American States, http://www.oas.org
Pilot Program to Conserve the Brazilian Rain Forest (PPG7) http://www.mma.gov.br/ppg7/
Radiobrás: http://www.radiobras.gov.br
The International Union for Conservation of Nature www.iucn.org
Traffic http://www.traffic.com
United Nations: http://www.un.org
United Nations Development Program (UNDP) www.undp.org/
United Nations Educational, Scientific and Cultural Organization (UNESCO) http://www.unesco.org
United Nations Environmental Program, http://www.unep.org
United Nations Forum on Forests (UNFF) http://www.un.org/esa/forests/
United Nations Framework Convention on Climate Change Secretariat: http://unfccc.int
World Bank (WB) http://www.worldbank.org/
World Bank (WB) Inter-American Commission on Human Rights: http://www.cidh.oas.org
World Trade Organization (WTO) www.wto.org
World Wildlife Fund for Nature http://www.wwf.org

Index

ACT, 34, 188
  ACTO, 191
    programmatic areas, 168
  Andean countries, 83
  Declaration of Manaus, 2004, 90
  operational agreements and understandings, 86
  Permanent Secretariat, 109, 216
  spatial domain of validity, 88
actio popularis, 286
African Convention on the Conservation of Nature and Natural Resources 1968, 269
aftosa fever, 167
Agreement on Trade-Related Aspects of Intellectual Property Rights 1995 (TRIPS), 218
Amazon, 1
  Amazon States, 1–3, 5, 19, 23. *See* basin States or Amazon countries
  area, 1, 23
  biodiversity, 1, 16, 36, 47, 49
  biome, 23
  boundaries, 24, 50–51, 54
    Brazil-Bolivia, 54
    Brazil-Colombia, 55
    Brazil-Guyana, 57
    Brazil-Peru, 58
    Brazil-Suriname, 57
    Brazil-Venezuela, 60
  Brazilian Amazon, 2
    agricultural frontier, 42
    agriculture, 40
    arc of deforestation, 128. *See* Arco do Desmatamento
    Associação Brasileira dos Supermercados, 46
    beef industry, 207
    climate change, 44
    deforestation, 2, 10, 38, 40, 42
    global warming, 239
    highways, 31
    indigenous peoples. *See* indigenous communities
    land disputes, 42
    Legal Amazon, 42–43. *See* Amazônia Legal
    Legal Reserve, 171
    logging, 45
    national patrimony, 266
    national system of protected areas, 171
    Pacto pela Valorização da Floresta e pelo Fim do Desmatamento na Amazônia, 229
    Plan of Action for the Prevention and Control of Deforestation in the Legal Amazon 2004, 170, 171
    Programa de Certificação de Produção Responsável na Cadeia Bovina, 46, 212
    soy moratorium, 45, 211
    soybeans, 40, 42, 207, 211
    stock farming, 42–44, 212. *See* cattle, beef
    Sustainable Amazon Plan, 2008, 170
  cerrado, 38
  climate change, 1, 36, 49
  continental Amazon, 1
  deforestation, 30, 47
    causes, 38
  distribution, 24
  economic growth, 33
  ecosystems, 1
  fauna and flora, 189
  foreign interference, 77–78
  foreign intervention, 60
  forests, 1, 26
    deforestation, 25

Amazon (*cont.*)
freshwater, 1, 16, 26. *See* water resources, watercourses
GDP, 33
independence, 50–51, 153
indigenous communities, 31
indigenous peoples. *See* indigenous communities
internationalization, 76–77
languages, 25
limits, 87
location, 23
pan-Amazonia, 1, 23, 32
population, 29
privatization, 12, 303
protected areas, 47
regionalization, 304
slaughterhouses, 207, 212
species, 1
trade, 34
Amazon Charter, 99
Amazon club, 77
Amazon Cooperation Treaty 1978, 12, 74
1989 Declaration of Manaus, 97
1989 Quito Declaration, 100
1992 Manaus Declaration, 12, 93, 97–98
1995 Lima Declaration, 94
1998 Protocol of Amendment, 97
2004 Declaration of Manaus, 12, 90
2005 Declaration of Iquitos, 99, 114
2009 Declaration of Heads of States, 116, 123, 228
absolute equality, 90
ACC, 96, 103
Regulation, 104
ACTO, 12, 87, 110
headquarters, 110
Regulation, 111
strategic axes, 114
Strategic Plan, 94
Ad Hoc Consultation Committee, 108
Ad Hoc Working Group, 109, 111–12, 117
biodiversity, 93
Bonn Report, 116
CCOOR, 112
Regulation, 112
decisions, 104
diplomatic relations, 81
draft text, 82, 84
draft treaty, 84
duty to cooperate, 91, 95
ecological balance, 91
effect. *See* conflicts of law
effectiveness, 118
entry into force, 75, 85
environmental conservation, 89
fauna and flora, 92
financial mechanism, 117
framework agreement, 19, 74
freedom of navigation, 90
French Guyana, 86
full incorporation, 78
implementation, 95
indigenous peoples, 92
ACTO Coordinating Office, 175
Special Commission, 175
Meetings of the Presidents, 100
MMFA, 12, 96, 101
Regulation, 101
national sovereignty over natural resources, 88
negotiation, 75, 83, 85, 97
objective, 74
official languages, 85
operational agreements and understandings, 19, 75, 86, 97
parties, 74
Permanent National Commissions, 104
Permanent Secretariat, 96, 111, 113–14, 118
physical integration, 84
priority goal, 78
Pro Tempore Secretariat
Regulation, 108
Pro Tempore Secretariats, 106–08, 113, 117
programmatic areas, 114
rationale, 75
regulations, 98
resolutions, 97–98
signature, 85
sovereignty, 77, 79, 83
spatial domain of validity, 86
Special Commissions, 105
sustainable development, 91
Tarapoto Criteria and Sustainability Indicators, 94
Tarapoto process, 121, 173
water pollution, 156
water resources, 92, 155
Amazon Fund. *See* Fundo Amazônia, 242
Amazonian Protection System. *See* Sipam, 173
American Convention on Human Rights, 1969 *See* Pact of San José
American Declaration of the Rights and Duties of Man, 1948, 176
American Hudson Institute, 76
Andean Community, 127, 132
Andean countries, 83
1996 Common Regime on Access to Genetic Resources

intangible components, 162
2005 Paracas Declaration, 12
Andean Integration System, 132, 134
Andean System on Agriculture, Food Safety and Environment Conservation. *See*
CAN law, 129
applicability, 134
direct effect, 136
CAN legal system, 134
chemical pesticides
common mechanism, 1998, 174
Commission, 133–34, 136
Committee of Environmental Authorities, 137
Common Regime on Access to Genetic Resources, 137
Council of Foreign Ministers, 133–34, 136
Council of Ministers of the Environment and Sustainable Development, 12, 137
Court of Justice of the Cartagena Agreement, 129, 133
1979 treaty, 133
Decisions and Resolutions, 136
disaster prevention, 174
Committee, 174
genetic resources
1996 Common Regime, 162
access procedure, 163
Committee, 162
indigenous peoples
Working Group, 2002, 175
member States, 132
new plant varieties, 163
1993 common regime, 163
Subregional Committee, 164
non-compliance procedure, 135
Parliament, 133
1997 Additional Protocol, 133
Presidential Council, 133
Protocol of Trujillo 1996, 132
Andean Development Corporation (CAF), 148
Environmental Strategy, 148
Sustainable Development Coordination Office, 148
Andean Pact or Cartagena Agreement, 129, 132, 134, 135, 164, 220
Andean Subregional Integration Agreement 1969, 84, 129, 216. *See* Cartagena Agreement or Andean Pact
animal health, 167
Antarctic, 21, 267, 270
Antarctic treaty 1959, 275
Antarctic treaty system, 275
Convention on the Conservation of Antarctic Living Resources 1980, 280
Convention on the Conservation of Antarctic Marine Living Resources 1980, 299
environmental impact assessment, 276
Environmental Protocol 1991, 276
Measures for the Conservation of Fauna and Flora 1964, 276
natural reserve, 276
precautionary principle, 276
sovereignty, 275
special conservation area, 276
Application of Forest Legislation in the Amazon, 173
Arctic, 3, 277
Arctic Environmental Protection Strategy 1991, 3
Arctic States, 3, 277
Council, 278
Declaration on the Protection of the Arctic Environment, 278
Environmental Protection Strategy 1991, 278
indigeous communities, 279
Monitoring and Assessment Program, 279
Area. *See* UNCLOS
arsenic, 28
Association of Southeast Asian Nations (ASEAN), 18
Atlantic Ocean, 27

Barcelona Traction, Light and Power Company case, 285
Belém–Brasília, 42
biodiversity, 46, 159
1992 Convention, 13, 47, 93, 115, 189
Hague Ministerial Declaration 2002, 283
jurisdictional scope, 282
2002 Regional Biodiversity Strategy for the Andean Tropical Countries, 159
mega-diverse countries, 46
Regional Plan of Action for the Amazonian Biodiversity, 159
biosphere, 21
Bolivian Syndicate of New York City, 55
Brasília, 78, 83, 85, 104
Brazilian Constitution 1988, 171, 266
Brazilian DETER/PRODES program, 172
Brazilian Development Bank, 242
Brazilian General Attorney Office (Ministério Público), 212
Brazilian Growth Acceleration Program (PAC), 158
Brazilian Institute of Environment and Renewable Natural Resources (IBAMA), 158

Brazilian Institute of Geography and Statistics (IBGE), 30
Brazilian Ministry of Foreign Affairs, 75, 82–83, 85
Brazilian National Indian Foundation, 209
Brazilian National Institute for Space Research (INPE), 38, 70, 171–72, 239, *see also* Instituto Nacional de Pesquisas da Amazônia (INPA)
Brazilian National Water Agency, 154
  Technical Chamber of Trans-boundary Water Resource, 157
Brazilian Superior War College, 78

campinaranas, 26
carbon dioxide, 2, 34
carbon market, 233
  Chicago Climate Exchange, 236
  forest carbon, 237
  mandatory or compliance carbon market, 233
  voluntary carbon market, 236
    voluntary or verified emissions reductions (VERs), 236
carbon offsets, 233
carbon sequestration, 254
carbon sinks, 34, 269, 281
carbon trading, 20
Caricom, 127, 137
  1973 treaty. *See* treaty of Chaguaramas
  2001 revised treaty, 138
  Caribbean Court of Justice, 138
  Conference of Heads of Government, 138
  Council of Ministers, 138
  decisions, 139
  Miniterial Councils, 138
  recommendations, 139
  single market economy, 138
cerrado, 38, 40, 239
certification, 230
certification schemes, 231
certified emission reductions, 233
Charter of Economic Rights and Duties of States 1974, 268
Charter of Economic Rights and Duties of States 1984, 6
Chernobyl accident, 287
CITES, 13, 185, 189
climate change, 46
  Eliasch Review, 240
  Latin America, 205
  post-2012 climate regime, 200
  stabilization target, 239
  Stern Review, 240
Climate Change Convention 1992, 15, 192
  Bali Action Plan, 243
    Ad hoc working groups, 244
  Climate Change Convention 1992,
    Copenhagen Accord
      Green Climate Fund, 245
  Copenhagen Accord, 229, 244
    bodies, 245
  Subsidiary Body for Scientific and Technological Advice (SBSTA), 243
cobiça international, 76
common areas, 267
common but differentiated responsibility. *See* Climate Change Convention 1992
common concern, 21, 279
  Amazon, 282
  definition, 280
  forests, 282
common concern of humankind, 270
common heritage of humankind, 273
  Amazon, 274
  benefit sharing, 274
  definition, 271
  non-exclusive use, 273
  ownership, 273
common interests, 4–5
community of interests, 268
community of States, 2
conferences of the parties (COPs). *See* meetings of the parties
conflicts of law, 216
  Andean Community Court of Justice, 217–18
  CAN law, 217
  Carribean Court, 218
  Mercosur Permanent Court of Appeal, 221
  self-contained regimes, 222
Conservation and Sustainable Use of Brazilian Biodiversity (PROBIO), 161
Convention on Nature Protection and Wildlife Preservation in the Western Hemisphere 1940, 68, 131, 185
Convention on the Conservation of Migratory Species of Wild Animals, 1979, 189
Convention on the Cooperation for the Sustainable Development of the Danube River 1994, 9
Convention on the High Seas 1958, 269
Convention on the Protection of the Archeological, Historical and Artistic Heritage of American Nations 1976, 131
cooperation, 3, 5, 11–12
  cooperative sovereignty, 6

duty to cooperate, 6–7, 10, 13
international economic and social cooperation, 6–7
principle, 6, 8
regional cooperation, 3, 48, 66, 79
environment, 67
Coordinator of Indigenous Organizations of the Amazon Basin, 128
customary international law, 10

debt-for-nature swap, 229
Declaration on Principles of International Law Concerning Friendly Relations and Cooperation among States 1970, 6–7
Declaration on the Establishment of a New International Economic Order 1974, 6
deep seabed, 272
droit de regard, 294, 295, 296, 297, 298, 299, 302

economic neo-colonialism, 252
ecosystem, 3
ecosystems, 17, 81
El Nino, 38
erga omnes obligations, 285
European Commission, 87
Joint Research Centre, 87
European Community, 117, 139
European Court of Justice, 139
European Union Emission Trading Scheme, 234
externality, 240
extractivist reserves, 240

FAO
World Forestry Congress, 15
fauna and flora, 164
financial assistance, 15–16
Food and Agriculture Organization (FAO), 15, 117, 196
Committee on Forestry, 197
Conference, 196
Forestry Department, 197
headquarters, 196
Latin America and the Caribbean regional office, 197
Regional Forestry Commissions, 197
Strategic Plan for Forestry, 197
Forest Antimary, 231
Forest Stewardship Council (FSC), 230
forests
Brazilian Public Forest Management Law 2006, 170
certification, 20, 230
deforestation, 16
Forest Principles, 1992, 191
Non-legally Binding Instrument on All Types of Forests 2007, 191, 228
tropical forests, 2
climate change, 2
deforestation, 2, 37
species, 2
fossil fuels, 2
French Guyana, 19
Friends of the Earth, 212
Fundo de Investimento da Amazônia, 78
future generations, 274

G77, 11
Declaration of Caracas 1989, 11
Galapagos islands, 300
General Agreement on Tariffs and Trade (GATT) 1994, 286
Restrictions on Import of Tuna case, 287
genetic resources, 162
GHG emissions, 45–46
global commons, 276, 299
Global Environment Facility (GEF), 156
Trust Fund, 206
Global Environmental Outlook Amazonia (GEO-Amazonia), 199
Global Forest Watch, 301
global warming. *See* climate change
good neighborliness, 223
governance, 260
Great American Chaco, 161
greenhouse gases (GHGs), 34, 44
Greenpeace, 206, 211
grilagem, 170, 260
Group of Seven (G7), 207

Hague Conference for Codification of International Law 1930, 271
Hague Recommendations on International Environmental Law, 284
hazardous substances, 174
historical responsibility, 8, 290
human genome, 296
human rights, 3, 4
monitoring. *See* right of surveillance

ICJ
Burkina Faso–Mali case, 53
IDB
Blue Ribbon Panel on the Environment, 146
ILA New Delhi Declaration 2002, 193, 194

illegal land grabbing, 170, 211. *See also* grilagem
Indigenous and Tribal Peoples Convention 1989, 128, 177
indigenous peoples. *See* indigenous communities or populations
2008 Manaus Declaration, 252
American Declaration, 176
Coordinating Body for the Indigenous Organizations of the Amazon Basin, 175
lands, 240
Institute of Environment and Renewable Natural Resources (IBAMA), 209
Inter Caetera, 50
Inter-American Charter on Social Guarantees, 1948, 176
Inter-American Commission on Human Rights (IACHR), 294
Inter-American Court of Human Rights, 128, 131, 176, 178
Aloeboetoe et al. vs. Suriname, 183
Matter of Pueblo indígena de Kankuamo regarding Colombia, 178, 184
Mayagna (Sumo) Awas Tingni Community vs. Nicaragua, 178
Moiwana Community vs. Suriname, 183
monitoreo continuo, 184
Pueblo indígena de Sarayaku vs. Ecuador, 179
Saramaka People vs. Suriname, 181–82
Sawhoyamaxa Indigenous Community vs. Paraguay, 182
Yakye Axa Indigenous Community vs. Paraguay, 178
Yatama vs. Nicaragua, 183
Inter-American Development Bank (IDB), 145
Environment and Safeguards Compliance Policy, 146
Environment Policy, 146
Environment Strategy, 146
Intergovernmental Panel on Climate Change (IPCC), 46, 233
International Climate and Forest Initiative, 201
international community, 1–3, 5, 188, 298
International Convention on Civil Liability for Oil Pollution Damage 1969, 10
International Court of Justice (ICJ), 53, 286
Burkina Faso and the Republic of Mali, 54
Gabčíkovo-Nagymaros Project, 286
Namibia Advisory Opinion, 98
international eagerness. *See cobiça international*
international environmental law, 4
International Institute of the Hylean Amazon, 69, 76
International Law Commission, 223
international law of coexistence, 4–5
international law of cooperation, 4–5, 7
international legal order, 4
international legal system, 7
international legal system of the Amazon, 18, 151–52
components, 151
international non-governmental organizations, 210
International Seabed Authority (ISA), 272
International Tropical Timber Agreement (ITTA), 2006, 190
International Tropical Timber Organization (ITTO), 190
Criteria and Indicators for the Sustainable Management of Tropical Forests, 2005, 190
IUCN, 191
IUCN Red List, 210

Jari project, 76
jus cogens. *See* conflicts of law

Kyoto Protocol
certified emission reductions (CERs). *See* Climate Change Convention 1992
Emissions Trading. *See also* Climate Change Convention 1992
Kyoto Protocol 1997, 192, 233
Annex 1 parties, 243. *See* Climate Change Convention 1992
Clean Development Mechanism, 192, 196. *See also* Climate Change Convention 1992
first commitment period, 233
reforestation and afforestation. *See* Climate Change Convention 1992

land ownership, 11. *See* land tenure
land tenure, 251
Large Scale Biosphere-Atmosphere Experiment (LBA), 208
Latin America and the Caribbean, 1, 47
forests, 1, 36–37
law enforcement, 260
lege feranda, 19
lex specialis. *See* conflicts of law
locus standi, 285

market failure, 240
market-based instruments, 20, 227, *See also* market-based mechanisms
definition, 229

markets for environmental services, 229
Mercosur, 127, 140
  1991 Protocol of Brasília, 141
  1992 Declaration of Canela, 142
  1994 Protocol of Ouro Preto, 140–41
  1995 Declaration of Taranco, 142
  2001 Framework Agrement on the Environment, 143
  2002 Protocol of Olivos, 141
  Ad Hoc Tribunal, 144
  Common Market Group, 141–42
  Council of the Common Market, 141
  dispute settlement mechanism, 141
  Framework Agreement on the Environment 2001, 128, 185, 224
  Joint Parliamentary Commission, 141
  legal sources, 141
  Mercosur law, 142
    duty to incorporate, 142
  Parliament, 141
    2005 Protocolo Constitutivo, 141
  Permanent Court of Appeal, 141, 144
  Reunión Especializada en Medio Ambiente, 142
  Treaty of Asunción, 140
  Trade Commission, 141
  Venezuela
    2006 Protocol of Accession, 140
  Working Sub-Group No. 6 on the Environment, 142
mercury, 28
Montreal Protocol 1987, 193, 227
Moon Treaty 1967, 269

National Economic and Social Development Bank (BNDES), 43
National Institute for Space Research, 127
natural resources, 6, 14, 266, 268
  domestic, 267
  legal status, 267
  shared or transboundary, 267
new international order, 6

Organization of American States (OAS), 68, 70, 127, 129
  1948 Charter, 129
  Committee on Sustainable Developement, 131
  Council for Integral development, 131
  Department of Sustainable Development, 131
  General Assembly, 130
  General Secretariat, 130
  inter-American conferences, 68
  Meeting of Consultation of Ministers of Foreign Affairs, 130
  Permanent Council, 130
  principles, 130
  projects, 131
Our Own Agenda, 14, 146, 277
outer space, 270, 299

pacta sunt servanda, 289
pactum de contrahendo, 19
Pan American Health Organization, 168
Pan American Union, 69, 185
paper parks, 260
Payments for environmental services (PES), 20, 253
  additionality, 259
  Amazon, 255
  baseline, 258
  Bolsa Floresta, 256
  conditionality, 253
  Costa Rica's PES Program, 254
  definition, 253
  permanence, 259
  poverty alleviation, 258
  Program for the Socio-Environmental Development of Rural Family Production (Proambiente), 255
Peace of Westphalia, 4
permanent sovereignty over natural resources, 13–14, 273
photosynthesis, 2
Pilot Program to Conserve the Brazilian Rain Forest (PPG7), 11, 207
  projects, 209
Plan of Action for the Sustainable Development of the Americas 1996, 119
positive agendas, 128
positive incentives, 20, 227–28
precautionary principle, 6
principle of effectiveness, 139
PROBIO, 161
Programa Brasileiro de Certificação Florestal – CERFLOR, 232
protected areas, 47
  transboundary, 48, 161
Protected Areas System (Sistema Estadual de Unidades de Conservação), 247
protective barrier. *See cordão protetivo*

RADAM project (Radar na Amazônia)., 171
Rainforest Alliance, 231

Rainforest Trust Fund, 208
Reducing Emissions from Deforestation and Forest Degradation (REDD), 200, 237
  additionality, 241
  baselines, 241
  Coalition for Rainforest Nations, 243
  Forest Partners Program (Programa Socio Bosque), 248
  fund-based approach, 242
  hot air credits, 250
  Juma Reserve, 247
  leakage, 241
  market approach, 242
  negotiations, 243
  Noel Kempff Climate Action Project, 246
  permanence, 241
right of surveillance. *See* droit de regard
  human rights, 294
  indigenous peoples, 295
  UNCLOS, 299
right to development, 6
Rio Declaration 1992, 7–8
rubber boom, 56

sanitary cooperation, 167
sea turtles, 287
shared responsibility, 284
slash-and-burn, 38
slave labor, 212
SmartWood, 231
soft law, 4
South American Community of Nations, 17
sovereign responsibility, 13–14, 17, 90, 303
special rapporteur, 294, 295
species, 2
  extinction, 2
state Law on Climate Change, Environmental Conservation and Sustainable Development (Law 3.135/2007), 247
state responsibility, 223
  2001 Artices on State Responsibility, 222
state sovereignty, 4, 12
Stockholm Conference 1972, 5, 67, 82, 282
Stockholm Declaration 1972, 8–10
Subregional Integration Agreement, 1969. *See* Andean Pact or Cartagena Agreement
Summit of the Americas, 160
  1996 Declaration of Santa Cruz de la Sierra, 160
Superintendência do Desenvolvimento Econômico da Amazônia, 78
  Amazonas Sustentável Foundation, 247, 256
System for the Vigilance of the Amazon (SIVAM), 171

terra nullius, 52, 54
territorial sovereignty, 5, 274
TRAFFIC, 211
Transamâzonica, 42, 79
transfer of technology, 14, 273–74, 285
treaties
  framework agreement, 19
  law-making treaties, 4
  multilateral environmental agreements, 8, 189
Treaty of Badajoz 1801, 53
Treaty of Madrid 1750, 51, 53
Treaty of San Ildefonso 1777, 51, 53
Treaty of Tordesillas 1494, 51
tropical diseases, 166
Tropical Forest Conservation Act 1988, 230
tropical timber, 45

UN Conference of the Law of the Sea 1958, 271
UN Declaration on the Rights of Indigenous Peoples 2007, 128
UN Economic and Social Council (ECOSOC), 283
UN Forum on Forests (UNFF), 114, 283
  Multi-year Program, 284
UN Millennium Declaration Goals 2000, 200
UNCED, 7, 12, 93, 98, 101. *See* Earth Summit. *See* Rio Declaration, Agenda 21
  1992 Forest Principles, 283
  Agenda 21, 170
UNCLOS, 10, 90, 194
  Area, 271
  New Agreement Relating to the Implementation of Part XI, 15
  seabed regime, 15
UNDP, 117, 200
  Climate Change Strategy 2008, 214
UNEP, 132, 198
  Draft Principles, 9, 268
  Environmental Law Guidelines and Principles on Shared Natural Resources 1978, 193
  environmntal governance system, 214
  headquarters, 198
  priority areas, 198
  regional offices, 198
UNESCO, 69–70, 191
Union of South American Nations (UNASUR), 149
United Nations, 5

Charter, 5–7, 13, 296
General Assembly, 7–8, 13, 270
Human Rights Commission, 6
Security Council, 98
UN REDD Initiative, 215
UN Delivering as One, 215
uti possidetis, 51–53, 58
uti possidetis de facto, 53, 83
uti possidetis iuris, 53

Vienna Convention on the Law of Treaties, 223
Vilcabamba–Amboró corridor, 162

watercourses
Acre River, 157
Amazon River, 26
discharge, 27
exclusive navigation rights, 61
freedom of navigation, 61, 63–66
hydropower, 154
navigation, 27, 60, 155
opening of, 60, 62–63, 65
pollution, 28
Riparian States, 27
shared watercourses, 153
source, 27
special concession, 61, 63, 64. *See also* gracious concession or special favor
tributaries, 27–28
Javary, 51
Jirau, 123, 158
La Plata River, 18, 28, 154
1969 treaty, 84
1969 treaty, 50, 70–71
CIC, 71
Madeira River, 123, 158
Orinoco, 1
Santo Antônio, 123, 158
Tocantins, 1
Xingu River, 42
Yapurá, 51
Watercourses Convention 1997, 193
wetlands, 300
Ramsar Convention 1971, 191, 228, 300
Montreux Record, 195
Ramsar List, 191, 195
Transboundary Ramsar Sites, 191
World Bank, 201
carbon funds, 202
Clean Technology Fund, 202
Climate Investment Funds, 202
environmental policy, 202
Forest Carbon Partnership, 201, 244
Forest Investment Program, 203
headquarters, 201
Strategic Climate Fund, 203
Strategic Framework on Climate Change and Development, 202
World Charter for Nature 1982, 8
World Commission on Environment and Development, 13
World Conservation Monitoring Centre, 192
World Conservation Strategy, 1980, 93
World Health Organization, 168
world heritage
List of World Heritage in Danger, 300
reactive monitoring, 300
UNESCO World Heritage Centre, 301
World Heritage Committee, 191, 194
World Heritage Convention 1972, 9, 21, 195
World Heritage Fund, 194
World Heritage List, 191, 194
World Meteorological Organization, 233
World Summit on Sustainable Development 2002, 283
World Trade Organization (WTO), 286
Appelate body
Import Prohibition of Certain Shrimp and Shrimp Products, 286
World Trade Organization(WTO), 218
World Wide Fund For Nature (WWF), 213
biodiversity goal, 213

Zero Deforestation in the Amazon, campaign, 211